Immunobiology of the
SHARK

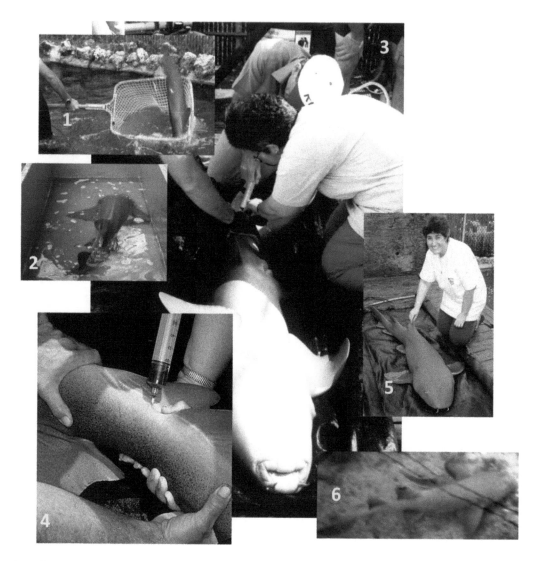

Frontispiece A shark bleed. 1. Netting a captive shark from the sea water channel. 2. Shark in anesthetizing tank. 3. Bleeding the anesthetized shark. 4. Blood withdrawn from the caudal sinus. 5. Recovered shark ready to return to the water. 6. Shark back in the channel. (Photographs courtesy of the Smith Collection.)

Immunobiology of the
SHARK

EDITED BY

Sylvia L. Smith
Department of Biological Sciences
Florida International University

Robert B. Sim
Department of Pharmacology
University of Oxford

Martin F. Flajnik
Department of Microbiology and Immunology
University of Maryland at Baltimore

CRC Press
Taylor & Francis Group
Boca Raton London New York

CRC Press is an imprint of the
Taylor & Francis Group, an **informa** business

CRC Press
Taylor & Francis Group
6000 Broken Sound Parkway NW, Suite 300
Boca Raton, FL 33487-2742

First issued in paperback 2019

ISBN-13: 978-1-4665-9574-3 (hbk)
ISBN-13: 978-0-367-37804-2 (pbk)

Visit the Taylor & Francis Web site at
http://www.taylorandfrancis.com

and the CRC Press Web site at
http://www.crcpress.com

Dedication

To my late husband, Professor David Spencer Smith (Oxford University, U.K.), for his ever present enthusiasm for my shark research and his invaluable encouragement and support as I re-established and restarted my research program following the devastating effects of Hurricane Andrew when all eight long-term study sharks, who had faithfully served us for ten years, lost their lives. I dedicate this book also to those magnificent animals.

Sylvia Lubna (Hyder) Smith
2014

Nurse sharks

Contents

Chapter 1

Gregor M. Cailliet and David A. Ebert

Chapter 2

Helen Dooley

Chapter 3

Lynn L. Rumfelt

Chapter 4

Carl A. Luer, Catherine J. Walsh, and A.B. Bodine

Chapter 5

Ashley N. Haines and Jill E. Arnold

Chapter 6

Catherine J. Walsh and Carl A. Luer

Chapter 7

C.J. Secombes, J. Zou, and S. Bird

Chapter 8

Sylvia L. Smith and Masaru Nonaka

Chapter 9
Simona Bartl and Masaru Nonaka

Chapter 10
Ellen Hsu

Chapter 11
Michael F. Criscitiello

Chapter 12
Michael F. Criscitiello

Chapter 13
Byrappa Venkatesh and Yuko Ohta

Chapter 14
Catherine J. Walsh and Carl A. Luer

Chapter 15
Nichole Hinds Vaughan and Sylvia L. Smith

Chapter 16
Liza Merly and Sylvia L. Smith

With the possible exception of the giant panda, perhaps no other animal has captured the interest of the general public as the shark. As general biologists, we recognize the cartilaginous fishes, such as sharks, skates, rays, and the *Chimera*, of which ratfish represent an extant form, to be the living representatives of a distant major radiation in the phylogenetic development of the vertebrate form. As such, this fascinating group potentially holds clues to understanding basic biological processes whose underlying genetic bases have been obscured during the passage of evolutionary time. As immunologists, we recognize the cartilaginous fishes to be the most phylogenetically distant group of species from man that possess an adaptive immune system in which immunoglobulin and T-cell antigen receptors function as mediators of humoral and cellular immunity. Intensive studies of immunity in the cartilaginous fish over the past three decades have resulted in remarkable discoveries that have contributed much to our understanding of how the highly complex immune system of the vertebrates has evolved.

Immunobiology of the Shark is a comprehensive resource for immunologists, evolutionary biologists, and other scientists interested in understanding how the integration of cellular and humoral processes has created a highly effective survival mechanism. This volume incorporates a rich, diverse literature into a well-reasoned, comprehensive package that will permit the reader to understand topics as diverse as the placement of the shark in phylogeny, the uniqueness of its physiology, and the mechanisms regulating how the rearranging immune genes make the somatic immune cell a selectable unit. An exceptional number of findings are described that are of immediate relevance to understanding immune regulation in the far more widely studied immune systems of man and mouse. The fundamental differences that exist between lymphoid architecture in cartilaginous fish and "higher" vertebrates, which create distinct microenvironments for immune cell differentiation and maturation, are the primary topics of several chapters. In addition to the unique adaptations characteristic of some components of immunity in the shark, other aspects of immunity closely resemble those seen in the higher vertebrates, presenting a remarkable story of evolutionary variation and stability. This volume is particularly notable in the diversity of approaches that were employed to study the immunobiology of the shark, ranging from basic microscopic observations to detailed genome annotation. A series of fascinating questions are raised throughout this volume that can be addressed experimentally with today's technology. If one message rings true, it is that alternative animal model systems, such as sharks and their relatives, are critical to our understanding of the scope of evolutionary change and the mechanisms underlying variation. *Immunobiology of the Shark* will prove to be an invaluable resource as well as guide for future investigations of how the systematic integration of pathways and processes give rise to successful biological adaptations.

Gary W. Litman
Associate Vice Dean for Research, Children's Research Institute
Hines Professor/Vice Chairman, Department of Pediatrics
Distinguished University Professor
University of South Florida Morsani College of Medicine
Director, Laboratory of Molecular Genetics
All Children's Hospital-Johns Hopkins Medicine

In the past three decades, our understanding of the biology of sharks has advanced significantly, particularly in the area of shark immunology. There is a lack and pressing need for a comprehensive reference book that specifically focuses on various aspects of shark immunobiology, a topic that, although touched upon in several excellent immunology text books under a general comparative immunology section, is not dealt with in depth. This book, *Immunobiology of the Shark*, attempts to fill the void and to provide readers with a resource in which major aspects of shark biology are covered in chapters written by authors that are acknowledged experts in the field. We are grateful for their contributions. Most published books on sharks address natural history, habitat, predation, ecology, classification, development, physiology, and occasionally evolution, but not specifically the immune system. This book is a reference text intended for students and experienced researchers alike engaged in (or considering research in) the study of shark immunity and immune systems. It will be a valuable reference source not only for comparative immunologists but also for main stream immunologists, comparative physiologists, geneticists, and biologists interested in comparative studies. The book also contains useful information for evolutionary and conservation biologists and ecologists. Students, postdoctoral scholars, and investigators actively engaged in shark research will find the contents of the book of immense value in obtaining a better understanding of various aspects of shark immunobiology and as an excellent resource of referenced material. Chapters are comprehensive and include up-to-date developments in the field including work published as recently as a few months preceding publication of the book. Contributing authors were asked to include an introductory overview of the chapter topic, a historical perspective of the development of the field, and an in-depth review of all that is currently known on the subject. Each chapter provides a comprehensive bibliography that includes early pioneering studies.

From an evolutionary perspective, study of the immune system of sharks and other elasmobranchs makes sense because the shark is the most primitive vertebrate that possesses an immune system that has key functional components of both the innate and adaptive immune system present in mammals. Studies on sharks have contributed to a better understanding of the evolution of our own immune system. Of particular interest to comparative immunologists will be the comparative approach taken by authors in writing chapters, in which comparisons are drawn to similar systems or elements in vertebrate species representing other classes. This is particularly notable in the chapter on shark complement (an integral component of innate immunity), in which the authors bring out in considerable molecular detail the genetic and phylogenetic and functional similarities and differences between the shark and human system.

For readers whose interest in sharks is recent and/or who are less familiar with shark species and their natural history, the first three chapters introduce the reader to general shark habitats and distribution, the diversity of species, their overall development and reproduction, and an overview of the elements of innate and adaptive immune system. In Chapter 4, the authors describe the organization and role of various tissue and organ systems that are key players in immune defense. Chapters 5 and 6 go a long way to clarify terminology, morphological features, and function of blood cell types. The authors provide a comprehensive classification and nomenclature of immune blood cells including correlation with essential immune functions such as phagocytosis, chemotaxis, and cytotoxicity. Chapter 7 is a comprehensive review at the cellular and molecular level of shark cytokines and chemokines and their essential role as messengers in intercellular communication and regulation of host immune response. The major humoral innate effector system, complement is described in molecular and functional detail in Chapter 8. Studies on the elements of the shark's adaptive immune system (immunoglobulins, T-cell antigen receptor, major histocompatibility antigens) are described in detail at the molecular level in the following four chapters. The underlying genetics and the novel mechanisms employed to generate diversity not seen in higher vertebrates are defined.

The mechanisms of gene recombination for generating diverse immunoglobulin and TCR repertoires are clearly presented. Furthermore, the arrangement of MHC genes involved in antigen processing and presentation have been mapped and typical structural features detailed. Chapter 13 summarizes results of recent genome sequencing of elephant shark, *Callorhinchus milii*, and makes useful comparison of the immunogenome of cartilaginous fish with bony vertebrates. The book also includes a chapter on the *in vitro* culture of elasmobranch cells. It reviews the problems, particularly related to the composition of the growth medium, that were overcome in the early attempts to successfully grow and maintain long-term cell cultures. The antimicrobial activity of proteins and peptides, found in various body fluids of vertebrates and invertebrates, is a critical immediate first line of defense, and sharks similarly have their share of such defense molecules. A review of antimicrobial factors is undertaken in Chapter 15. A book on shark immunobiology cannot ignore the commercial exploitation of sharks for products that are globally marketed for medicinal and/or dietary purposes. Thus, in the last chapter, the authors review shark-derived immunomodulators and discuss their bioactive properties, their alleged and proven usefulness in complementary medicine, and the impact this industry is having on the survival chances of shark species. We hope the reader will find the chapters clearly written with useful supporting illustrations. The book's full color illustrations should make it easier for the reader to comprehend the complex figures of cellular and molecular structures and to correlate them with information provided in the text.

It is hoped that this book illustrates the magnificence of these animals as model systems and brings out the importance to study them in order to further understand their complex and often enigmatic biology. The book also serves as recognition of the valuable contribution shark research has made in revealing evolutionary changes, system diversity, and the potential harmful effect of ecological imbalance from commercial exploitation and shows the need and importance of continued research. Once again, with apologies to Ogden Nash, we take the liberty of sharing a previously published quote (Smith, *Immunological Reviews*, 1998; 166:67–78), which still holds true today.

Two things I know about the Shark:
Its bite is worser than its bark.
And in its plasma we may see
The forerunners of things to be.

Sylvia L. Smith
Florida International University

Robert B. Sim
University of Oxford
University of Leicester

Martin F. Flajnik
University of Maryland at Baltimore

Acknowledgments

We thank Ms. Rhia Rae Jones and Mr. Ivan Santiago of Florida International University's Academic Imaging Services for enhancing the images that appear in the frontispiece collage and the dedication. The cover picture of a nurse shark is courtesy of Professor Max Telford, University College London, London, UK.

Sylvia L. Smith, PhD, SM (ASCP, AAM)
Professor Emeritus
Department of Biological Sciences
Florida International University
Miami, Florida

Martin F. Flajnik, PhD
Professor
Department of Microbiology and
 Immunology
University of Maryland at Baltimore
Baltimore, Maryland

Robert B. Sim, DPhil
Senior Research Fellow
Department of Pharmacology
University of Oxford
Oxford, UK

and

Honorary Professor
Department of Infection, Immunity,
 Inflammation
University of Leicester
Leicester, UK

Contributors

Jill Arnold, MS, MT (ASCP)
Director of Laboratory Services
National Aquarium in Baltimore
Baltimore, Maryland

Simona Bartl, PhD
Moss Landing Marine Laboratories
Moss Landing, California

Steve Bird, PhD
Department of Biological Sciences
School of Science and Engineering
University of Waikato
Hamilton, New Zealand

A.B. Bodine, PhD
Department of Animal and Veterinary Sciences
Clemson University
Clemson, South Carolina

Gregor M. Cailliet, PhD
Pacific Shark Research Center
Moss Landing Marine Laboratories
Moss Landing, California

Michael F. Criscitiello, PhD
Comparative Immunogenetics Laboratory
Department of Veterinary Pathobiology
College of Veterinary Medicine and
  Biomedical Sciences
Texas A&M University
College Station, Texas

Helen Dooley, PhD
Scottish Fish Immunology Research Centre
School of Biological Sciences
University of Aberdeen
Aberdeen, UK

David A. Ebert, PhD
Pacific Shark Research Center
Moss Landing Marine Laboratories
Moss Landing, California

Ashley N. Haines, PhD
Department of Biology
Norfolk State University
Norfolk, Virginia

Ellen Hsu, PhD
Department of Physiology and Pharmacology
State University of New York
Health Science Center at Brooklyn
Brooklyn, New York

Carl A. Luer, PhD
Marine Biomedical Research Program
Center for Shark Research
Mote Marine Laboratory
Sarasota, Florida

Liza Merly, PhD
Marine and Atmospheric Science Program
Division of Marine Biology and Fisheries
Rosentiel School of Marine and Atmospheric
  Science
University of Miami
Coral Gables, Florida

Masaru Nonaka, PhD
Department of Biological Sciences
Graduate School of Science
The University of Tokyo
Tokyo, Japan

Yuko Ohta, PhD
Department of Microbiology
  and Immunology
School of Medicine
University of Maryland
Baltimore, Maryland

Lynn L. Rumfelt, PhD
Department of Biology
School of Arts and Sciences
Gordon State College
Barnesville, Georgia

Christopher J. Secombes, PhD
Scottish Fish Immunology Research Centre
University of Aberdeen
Aberdeen, UK

Sylvia L. Smith, PhD
Department of Biological Sciences
Florida International University
Miami, Florida

Nichole Hinds Vaughan, PhD
Biological Sciences, Physical Sciences, &
  Wellness
Broward College Central Campus
Davie, Florida

Byrappa Venkatesh, PhD
Comparative Genomics Laboratory
Institute of Molecular and Cell Biology
Biopolis, Singapore

Catherine J. Walsh, PhD
Marine Immunology Program
Center for Shark Research, Mote Marine
  Laboratory
Sarasota, Florida

Jun Zou, PhD
Scottish Fish Immunology Research Centre
University of Aberdeen
Aberdeen, UK

CHAPTER 1

The Diversity and Natural History of Chondrichthyan Fishes

Gregor M. Cailliet and David A. Ebert

CONTENTS

SUMMARY

The purpose of this introductory chapter is to (1) describe the worldwide diversity of chondrichthyan fishes, their relationships to each other, their species richness (i.e., biodiversity), and general habitat utilization pattern and (2) characterize their life histories, briefly review fishery impacts, and summarize their conservation status.

INTRODUCTION

There are three major groups of fishes in the world, totaling to date ~33,000 species (Eschmeyer, 2013). These include the jawless fishes (Agnatha), which comprise only 0.4% of the species (Figure 1.1). Another, much larger group is the bony fishes (Osteichthyes), which dominate the world's fishes comprising ~96% of the total. And finally, the subject of this chapter, the cartilaginous fishes (Chondrichthyes) make up the remaining ~3.7% of the species of fishes worldwide.

Chondrichthyans are a relatively successful group of fishes and inhabit most marine ecosystems. Worldwide, a little more than half (~55%) of chondrichthyans occur on continental shelves from the intertidal zone down to 200 m depth, and to a lesser extent on insular shelves (Compagno, 1990). The distribution of marine organisms is influenced by a variety of factors, with temperature the principal factor determining their geographic ranges. Most marine species exhibit stratified distribution patterns, with species diversity decreasing from warm equatorial to cool high-latitude seas (Barnes and Hughes, 1999; Briggs, 1995; Luning and Asmus, 1991).

The cartilaginous fishes occupy a wide range of habitats, including freshwater riverine and lake systems, inshore estuaries and lagoons, coastal waters, the open sea, and the deep ocean. Most species have a relatively restricted distribution, occurring mainly along continental shelves and slopes and around islands, with some endemic to small areas or confined to narrow depth ranges. Others are disjunct in their distribution, represented by many populations occurring in widely separated areas around the world. Only a relatively small number of species are known to

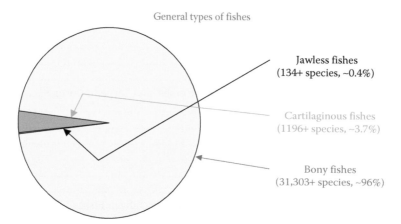

Figure 1.1 A diagram representing the contribution of jawless, cartilaginous, and bony fishes of the world's total fish diversity of 32,633 species. (D.A. Ebert, personal database, May 1, 2013; updated and modified from Compagno, L.J.V., *Environ. Biol. Fish.*, 28, 33–75, 1990; McEachran, J.D., and Aschliman, N., *Biology of Sharks and Their Relatives*, CRC Press, Boca Raton, FL, 2004; Nelson, J.S., *Fishes of the World*, Wiley, Hoboken, NJ, 2006; Eschmeyer, W.N., editor, *Catalog of Fishes*, California Academy of Sciences. http://research.calacademy.org/research/ichthyology/catalog/fishcatmain.asp. Electronic version. Accessed April 15, 2013.)

be genuinely wide ranging. The best studied of these are the large pelagic species, which make extensive migrations across ocean basins. However, at least some of the deepwater species, such as the deeper-dwelling squaloids, may exhibit similar wide-ranging movements, although very few of these have been studied.

The diversity of chondrichthyans has likely been underestimated, because species complexes, especially among the skates (Order Rajiformes) and catsharks (Family Scyliorhinidae), are poorly known in several regions. Only within the past decade has significant progress been made toward understanding and identifying the species-rich skate faunas that occur on continental slopes and shelf areas. Other groups, such as the catsharks and chimaeroids (Order Chimaeriformes), are still poorly known taxonomically and very little is known about their life histories.

The biology of the chondrichthyan fishes is among the most poorly known and least understood of all major marine vertebrate groups. Detailed information on life history and reproductive dynamics is available only for a few species, especially those that are of importance for directed fisheries. This is the result of both the low priority placed on cartilaginous fish research and the considerable difficulty of data collection for many species (particularly those restricted to deepwater habitats, or that are sampled only at certain times of year or at some stages in the life cycle).

Cartilaginous fishes are predominantly predatory; however, some are also scavengers and some of the largest (whale, basking and megamouth sharks, and manta rays) are suction or filter feeders on plankton and small fish, similar to the great whales. None are herbivorous. The predatory sharks are at, or near, the top of marine food chains. Therefore, wherever they occur their numbers are relatively small compared to those of most teleost fishes.

Several major teleost and invertebrate fisheries take chondrichthyans as bycatch (i.e., accidental capture), but they are usually discarded, and little information is gathered regarding species complexes or abundance. One exception is for larger pelagic shark species (e.g., *Lamna* spp.) that have been intensely targeted by directed fisheries in some regions, both for their flesh and fins for soup in some nations. There is global concern of the status of these pelagic sharks (Dulvy *et al.*, 2008; see Chapter 16).

CLASSIFICATION, BIODIVERSITY, AND HABITAT ASSOCIATIONS

Classification

The class Chondrichthyes is generally divided into two major groups, the large subclass Neoselachii, which is subdivided into two cohorts, the Selachii (sharks) and the Batoidea (rays) (see Table 1.1), and the smaller subclass Holocephalii, which includes all of the chimaeras. The Selachii is further subdivided into two superorders, the Squalomorphii and Galeomorphii. The superorder Squalomorphii includes the orders (Table 1.1) Hexanchiformes, Squaliformes, Squatiniformes, and Pristiophoriformes, whereas the superorder Galeomorphii includes the Heterodontiformes, Orectolobiformes, Lamniformes, and Carcharhiniformes.

The cohort Batoidea as defined here includes all of the batoids and recognizes four orders: Torpediniformes, Pristiformes, Rajiformes, and Myliobatiformes (Table 1.1). The ordinal classification used here follows the studies of Compagno (2001, 2005), McEachran and Aschliman (2004), and Nelson (2006), but recognizes that there is still considerable disagreement among taxonomists (de Carvalho, 1996; Naylor *et al.*, 2005; Shirai, 1992) regarding the relationships of these groups.

There are currently 13 recognized orders of Chondrichthyan fishes (Table 1.1). More recently, the term *sharks* has been used, in the broad sense, to include the sharks (Selachii: eight orders), rays (Batoidea [rays or *flat sharks*]: one or four orders, depending on your source of information), and chimaeras (ghost or silver sharks: one order). The elasmobranchs (cartilaginous fishes comprising sharks, rays, and skates) are the dominant chondrichthyan group, with approximately 57 families representing 96% of extant species. The remaining 4% includes the three chimaera families.

In higher taxonomic groupings (genera and above), the sharks are more diverse than the batoids, but among all elasmobranch species there are more batoids (56%) than sharks (44%).

Worldwide approximately 1196 species of cartilaginous fishes (D.A. Ebert, personal database, May 1, 2013; all taxonomic details are as of this date, even though some species have been changed since then) have been described (Table 1.1). New species are still being described at a fairly prodigious rate (Figure 1.2). Approximately 217 new species were named within the past decade. This is more than

Table 1.1 **The Taxonomic Composition and Richness of the Eight Orders of Selachii (Sharks), and Either One Major Order or Four Minor Orders of Batoidea (Rays), and the One Order of Chimaera (Ratfishes) Worldwide, Which Make Up the Elasmobranchs**

Order	Family	(%)	Genera	(%)	Species	(%)
Hexanchiformes (cow and frilled sharks)	2	3.5	4	2.0	6	0.5
Squaliformes (dogfish sharks)	7	12.3	24	11.9	126	10.5
Pristiophoriformes (sawsharks)	1	1.8	2	1.0	7	0.6
Squatiniformes (angel sharks)	1	1.8	1	0.5	20	1.7
Heterodontiformes (horn or bullhead sharks)	1	1.8	1	0.5	9	0.8
Orectolobiformes (carpet sharks)	7	12.3	14	7.0	46	3.8
Lamniformes (mackerel sharks)	7	12.3	10	5.0	15	1.3
Carcharhiniformes (ground sharks)	8	14.0	51	25.4	279	23.3
Torpediniformes (electric rays)	2	3.5	13	6.5	69	5.8
Pristiformes (sawfishes)	1	1.8	2	1.0	5	0.4
Rajiformes (skates and guitarfishes)	7	12.3	46	22.9	348	29.1
Myliobatiformes (eagle rays)	10	17.5	27	13.4	217	18.1
Chimaeriformes (chimaeras or silver sharks)	3	5.3	6	3.0	49	4.1
Total	57		201		1196	

Source: D.A. Ebert, Personal database, May 1, 2013.

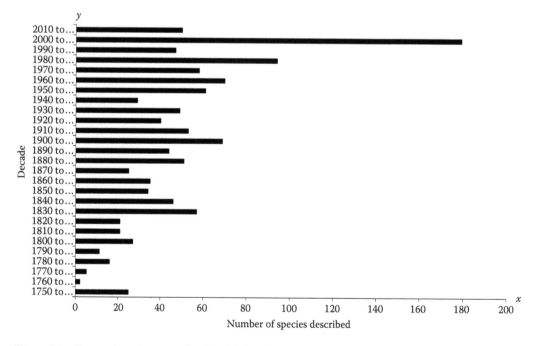

Figure 1.2 The number of new species (x-axis) described by decade (y-axis) from 1750 to 2012. (From D.A. Ebert, Personal database, May 1, 2013.)

the total number of chondrichthyans ($n = 199$) named over the previous three decades (1970–1999). In addition, it is estimated that at least 100 or more species are still awaiting formal description by researchers.

Marine Habitats Used by Chondrichthyans

For the purposes of describing habitats utilized by sharks, rays, and skates, we first distinguish between those in freshwater and marine. Then, marine habitats can be divided into (1) coastal habitats, which include those that extend offshore from the intertidal to the edge of the continental shelf, and (2) open ocean habitats, which are those in deeper water past the continental shelf edge.

Furthermore, within the coastal and open ocean, two kinds of habitats can be distinguished: pelagic habitats are those in the open water column up to the surface, not on the bottom, whereas demersal (or benthic) habitats are those on or near the bottom.

The water column can be divided into the upper zone (epipelagic: typically 0–200 m), middle zone (mesopelagic: typically 200–1000 m), and deepest zones (bathypelagic: 2000–4000 m and abyssopelagic: >4000 m).

The demersal habitats can also be divided into depth zones, starting shallow with inshore waters including bays and estuaries, and then further out on the floor of the continental shelf. Deeper demersal fish habitats occur at and beyond the edge of the continental shelf and slope, where the bottom goes to abyssal depths. This edge marks the boundary between coastal, relatively shallow, benthic, and the deep benthic habitats. Deepwater (or "bathydermersal") habitats are those at depths greater than 200 m.

In addition, epibenthic is used to describe organisms that occupy the top of the ocean floor, compared to those that burrow into the seafloor itself. And benthopelagic refers to fishes that swim just above the seafloor at depths >200 m (the typical edge of the continental shelf).

The majority of chondrichthyans (94%) occupy a benthic or benthopelagic habitat, with pelagic species representing only about 6% of the fauna (Compagno, 1990). Most benthic and benthopelagic species belong to one of the five major chondrichthyan groups, including the skates, squaloids, scyliorhinids, triakids, and chimaeroids, occurring in high-latitude seas. Approximately 70% of benthic and benthopelagic chondrichthyan species occur along the outer continental shelf and upper slopes, whereas 30% are considered shelf and nearshore species.

Although chondrichthyans may be characterized by broad assumptions about their habitat utilization or type, very little is known regarding ontogenetic, spatial, and temporal changes for most species (Ebert *et al.*, 2006). For example, several genera of squaloids (e.g., *Somniosus, Parmaturus*, and some *Squalus* spp.; Ebert, 2003) are known to occupy a midwater habitat as neonates, but develop a more benthic lifestyle as they grow in size and mature. Skates may also change habitat as they grow and mature, as demonstrated by adult *Bathyraja aleutica*, which tend to inhabit the area along the shelf-slope break where their nurseries are found, whereas neonates of the species migrate down the slope and occupy a deeper habitat. As this species grows in size, it migrates back up the slope and eventually occupies the slope shelf break. Additional details are provided in section "Biodiversity of sharks" for specific groups of sharks and rays.

Biodiversity of Sharks

Hexanchiformes

The Hexanchiformes (cow and frilled sharks) comprises two families, four genera, and six species of moderate-sized to very large sharks (Table 1.1; Figure 1.3). These sharks are unique in that they are the only group to possess six or seven paired gill slits, a single dorsal fin, and an anal fin. The hexanchoids are usually considered to be one of the more primitive groups of modern day sharks. Members of this group occupy a benthopelagic habitat and range from tropical

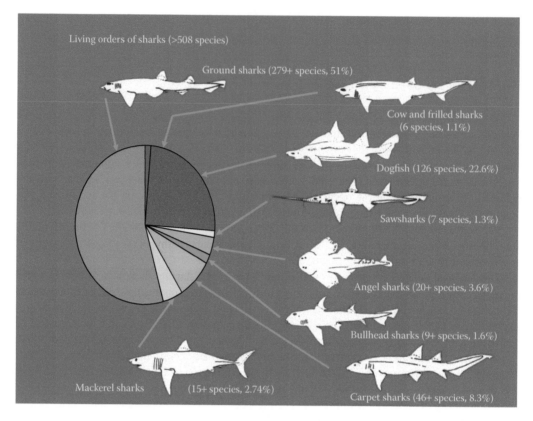

Figure 1.3   Diagram of the relative proportion of total species of sharks in each of the first eight orders in Table 1.1. (D.A. Ebert, Personal database, May 1, 2013; updated and modified from Compagno, L.J.V., *Environ. Biol. Fish.*, 28, 33–75, 1990.)

to cold-temperate seas. Despite the low number of species within this group, their wide-ranging distribution is comparable to that of more species-rich groups.

All members of this order exhibit yolk-sac viviparity, with litter sizes ranging from less than a dozen to more than 100 depending on the species (Ebert, 1990). The Chlamydoselachidae are sporadically distributed worldwide.

The Hexanchidae are also wide ranging, occurring mostly along upper continental slopes and the outer continental shelf, with the exception of *Notorynchus cepedianus*, a primarily coastal continental shelf species. *Hexanchus griseus* is perhaps one of the most abundant and widest ranging shark species, with a distribution comparable to *Squalus acanthias* (see Ebert *et al.*, 2010) and *Prionace glauca*. The hexanchoids are apex predators in those environments in which they occur. The trophic level (TL, position in the food chain, with 1 corresponding to plants and 5 corresponding to apex predators) of this shark group is one of the highest of any group, ranging between 4.2 and 4.7 (Cortés, 1999), with members feeding oncephalopods, large teleosts, including billfish, pinnipeds, and even small cetaceans on occasion (Ebert, 1990, 1991, 1994).

Squaliformes

The Squaliformes (dogfish sharks) is the third largest group of chondrichthyans, with seven families, 24 genera, and ~126 described species (Table 1.1; Figure 1.3). The squaloids are a morphologically diverse group that contains some of the smallest and largest known shark species. This order is characterized by two small to moderately large dorsal fins, which may be preceded by a fin spine in some species, the absence of an anal fin, eyes without a nictitating membrane, and five paired gill slits located anterior to the pectoral fin.

Members of this group inhabit both shallow and deepwater environs in temperate waters, but are generally replaced by requiem (Carcharhiniformes: Carcharhinidae) and hammerhead (Carcharhiniformes: Sphyrnidae) sharks in shallow warm-temperate and tropical seas.

Most of the species in this group occur in four families, the Centrophoridae, Etmoperidae, Squalidae, and Somniosidae, with one genus (*Somniosus* spp.) containing three very large species. The other nine species represented in this group are distributed among three relatively small families, the Echinorhinidae, Dalatiidae, and Oxynotidae whose members have scattered but wide-ranging distributions.

All members of this shark group exhibit yolk-sac or limited histotrophy (please refer to definitions of reproductive modes in the Life Histories section under Reproduction; Musick and Ellis, 2005), with litter sizes as small as one for some species (e.g., *Centrophorus* spp.) but upward of 300 or more in some of the large *Somniosus* species. The overall mean TL for this group is 4.1, but ranges widely from 3.5 to 4.4 depending on the individual species (Cortés, 1999). The number of species in this group is expected to increase with continued exploration of the deep sea and improved taxonomic resolution incorporating molecular tools.

The members of this group occupy a wide variety of habitats and are found from inshore to over 6000 m. Individual species may be demersal or epipelagic with some species exhibiting pronounced ontogenetic changes in habitat during their lifetime.

The Squalidae are a wide-ranging, benthopelagic family with two genera and 26 species recognized globally.

The Centrophoridae are a taxonomically challenging group with two genera and 15 species currently being recognized worldwide.

The Etmopteridae are the largest family of squaloid sharks, with four genera and 45 or more species occurring worldwide. Members are benthopelagic, deepwater sharks that mostly occur between 200 and 1500 m depth, though some species may range down to 4500 m (Compagno *et al.*, 2005).

The Somniosidae are one of the widest-ranging groups, occurring in most seas from the tropics to polar seas. Members of this group occupy a benthopelagic or pelagic habitat. The group consists of seven genera and 17 described species, containing some of the smallest and largest known shark species.

The Echinorhinidae are a small group consisting of a single genus with two wide-ranging benthopelagic species (*Echinorhinus brucus* and *E. cookei*) and a scattered worldwide distribution.

The Oxynotidae are a morphologically distinct group of five small- to medium-sized sharks within a single genus. Members are benthic dwelling sharks that occur between 50 and 650 m deep.

The Dalatiidae are a small group of seven genera and 10 species of dwarf to medium-sized, wide-ranging sharks. Members of this group are mostly pelagic or benthopelagic and occur from near-surface waters to over 1500 m deep.

Squatiniformes

The Squatiniformes (angel sharks) comprises a single family (Squatinidae) and genus (*Squatina*) of small, dorsoventrally flattened sharks (Table 1.1; Figure 1.3). The members of this group are all very similar morphologically and are often misidentified. Recent systematic work has helped clarify the regional status of the group and has led to the description of six new species in recent years. Members of this group are benthic inhabitants and are often seen partially buried in the sediment. There are approximately 20 described species worldwide, mostly in temperate to subtropical seas.

Members of this group exhibit yolk-sac viviparity. Litter sizes for individual species in this group are poorly known, but may be as high as 25 per reproductive cycle, with an average range between 6 and 12 (Ebert, 2003; personal database). These sharks are voracious benthic predators, feeding on crustaceans, cephalopods, and teleosts, and have a mean TL of 4.1 (Cortés, 1999).

Pristiophoriformes

The Pristiophoriformes (saw sharks) is a small, little known, morphologically distinct, shark group with a single family (Pristiophoridae), two genera, and seven described species (Table 1.1; Figure 1.3). These slender-bodied sharks have a long, flattened, saw-like snout, lateral rostral teeth, a pair of long string-like barbels in front of the nostrils, two dorsal fins, and no anal fin. Most species have five paired gill slits, though one species (*Pliotrema warreni*) endemic to southern Africa has six. These benthopelagic sharks primarily occur on continental shelves in temperate areas and in deeper seas, usually along the upper continental slope, in tropical regions.

Heterodontiformes

The Heterodontiformes (bullhead or horn sharks) is a small shark group with one family (Heterodontidae) and a single genus (*Heterodontus*) of nine similar-looking species (Table 1.1; Figure 1.3). These are the only living sharks with a dorsal spine preceding each dorsal fin and an anal fin. Members of this group are primarily benthic-dwelling, nocturnal sharks that tend to rest in caves and rocky crevices during the day, but become quite active at night. They are mostly nearshore sharks, occurring from the intertidal down to about 100 m, but at least one species is known to occur down to 275 m. The members of this group are temperate to tropical sharks.

Members of this group all have an oviparous reproductive mode. Although fecundity is not well known for this group, estimates for those species where some information is available suggest that they may deposit between 12 and 24 egg cases per season (Compagno, 2001, 2005; Ebert, 2003). These sharks occupy a relatively low mean TL of 3.2 (Cortés, 1999) because they feed mostly on gastropods, clams, crabs, shrimps, sea urchins, polychaete worms, and small fishes.

Orectolobiformes

The Orectolobiformes (carpet sharks) is composed of seven families, 14 genera, and 46 species globally (Table 1.1; Figure 1.3). Members of this group can be distinguished by the presence of nasal barbels, two spineless dorsal fins, and an anal fin. They are primarily benthic, tropical to warm-temperate sharks found from intertidal waters to about 200 m deep and usually on rocky or coral reefs, but one notable species (*Rhincodon typus*) occupies an oceanic, pelagic habitat.

Members of this group may exhibit either an oviparous or yolk-sac viviparous mode of reproduction. Fecundity estimates for most species in this group are poorly known except for *R. typus*, which has one of the largest litters known among modern chondrichthyans, producing at least 300 young per reproductive cycle (Joung *et al.*, 1996). The mean TL for this group is 3.6, though it ranges from 3.1 to 4.0 among the various families (Cortés, 1999).

Lamniformes

The Lamniformes (mackerel sharks) is a small but diverse group containing seven very distinctive families, 10 genera, and 15 species (Table 1.1; Figure 1.3). Although each of these families appears to be morphologically distinct, they are all united by a number of unique features, including a pointed snout, lack of dorsal fin spines, a similar style of dentition, and oophagous reproduction. These are mostly large, actively swimming, voracious sharks and are among the most media-publicized species, including white, salmon, mako, and thresher sharks. These sharks occupy a wide range of nearshore, coastal, deep-sea, and pelagic habitats from polar and cold-temperate to tropical seas.

Two families, the Cetorhinidae and Lamnidae, are large, wide-ranging groups that include three species (*Cetorhinus maximus*, *Lamna ditropis*, and *L. nasus*) that occur primarily in cold-temperate and polar seas.

The Odontaspididae, Mitsukurinidae, Pseudocarcharinidae, and Megachasmidae are small families, the latter three of which are monotypic (having only one species per family).

Reproduction in these sharks, for which information is available, is a unique form of viviparity, in which embryonic sharks feed on unfertilized eggs (oophagy) and in one species their fellow embryos within the uterus (intrauterine cannibalism).

Information on fecundity in this group of sharks is sparse, but is generally low, with estimates of only one or two young per cycle to perhaps 16 in some species. TL estimates for this group are high, ranging from 4.2 to 4.5, with the exception of the filter-feeding *C. maximus* (TL = 3.2). Depending on the species, diets may consist of marine mammals, other chondrichthyans, large teleosts, and cephalopods (Cortés, 1999).

Carcharhiniformes

The Carcharhiniformes (ground or requiem sharks) is the most diverse and largest group of shark-like cartilaginous fishes, comprising eight families, 51 genera, and at least 279 described species (Table 1.1; Figure 1.3).

The Scyliorhinidae (catsharks) are the largest family of sharks, comprising 17 genera and at least 148 species worldwide. The reproductive mode of this group is mostly oviparous, although a few species exhibit yolk-sac viviparity. Few fecundity estimates for scyliorhinids exist, but depending on the species, fecundity may range from 29 to 190, with an average of 60 (Musick and Ellis, 2005). The mean TL of these generally small demersal sharks is 3.9 (Cortés, 1999). Diets include crustaceans, cephalopods, and small bony fishes.

The Triakidae (houndsharks) are a large family consisting of nine genera and 47 species worldwide. The triakids exhibit a viviparous reproductive mode and are either mucoid histotrophs or placental (Musick and Ellis, 2005), with litter sizes ranging from 2 to over 50 (Ebert, 2003). The members of this group have a mean TL of 3.8 (Cortés, 1999) and are voracious predators on the local crustacean fauna, although some species (e.g., *G. galeus*) predominantly prey upon teleosts.

The Pseudotriakidae (false catsharks) are a small group of only four species, one of which, *Pseudotriakis microdon*, is fairly wide ranging. All false catshark species exhibit a special form of oophagous reproduction.

The Carcharhinidae (requiem sharks) are one of the larger shark families, comprising 12 genera and about 57 species worldwide. The family is most diverse in tropical to warm-temperate seas, but one species, *P. glauca*, is perhaps one of the most wide-ranging pelagic shark species occurring in all cold-temperate seas.

The Sphyrnidae (hammerhead sharks) are a small group with two genera and eight species, of mostly tropical to warm-temperate species. The members of these families are live bearers exhibiting placental viviparity, and depending on the species, litters may range from 2 to over 80. The mean TL for these shark groups is 4.0 with a range of 3.9–4.3 (Cortés, 1999). Most species feed on cephalopods, teleost fishes, and in some instances, other chondrichthyans.

Biodiversity of Rays

Batoids: Rajiformes, Myliobatiformes, Pristiformes, and Torpediniformes

Worldwide, the batoids are the largest cartilaginous fish group composed of four orders (Table 1.1; Figure 1.4), 20 families, 88 genera, and at least 639 species, a number likely to increase as new species are described (Table 1.1; Figure 1.4). The members of this group are morphologically diverse in appearance, but are united by their dorsoventrally flattened bodies and ventrally located gill slits. As with sharks, biodiversity estimates in the batoids is from D.A. Ebert's personal database, May 1, 2013, in addition to citations listed.

The Rajiformes includes both the skates and guitarfishes. This order has six families (Rhinidae, Rhynchobatidae, Rhinobatidae [guitarfishes], Anacanthobatidae, Arhynchobatidae, and Rajidae), comprising 46 genera and at least 348 described species (Table 1.1; Figure 1.4). The suborder Rajoidei consists of the *true* skates (Anacanthobatidae, Arhynchobatidae, and Rajidae), which are the most specious group of chondrichthyans, with 34 genera and at least 287 species (Ebert and Compagno, 2007; Last *et al.*, 2008).

Most skates are primarily benthic species occurring from the nearshore to depths of over 4000 m. All skate species are oviparous, with highly variable estimates of litter size ranging from about 18 to 350 egg cases per year for those species for which limited information is available (Ebert *et al.*, 2008; Lucifora and Garcia, 2004; Musick and Ellis, 2005). The mean TL for the Arhynchobatidae and Rajidae is 3.9 and 3.8, respectively (Ebert and Bizzarro, 2007).

The remaining three batoid orders include the Myliobatiformes (stingrays), which comprises eight families, 27 genera, and 217 species (Table 1.1; Figure 1.4). A second order is the Pristiformes

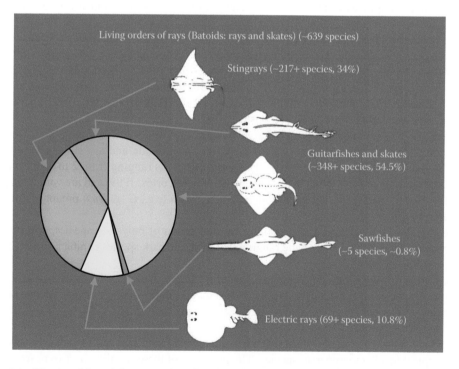

Living orders of rays (Batoids: rays and skates) (~639 species)

Stingrays (~217+ species, 34%)

Guitarfishes and skates
(~348+ species, 54.5%)

Sawfishes
(~5 species, ~0.8%)

Electric rays (69+ species, 10.8%)

Figure 1.4   Diagram of the relative proportion of total species of rays in each of the five orders currently recognized. (D.A. Ebert, Personal database, May 1, 2013; updated and modified from Compagno, L.J.V., *Environ. Biol. Fish.*, 28, 33–75, 1990; McEachran, J.D., and Aschliman, N., *Biology of Sharks and Their Relatives*, CRC Press, Boca Raton, FL, 2004; Nelson, J.S., *Fishes of the World*, Wiley, Hoboken, NJ, 2006.)

(sawfishes), which has only one family Pristidae, two genera, and a total of only five species world-wide. Last is the order Torpediniformes (electric rays), which has two families, 13 genera, and 69 widely scattered species worldwide.

Chimaeriformes

The Chimaeriformes (chimaeras, ratfish, or silver sharks) is a small, primitive group of chon-drichthyans comprising three families, six genera, and 49 described species (Table 1.1). The chi-maeras are readily distinguished by their elongated tapering body, filamentous whip-like tails, smooth scales, large venomous dorsal fin spine, broad wing-like pectoral fins, and open lateral line canals, which appear as grooves on the head and along the trunk.

Chimaeras are sluggish-swimming fishes usually found along outer continental shelves and upper slopes, although some species, especially members of the Callorhinchidae, are common in nearshore coastal waters. All members of this group are oviparous, but estimates of fecundity, rang-ing from 6 to 24 depending on the species, are sketchy (Barnett *et al.*, 2009).

LIFE HISTORIES

Introduction

The biology of the chondrichthyan fishes is among the most poorly known and least understood of all major marine vertebrate groups. Detailed information on life history and reproductive dynam-ics is available only for a few of the species that are of importance for directed fisheries. This is the result of both the low priority placed on cartilaginous fish research and the considerable difficulty of data collection for many species (particularly those restricted to deepwater habitats, or which are sampled only at certain times of year or at some stages in the life cycle).

Among the ~1200 species of known chondrichthyans, there is considerable variation in life history parameters (e.g., litter sizes among viviparous species vary from 1 to 300 (Compagno, 1990; Dulvy and Reynolds, 2002). Studies on life history parameters such as age and growth, along with basic information on distribution, abundance, movements, feeding, reproduction, and genetics, are essential for biologists to understand and predict how populations will grow and how they will respond to fish-ing pressure. Accurate age estimates provide valuable information on recruitment, age at maturity, age-specific reproduction and mortality rates, longevity, and growth rates of fished populations.

Many recommendations have been made to improve our knowledge of shark life histories (Cailliet and Tanaka, 1990; Cailliet *et al.*, 1990). These mainly center on getting more and bet-ter information about those sharks for which few biological data are available. Verification of growth zones in calcified structures is certainly among the most important, but the list also includes improved precision and accuracy, better growth models, increased sample sizes (of size classes, sexes, and geographic locations), more tag-recapture studies, and development and application of new methods of age determination, verification, and validation (Cailliet and Goldman, 2004; Cailliet *et al.*, 2006; Goldman *et al.*, 2012).

The chondrichthyans for which age and growth have been estimated and verified generally exhibit strongly K-selected life history strategies (few offspring) (Holden, 1974), especially when compared with the vast majority of r-selected, highly fecund teleost fishes (many offspring). With few exceptions (e.g., Simpfendorfer, 1992, 1993, 1999), these cartilaginous fishes exhibit, to a greater or lesser degree,

- Slow growth
- Late age at maturity

- Low fecundity and productivity (small, infrequent litters)
- Long gestation periods
- High natural survivorship for all age classes
- Long life

This suite of biological traits, which has developed over some 400 million years of evolution, results in a low reproductive potential for most species. This is an appropriate and successful strategy for an environment where the main natural predators of these fish (even as juveniles) are larger sharks. These top predators need only to produce a very few young capable of reaching maturity in order to maintain population levels under natural conditions. However, these K-selected life history characteristics, combined with the tendency of many species to aggregate by age, sex, and reproductive stage, have serious implications for the sustainability of fisheries for cartilaginous species, particularly for apex predators with few or no natural enemies and naturally small populations, even at their centers of distribution. Their limited reproductive productivity and, for many species, restricted geographical distribution severely limit the capacity of populations to sustain and recover from declines resulting from human activities (Stevens *et al.*, 2000, 2005).

Age and Growth

In the past, age and growth studies have mainly utilized size frequencies and size-at-birth and maturity estimates to predict how shark populations will fluctuate. In many cases, growth information was predicted from zones in calcified structures such as spines and vertebrae (Cailliet and Goldman, 2004; Cailliet *et al.*, 1986, 1990; Goldman *et al.*, 2012). The basic process was to remove calcified structures, such as vertebral centra, and process them so that their growth zones could be counted. Once sufficient samples of all size classes were analyzed, the result was a growth curve that represented the rate at which size (usually total length) increased with increasing age (estimated from the number of bands). In those chondrichthyans for which vertebrae were not useful aging structures, other hard parts were used (i.e., spines or caudal thorns; Gallagher and Nolan, 1999; Rago and Sosabee, 1997).

Unfortunately, most sharks and rays have not yet been reliably aged (Cailliet and Goldman, 2004; Goldman *et al.*, 2012). For some species, using the current scientific techniques on calcified structures may not even be possible, and time-consuming and expensive tagging and recapture studies are needed (Cailliet *et al.*, 1992; Kusher *et al.*, 1992).

In addition, in only very few cases have the growth zones been temporally validated, resulting in potential imprecision and inaccuracy in estimates of age at a given size, which can seriously affect our ability to manage these populations. Numerous verification and validation techniques have been reviewed by Cailliet *et al.* (1986, 1990) and Cailliet (1990). The best approach is to mark these growth zones at an initial time, then analyze growth zone deposition subsequent to a period of time the shark remained alive and at large in the field or in captivity. This technique has only been successfully applied to a few species of sharks (see Cailliet, 1986, 1990; Kusher *et al.*, 1992 for chondrichthyans and Campana, 2001 for fishes in general). It is easily seen that out of the hundreds of species of sharks and rays that exist in the world's oceans, very few have been convincingly studied.

Age estimates are used in calculations of a species' growth, mortality, and fecundity rates and, therefore, are fundamental to the assessment of a population's status and vulnerability to exploitation. As with other fishes, age estimates for chondrichthyans are produced primarily through counts of growth bands in calcified structures. To date, most chondrichthyan aging studies have relied on counts of pairs of opaque and translucent vertebral bands, which are usually interpreted as representing one year of growth (Cailliet and Goldman, 2004; Goldman *et al.*, 2012).

A common problem with aging in many deep-dwelling species is that banding patterns are difficult to discern due to poor vertebral calcification. In past studies, a variety of staining techniques have been used in attempts to improve band clarity, with differential success (Cailliet and Goldman, 2004; Goldman *et al.*, 2012). Methods involving the decalcification of vertebral centra may prove particularly useful in interpreting banding patterns in deepwater species, as has been demonstrated for *Malacoraja senta* and *Galeus melastomus* (Correia and Figueiredo, 1997; Natanson *et al.*, 2007). Several studies have used alternative structures to obtain age estimates for species in which vertebral banding patterns are unclear.

Counts of growth bands in the dorsal fin spines of squaloid sharks and chimaeras and in the neural arches of hexanchiformes have proven useful, though they have yet to be validated for most species (Cailliet and Goldman, 2004; Goldman *et al.*, 2012). Gallagher and Nolan (1999) investigated the suitability of caudal thorns as an aging structure for four bathyrajid species and found a high degree of agreement between age estimates based on counts of thorn and vertebral bands. The use of this alternative method, although successful in some skate species (Matta and Gunderson, 2007), has been found to result in a high degree of imprecision between thorn and vertebral band counts (Davis *et al.*, 2007; Perez *et al.*, 2010). Therefore, although caudal thorns may provide a nonlethal approach to aging, their use may not be appropriate for some skate species and should be investigated further before being widely applied.

Of the chondrichthyan species studied, several have proven difficult or impossible to age using conventional methods, even for species in which vertebral banding is evident. Natanson and Cailliet (1990) found that vertebral growth zones in *Squatina californica* reflected somatic growth but were not deposited annually. This was also found to be the case for *C. maximus* (Natanson *et al.*, 2008). Age estimates for deepwater catsharks (Scyliorhinidae) are also lacking, though a modified decalcification technique developed for *G. melastomus* may have application to other deepwater scyliorhinid species (Correia and Figueiredo, 1997).

Although chondrichthyans are widely believed to be a slow-growing, late-maturing group exhibiting typically low fecundities, available estimates of growth coefficients (k) in the von Bertalanffy growth model (see Cailliet and Goldman, 2004; Cailliet *et al.*, 2006; Goldman *et al.*, 2012) for species occurring predominantly in high-latitude seas vary widely (from 0.036 for *Squalus suckleyi* [see Ebert *et al.*, 2010] to 0.370 for *Beringraja binoculata*), reflecting the diversity of life history modes present. k values range from 0.036 to 0.12 in the Squaliformes, from 0.061 to 0.17 in the Lamniformes, and from 0.073 to 0.369 in the Carcharhiniformes (Ebert and Winton, 2010).

Available estimates of growth coefficients for the Rajiformes cover the broadest range, varying from 0.057 for *Dipturus batis* to 0.370 for *B. binoculata*. Estimates of k values for chimaeras range from 0.067 for *Chimaera monstrosa* to 0.224 for *Callorhynchus milii*. Though estimates for all groups vary widely, differences in k values may also reflect differences in aging methodologies, growth models, or insufficient sample sizes (Cailliet and Goldman, 2004; Goldman *et al.*, 2012).

Estimates of age at maturity and longevity of chondrichthyans vary widely within and among chondrichthyan groups. Ages at maturity range from a low of three for *Mustelus henlei* to a high of 45 for the deepwater gulper shark *Centrophorus squamosus*. Higher ages at maturity generally correspond to higher estimates of longevity, though there are some exceptions. For sharks, longevity estimates range from 14 to 70 years in the Squaliformes, from 20 to 65 years in the Lamniformes, and from 8 to 40 years in the Carcharhiniformes. Maximum age estimates for skates range from 7 to 50 years, with the two available estimates for chimaera species of 9 and 29 years. It is important to note that age estimates produced from different regions of a species' range may provide different estimates of age at maturity and longevity, whether due to latitudinal differences in population parameters or differences in sampling methodology (Cailliet and Goldman, 2004; Frisk and Miller, 2006; Goldman *et al.*, 2012; Lombardi-Carlson *et al.*, 2003).

Few studies reporting on the age and growth of chondrichthyan species have verified or validated their results, which most simply assume annual patterns of vertebral band deposition based on the few validated studies to date. Validation of the temporal periodicity of vertebral banding is essential in order to ensure the accuracy of age estimates since several studies have demonstrated that band deposition is not directly related to time in all chondrichthyan species (Cailliet and Goldman, 2004; Goldman *et al.*, 2012; Natanson and Cailliet, 1990; Natanson *et al.*, 2008). Age estimates for several shark and skate have been validated. Kusher *et al.* (1992) and Smith *et al.* (2003) both validated annual band deposition patterns in *Triakis semifasciata* through tag-recapture of oxytetracycline (OTC)-injected individuals, with Smith *et al.* (2003) reporting on a recapture of an individual after 20 years at large. Age estimates for *L. nasus* have also been validated by OTC tag-recapture and bomb radiocarbon in the western North Atlantic (Campana, 2001; Campana *et al.*, 2002; Natanson *et al.*, 2002).

Although rates of band deposition in dorsal spines have only been validated for *S. (acanthias) suckleyi* using OTC injection and bomb radiocarbon (Beamish and McFarlane, 1985; Campana *et al.*, 2006), banding patterns in the spines of *Centroselachus crepidater* have been verified by correlation between band counts and the results of radiometric analysis (Irvine *et al.*, 2006a and b). McPhie and Campana (2009) reported the only direct validation for a skate species using bomb radiocarbon to provide evidence of annual band-pair formation in *Amblyraja radiata*. Gallagher and Nolan (1999) tagged 550 bathyrajids in the waters off the Falkland Islands but only reported one return for an undescribed species not investigated in their study. Though the individual was only at liberty for 10 months, OTC was incorporated into an opaque band with a subsequent translucent band forming during that time period, suggesting an annual banding pattern.

Of those chondrichthyan fishes that have been aged, most are relatively long lived (up to about 75 years; McFarlane and Beamish, 1987) and very slow to reach maturity. Age to maturity ranges from the unusually short 1–2 years in the Australian sharpnose shark, *Rhizoprionodon taylori* (Simpfendorfer, 1992, 1993) to 20–25 years in the spiny dogfish (McFarlane and Beamish, 1987) and the dusky shark, *Carcharhinus obscurus* (Natanson *et al.*, 1995). Because of the paucity of validated age and growth studies coupled with comprehensive information on reproductive habits, such information is not known for most chondrichthyan species.

Reproduction

Chondrichthyans exhibit two main reproductive modes, oviparity and viviparity, with variations that have evolved within each primary mode. Oviparous species are those that deposit egg cases, usually a leathery external shell that protects developing embryos, on the seafloor bed. These egg cases are useful taxonomic tools because most egg cases have morphological characteristics that are unique to each species (Ebert, 2005; Ebert and Davis, 2007; Treloar *et al.*, 2006, 2007). This reproductive mode is exhibited by all members of the order Chimaeriformes and the families Arhynchobatidae, Rajidae, and most of the Scyliorhinidae.

Viviparous species, in which the developing embryos are retained within the mother's uterus, exhibit two primary developmental categories, lecithotrophy and matrotrophy. And, in these developmental categories, there is a variety of reproductive modes (Musick and Ellis, 2005). Lecithotrophic, or yolk-sac, viviparous species are those chondrichthyans whereby the young are nourished by a yolk sac and receive no maternal nourishment during development. Embryos of matrotophic species receive maternal nourishment through ingestion of lipids or mucous (histotroph) produced by the uterine wall or of extra unfertilized eggs produced by the mother (Musick and Ellis, 2005).

It is commonly believed that the females of most chondrichthyan species mature at larger sizes than males. However, this does not hold true for most chondrichthyans that exhibit an oviparous reproductive mode. Contrary to this paradigm, the sexes of most oviparous chondrichthyans mature at a similar size and in many species the males may mature at a slightly larger size (Ebert, 2005; Ebert *et al.*, 2006, 2007).

A notable exception among oviparous chondrichthyans, however, can be found among those skate species that attain a maximum total length exceeding 1.5 m. In these species, the females tend to grow to a notably larger size than males (Ebert *et al.*, 2008). Reasons for this are unclear, but may include (1) no advantage for oviparous females to attain a larger size to produce larger young, as is commonly found in viviparous species, or (2) the time an egg case might remain *in utero* is relatively minimal (as little as every 1–6 days) when compared to most viviparous species, which may carry their young *in utero* for up to two years or more (Carrier *et al.*, 2004; Ebert, 2003; Ebert *et al.*, 2008; Holden *et al.*, 1971; Ishihara *et al.*, 2002).

Fecundity estimates for most oviparous chondrichthyans are poorly known except for captive studies that suggest fecundity in some species may be relatively high both annually and throughout the entire life span of an individual skate (Ebert, 2005; Holden *et al.*, 1971; Ishihara *et al.*, 2002).

Depending on the species, females may bear from one or two (in the case of the sand tiger shark, *Carcharias taurus* and manta ray, *Manta birostris*) to 300 young (in the whale shark, *R. typus*). Gestation rates are unknown for most species, but range from around three months (e.g., rays in *Dasyatis* and *Urolophus halleri*; Hamlett and Koob, 1999) to more than 22 months for the ovoviviparous spiny dogfish (Pratt and Casey, 1990), which has the longest gestation period known for any living vertebrate. Breeding does not always occur annually in females: some species have one or more *resting* years between pregnancies.

Although data are limited, it appears that though oviparous species possess a reproductive strategy that produces smaller offspring, they may have a notably higher fecundity than observed in their viviparous relatives (Ebert *et al.*, 2008; Lucifora and Garcia, 2004; Musick and Ellis, 2005). Fecundity estimates in skates range from a low of 18–350 egg cases per year, depending on the species.

Of particular interest is the potential fecundity of two skate species, *B. binoculata* and *B. pulchra*, each of which has multiple embryos per egg case; a unique mode even among skates. The fecundity of *B. binoculata* is known to exceed 350 egg cases per year and has been estimated to produce upward of 48,000 embryos during its lifetime, a number far exceeding that of any other chondrichthyan species (Ebert *et al.*, 2008). Ishihara *et al.* (2002) reported that *Okamejei kenojei* matured within three years and produced from 300 to 600 egg cases during its life span. These authors also found that these skates grew very little after attaining sexual maturity, putting most of their energy into reproduction.

It has been suggested that chondrichthyans may become less fecund and may even senescence with age. This has been observed in *G. galeus* and in several eastern Bering Sea and western South Pacific skate species (Ebert, 2005; Peres and Vooren, 1991).

Ebert (2005) found evidence of senescence in three species of Bering Sea skates and found that the number of ovarian eggs appeared to decline slightly in larger, and presumably older, individuals in at least one species, *B. aleutica*. Frisk (2004) found that the net reproductive rate of *Leucoraja erinacea* and *Leucoraja ocellata* peaked at 7 and 13 years, respectively, but declined in fecundity beyond these ages.

These observations indicate that younger mature fish are more fecund than older, and in some instances, larger fish. This is in contrast to several species of viviparous elasmobranchs, especially those in the family Triakidae, in which size appears to be a more important indicator of fecundity than age (Ebert and Ebert, 2005). These life history traits have both ecological and fisheries management implications. The removal of older, less fecund oviparous individuals that tend to consume more resources may in effect provide more resources and habitat for younger, more virile fish. This may create a much healthier, more robust population of these important demersal elasmobranchs.

There are three main patterns of embryonic development in chondrichthyans, all of which involve considerable parental investment to produce small numbers of large, fully developed young that have a relatively high natural survival rate (Hamlett, 1997; Hamlett and Koob, 1999). Internal fertilization of relatively few eggs is followed by either one of the following:

- Attachment of the embryo by a yolk-sac placenta (placental viviparity)
- Development of unattached embryos within the uterus, with energy supplied by large egg yolks (ovoviviparity or aplacental yolk-sac viviparity), ingestion of infertile eggs (oophagy), ingestion of

eggs and smaller embryos (adelphophagy) or fluids secreted by the uterus (the last three are all forms of matrophagy)
- Development of the young within large leathery egg cases that are laid and continue to develop and hatch outside the female (oviparity)

Following their high initial investment in pup production, many sharks and rays subsequently give birth in sheltered coastal or estuarine nursery grounds, where predation risks to the pups (primarily from other sharks) are reduced (Branstetter, 1990), or deposit eggs in locations where they are most likely to survive undamaged until the pups emerge. There is no known post-birth parental care. Nevertheless, it is thought that most chondrichthyans have relatively low natural mortality coefficients (M). However, accurately estimating M is one of the most difficult things to do in marine fishes and usually indirect methods are used (Gunderson, 1980; Gunderson and Dygert, 1988; Hoenig, 1983; Jensen, 1996; Pauly, 1980; Vetter, 1987). Few direct estimates of M have been generated for chondrichthyan fishes (see Gruber et al., 2001; Hoenig and Gruber, 1990; Manire and Gruber, 1993; Simpfendorfer, 1999).

Although the large majority of chondrichthyan species are slow growing with low productivity, a few species of sharks, especially many of the smaller species, are not as extreme in their life histories as the larger, K-selected species (Smith et al., 1998). For example, the Australian sharpnose shark matures at age one, lives to age six or seven, and has an average natural mortality rate of about 0.6 (Simpfendorfer, 1999). This is in contrast to the sandbarshark Carcharhinus plumbeus with an average natural mortality rate of only about 0.10–0.05 (Sminkey and Musick, 1996). Species that have shorter life spans are likely to have higher productivity and are better able to sustain commercial fisheries, although they still require careful and conservative management (e.g., the gummy shark, Mustelus antarcticus in Southern Australia, with a maximum age of 16 years).

Demography

Chondrichthyans generally exhibit long life spans, late ages at maturity, and relatively low fecundities, characteristics that make age-structured models, such as life history tables and matrix population models, more appropriate for analyzing population dynamics than biomass dynamic models traditionally applied to teleost stocks (Cortés, 1998). These types of models can provide insight into relationships among life history traits, population productivity, and a species' consequent ability to withstand exploitation. However, as these types of analyses are dependent on detailed, age-specific life history data, it is not surprising that little exists in the way of published demographic studies for high-latitude chondrichthyans, given the relative lack of life history information available.

Several studies have recognized body size as an indicator of a species' vulnerability to exploitation in skate assemblages, based on the slower growth, later maturity, and lower productivity of larger species (Dulvy and Reynolds, 2002; Frisk, 2010; Frisk et al., 2002; Walker and Hislop, 1998). Though this observation may hold true for some species assemblages, body size should not be considered a reliable indicator in all cases. For example, B. binoculata is the largest skate in the eastern North Pacific skate complex but has among the highest growth and reproductive rates estimated for a skate species to date.

Smith et al. (1998) provided estimates of intrinsic rebound potentials based on reported life history traits for 26 Pacific Ocean shark species and predicted one of the smallest species included in their analysis, S. suckleyi (=S. acanthias in Smith et al., 1998), to have the highest vulnerability to exploitation.

Cortés (2002) found no relationship between body size and population growth rates in a study investigating 38 shark species. Although body size is a convenient measure that is easily obtainable, in reality other traits, such as growth rate, age at maturity, and fecundity, are driving this observation. Therefore, although it may correlate with other demographic parameters in many species (Frisk, 2010), body size alone should not be prioritized as an indicator of vulnerability; many relatively small deepwater species are among the most unproductive and consequently most vulnerable chondrichthyan species (García *et al.*, 2008; Kyne and Simpfendorfer, 2010; Smith *et al.*, 1998; Simpfendorfer and Kyne, 2009).

Estimates of population growth rates have only been published for several species of chondrichthyans. Demographic analyses have calculated intrinsic rates of population increase (r) ranging from 0.067 for *T. semifasciata* (Cailliet, 1992) to 0.307 for *Bathyraja interrupta* (Barnett *et al.*, 2013), with corresponding population doubling times of 10.345 and 2.350 years in the absence of fishing pressure, respectively. Intrinsic rates of increase for squaloid sharks range from 0.071 for *S. suckleyi* to 0.118 for *S. acanthias*, for lamnids from 0.081 to 0.086, and for the carcharhinids from 0.067 for *T. semifasciata* to 0.143 for *Mustelus henlei*. Predicted r values for the Rajiformes were at the higher end of the range, varying from 0.13 for *L. ocellata* to 0.307 for *B. interrupta* (Ainsley *et al.*, 2014; Barnett *et al.*, 2013).

For two species of skate, the only estimates of population growth incorporated both natural and fishing mortality, reporting low or declining r values for *A. radiata* and *Raja clavata*, respectively (Walker and Hislop, 1998). To date, no studies investigating the demography of high-latitude chimaeras have been published. Gedamke *et al.* (2007) highlighted problems with the methodology and interpretation of the population models used to produce such estimates. As advances are made in this field, the above estimates and the management recommendations based on them will doubtlessly be revised.

Stock assessments are only as good as the data used in their production; the better the life history estimates input, the more reliable the management strategy based on the results of the assessment. Because latitudinal variation in growth rates is often reported in marine species, with members of the same species from higher latitude seas attaining greater ages and ages at maturity, it is important to obtain life history information and conduct demographic analyses in all regions of a species' range, particularly in areas where the population is subject to exploitation (Frisk, 2010; Frisk and Miller, 2006; Lombardi-Carlson *et al.*, 2003).

When life histories are compared across taxa (Camhi *et al.*, 1998; Cortés, 1998, 2002), it is immediately apparent that many sharks and rays are among the latest maturing and slowest reproducing of vertebrates. Their reproductive strategies, along with the relatively close relationship between parent stock and subsequent recruitment from their live-borne or egg-borne early development, contrast markedly with those employed by all but a few examples of the teleosts, which support most fisheries (sharks and rays provide around 1% of the total world catch [Stevens, 2010]). In general, cartilaginous fishes are much slower growing and live longer than teleosts. Thousands to tens of millions of tiny eggs are produced annually by large teleost fishes and, although only very few of the young produced survive to maturity, recruitment to the adult population is broadly independent of the size of the spawning stock (at least until the latter declines to extremely low levels). This is partly due to the operation of density-dependent factors that compensate for adult population decline.

Once basic life history information, such as age, size, mortality (age-specific death rates) and natality (age-specific birth rates), is available, demography can be applied to better understand the population dynamics of sharks. Using life tables constructed of survivorship and reproductive schedules (Krebs, 1985; Mertz, 1970; Mollet and Cailliet, 2002), one can calculate the following reproductive demographic parameters:

- *Net reproductive rate* (R_o or multiplication rate per generation)
- *Generation length* ($G =$ the average time between the birth of an individual and the birth of her first offspring; also defined as the mean age of living, reproductive females in the population by International Union for Conservation of Nature)
- *Intrinsic (instantaneous) rate of increase* or growth coefficient of the population (r)
- *Finite (usually annual) rate* of population increase (e^r)
- *Doubling time* (time, in years, it takes for a population to double)

This approach has only been utilized successfully for a few shark species (Brewster-Geisz and Miller, 2000; Cailliet, 1992; Cailliet *et al.*, 1992; Cortés, 1995, 1998; Cortés and Parsons, 1996;

Simpfendorfer, 1999; Sminkey and Musick, 1996). Demographic analyses have helped manage some of these species. For example, the leopard shark, *T. semifasciata* from California waters (Cailliet, 1992; Kusher *et al.*, 1992) was estimated to have an R_0 of 4.47, a G of 22.35 years, and an r of 0.067, in the absence of fishing pressure. However, when fishing mortality is included, these population parameters are radically reduced and suggest the need for management procedures such as size and bag limits.

These results are even more graphic for the longer-lived sandbar shark, for which demographic analyses using both life history tables and stochastic matrix modeling indicate that their annual rate of population increase is only between 2.5% and 11.2% (most likely 5.2% maximum; Brewster-Geisz and Miller, 2000; Cortés, 1999; Sminkey and Musick, 1996). Thus, one can readily see how a shark population, with relatively slow growth, late age at maturity, long gestation period, and low fecundity, can be very vulnerable to overfishing.

As a result of few and inadequate age and growth estimates, the use of stock replacement and yield per recruitment models (Smith and Abramson, 1990) and demographic analyses has not been widely applied. Because of these gaps in ecological knowledge, shark populations have continued to suffer from overexploitation without the benefit of reasonable management strategies.

A relatively recent analytical technique (Au and Smith, 1997), termed *intrinsic rebound potentials* by Smith *et al.* (1998), requires less basic life history information and may prove very useful in early management efforts on newly developing shark and ray fisheries. Their method incorporated density dependence as r depended on the level of fishing mortality and the resulting decrease in population size. Productivity was strongly affected by age at maturity and little affected by maximum age. Sharks with the highest recovery potential tend to be smaller, early maturing, relatively short-lived inshore coastal species such as *Mustelus* and *Rhizoprionodon*. Those with the lowest recovery potential tend to be larger-sized, slow-growing, late-maturing, and long-lived coastal sharks such as *C. obscurus*, *C. plumbeus*, *C. leucas*, *Sphyrna lewini*, and others. The smaller-sized *S. (acanthias) suckleyi* and *Galeorhinus* were also in this group (Cortés *et al.*, 1998, 2002; Smith *et al.*, 1998).

Alternative approaches to determining vulnerability have looked for other life history traits such as body size, which are correlated with response to exploitation. In skates, body size appears to be a good predictor of vulnerability to exploitation (Dulvy and Reynolds, 2002; Dulvy *et al.*, 2000) with larger species having lower replacement rates than smaller species (Dulvy and Reynolds, 2002; Dulvy *et al.*, 2000; Walker and Hislop, 1998). However, this trend is less clear in western Atlantic skates and in Pacific sharks; there is no body size correlation with Smith *et al.*'s (1998) rebound potential (Stevens *et al.*, 2000). Although the detection of species that are potentially vulnerable to exploitation is in its infancy, refinements of these new approaches may well lead to useful tools for the assessment of vulnerability.

All traditional fisheries management strategies are based on typical teleost reproductive strategies and life history characteristics. In contrast, recruitment of cartilaginous fishes to the adult population is very closely linked to the number of breeding females (Rago and Sosabee, 1997). This suggests that as mature individuals are fished out, the number of younger fish that will support future generations will also decline, which in turn limits future productivity of the fishery and the capacity of shark populations to recover from overfishing. In this respect, the reproductive potential and strategies of the cartilaginous fishes, particularly the larger species, are more closely related to those of the cetaceans, sea turtles, and large land mammals and birds than to the teleost fishes (Musick, 1997, 1999; Musick *et al.*, 2000). As a result, a very different approach to management than that currently employed for teleosts is required for chondrichthyan fisheries to be sustainable (Stevens, 2010).

It should, however, be noted that some density-dependent factors do operate for elasmobranch stocks, notably the increase in survivorship of juveniles and smaller species as adults and larger species are fished down.

Chondrichthyans exhibit a wide variety of life history characteristics and subsequent variability in vulnerability to exploitation, with several species among the least productive and most suscep-tible to exploitation of all chondrichthyans. Problems with taxonomic resolution remain for many of the deep-dwelling, less economically valuable and hence less well-known species complexes taken primarily as bycatch. In fact, far more high-latitude seas chondrichthyans are likely taken by indi-rect fisheries, which tend to underestimate their biomass, than those that are the focus of targeted fisheries.

Diets, Feeding Ecology, and Ecological Role of Chondrichthyan Fishes

Trophically, chondrichthyans occupy a similar role to that of other high-level predators (e.g., birds and marine mammals) and as such may influence and shape those marine communities in which they occur. The major demersal groups, the squaloids, skates, scyliorhinids, and triakids, can be broadly separated by TL, with mean averages of 4.2, 3.9, 3.8, and 3.8, respectively.

However, many individual species estimates of TL were based on a single regional study with few data points, thus not taking into account that TLs may vary ontogenetically, spatially, or tem-porally. Furthermore, most of these studies do not take into account possible shifts in habitat prefer-ence due to changes in life stage. In addition, the diets of many larger, potentially trophically higher, demersal species (e.g., skates) are virtually unknown. The feeding ecology and, hence, trophic role in marine communities of the chimaeroids are by and large unknown.

The ecological role of chondrichthyans, for example, their influence on the structure of com-plex fish communities, has only recently been recognized as intensive fisheries have disturbed ecological systems. For example, the abundance of spiny dogfish, *S. (acanthias) suckleyi* and several species of skates (Rajidae) was observed to increase drastically off New England after stocks of demersal teleosts such as cod, *Gadus morhua*, and haddock, *Melanogrammus aegle-finus*, collapsed from overfishing (Anonymous, 2005). The increase in chondrichthyans was purported to be due to the decrease in their teleost competitors and predators on young, but this hypothesis has yet to be supported by additional data. More recently, with declining avail-ability of the traditional teleost species, the fishing industry has been targeting spiny dogfish in the Northwest Atlantic, with the consequence that these stocks are now in serious jeopardy (Fordham *et al.*, 2006).

Others studying fisheries that catch skates (Casey and Myers, 1998; Dulvy *et al.*, 2000; Walker and Hislop, 1998) have noted declines in the larger species of skate in the North Atlantic such as *Dipturus laevis*, *D. batis*, *D. oxyrhyncus*, and *Rostroraja alba*, whereas smaller species increased in abundance. Increased fishing mortality has altered the species composition so that the species with the earliest age at maturity now dominate. Dulvy *et al.* (2000) suggested that the removal of larger skates may have led to the increase in smaller species through increased food availability. Competitive release such as this has also been implicated in a community shift from a teleost-dominated to a chondrichthyan-dominated community on Georges Bank (Murawski and Idoine, 1992).

Inferential evidence of predation by large sharks on small juvenile sharks was provided by Van Der Elst (1979), who reported that the abundance of young dusky sharks increased off South Africa after the large, predatory sharks had been reduced in numbers by protective beach mesh-ing. However, Dudley and Cliff (1993) pointed out that small sharks are not as important in the diet of large sharks, as suggested by the 1979 study. Musick *et al.* (1993) suggested that juvenile sandbar sharks increased substantially in the Chesapeake Bight off the US mid-Atlantic coast after a 70%–80% reduction in large shark populations. A small species, the Atlantic sharpnose shark, *Rhizoprionodon terraenovae*, also increased in abundance at the same time.

Although the diets of many shark species have been studied in detail (Cortés, 2000; Simpfendorfer *et al.*, 2001), the trophic effects of sharks on ecosystems other than those mentioned earlier in this section are largely unknown. However, it is widely considered that, as apex predators, the larger species are likely to significantly affect the population size of their prey species and the structure and species composition of the lower trophic levels of the marine ecosystem. This, of course, is dependent upon the rate at which chondrichthyans consume prey. Gastric evacuation studies indicate that many sharks take a relatively long period of time to process their meals (Cortés, 1997; Cortés and Gruber, 1990; Wetherbee *et al.*, 1987), and, therefore, do not eat often, reducing the potential effect on their prey. Anderson and Hafiz (2002) report that Maldivian fishermen believe that if they remove all pelagic sharks from an area, they will no longer catch tuna, implying that the removal of their predators will alter the behavior of tuna so that shoaling no longer occurs. More research is required to confirm that these species do, indeed, play a similar role to that of apex predators in the terrestrial environment.

Stevens *et al.* (2000) reviewed trophic interactions resulting from the effects of fishing on chondrichthyans and also modeled selected ecosystems to examine the effects of chondrichthyan removal. They found that ecosystem responses to the removal of sharks are complex and fairly unpredictable, though may well be ecologically and economically significant and should be studied further. For example, removal of tiger sharks from a tropical ecosystem resulted in a decline in numbers of some important commercial fish species, such as tuna, even though the latter were not important prey for the sharks and might therefore have been expected to increase in abundance following loss of sharks from the ecosystem. The tuna decline, in fact, occurred because the sharks kept populations of other predators of these fishes in check.

FISHERIES AND CONSERVATION

Historically, fisheries targeting chondrichthyans have proven unsustainable. As fisheries begin to exploit populations in deeper, more remote waters due to continued global population declines in traditionally targeted, shallow nearshore stocks (Roberts, 2002), there is an immediate need to direct funding and research effort toward the collection of *baseline* population data for poorly known high-latitude species so that management strategies can be developed and implemented.

In addition, because more emphasis is now placed on ecosystem-based management, this requires more research on multispecies interactions. An example is the study by Agnew *et al.* (2000), which investigated multispecies skate and ray off the Falkland Islands (Agnew *et al.*, 2000). Others (Link *et al.*, 2002) are doing similar community-based studies in other areas.

Thus, with the information at hand, the vulnerability of chondrichthyans to exploitation varies with life history, size attained, depth of habitat, and other aspects of their biology. It will be interesting to see in the chapters in this book how shark and ray immunology might fit into this consideration. If species, genera, families, and even orders of chondrichthyans are different in their immune systems and how well they protect individual fish, then some might be able to handle environmental perturbations better than others. The answer remains to be seen.

ACKNOWLEDGMENTS

We thank Megan Winton (Pacific Shark Research Center/Moss Landing Marine Laboratories), Joe Bizzarro (Pacific Shark Research Center/Moss Landing Marine Laboratories and University of Washington), Jack Musick (Virginia Institute of Marine Science), Will White and John Stevens (CSIRO, Hobart, Tasmania), Malcolm Francis (National Institute of Water and Atmospheric Research, Wellington, New Zealand), Hajime Ishihara (W&I Associates Corporation, Japan), Pete Kyne (Charles Darwin University, Australia), Colin Simpfendorfer (James Cook University, Australia), Dave Kulka (formerly Department of Fisheries and Ocean, Canada), Daniel Figueroa

(Departamento de Ciencias Marinas, Universidad Nacional de Mar del Plata), and Lisa Natanson (NOAA Fisheries Service, Northeast Fisheries Science Center), who provided invaluable assistance on various portions of this study. Funding for this project was provided by NOAA/NMFS to the National Shark Research Consortium and Pacific Shark Research Center, the David and Lucile Packard Foundation, and the North Pacific Research Board.

REFERENCES

Agnew, D.J., Nolan, C.P., Beddington, J.R., and Baranowski, R. Approaches to the assessment and management of multispecies skate and ray fisheries using the Falkland Islands fishery as an example. *Canadian Journal of Fisheries and Aquatic Sciences* 2000; 57:429–440.

Ainsley, S.M., Ebert, D.A., Natanson, L.J., and Cailliet G.M. A comparison of age and growth of the Bering Skate, *Bathyraja interrupta* (Gill and Townsend, 1897), from two Alaskan large marine ecosystems. *Fisheries Research* 2014; 154:17–25.

Anderson, R.C., and Hafiz, A. Elasmobranch fisheries in the Maldives. In: *Elasmobranch Biodiversity, Conservation and Management: Proceedings of the International Seminar and Workshop*, Sabah, Malaysia, July 1997 IUCN/SSC Shark Specialist Group. IUCN, Gland, Switzerland; Cambridge, UK, 2002; 114–121.

Anonymous. Status of the Fishery Resources off the Northeastern United States for 1994. NOAA Technical Memorandum NMFS-NE-NMFS. National Marine Fisheries Service, Springfield, VA, 2005; 108.

Au, D.W., and Smith, S.E. A demographic model with population density compensation for estimating productivity and yield per recruit of the leopard shark (*Triakis semifasciata*). *Canadian Journal of Fisheries and Aquatic Science* 1997; 54:415–420.

Barnes, R.S.K., and Hughes, R.K. *An Introduction to Marine Ecology*. Blackwell Science Ltd, Malden, MA, 1999; 296 p.

Barnett, L.A.K., Earley, R.L., Ebert, D.A., and Cailliet, G.M. Maturity, fecundity, and reproductive cycle of the spotted ratfish, *Hydrolagus colliei*. *Marine Biology* 2009; 156:301–316.

Barnett, L.A.K., Winton, M.V., Ainsley, S.M., Cailliet, G.M., and Ebert, D.A. Comparative demography of skates: Life-history correlates of productivity and implications for management. *PLoS ONE* 2013; 8(5):e63421. doi:10.1371/journal.pone.0063421.

Beamish, R.J., and McFarlane, G.A. Annulus development on the second dorsal spine of the spiny dogfish (*Squalus acanthias*) and its validity for age determination. *Canadian Journal of Fisheries and Aquatic Sciences* 1985; 42(11):1799–1805.

Branstetter, S. Early life history implications of selected carcharhinoid and lamnoid sharks of the Northwestern Atlantic. In: Pratt, H.L., Jr., Gruber, S.H., and Taniuchi, T., editors. *Elasmobranchs as Living Resources: Advances in the Biology, Ecology, Systematics, and the Status of the Fisheries*. NOAA Technical Report NMFS. NMFS, Springfield, VA, 1990; 90:17–28.

Brewster-Geisz, K.K., and Miller, T.J. Management of the sandbar shark, *Carcharhinus plumbeus*: Implications of a stage-based model. *Fishery Bulletin* 2000; 98:236–249.

Briggs, J.C. *Global Biogeography*. Elsevier, Amsterdam, the Netherlands, 1995.

Cailliet, G.M. Elasmobranch age determination and verification: An updated review. In: Pratt, H.L., Jr., Gruber, S.H., and Taniuchi, T., editors. *Elasmobranchs as Living Resources: Advances in the Biology, Ecology, Systematics, and the Status of the Fisheries*. NOAA Technical Report NMFS. NMFS, Springfield, VA, 1990; 90:157–165.

Cailliet, G.M. Demography of the central California population of the leopard shark (*Triakis semifasciata*). *Australian Journal of Marine and Freshwater Research* 1992; 43:183–193.

Cailliet, G.M., and Goldman, K. Age determination and validation in chondrichthyan fishes. In: Carrier, J.C., Musick, J.A., and Heithaus, M.R., editors. *Biology of Sharks and Their Relatives*, CRC Press, Boca Raton, FL, 2004; 399–447.

Cailliet, G.M., Martin, L.K., Kusher, D., Wolf, P., and Welden, B.A. Techniques for enhancing vertebral bands in age estimation of California elasmobranchs. In: Prince, E.D., and Pulos, L.M., editors. *Tunas, Billfishes, Sharks. Proceedings of an International Workshop on Age Determination of Oceanic Pelagic Fishes*. NOAA Technical Report NMFS. NMFS, Springfield, VA, 1983; 8:157–165.

Cailliet, G.M., Mollet, H.F., Pittenger, G.G., Bedford, D., and Natanson, L.J. Growth and demography of the Pacific angel shark (*Squatina californica*), based upon tag returns off California. *Australian Journal of Marine and Freshwater Research* 1992; 43:1313–1330.

Cailliet, G.M., Radtke, R.L., and Welden, B.A. Elasmobranch age determination and verification: A review. In: Uyeno, T., Arai, R., Taniuchi, T., and Matsuura, K., editors. *Indo-Pacific Fish Biology: Proceedings of the Second International Conference on Indo-Pacific Fishes*. Ichthyological Society of Japan, Tokyo, 1986; 345–360.

Cailliet, G.M., Smith, W.D., Mollet, H.F., and Goldman, K.J. Age and growth studies of chondrichthyan fishes: The need for consistency in terminology, verification, validation, and growth function fitting. *Environmental Biology of Fishes* 2006; 77:211–228.

Cailliet, G.M., and Tanaka, T. Recommendations for research needed to better understand the age and growth of elasmobranchs. In: Pratt, H.L., Jr., Gruber, S.H., and Taniuchi, T., editors. *Elasmobranchs as Living Resources: Advances in the Biology, Ecology, Systematics, and the Status of the Fisheries*. NOAA Technical Report NMFS. NMFS, Springfield, VA, 1990; 90:505–507.

Cailliet, G.M., Yudkin, K.G., Tanaka, S., and Taniuchi, T. Growth characteristics of two poulations of Mustelus Manazo from Japan based on cross readings of vertebral bands. In: Pratt, H.L., Jr., Gruber, S.H., and Taniuchi, T., editors. *Elasmobranchs as Living Resources: Advances in the Biology, Ecology, Systematics, and the Status of the Fisheries*. NOAA Technical Report NMFS. NMFS, Springfield, VA, 1990; 90:167–176.

Camhi, M., Fowler, S.H., Musick, J., Brautigam, A., and Fordham, S. Sharks and Their Relatives: Ecology and Conservation. Occasional Paper of the IUCN Species Survival Commission No. 20, IUCN, Gland, Switzerland; Cambridge, UK, 1998.

Campana, S. Accuracy, precision and quality control in age determination, including a review of the use and abuse of age validation methods. Review Paper, *Journal of Fish Biology* 2001; 59:197–242.

Campana, S.E., Jones, C., McFarlane, G.A., and Myklevoll, S. Bomb dating and age validation using the spines of spiny dogfish (*Squalus acanthias*). *Environmental Biology of Fishes* 2006; 77:327–336.

Campana, S.E., Natanson, L.J., and Myklevoll, S. Bomb dating and age determination of large pelagic sharks. *Canadian Journal of Fisheries and Aquatic Sciences*. 2002; 59:450–455.

Carrier, J.C., Pratt, H.L., and Castro J.I., Reproductive biology of elasmobranchs. In: Carrier, J.C., Musick J.A., and Heithaus M.R., editors. *Biology of Sharks and Their Relatives*. CRC Press, Boca Raton, FL, 2004; 269–286.

Casey, J.M., and Myers, R.A. Near extinction of a large, widely distributed fish. *Science* 1998; 28:690–692.

Compagno, L.J.V. Alternate life history styles of cartilaginous fishes in time and space. *Environmental Biology of Fishes* 1990; 28:33–75.

Compagno, L.J.V. Sharks of the world. An annotated and illustrated catalogue of shark species known to date. Volume 2. bullhead, mackerel and carpet sharks (Heterodontiformes, Lamniformes and Orectolobiformes). FAO Species Catalogue for Fishery Purposes. FAO, Rome, Italy, 2001; 1(2).

Compagno, L.J.V. Checklist of living chondrichthyans. In: Hamlett, W.C., and Jamieson, B.G.M., editors. *Reproductive Biology and Phylogeny of Chondrichthyes: Sharks, Rays, and Chimaeras*. Science Publishers, Enfield, NH, 2005; 503–538.

Compagno, L., Dando, M., and Fowler, S. *Sharks of the World: Collins Field Guide*. HarperCollins Publishers, Ltd., London, 2005; 368 p.

Correia, J.P., and Figueiredo, I.M. A modified decalcification technique for enhancing growth bands in deep-coned vertebrae of elasmobranchs. *Environmental Biology of Fishes* 1997; 50:225–230.

Cortés, E. Demographic analysis of the Atlantic sharpnose shark, *Rhizoprionodon terraenovae*, in the Gulf of Mexico. *Fishery Bulletin* 1995; 93:57–66.

Cortés, E. A critical review of methods of studying fish feeding based on analysis of stomach contents: Application to elasmobranch fishes. *Canadian Journal of Fisheries and Aquatic Science* 1997; 54:726–738.

Cortés, E. Demographic analysis as an aid in shark stock assessment and management. *Fisheries Research* 1998; 39:199–208.

Cortés, E. Standardized diet compositions and trophic levels of sharks. *ICES Journal of Marine Science* 1999; 56:707–719.

Cortés, E. Incorporating uncertainty into demographic modeling: Application to shark populations and their conservation. *Conservation Biology* 2002; 16:1048–1062.

Cortés, E., and Gruber S.H. Diet, feeding habits and estimates of daily ration of young lemon sharks, *Negaprion brevirostris. Copeia* 1990; 1:204–218.

Cortés, E., and Parsons, G.R. Comparative demography of two populations of the bonnethead shark (*Sphyrna tiburo*). *Canadian Journal of Fisheries and Aquatic Sciences* 1996; 53:709–717.

Davis, C.D., Cailliet, G.M., and Ebert, D.A. Age and growth of the roughtail skate *Bathyraja trachura* (Gilbert 1892) from the eastern North Pacific. *Environmental Biology of Fishes* 2007; 80:325–336.

de Carvalho, M.R. Higher-level elasmobranch phylogeny, basal squaleans, and paraphyly. In: Stiassny, M.L.J., Parenti, L.R., and Johnson, G.D., editors. *Interrelationships of Fishes*. Academic Press, San Diego, CA, 1996; 35–62.

Dudley, S.F.J., and Cliff, G. Some effects of shark nets in the Natal nearshore environment. *Environmental Biology of Fishes* 1993; 36:243–255.

Dulvy, N.K., Baum, J.K., Clarke, S., Compagno, L.J.V., Cortés, E., Domingo, A., Fordham, S. *et al.* You can swim but you can't hide: The global status and conservation of oceanic pelagic sharks and rays. *Aquatic Conservation: Marine and Freshwater Ecosystems* 2008; 18(5):459–482.

Dulvy, N.K., Metcalfe, J.D., Glanville, J., Pawson, M.G., and Reynolds, J.D. Fishery stability, local extinctions, and shifts in community structure in skates. *Conservation Biology* 2000; 14(1):283–293.

Dulvy, N.K., and Reynolds, J.D. Predicting extinction vulnerability in skates. *Conservation Biology* 2002; 16(2):440–450.

Ebert, D.A. The Taxonomy, Biogeography and Biology of Cow and Frilled Sharks (Chondrichthyes: Hexanchiformes). Doctoral Dissertation, Rhodes University, Grahamstown, South Africa, 1990.

Ebert, D.A. Diet of the seven gill shark *Notorynchus cepedianus* in the temperate coastal waters of southern Africa. *South African Journal of Marine Science* 1991; 11:565–572.

Ebert, D.A. Diet of the six gill shark *Hexanchus griseus* off southern Africa. *South African Journal of Marine Science* 1994; 14:213–218.

Ebert, D.A. *Sharks, Rays, and Chimaeras of California*. University of California Press, Berkeley, CA, 2003.

Ebert, D.A. Reproductive biology of skates, *Bathyraja* (Ishiyama), along the eastern Bering Sea continental slope. *Journal of Fish Biology* 2005; 66:618–649.

Ebert, D.A., and Bizzarro, J.J. Standardized diet composition and trophic levels in skates. *Environmental Biology of Fishes* 2007; 80:221–237.

Ebert, D.A., and Compagno, L.J.V. Biodiversity and systematics of skates (Chondrichthyes: Rajiformes: Rajoidei). *Environmental Biology of Fishes* 2007; 80:111–124.

Ebert, D.A., Compagno, L.J.V., and Cowley, P.D. Reproductive biology of catsharks (Chondrichthyes: Scyliorhinidae) off the west coast of southern Africa. *ICES Journal of Marine Science* 2006; 63:1053–1065.

Ebert, D.A., and Davis, C.D. Description of skate egg cases (Chondrichthyes: Rajiformes: Rajoidei) from the eastern North Pacific. *Zootaxa* 2007; 1393:1–18.

Ebert, D.A., and Ebert, T.B. Reproduction, diet and habitat use of leopard sharks, *Triakis semifasciata* (Girard), in Humboldt Bay, California, USA. *Marine and Freshwaer Research* 2005; 56:1089–1098.

Ebert, D.A., Smith, W.D., and Cailliet, G.M. Reproductive biology of two commercially exploited skates, *Raja binoculata* and *R. rhina*, in the western Gulf of Alaska. *Fisheries Research* 2008; 94:48–57.

Ebert, D.A., Smith, W.D., Haas, D.L., Ainsley, S.M., and Cailliet, G.M. Life History and Population Dynamics of Alaskan Skates: Providing Essential Biological Information for Effective Management of by Catch and Target Species. North Pacific Research Board Project 510 Final Report, North Pacific Research Board, Anchorage, AL. 2007.

Ebert, D.A., White, W.T., Goldman, K.J., Compagno, L.J.V., Daly-Engel, T.S., and Ward, R.D. Reevaluation and redescription of *Squalus suckleyi* (Girard, 1854) from the North Pacific, with comments on the *Squalus acanthias* subgroup (Squaliformes: Squalidae). *Zootaxa* 2010; 2612:22–40.

Ebert, D.A., and Winton, M.V. Chondrichthyans of high latitude seas. In: Carrier, J.C., Musick, J.A., and Heithaus, M.R., editors. *Sharks and Their Relatives II: Biodiversity, Adaptive Physiology and Conservation*. CRC Press, Taylor & Francis Group, Boca Raton, FL, 2010; Chapter 3; 115–158.

Eschmeyer, W.N., editor. Catalog of Fishes. California Academy of Sciences. http://research.calacademy.org /research/ichthyology/catalog/fishcatmain.asp. Electronic version. Accessed April 15, 2013.

Fordham, S., Fowler, S.L., Coelho, R., Goldman, K.J., and Francis, M. *Squalus acanthias*. 2006 IUCN Red List of Threatened Species. www.iucnredlist.org.

Frisk M.G. Biology, life history and conservation of elasmobranchs with an emphasis in western Atlantic skates. Dissertation, University of Maryland, College Park, MD. 2004.

Frisk, M.G. Life history strategies of batoids. In: Carrier, J.C., Musick, J.A., and Heithaus, M.R., editors. *Biology of Sharks and Their Relatives: Physiological Adaptations, Behavior, Ecology, Conservation and Management of Sharks and Their Relatives*. CRC Press, Boca Raton, FL, 2010; 2:283–316.

Frisk, M.G., and Miller, T.J. Age, growth and latitudinal patterns of two Rajidae species in the northwestern Atlantic: Little skate (*Leucoraja erinacea*) and winter skate (*Leucoraja ocellata*). *Canadian Journal of Fisheries and Aquatic Sciences* 2006; 63:1078–1091.

Frisk, M.G., Miller, T.J., and Fogarty, M.J. The population dynamics of the little skate *Leucoraja erinacea*, winter skate *Leucoraja ocellata*, and barndoor skate *Dipturus laevis*: Predicting exploitation limits using matrix analyses. *ICES Journal of Marine Science* 2002; 59:576–586.

Gallagher, M.J., and Nolan, C.P. A novel method for the estimation of age and growth in rajids using caudal thorns. *Canadian Journal of Fisheries and Aquatic Sciences* 1999; 56(9):1590–1599.

García, V.B., Lucifora, L.O., and Myers, R.A. The importance of habitat and life history to extinction risk in sharks, skates, rays and chimaeras. *Proceedings of the Royal Society B* 2008; 275:83–89.

Gedamke, T., Hoenig, J.M., Musick, J.A., DuPaul, W.D., and Gruber, S.H. Using demographic models to determine intrinsic rate of increase and sustainable fishing for elasmobranchs: Pitfalls, advances and applications. *North American Journal of Fishery Management* 2007; 27:605–618.

Goldman, K.J., Cailliet, G.M., Andrews, A.H., and Natanson, L.J. Assessing the age and growth of chondrichthyan fishes. In: Carrier, J., Musick, J.A., and Heithaus, M.R., editors. *Biology of Sharks and Their Relatives*, Second Edition. CRC Press, Boca Raton, FL, 2012; Chapter 14, 419–447.

Gruber, S.H., de Marignac, J.R.C., and Hoenig, J.M. Survival of juvenile lemon sharks at Bimini, Bahamas, estimated by mark-depletion experiments. *Transactions, American Fisheries Society* 2001; 130:376–384.

Gunderson, D.R. Using r-K selection theory to predict natural mortality. *Canadian Journal of Fisheries and Aquatic Sciences* 1980; 37:2266–2271.

Gunderson, D.R., and Dygert, P.H. Reproductive effort as a predictor of natural mortality rate. *Journal du Conseil, International Exploration de la Mer* 1988; 44:200–209.

Hamlett, W.C. Reproductive modes of elasmobranchs. *Shark News* 1997; 9:1–3.

Hamlett, W.C., and Koob, T.J. Female reproductive system. In: Hamlett, W.C., editor, *Sharks, Skates, and Rays: The Biology of Elasmobranch Fishes*, The Johns Hopkins University Press, Baltimore, MD 1999; Chapter 15, 398–443.

Hoenig, J.M. Empirical use of longevity data to estimate mortality rates. *Fishery Bulletin* 1983; 82(1):898–903.

Hoenig, J.M., and Gruber, S.H. Life history patterns in the elasmobranchs: Implications for fisheries management. In: Pratt, H.L., Jr., Gruber, S.H., and Taniuchi T., editors. *Elasmobranchs as Living Resources: Advances in the Biology, Ecology, Systematics, and the Status of the Fisheries*. NOAA Technical Report NMFS. NMFS, Springfield, VA, 1990; 90:1–16.

Holden, M.J. Problems in the rational exploitation of elasmobranch populations and some suggested solutions. In: Harden, F.R., and Jones, C., editors. *Sea Fisheries Research*. Logos Press, London, 1974; 117–138.

Holden, M.J., Rout, D.W., and Humphreys, C.N. The rate of egg laying by three species of ray. *Journal du Conseil, International Exploration de la Mer* 1971; 33(3):335–339.

Irvine, S.B., Stevens, J., and Laurenson, L. Comparing external and internal dorsal-spine bands to interpret the age and growth of the giant lantern shark, *Etmopterus baxteri* (Squaliformes: Etmopteridae). *Environmental Biology of Fishes* 2006a; 77(3–4):253–264.

Irvine, S.B., Stevens, J.D., and Laurenson, L.J.B. Surface bands on deepwater squalid dorsal-fin spines: An alternative method for aging *Centroselachus crepidater*. *Canadian Journal of Fisheries and Aquatic Sciences* 2006b; 63(3):617–627.

Ishihara, H., Mochizuki, T., Homma, K., and Taniuchi, T. Reproductive strategy of the Japanese common skate (Spiny rasp skate) *Okameji kenojei*. In: Fowler, S.L., Reed, T.M., and Dipper, F.A., editors. *Elasmobranch Biodiversity, Conservation and Management: Proceedings of the International Seminar and Workshop*, Sabah, Malaysia, July 1997. IUCN SSC Shark Specialist Group. IUCN, Gland, Switzerland; Cambridge, 2002; 236–240.

Jensen, A.L. Beverton and Holt life history invariants result from optimal trade-off of reproduction and survival. *Canadian Journal of Fisheries and Aquatic Sciences* 1996; 53:820–822.

Joung, S.J., Chen, C.T., Clark, E., Uchida, S., and Huang, W.Y.P. The whale shark, *Rhincodon typus*, is a livebearer: 300 embryos found in one "megamamma supreme." *Environmental Biology of Fishes* 1996; 46:219–223.

Krebs, C. *Ecology, the Experimental Analysis of Distribution and Abundance*, Third Edition, Harper and Row, New York, NY, 1985.

Kusher, D.I., Smith, S.E., and Cailliet, G.M. Validated age and growth of the leopard shark, *Triakis semifasciata*, from central California. *Environmental Biology of Fishes* 1992; 35:187–203.

Kyne, P.M., and Simpfendorfer, C.A. Deepwater chondrichthyans. In: Carrier, J.C., Musick, J.A., and Heithaus, M.R., editors. *Sharks and Their Relatives II: Biodiversity, Adaptive Physiology, and Conservation.* CRC Press, Boca Raton, FL, 2010; 37–114.

Last, P.R., White, W.T., and Pogonoski, J.J., editors. *Descriptions of New Australian Chondrichthyans.* CSIRO Marine and Atmospheric Research Paper No. 022.2008. CSIRO, Hobart, Tasmania, 2008.

Link, J.S., Garrison, L.P., and Almeida, F.P. Ecological interactions between elasmobranchs and groundfish species on the northeastern U.S. continental shelf. I. Evaluating predation. *North American Journal of Fishery Management* 2002; 22:550–562.

Lombardi-Carlson, L.A., Cortés, E., Parsons, G.R., and Manire, C.A. Latitudinal variation in life-history traits of bonnethead sharks, *Sphyrna tiburo* (Carcharhiniformes: Sphyrnidae), from the eastern Gulf of Mexico. *Marine and Freshwater Research* 2003; 54:875–883.

Lucifora, L.O., and Garcia, V.B. Gastropod predation on egg cases of skates (Chondrichthyes, Rajidae) in the southwestern Atlantic: Quantification and life history implications. *Marine Biology* 2004; 145:917–922.

Luning, K., and Asmus, R. Physical characteristics of littoral ecosystems, with special reference to marine plants. In: Mathieson, A.C., and Nienhaus, P.H., editors. *Ecosystems of the World 24: Intertidal and Littoral Ecosystems.* Elsevier, Amsterdam, the Netherlands, 1991.

Manire, C.H., and Gruber, S.H. A Preliminary Estimate of Natural Mortality of 0-age Lemon Sharks. NOAA Technical Report, NMFS, Springfield, VA, 1993; 115:65–71.

Matta, M.E., and Gunderson, D.R. Age, growth, maturity, and mortality of the Alaska skate, *Bathyraja parmifera*, in the eastern Bering Sea. *Environmental Biology of Fishes* 2007; 80:309–323.

McEachran, J.D., and Aschliman, N. Phylogeny of batoidea. In: Carrier, J.C., Musick, J.A., and Heithaus, M.R., editors. *Biology of Sharks and Their Relatives.* CRC Press, Boca Raton, FL, 2004; 79–113.

McFarlane, G.A., and Beamish, R.J. Validation of the dorsal spine method of age determination for spiny dogfish. In: Summerfelt, R.C., and Hall, G.E. editors, *Age and Growth of Fish.* Iowa State University Press, Ames, IA, 1987; 287–300.

McPhie, R.P., and Campana, S.E. Bomb dating and age determination of skates (family Rajidae) off the eastern coast of Canada. *ICES Journal of Marine Science* 2009; 66(3):546–560.

Mertz, D.B. Notes on methods used in life history studies. In: Connell, J.H., Mertz, D.B., and Murdoch, W.W., editors. *Readings in Ecology and Ecological Genetics.* Harper and Row, New York, NY, 1970; 4–17.

Mollet, H.F., and Cailliet, G.M. Comparative population demography of elasmobranchs using life history tables, Leslie matrices, and stage-based matrix models. *Marine and Freshwater Research* 2002; 53(2):503–516.

Murawski, S.A., and Idoine, J.S. Multispecies size composition: A conservative property of exploited fishery systems? *Journal of Northwest Atlantic Fishery Science* 1992; 14:79–85.

Musick, J.A. Restoring stocks at risk. *Fisheries* 1997; 22(7):31–32.

Musick, J.A. editor. *Life in the Slow Lane: Ecology and Conservation of Long-lived Marine Animals.* American Fisheries Society Symposium 23, Bethesda, MD, 1999.

Musick, J.A., Branstetter, S., and Colvocoresses, J.A. *Trends in Shark Abundance from 1974 to 1991 for the Chesapeake Bight Region of the US mid-Atlantic Coast.* NOAA Technical Report NMFS. NMFS, Springfield, VA, 1993; 115:1–19.

Musick, J.A., Burgess, G., Cailliet, G.M., Camhi, M., and Fordham, S. Management of sharks and their relatives (Elasmobranchii). *Fisheries* 2000; 25(3):9–13.

Musick, J.A., and Ellis, J.K. Reproductive evolution of chondrichthyans. In: Hamlett, W.C., and Jamieson, B.G.M., editors. *Reproductive Biology and Phylogeny of Chondrichthyes: Sharks, Batoids, and Chimaeras.* Science Publishers, Enfield, NH, 2005; 45–79.

Natanson, L.J., and Cailliet, G.M. Vertebral growth zone deposition in Pacific angel sharks. *Copeia* 1990(4):1133–1145.

Natanson, L.J., Casey, J.G., and Kohler, N.E. Age and growth estimates of the dusky shark, *Carcharhinus obscurus*, in the western north Atlantic Ocean. *Fishery Bulletin* 1995; 93:116–126.

Natanson, L.J., Mello, J.J., and Campana, S.E. Validated age and growth of the porbeagle shark (*Lamna nasus*) in the western North Atlantic Ocean. *Fishery Bulletin* 2002; 100:266–278.

Natanson, L.J., Sulikowski, J.A., Kneebone, J.R., and Tsang, P.C. Age and growth estimates for the smooth skate, *Malacoraja senta*, in the Gulf of Maine. *Environmental Biology of Fishes* 2007; 80:293–308.

Natanson, L.J., Wintner, S.P., Johansson, F., Piercy, A., Campbell, P., De Maddalena, A., Gulak, S.J.B. *et al.* Ontogenetic vertebral growth patterns in the basking shark *Cetorhinus maximus*. *Marine Ecology Progress Series* 2008; 361:267–278.

Naylor, G.J.P., Ryburn, J.A., Fedrigo, O., and López, A. Phylogenetic relationships among the major lineages of sharks and rays. In: Hamlett, W.C., and Jamieson, B.G.M., editors. *Reproductive Biology and Phylogeny of Chondrichthyes: Sharks, Batoids, and Chimaeras*. Science Publishers, Enfield, NH, 2005; 1–25.

Nelson, J.S. *Fishes of the World*. Wiley, Hoboken, NJ, 2006.

Pauly, D. On the interrelationship between natural mortality, growth parameters, and mean environmental temperature in 175 fish stocks. *Journal du Conseil, International Exploration de la Mer* 1980; 39(2):175–192.

Peres, M.B., and Vooren, C.M. Sexual development, reproductive cycle, and fecundity of the school shark *Galeorhinus galeus* off southern Brazil. *Fishery Bulletin* 1991; 89(4):655–667.

Perez, C.R., Cailliet, G.M., and Ebert, D.A. Age and growth of the sandpaper skate. *Bathyraja kincaidii* (Garman, 1908), using vertebral centra with an investigation of caudal thorns. *Journal of the Marine Biological Association of the United Kingdom* 2010; 91(6):1149–1156.

Pratt, H.L., Jr., and Casey, J.G. Shark reproductive strategies as a limiting factor in directed fisheries, with a review of Holden's method of estimating growth parameters. In: Pratt, H.L., Jr., Gruber, S.H., and Taniuchi, T., editors. *Elasmobranchs as Living Resources: Advances in the Biology, Ecology, Systematics, and the Status of the Fisheries*. NOAA Technical Report NMFS. NMFS, Springfield, VA, 1990; 90:97–109.

Rago, P., and Sosabee, K. Spiny Dogfish (*Squalus acanthias*) Stock Assessment. NMFS SAW-26, SARC Working Paper D1, NOAA (National Oceanic and Atmospheric Administration) Fisheries (National Marine Fisheries Service), Silver Spring, MD, 1997; 1–83.

Roberts, C.M. Deep impact: The rising toll of fishing in the deep sea. *Trends in Ecology and Evolution* 2002; 17(5):242–245.

Shirai, S. *Squalean Phylogeny: A New Framework of "Squaloid" Sharks and Related Taxa*. Hokkaido University Press, Sapporo, Japan, 1992.

Simpfendorfer, C.A. Reproductive strategy of the Australian sharpnose shark, *Rhizoprionodon taylori* (Elasmobranchii: Carcharhinidae), from Cleveland Bay, northern Queensland. *Australian Journal of Marine and Freshwater Research* 1992; 43:67–75.

Simpfendorfer, C.A. Age and growth of the Australian sharpnose shark, *Rhizoprionodon taylori*, from north Queensland, Australia. *Environmental Biology of Fishes* 1993; 36:233–241.

Simpfendorfer, C.A. Mortality estimates and demographic analysis for the Australian sharpnose shark, *Rhizoprionodon taylori*, from northern Australia. *Fishery Bulletin* 1999; 97:978–986.

Simpfendorfer, C.A., Goodreid, A.B., and McAuley, R.B. Size, sex and geographic variation in the diet of the tiger shark, *Galeocerdo cuvier*, from western Australian waters. *Environmental Biology of Fishes* 2001; 61:37–46.

Simpfendorfer, C.A., and Kyne, P.M. Limited potential to recover from overfishing raises concerns for deep-sea sharks, rays and chimaeras. *Environmental Conservation* 2009; 36(2):97–103.

Sminkey, T.R., and Musick, J.A. Demographic analysis of the sandbar shark, *Carcharhinus plumbeus* in the western North Atlantic. *Fishery Bulletin* 1996; 94:341–347.

Smith, S.E., and Abramson, N.E. Leopard shark (*Triakis semifasciata*) distribution, mortality rate, yield, and stock replenishment estimates based on a tagging study in San Francisco Bay. *Fishery Bulletin* 1990; 88:371–381.

Smith, S.E., Au, D.W., and Show, C. Intrinsic rebound potentials of 26 species of Pacific sharks. *Marine and Freshwater Research* 1998; 49:663–678.

Smith, S.E., Mitchell, R.A., and Fuller, D. Age validation of a leopard shark (*Triakis semifasciata*) recaptured after 20 years. *Fishery Bulletin* 2003; 101:194–198.

Stevens, J.D. Epipelagic oceanic elasmobranchs. In: Carrier, J.C., Musick, J.A., and Heithaus, M.R., editors. *Sharks and Their Relatives II: Biodiversity, Adaptive Physiology, and Conservation*. CRC Press, Boca Raton, FL, 2010; 3–35.

Stevens, J.D., Bonfil, R., Dulvy, N.K., and Walker, P.A. The effects of fishing on sharks, rays, and chimaeras (chondrichthyans), and the implications for marine ecosystems. *ICES Journal of Marine Science* 2000; 57:476–494.

Stevens, J.D., Walker, T.I., Cook, S.F., and Fordham, S.V. Threats faced by chondrichthyan fish. In: Fowler, S.L., Cavanagh, R.D., Camhi, M., Burgess, G.H., Cailliet, G.M., Fordham, S.V., Simpfendor, C.A., and Musick, J.A., editors. *Sharks, Rays and Chimaeras: The Status of the Chondrichthyan Fishes*. Status survey. IUCN/SSC Shark Specialist Group, Gland, Switzerland, 2005; Chapter 5, 48–57.

Treloar, M.A., Laurenson, L.J.B., and Stevens, J.D. Descriptions of rajid egg cases from southeastern Australian waters. *Zootaxa* 2006; 1231:53–68.

Van Der Elst, R.P. A proliferation of small sharks in the shore-based Natal sport fishery. *Environmental Biology of Fishes* 1979; 29:349–362.

Vetter, E.F. Estimation of natural mortality in fish stocks: A review. *Fishery Bulletin* 1987; 86:25–43.

Walker, P.A., and Hislop, J.R.G. Sensitive skates or resilient rays? Spatial and temporal shifts in ray species composition in the central and north-western North Sea between 1930 and the present day. *ICES Journal of Marine Science* 1998; 55:392–402.

Wetherbee, B.M., Gruber, S.H., and Ramsey, A.L. X-radiographic observation of food passage through the digestive tract of the juvenile lemon shark. *Transactions of the American Fisheries Society* 1987; 116:763–767.

CHAPTER 2

Athena and the Evolution of Adaptive Immunity

Helen Dooley

CONTENTS

SUMMARY

Being mammals ourselves, it is not surprising that the immune system of this phylogenetic group holds our special interest and has been studied to the greatest extent. However, the immensely complex immune system we observe in mammals arose through the stepwise accumulation of small changes over billions of years. Although some form of innate immunity has been found in all species thus far examined, adaptive immunity, centered upon highly antigen-specific receptors generated via somatic recombination, emerged after the divergence of the vertebrates. The cartilaginous fishes (e.g., chimera, sharks, skates, and rays) are the oldest phylogenetic lineage to have a mammalian-like adaptive immune system based upon immunoglobulin super-family domains (immunoglobulins, T-cell receptors, and major histocompatibility complex) that are somatically recombined by the RAG proteins. Their location at this pivotal point in phylogeny means that the cartilaginous fishes are vital in furthering our understanding regarding the evolution of the immune system in general and the emergence of adaptive immunity specifically. In this chapter, I will discuss some of the findings regarding immunity in the cartilaginous fishes and other nonhuman species, which have improved our understanding regarding the evolution of the immune system.

INTRODUCTION

Being mammals ourselves, it is not surprising that the immune system of this phylogenetic group holds our special interest and has been studied to the greatest extent. Even within this group, the weight of the work has focused upon immune function in humans and our experimental surrogates among the primates and rodents. However, the immensely complex immune system we observe in mammals did not spring up fully armed, as the ancient war goddess Athena did from the forehead of Zeus; rather, it arose through the stepwise accumulation of small changes over billions of years. The subsequent honing by selection has led to an immune system which, in this case not unlike Athena, fights only with reasonable cause and uses discipline and (evolutionary) strategy in order to avoid widespread destruction.

Although much of the information regarding the functioning of the immune system has been established from studies in mammals, some of the major leaps forward in the field actually came about through the study of other, "less conventional" species; for example, in the 1960s, the T (*thymus*)- and B (*bursa*)-cell lineages fundamental to the adaptive immune system (AIS) were first described in chicken (Cooper *et al.*, 1965), and, more recently, studies of insects revealed toll-like receptors (TLRs) as crucial mediators of innate immunity (Hoffmann, 2003). Luckily for us, the vast amounts of data being generated from genomic and transcriptomic projects in species across phylogeny are facilitating comparative studies and allowing us to trace the evolutionary origins of the various features of the mammalian immune system. From such studies, it has become apparent that the innate immune system (IIS), which uses broadly specific, germline-encoded pattern recognition receptors (PRRs) and has been found in some form in every animal species thus far examined, was present long before the AIS. In contrast, the AIS, centered on highly antigen-specific receptors generated via somatic recombination, emerged after the divergence of the vertebrates and thereafter became intrinsically linked with the innate system already in place.

Although the jawless fishes (lamprey and hagfish) have rearranging receptors based upon leucine-rich repeat proteins (LRRs) (see the following text), the cartilaginous fishes are the most phylogenetically distant group relative to mammals that have an AIS, which is also based upon immunoglobulin super-family domains (immunoglobulins [Igs], T-cell receptors [TCRs], and major histocompatibility complex [MHC]) that are somatically recombined in their lymphocytes. The cartilaginous fishes last shared a common ancestor with other jawed vertebrates ~500 million years ago (MYA) (Figure 2.1) and (at best guess) comprise over 1200 extant species separated between two subclasses, the Holocephali (chimeras and ratfish) and Elasmobranchii (sharks, skates, and rays), which diverged ~350 MYA. As we hope to illustrate in this work, their location at this pivotal point in phylogeny means that the cartilaginous fishes are vital in furthering our understanding regarding the evolution of the immune system and the emergence of adaptive immunity.

ASPECTS OF INNATE IMMUNITY

Probably the most ancient and universal tool of host defense is phagocytosis; used by single-celled organisms to obtain nutrition, in multicellular animals, the process has been adapted for the clean-up of dead or dying host cells (self) as well as for the engulfment and destruction of invading pathogens (nonself) (Desjardins *et al.*, 2005). Indeed, it was the observation of phagocytosis by mesenchymal cells in echinoderms (such as starfish and sea urchins) that led to the original concept of self- versus nonself-recognition (Litman and Cooper, 2007). In mammals, phagocytosis is mainly carried out by the so-called professional phagocytes, cells of myeloid origin that include macrophages, monocytes, and polymorphonuclear leukocytes (PMNs). These cells (along with various other immune cells) carry PRRs that recognize the so-called pathogen-associated molecular patterns (PAMPs) produced by microbes but not host cells. Phagocytes can process these targets in different ways; in the case of internalized apoptotic cells, no inflammatory response is initiated and an autoimmune reaction is avoided. However, when microbes are bound by these receptors, they become encircled by the cell membrane, enclosing them within a vesicle (phagosome), where two things happen: (1) in an oxygen-dependent process known as *respiratory burst*, phagocytes produce

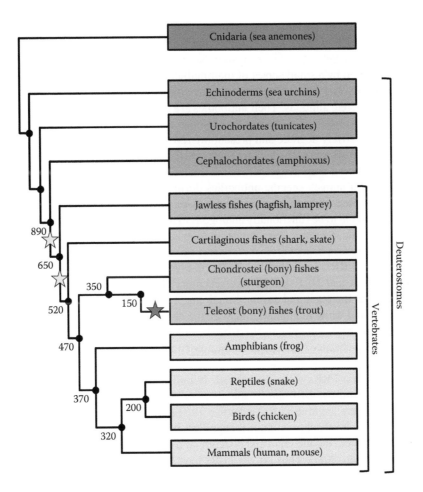

Figure 2.1   Phylogenetic relationship among selected Eumetazoa phyla and classes. Divergence times of the vertebrate lineages (shown as MYA) are based upon the data of Blair and Hedges. (From Blair, J.E. and Hedges, S.B., *Mol. Biol. Evol.*, 22 (11), 2275–2284, 2005.) The 2R hypothesis proposes two rounds of genome-wide duplication (GWD): the first prior to the emergence of the vertebrates and the second prior to the emergence of the jawed vertebrates (indicated by yellow stars). A third, lineage-specific GWD (red star), seems to have occurred sometime between the divergence of the chondrostei and teleost fishes.

reactive oxygen species (ROS), such as hydrogen peroxide (H_2O_2) and superoxide anion (O_2^-), which are toxic to the engulfed microbe; and (2) the phagosome fuses with a lysosome, an intracellular vesicle containing enzymes, proteins, and peptides, which can kill and degrade the internalized microbes. In jawed vertebrates, the microbe-derived peptides resulting from this process will ultimately be loaded onto MHC molecules for presentation to T cells, leading, in due course, to the production of neutralizing antibodies by B cells (Desjardins *et al.*, 2005) (see the following text). Tissue-resident macrophages and dendritic cells (DCs) are usually the first to encounter microorganisms that have breached the epithelia; as well as helping clear the pathogen through phagocytosis, once activated they also release cytokines and chemokines, setting up a state of inflammation and recruiting other immune cells to the site of infection.

It has long been known that B cells can internalize soluble antigens, which are specifically bound by B-cell receptor (BCR) complexes on their surface, allowing the eventual presentation of antigen-derived peptides to T cells in the context of MHC. However, until recently, it was believed that primary B cells were incapable of non-BCR-mediated phagocytosis. Studies in bony fishes, amphibians, and reptiles showed that certain B-cell subsets in these species have a high capacity for phagocytosis, ingesting both large inert particles and live bacteria, killing the latter following their uptake (Li *et al.*, 2006). Subsequent investigation has shown that these phagocytic and bactericidal abilities are also shared by a small subset of mammalian B cells, specifically the B-1 lymphocytes of the peritoneal cavity. B-1 cells are a unique, innate-like subset of B cells which are self-renewing and spontaneously secrete antibody. Their antibodies are often polyspecific and the binding regions appear to have been evolutionarily selected to recognize common PAMPs. Following phagocytosis, the B-1 cells can effectively degrade and present microbe-derived peptides to host T cells, thereby priming the adaptive response to the invader (Parra *et al.*, 2012). That this and other innate abilities are possessed only by B-1 cells suggests they are evolutionarily the oldest B-cell subset, with B-2 cells, the producers of antigen-specific antibodies during an adaptive response (see the following text), emerging later. Recent evidence supports the existence of distinct B-cell lineages in the cartilaginous fishes that seem to be functionally analogous to mammalian B-1 and B-2 cells (Dooley and Flajnik, 2005); it would, therefore, be interesting to examine if the B-1-like cells of cartilaginous fishes share a capacity for phagocytosis and/or some of the innate abilities observed in mammals.

A whole host of PRRs, including the TLRs, C-type lectin receptors (CLRs), nod and nod-like receptors (NODs and NLRs), RIG-like receptors (RLRs), and scavenger receptors (SRs), are found throughout the plant and animal kingdoms. The diversity of these receptors is entirely germline encoded, and while in humans and mice the number of these *innate* receptors is limited, in some invertebrates and many plants, there has been a great expansion of various PRR families, suggesting that in the absence of adaptive immunity complex, innate mechanisms evolve for host defense.

TLRs were the first PRRs to be discovered. Initially identified as key players in embryonic development in the fruit fly, *Drosophila*, members of this PRR family have now been found in numerous species. There are nine TLRs (including *Toll*) encoded in the *Drosophila* genome, and at least some of these are involved in immune recognition, being activated when they bind to their particular PAMP(s) and inducing the rapid and differential transcription of antibacterial and antifungal peptides (Valanne *et al.*, 2011). Thus far, 10 TLRs have been identified in humans and equivalent forms of many of these have been identified in other vertebrate species, including the elephant shark (Venkatesh *et al.*, 2014). However, it is also very apparent that there are differences in the repertoire of TLRs possessed by different species, for example, TLRs 11, 12, and 13 are found in mice but not humans, whereas lower vertebrates (lampreys, bony fishes, and amphibians) encode an additional set of TLRs distinct from those found in mammals (the so-called fish-type TLRs) (Kasamatsu *et al.*, 2010; Palti, 2011); the TLR4 receptor and its associated components are missing from the elephant shark genome (Venkatesh *et al.*, 2014), explaining the apparent lack of lipopolysaccharide (LPS) responsiveness in this group (Dooley and Flajnik, unpublished data). Therefore, as would be expected, the TLR repertoire possessed by a host evolves to fit its environment and enable recognition of pathogens that it is likely to encounter.

TLRs are type I transmembrane proteins and comprise an ectodomain that mediates the recognition of a specific set of PAMPs, a transmembrane region, and a cytosolic TIR domain that activates downstream signaling pathways. Some TLRs act as cell-surface receptors, whereas others are located on the surface of intracellular vesicles, where they can detect pathogen components that have been taken up by the cell. Ligand binding by the LRR domain of the TLR induces a conformational change that is transmitted across the membrane and allows the intracellular TIR domain to bind to an adaptor protein. In humans, four adaptor proteins have been identified so far (MyD88, TRIF, TRAM, and TIRAP), and the type of response raised depends upon the recruitment of a single or a specific combination of these adaptor proteins. All of the signaling pathways culminate in the activation of NFκB and/or other pathway-dependent kinases (e.g., MAP kinase, IRF-3, and IRF-7) and the subsequent induction of various inflammatory cytokines or type I interferons (Kawai and Akira, 2011). These signaling pathways are highly conserved throughout phylogeny, and in all species thus far examined, activation of TLRs leads to the production of important mediators of innate immunity, such as antimicrobial peptides, cytokines, and chemokines.

The more recent discovery of a multitude of non-TLR PRR families (as listed earlier) suggests that this aspect of innate immunity is actually much more sophisticated than first thought. Different types of PRR appear to work in concert to sense infection and protect the host; for example, NOD and NOD-like receptors are found in the cellular cytosol, where they bind fragments of bacterial cell-wall proteoglycans. They are used to help protect cells in which TLR expression is low and can also act synergistically with TLRs to help clear infections. Also, as mentioned earlier, some organisms have greatly expanded their PRR repertoire, presumably in order to increase their innate protection; this is exemplified by the purple sea urchin, *Strongylocentrotus purpuratus* (an echinoderm; Figure 2.1), which has over 200 different TLR genes and as many NLR and SRCR scavenger receptor genes (Rast and Messier-Solek, 2008). Whether the members of these gene families are expressed in a monospecific (one receptor specificity per cell, such as antibodies) or variegated manner (multiple receptor specificities per cell, such as NK receptors) has yet to be determined. This information will not only go some way to aid our understanding of how the sea urchin immune system functions but also inform us as to whether such an expanded repertoire of PRRs function in a manner similar to that of the AIS.

As outlined earlier, receptor-mediated phagocytosis emerged early during evolution; however, in the intervening period it has become overlaid by additional pathways and systems, which act to increase the specificity of self-/nonself-recognition, increase phagocytic efficiency, or, uniquely in vertebrates, initiate an adaptive immune response. The complement system is one such system that has evolved over millions of years to fulfill some of these roles; it is a complex network of plasma proteins that function through the opsonization of pathogens and induction of powerful inflammatory responses that help fight infection. In mammals, there are three complement activation pathways: (1) the classical pathway, triggered by C1q binding directly to microbial targets or to immune (antigen–antibody) complexes and mediated by the serine proteases C1r and C1s; (2) the lectin pathway, initiated by the binding of mannose-binding lectin (MBL) or ficolins to carbohydrate arrays on the surface of microbes and mediated by mannan-binding lectin-associated serine proteases (MASPs); and (3) the alternative pathway in which spontaneously activated C3 binds directly to the surface of the pathogen and recruits factor B (Bf). The three pathways converge with the cleavage of C3. The fragments released bind directly to the invading pathogen, enhancing phagocytosis and initiating the assembly of the membrane-attack complex (MAC), which forms pores that disrupt the surface of the pathogen.

Despite its complexity in mammals, accumulating data indicates the origins of the complement system are also very ancient; C3, Bf, and MASP have all been found in the sea anemone, *Nematostella vectensis* (Kimura *et al.*, 2009) (a cnidarian; Figure 2.1). The fact that all three molecules are expressed from the endoderm suggests they protect the primitive gut (coelenteron) and the circulatory system of this group. Echinoderms also express C3 and Bf, but in this case they are secreted by phagocytes directly onto the surface of pathogens, increasing their uptake by phagocytic cells (Smith *et al.*, 2001). Thus, the molecules required for the initiation of the alternative and lectin pathways were present long before the emergence of the jawed vertebrates. Although a homologue of C1q is found in the tunicate *Ciona* (a urochordate), the gene duplications leading to C3/C4/C5, Bf/C2, and MASP1/MASP2/C1r/C1s, and establishing the classical pathway of activation, seem to have occurred much later. The presence of MASP1 in the jawless fishes but not MASP2, C1r, or C1s (Kimura *et al.*, 2008) indicates that the duplications giving rise to these genes occurred sometime after their divergence. Although shark serum proteins with functional analogy to C1r and C1s were described over 30 years ago by Jensen and colleagues (1981), it was only very recently that the genes for these classical pathway-initiating components were identified in the elephant shark genome (two potential genes for each) (Venkatesh *et al.*, 2014). Thus, all three complement pathways had been established before the divergence of the cartilaginous fishes from the common vertebrate ancestor.

Thus far, the terminal components of the complement cascade, those which form the MAC (C6/C7/ C8a/C8b/C9), have only been found in jawed vertebrates; this fits with functional data showing that lamprey serum has opsonizing but not cytolytic activity. The previous observed lytic activity exhibited by shark serum (Jensen *et al.*, 1981) in combination with the characterization of C5, C6, and C8 in sharks (Graham *et al.*, 2009; Kimura and Nonaka, 2009) and *in silico* identification of other MAC components in the recently published elephant shark genome (Venkatesh *et al.*, 2014) indicates the cytolytic pathway in cartilaginous fishes is comparable to that of mammals.

AT THE JUNCTION OF INNATE AND ADAPTIVE IMMUNITY

Although only jawed vertebrates have MHC, cells that mediate self-/nonself-recognition are found even in primitive animals such as sponges (porifera) (McKitrick and De Tomaso, 2010). In mammals, natural killer (NK) cells distinguish *abnormal* (virally infected or malignantly transformed) host cells using a complex repertoire of germline-encoded receptors, hence their historical classification is part of the IIS. Although NK cells share a common progenitor and certain functional features with T cells, they differ in that NK cells detect *missing-self* rather than *nonself.* Both activating and inhibitory NK receptors are found on the surface of each cell, and it is through these that the NK cells engage MHC class I molecules (see the following text) on target cells. NK-cell receptors evolve very rapidly, with even mouse and human NK receptors being encoded by different gene families; Ly49, a family of receptors homologous to C-type lectins, is expressed by mouse NK cells; KIRs (killer cell immunoglobulin-like receptors), a family of Ig-like receptors, are expressed by human NK cells; and NKG2/CD94, a heterodimeric, lectin-like receptor, is expressed on both human and mouse NK cells. Despite being structurally distinct, both Ly49 receptors and KIRs bind to classical class I MHC molecules, which are displayed upon the surface of all normally functioning, nucleated cells, whereas NKG2/CD94 interacts with nonclassical class I MHC molecules that are generally expressed by cells in response to metabolic or cellular stress. As NK-cell receptors are expressed in a variegated fashion, each cell expresses a different (random) set of receptors. This means activation thresholds vary from cell to cell and are dependent upon the number of both activating and inhibitory receptors engaged, as well as the qualitative differences in the signals transduced by each receptor (Lanier, 2005). Signals generated by inhibitory receptors act to dampen those from activating receptors, thus preventing inadvertent stimulation and potential harm to the host. However, in cases where a target cell has *altered* class I MHC (e.g., downregulated classical and/or upregulated nonclassical class I), the signaling balance is disrupted, and the NK cell is triggered to release the contents of intracellular cytotoxic granules (mainly perforin and granzymes) directly onto the surface of the abnormal cell, thereby initiating its destruction. The NK cells also begin to secrete IFN-γ, IL-10, and TNFα, which have immunoregulatory, antiviral, and antitumor properties (Lanier, 2005).

Despite their classification as *innate*, NK cells actually require priming by various factors produced by DCs or macrophages, such as IL-12, IL-15, IL-18, and type I interferons, to achieve their full effector potential; this priming can increase the killing activity of NK cells up to 100-fold. In turn, NK cells are themselves major producers of cytokines, including TNFα, which promotes tumor cell killing, and IFN-γ, which induces DC maturation (Vivier *et al.*, 2011). Recent evidence also shows that some mammalian NK cells are long lived, undergo expansion, and are able to mount a robust recall (memory) response (Sun and Lanier, 2009). Thus, NK cells actually appear to straddle the line between innate and adaptive immunity, showing features typical of both.

ASPECTS OF ADAPTIVE IMMUNITY

It has long been known that jawless fishes (hagfish and lamprey) showed signs characteristic of an adaptive immune response (i.e., accelerated rejection of second skin grafts, delayed-type hypersensitivity to tuberculin, and antigen-specific agglutination following immunization) (Finstad and

Good, 1964). However, despite focused studies that identified many homologues of genes expressed in mammalian lymphocytes, no evidence was found of the *usual* adaptive immune molecules (Igs, TCRs, or MHC) (Mayer *et al.*, 2002; Uinuk-Ool *et al.*, 2002). This apparent paradox was resolved when antigen-binding proteins based upon highly variable LRRs were found to be expressed as cell membrane bound and soluble serum proteins in lamprey. These proteins were christened variable lymphocyte receptors (VLRs).

There are three known germline *VLR* genes in lamprey (*VLRA*, *VLRB*, and *VLRC*); however, none of them is capable of encoding functional proteins; each gene contains the coding sequence for the invariant caps found at either end of the mature protein separated by noncoding DNA. However, each gene is also flanked by hundreds of copies of highly diverse LRR modules that, during cell development, are sequentially incorporated between the invariant VLR caps by a process called *copy choice* (Figure 2.2). The recombination-activating genes (RAG), which rearrange Igs and TCRs in jawed vertebrates (see the following text), are not found in the jawless fishes. Rather, the somatic recombination events required to generate a functional VLR are mediated by cytidine deaminases *CDA1* and *CDA2* (Rogozin *et al.*, 2007), which are members of the same family of enzymes responsible for post-rearrangement diversification in jawed vertebrates (see the following text) (Muramatsu *et al.*, 2000; Arakawa *et al.*, 2002). During the *copy choice* process, short regions of homology are used to prime the copying of donor LRR modules into the recipient *VLR* gene (Nagawa *et al.*, 2007), in a process highly reminiscent of immunoglobulin gene conversion. As the individual LRR modules are highly diverse, and the number of modules copied across into the gene varies, the potential combinatorial diversity of a single *VLR* gene is comparable to that of mammalian Igs (estimated at $>10^{14}$ combinations). The crystal structures of recombinant VLR monomers show they adopt a horseshoe-shaped solenoid that is similar in structure to other LRR proteins, such as the TLRs, and that the antigen-binding site spans the concave surface where the highly variable residues are located. VLR monomers have been shown to bind a wide variety of antigens with affinities equivalent to those of affinity-matured mammalian Igs (Tasumi *et al.*, 2009; Deng *et al.*, 2010).

Like vertebrate Igs, the VLRs are expressed in a monospecific manner, with each lymphocyte expressing only one VLR gene (Nagawa *et al.*, 2007; Hirano *et al.*, 2013). Indeed, based upon the nature of the VLR and cytidine deaminase expressed (VLRA and VLRC/CDA1 or VLRB/CDA2), the lymphoid cells found in lamprey could be separated into two distinct populations that seem to be functionally equivalent to the T- and B-lymphocytes of jawed vertebrates (Guo *et al.*, 2009). Although both populations of cells proliferate following antigenic stimulation, only the VLRB+ cells are able to differentiate into VLR-secreting cells after binding to their specific antigen. Secreted VLRB molecules occur as tetramers or pentamers of dimers, thus having 8–10 antigen-binding sites per multimer-like mammalian IgM (Herrin *et al.*, 2008). This allows VLRB molecules to undertake high-avidity, highly specific binding to repetitive antigens. VLRB+ cells were also shown to express homologues of genes typically found in mammalian B lymphocytes, including signal transduction components associated with the BCR (Syk and BCAP), various TLRs, the chemokine CXCL8, and the receptor for the proinflammatory cytokine interleukin-17 (IL-17R). In contrast, VLRA+ and VLRC+ cells respond preferentially to phytohaemagglutinin (PHA), a classical T-cell mitogen, and can be distinguished by transcription of the transmembrane receptor NOTCH-1, a factor shown to be essential for T-cell lineage commitment in mammals (Guo *et al.*, 2009). Further, a recent study of the gene expression profiles of the different cell populations by Hirano and colleagues (2013) showed that although VLRA+ cells express orthologues of genes associated with αβ T cells in mammals (e.g., the transcription factor TCF1 and the regulatory co-receptor CTLA-4), VLRC+ cells preferentially expressed those associated with mammalian γδ T cells, notably the γδ fate-determining transcription factor SOX13. Thus, the accumulated data suggest that the genetic programs required for the establishment of a primordial B-like and two T-like lineages were already present in the common ancestor of all vertebrates (Hirano *et al.*, 2013). After PHA stimulation, VLRA+ and VLRC+ cells both up-regulate the transcription of the

(a) VLRs

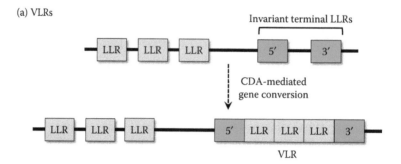

(b) Igs and TCRs

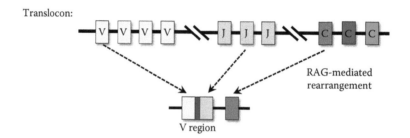

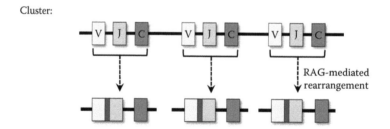

Figure 2.2   Structure of vertebrate lymphocyte receptor genes. (a) In jawless fishes, multiple leucine-rich repeat (LRR) modules are located next to each incomplete VLR gene. A gene conversion-like process sequentially incorporates multiple LRR modules into the gene, between the 3' and 5' invariant caps, to generate a functional VLR. This process is mediated by members of the cytidine deaminase (CDA) family of enzymes. (b) Most Ig and all TCR loci examined so far are arranged in a translocon configuration, where multiple variable (V) and joining (J) segments are located adjacent to the constant (C) region exons of multiple different isotypes. In Ig heavy chains and TCR β/δ chains, multiple diversity (D) segments are located between the V and J segments (not shown). RAG-mediated somatic rearrangement fuses together a single V and a single J segment (or a single V, D, and a J where D segments are present) to create a functional V region. The action of enzymes such as TdT during the rearrangement process diversifies the join between the segments (indicated in red), thus increasing the potential binding repertoire. In contrast, the cluster configuration can be thought of as a string of mini-translocons, with a single V segment, one or more D segments, a single J segment, and the constant region exons for a single isotype in each cluster. Rearrangement can occur only within a cluster but again enzymes such as TdT act to diversify the joining region and increase the potential binding repertoire. (Adapted from Boehm, T., *Nat. Rev. Immunol.*, 11 (5), 307–317, 2011.)

lamprey IL-17 homologue, which is assumed (but not yet proven) to engage its receptor present on the surface of VLRB cells. Conversely, VLRA/C cells express receptors for the cytokines produced by VLRB cells, such as IL-8, suggesting the elaborate communication networks used to coordinate B-cell and T-cell responses in mammals also arose prior to the divergence of the jawless and jawed vertebrates (Guo *et al.*, 2009; Hirano *et al.*, 2013).

In jawed vertebrates, B- and T-lymphocytes develop at distinct anatomical sites, and this spatial separation also seems to occur for VLRA/C and VLRB cells in lamprey. Studies of gene expression patterns show that VLRA/C (T-like) cells develop in discrete FOXN1+ *thymoids*, located at the tips of the lamprey gill filaments, whereas VLRB (B-like) cells develop in the haematopoietic tissue of the typhlosole and kidney (Bajoghli *et al.*, 2011). Thus, by the time the jawless fishes diverged from the common vertebrate ancestor, many aspects associated with the AIS were already in place, notably, distinct populations of receptor-bearing lymphocyte-like cells, whose development occurs in segregated organs and is controlled by the coordinated action of lineage-restricted transcription factors, as well as cytidine deaminases that facilitate somatic diversification. Despite this, the mammalian-like AIS, based upon somatically rearranging immunoglobulin super-family (IgSF) genes, did not appear until just before the emergence of the cartilaginous fishes (Flajnik and Kasahara, 2010).

It is widely believed that, for the emergence and development of the IgSF-based immune system to occur, two specific macroevolutionary events needed to happen; first, the vertebrate genome needed to undergo a round of duplication. It is now mostly accepted that the vertebrate genome experienced two rounds of whole genome duplication, one that is thought to have taken place sometime prior to the emergence of the jawless fishes and the second before the emergence of the jawed vertebrates (the 2R hypothesis; Figure 2.1) (Ohno, 1999; Kasahara, 2007). Such large-scale gene duplication events provide both the raw material and evolutionary opportunity for the development of new functions and systems and appear to have been crucial in the emergence of jawed vertebrate-type adaptive immunity.

The second event was the insertion of a mobile DNA element (transposon) into an ancestral IgSF domain (Sakano *et al.*, 1979). A transposon encodes two essential elements: the gene(s) encoding an enzyme capable of cleaving double-stranded DNA and a set of terminal repeat sequences that are recognized by this enzyme. The insertion and subsequent excision of a transposon by its transposase leaves short repeat sequences in the DNA with a gap between them. Such repeat sequences, the so-called recombination signal sequences (RSSs), flank the rearranging segments that make up the variable domains of Ig and TCR genes meaning they cannot be expressed until stitched back together through the action of the RAG enzymes. After the *RAG* genes were discovered and shown to encode enzymes that cleave DNA specifically at these RSSs (Oettinger *et al.*, 1990), it was proposed that a microbial transposon containing both the *RAG* genes and RSSs was inserted into an ancient IgSF gene leading to a new mechanism for immune diversification (Bernstein *et al.*, 1996). Surprisingly, homologues of the *RAG* genes have also been found in the genome of the purple sea urchin (Fugmann *et al.*, 2006), implying that this transposon invaded the genome 100 million years before the emergence of adaptive immunity and lay dormant, or that it has invaded the genome multiple times. There are many genes in lower chordates that contain IgSF variable (V)-type domains, so there are many candidate genes that could have been invaded by the RAG transposon in a vertebrate ancestor to generate the split V region, something akin to the V region of Ig light chains (see the following text). The split V gene could then multiply, thereby increasing repertoire diversity.

The molecules of the MHC are important in self-/nonself-discrimination and are encoded by a large cluster of genes that were first identified due to the role they played in the rejection of transplanted tissues. The MHC gene family has now been found in all vertebrate species from the cartilaginous fishes onward (Bartl, 1998) and is the means by which peptides derived from self- and nonself sources are displayed for recognition by T cells. Usually divided into three subgroups (MHC class I, II, and III), the MHC genes of most species are highly variable, being both polygenic (many genes in each class) and polymorphic (many variants in the population), thus allowing the display of a vast range of peptides, both by an individual and across a species.

MHC class I is expressed on the surface of all nucleated cells and presents peptides from the cellular cytosol (mainly self-peptides but also peptides derived from intracellular viruses and bacteria) to CD8+ *killer* T cells, which can then induce apoptosis of infected cells. Most cytosolic degradation is performed by the proteasome, a large, multicatalytic protease complex, which randomly cleaves proteins. A proportion of the proteasome output is able to be transported by the TAP (*transporters associated with antigen processing*) proteins and subsequently loaded onto MHC molecules. However, the proteasome actually exists in two forms, the constitutive form and an induced form, the so-called immunoproteasome, in which certain subunits have been replaced by their cytokine-induced counterparts (LMP2, LMP7, and MECL-1). The incorporation of LMP2 and LMP7 modulates the specificity and/or proteolytic activity of the complex, allowing the preferential production of peptides that are better transported by TAP or more efficiently loaded onto the MHC molecules. In almost every nonmammalian vertebrate studied to date, the TAP1 and TAP2 proteins, LMP2, and LMP7 are all found to be closely linked within the MHC class I region (Ohta *et al.*, 2002) (with the exception of birds that have MHC-linked TAP1/2 but not LMP2/7 [Shiina *et al.*, 2004]), indicating that the linkage of these class I-processing genes with class I is a primordial feature.

MHC class I molecules are composed of an α-chain, with three domains and a transmembrane region, associated with a β2 microglobulin (β2M) chain. Newly synthesized class I molecules are held in the cell's endoplasmic reticulum (ER) until the peptide-binding groove of the α-chain is occupied. Once loading is complete, the MHC complexes are released for display on the cell surface. Many viruses interfere with the loading or release process in order to prevent display of virally derived peptides and subsequent immune recognition. However, this results in low surface expression of classical class I MHC, which in itself is suggestive of abnormal cell function, and is detected by NK cells that can then destroy the infected cell (see earlier).

MHC class II molecules can be expressed by all cells but are normally found only on professional antigen-presenting cells (APCs); following pathogen or antibody–antigen complex uptake by APCs, the peptides generated by normal degradation pathways in intracellular vesicles are loaded onto class II MHC for recognition by CD4+ *helper* T cells. Class II molecules are formed of two chains, α and β, each having two domains and a transmembrane region. The peptide-binding groove is formed between the two chains and is protected prior to peptide loading by the invariant chain, which also directs the newly synthesized molecules to endosomes so that peptide loading can occur. If the MHC–peptide complex is recognized by a CD4+ T cell, an adaptive immune response is initiated that includes the activation of B cells to generate an appropriate antibody response.

Although MHC class I and II genes encode structurally related molecules that are responsible for antigen presentation to T cells, those of the class III region are heterogeneous in both structure and function. In fact, their definition as class III is based purely upon their location between class I and II in placental mammals, and they encode a miscellany of nonimmune proteins as well as immune proteins including various complement proteins and cytokines.

Clusters of genes resembling a *proto-MHC region* have been found in the genomes of protochordates such as tunicates (urochordates) and amphioxus (cephalochordates) (Figure 2.1); however, the genes encoding MHC class I and II are found only in the jawed vertebrates and therefore must have arisen abruptly sometime between the emergence of the jawless and jawed fishes. This ties in with the discovery of four paralogous MHC clusters in the genomes of the jawed vertebrates; although only one of these regions carries MHC class I and II genes, all four regions contain many of the accessory molecules important for the functioning of the AIS (Flajnik and Kasahara, 2001). Thus, although the exact origin of the ancestral MHC gene is still open to debate, the two genome-wide duplication (GWD) events that occurred around this time appear to have been fundamental for the emergence of MHC region as a whole.

Although MHC regions are found in all vertebrates, they vary widely in size and complexity between species; for example, in humans the MHC region contains ~140 genes (spanning ~4 Mb) (The MHC sequencing consortium, 1999); in contrast, chickens are considered to have a *minimal essential* MHC that is approximately 20-fold smaller and contains only 19 genes (spanning ~19 Kb) (Kaufman *et al.*, 1999). Unlike other vertebrates, the class I and class II genes in teleosts are not linked, and the class III genes are scattered throughout the genome. However, as class I and II genes are closely linked in shark (Ohta *et al.*, 2000), this appears to be a derived feature resulting from differential silencing and gene rearrangement following the teleost-specific GWD (Figure 2.1). Although the MHC locus has not been completely characterized in the cartilaginous fishes, the genes studied thus far have large introns as well as large intergenic distances (Ohta *et al.*, 2002), suggesting it will likely prove much larger than that of humans. Despite this, studies of shark MHC, with subsequent comparison to that of other vertebrates, have been instrumental in defining the emergence and evolution of the vertebrate MHC complex (Ohta *et al.*, 2000).

As described earlier, distinct subpopulations of lymphocyte-like cells have been found recently in the jawless fishes. However, although these cells are functionally analogous to T cells and B cells, they carry VLR proteins on their cell surface rather than TCRs or Igs found on the surface of lymphocytes in jawed vertebrates. In mammals, where lymphocyte development is best understood, cells of the lymphoid lineage derive from pluripotent hematopoietic stem cells (HSCs) in the bone marrow (and in the fetal liver/spleen). However, in the jawless fishes and *basal* jawed vertebrates, other organs fulfill this role; lamprey and hagfish do not have bone marrow and so their lymphocyte-like cells are thought to develop in the kidney or typhosole (Zapata and Amemiya, 2000). In the bony fishes and at least some amphibians, the kidney is the major hematopoietic organ, whereas in cartilaginous fishes, the Leydig organ (associated with the esophagus) and epigonal (associated with the gonad) serve this function. These organs have a histological organization remarkably similar to the bone marrow of higher vertebrates (Zapata and Amemiya, 2000) and express high levels of genes associated with somatic rearrangement (e.g., RAG and TdT) (see the following text) (Hansen, 1997; Miracle *et al.*, 2001).

Mammalian HSCs receive both soluble and membrane-bound signals from the bone marrow stromal cell network enabling them to commit to the lymphoid lineage, rather than the myeloid lineage; key among these is the cytokine IL-7, which stimulates the eventual differentiation of the common lymphocyte progenitors (CLPs). It is these cells that give rise to the early T-lineage precursors, which migrate from the bone marrow to the thymus to fully differentiate into T cells, and the B-lineage precursors, which remain in the bone marrow and differentiate into B cells. In humans and mice, as well as in bony and cartilaginous fishes, B cells continue to be generated during the entire life of the animal (although this does not appear to be true in GALT (gut-associated lymphoid tissue) species such as chickens, rabbits, and sheep) (Flajnik, 2002), whereas the generation of T cells generally slows in adults, and T-cell numbers are maintained by slow cellular division in the periphery.

The cell-surface receptors for both lymphocyte subsets are generated through somatic recombination of variable gene segments. These segments are *stitched* together in an orderly and stepwise manner so as to maximize the functional diversity of the receptor repertoire. Once a functional receptor is formed, it is displayed on the lymphocyte surface and is subject to rigorous testing to ensure it is capable of recognizing foreign antigens but does not react against self-molecules. Igs were the first molecules involved in specific immunity to be characterized and are found in all jawed vertebrates both as secreted, soluble proteins (antibodies) and on the surface of B cells in complex with accessory molecules as BCRs. Antibodies are composed of four chains, two heavy chains each of which is paired with a light chain. Each chain has two parts: a single variable (V) region, which determines the binding specificity of the molecule, and a number of constant (C) regions, which determine the antibody isotype and effector function(s) of the molecule. The V regions of the antibody are encoded by multiple gene segments, V(ariable), D(iversity), and J(oining) segments for the heavy chain, V and J segments only for the light chain, which are somatically recombined to generate the mature V region exon. In most species, the elements required to encode an antibody are arranged in the so-called translocon configuration where multiple V, (D), and J segments are located upstream of the exons that encode the constant regions for each of the different isotypes (Figure 2.2). In contrast, both the heavy and light chain genes of cartilaginous fishes are encoded as multiple mini-loci or *clusters*, with each cluster containing a single V segment, one or more D segments, a J segment, and one set of C region exons (of a single isotype) (Kokubu *et al.*, 1988; Rast *et al.*, 1994); rearrangement occurs only within a cluster but dependent upon the species and isotype examined the number of clusters present can range from a handful to many hundreds. The Ig genes of bony fishes are different again, with the heavy chains being present in the translocon organization but the light chains mostly being present in clusters (Bengten *et al.*, 2006).

In mammals, there is one heavy chain locus, encoding five isotypes (μ, δ, γ, α, and ε), and two light chain loci (κ and λ). As orthologues of IgM and IgD heavy chains (using the μ and δ constant regions, respectively) have now been identified from the cartilaginous fishes through to mammals, it appears that these two isotypes were present in the ancestor of all the jawed vertebrates (Ohta and Flajnik, 2006) (see Figure 2.3). Although IgM is very evolutionarily stable with regard to constant domain number, IgD is more evolutionarily labile showing lineage- or even species-specific domain duplications and deletions (Chen and Cerutti, 2011). A number of additional (lineage-specific) heavy chain isotypes (IgNAR in cartilaginous fishes, IgT/Z in bony fishes, and IgX in amphibians) and two further light chain isotypes (σ in all cold-blooded vertebrates and σ-cart/σ2 thus far only identified in cartilaginous fishes and coelacanth [Criscitiello and Flajnik, 2007; Saha *et al.*, 2014]) have also been identified in nonmammalian species.

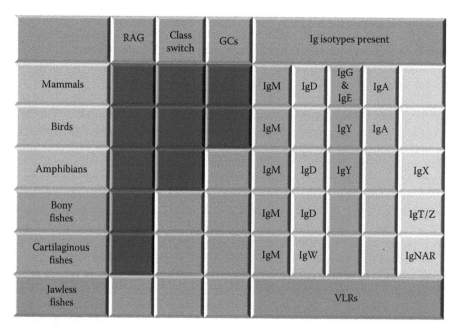

	RAG	Class switch	GCs	Ig isotypes present				
Mammals				IgM	IgD	IgG & IgE	IgA	
Birds				IgM		IgY	IgA	
Amphibians				IgM	IgD	IgY		IgX
Bony fishes				IgM	IgD			IgT/Z
Cartilaginous fishes				IgM	IgW		.	IgNAR
Jawless fishes				VLRs				

Figure 2.3 Emergence of adaptive immune features and the different immunoglobulin isotypes in vertebrates. Various functional features of the adaptive immune system, such as RAG-mediated recombination, antibody class switching, and germinal center (GC) formation, emerged at different times during the evolution of the jawed vertebrates; where a feature is present in a particular vertebrate lineage the box is shaded red and where absent shaded gray. None of these functional features has been detected in the jawless fishes. Multiple immunoglobulin (Ig) isotypes are also present in each of the jawed vertebrate lineages; those present in each lineage are listed and blue shading indicates that the isotypes in that column diverged from a common ancestor. As can be seen, orthologues of both IgM and IgD are found throughout phylogeny and likely arose in the common ancestor of all jawed vertebrates. The final column shows lineage-specific isotypes and these do not share a common ancestor. The jawless fishes do not encode Igs (or, for that matter, TCRs or MHC) in their genomes; however, they can generate an adaptive immune response based upon leucine-rich repeat proteins called variable lymphocyte receptors (VLRs). (Adapted from Flajnik, M.F., *Nat. Rev. Immunol.*, 2 (9), 688–698, 2002.)

In mammals, V gene recombination is a highly ordered process, with each stage needing to be successfully completed before progression to the next (heavy chain D-J then V-DJ rearrangement; light chain V-J rearrangement). Studies in cartilaginous fishes show their rearrangement process is slightly different, with random heavy chain V gene segments being recombined simultaneously; despite this, and the presence of multiple clusters, only one or, at most, two V regions are fully rearranged in each IgM+ shark B cell (Malecek *et al.*, 2008). The arrangement of the V gene segments combined with the action of enzymes such as TdT (which extends DNA ends during the rearrangement process) targets the greatest sequence diversity at the very center of the antigen binding site, thus generating a highly diverse preimmune (primary) repertoire. Due to the obvious risks associated with the gene rearrangement process, RAG gene expression is very tightly controlled, being expressed only in lymphocytes and only at certain developmental stages. Interestingly, some shark Ig clusters have their V region segmental elements partly or fully fused in the germline, likely as a result of a RAG-mediated rearrangement event in germ cells (Lee *et al.*, 2000), thus fixing the sequence and, by inference, the binding specificity of the antibody. These *germline-joined* genes seem to have a transcriptional advantage early in development but later become overwhelmed as antibody expression from rearranging cluster increases (Rumfelt *et al.*, 2001). It has been hypothesized that these antibodies have been evolutionarily selected to bind common pathogens or waste products and so confer protection on the animal before its AIS has fully matured.

Although humans, mice, amphibians, bony fishes, and cartilaginous fishes generate their primary antibody repertoire principally through recombination, other species generate theirs by means of somatic hypermutation (SHM) and/or gene conversion (see the following text), either in isolation or in combination with recombination (Flajnik, 2002). Sheep, for example, rearrange only a few invariant V region genes then use SHM to generate a diverse primary repertoire in their ileal Peyer's patches (Reynaud *et al.*, 1991), part of their mucosal-associated lymphoid tissue (MALT). Other species, such as chickens, have only a single gene, which is able to generate a functional V region by rearrangement; this V is then diversified through the process of gene conversion using upstream pseudogenes as donor templates (Reynaud *et al.*, 1989). This process occurs in the bursa, where B-cell proliferation and diversification continues until the cells receive some, as yet unknown, signal to migrate to the secondary lymphoid tissues. As the bursa degenerates with age, the adult chicken must *make do* with the B cells that it has produced early in life (Flajnik, 2002). However, as with many species, the antibody binding repertoire is not usually limited by the diversity that can be generated but rather by the number of B cells the individual animal can maintain. Regardless of the mechanism used to generate their primary repertoire, all species examined so far operate a system of allelic and isotypic exclusion, which ensures only antibodies of a single specificity are expressed by a B cell and thus processes such as self-tolerance and clonal deletion/expansion can operate effectively.

Once a functional BCR is expressed on the surface of a mammalian B cell and has passed checks testing for autoreactivity, it exits the bone marrow and enters the circulation. Most naïve B cells die within a few days of their release to the periphery; in order to survive, a B cell needs to enter follicles within the secondary lymphoid tissues and receive the signals (through its BCR and various cytokine receptors) required for its continued survival. Although some microbial antigens, typically repeated polymers such as bacterial LPS, can stimulate some B cells directly (giving a so-called T-independent primary response), a B cell typically needs to interact both with an antigen via its BCR and with helper T cells, through antigen-derived peptide: MHC class II complexes displayed on the B cell surface, to enable production of high-affinity antibodies of different isotypes (a T-dependent response) (see the following text).

In most jawed vertebrates, the thymus is a bilateral, lobed organ; histologically, it is organized into two well-defined areas, a tightly packed outer cortex where immature, proliferating thymocytes are found and an inner medulla region containing mature cells. Its importance was first discovered in chickens where surgical removal (thymectomy) resulted in immunodeficiency (Cooper *et al.*, 1965) and the recent discovery of a thymus equivalent in the lamprey (Bajoghli *et al.*, 2011) shows this primary lymphoid organ is present in all vertebrates, providing a discrete site for the development of T (or T-like) cells. As TCRs function through cell–cell contact, they are only made as membrane bound molecules and are classified into two functionally distinct sublineages (α:β or γ:δ), based upon the nature of the chains used to form the TCR. Both types of TCR are heterodimers, with each chain having a membrane-proximal constant region and a variable region which determines its binding specificity. The vast majority of T cells carry the α:β TCR and only recognize peptides in the context of MHC molecules; this does not appear to be the case for γ:δ T cells, and although their exact function is not entirely clear, it seems they may be able to bind free antigen and/or antigens presented by nonclassical MHC-like molecules (Champagne, 2011).

The TCR gene segments are arranged in a similar manner to Ig gene segments, although in all species examined so far (including the cartilaginous fishes [Chen *et al.*, 2009]) they are present only in the translocon arrangement (Figure 2.2), and are rearranged in the same manner, using the same enzymes, as Igs. The TCR α and γ loci, like that of Ig light chains, encode only V and J segments, whereas the TCR β and δ loci, like Ig heavy chains, encode V, D, and J segments. The structural diversity of TCRs is created at the junctions during rearrangement of these segments and, as in Igs, is focused upon the center of the binding site where most of the interaction with peptides occurs.

The development of T cells parallels that of B cells in many ways; there is an orderly and stepwise rearrangement of TCR gene segments combined with the sequential testing of generated receptors. When the lymphoid progenitors from the bone marrow first reach the thymus, they receive signals from the cortical stroma that trigger signaling through the master transcriptional regulator NOTCH-1 and commitment to the T lineage. After a period of proliferation, these cells, which do not express either of the co-receptors CD4 or CD8 and so are known as double-negative thymocytes, begin to transiently express T-cell surface markers such as CD44 and CD25 (part of the IL-2 receptor) and rearrange their TCR β locus. The cells that successfully make a functional β chain can display a pre-TCR (the β chain with the surrogate α chain, pre-T cell α) associated with the signaling complex CD3 on their surface. The display of pre-TCR complex halts further β chain rearrangement, triggers cell proliferation, and initiates the expression of both the CD4 and CD8 co-receptors. These cells are now known as double-positive thymocytes and once they cease proliferating, rearrangement at the α locus begins. The structure of α locus is such that it allows multiple attempts at producing a functional α chain, meaning that most double-positive cells will eventually produce a functional α:β TCR. Interestingly, in mammals the genes encoding the TCR δ chain are located within the α locus, meaning that if the α chain rearranges the genes for δ are deleted, ensuring that a committed α:β cell cannot also express a γ:δ TCR. However, due to the fact that the β, γ, and δ loci rearrange first, and do so almost simultaneously, T-cell lineage fate seems to be determined by whether a functional β chain is produced before those for γ and δ chains.

The epithelial cells of the cortical stroma express both MHC class I and II molecules, and contact with these cells is crucial for T-cell positive selection; most of the TCRs generated will not be able to recognize self-peptide:MHC complexes and will fail positive selection and die. Those cells that pass positive selection migrate to the medullary region where negative selection then occurs; specialized epithelial cells (medullary thymic epithelial cells or MTECs) display peptides from peripheral host proteins under the control of the transcription factor AIRE (Anderson *et al.*, 2005); T-cell recognition of self-antigens results in the deletion of these potentially autoreactive cells. Only ~2% of double-positive thymocytes pass both tests and continue to mature, expressing higher levels of TCR and ceasing to express one of the two co-receptor molecules, thereby becoming single-positive (CD4+ or CD8+) thymocytes. The class of MHC bound by the TCR determines which co-receptor is displayed; CD4 binds to an invariant site on MHC class II, whereas CD8 binds a similar site on MHC class I. Both of these co-receptor binding sites are located away from the peptide-binding groove, and full maturation requires integration of the signals from both the TCR and appropriate co-receptor.

All four TCR loci (α, β, γ, and δ) have been found in cartilaginous fishes, showing that they were all present in the ancestor of all the jawed vertebrates (Rast *et al.*, 1997). Interestingly, two forms of TCR δ chain are found in cartilaginous fishes; the first is the typical form also found in other species. However, in the second form, the δ chain contains an extra domain, atop the δV domain, and this extra domain is most closely related to the V region of one of the cartilaginous fish Igs (specifically the single-domain antigen-binding region of the novel antibody isotype IgNAR [Roux et al., 1998]). This *doubly rearranging* receptor makes up a significant proportion of the δ chain transcripts and is hypothesized to bind soluble antigen in a manner akin to the IgNAR V region (Criscitiello et al., 2006). An unusual form of TCR with similarity to NAR-TCR has also been found to have arisen by convergent evolution in marsupials; although the origin of the V region of this molecule is more ambiguous than that of NAR-TCR, the constant region of the so-called TCRμ also appears to be most closely related to those of TCRδ (Parra et al., 2007). In addition, recent studies have shown that shark γ chains carry somatically derived mutations and, as in Igs, the incorporated mutations are targeted to hotspots (see the following text). As the shark γ locus is much less complex than that of mammals, it is hypothesized that in this instance SHM is used to diversify the repertoire (Criscitiello et al., 2010; Chen et al., 2012). Thus, it appears the lack of MHC restriction upon γ:δ T cells may afford them more evolutionary freedom to explore different ways of binding.

As outlined earlier, once a pathogen is encountered by tissue-resident phagocytic cells (such as macrophages and DCs), it is internalized and degraded, and at the same time, the cell releases cytokines, setting up a state of inflammation in the infected tissue. The increased flow of lymph through the tissue enables the activated phagocytes and bacterial debris to reach nearby secondary lymphoid organs (spleen, lymph nodes, and MALT). It is here that the phagocytes mature into professional APCs, displaying pathogen fragments and co-stimulatory molecules on their surface in order to stimulate T-cell proliferation and differentiation.

Once B cells and T cells have fully matured in their respective primary lymphoid organs, they move into the blood stream, repeatedly passing through the secondary lymphoid organs until they either encounter their specific antigen or die of neglect. The mature naïve T cells enter the T-cell zones of the secondary lymphoid organs; if a particular T cell does not encounter its antigen, it receives survival signals through IL-7 and interaction with self-peptide:MHC complexes; this is not sufficient for full T-cell activation but is able to induce low-level cellular division thus maintaining T-cell numbers. However, if the T cell encounters a properly activated APC displaying the appropriate peptide:MHC ligand, it is prevented from leaving the T-cell zone. After a period of proliferation, the cell differentiates into a *primed* effector T cell, which no longer requires further co-stimulation to respond upon reencounter of its antigen.

When circulating mature naïve B cells enter the secondary lymphoid tissues, they have to pass through the T-cell zone on their way to the primary follicles, where they usually reside. However, B cells that have already encountered their specific antigen, and now display peptides derived from it in complex with MHC class II on their surface, become stalled at the border of the T- and B-cell zones. Here the cell is perfectly located to find and interact with primed T cells with specificity for the same antigen. After their initial encounter, the B and T cells migrate to the border of the T-cell zone and red pulp to form a primary focus. In mammals, primary foci appear about five days after infection, and it is here that B cells proliferate and form short-lived plasma cells that secrete antibody; this first phase of the humoral response generally provides protection through low-affinity IgM antibodies.

In the second phase of the response, some of the antigen-specific lymphocytes that have already proliferated migrate into the primary lymphoid follicles and form a germinal center (GC). The GC is essentially an island of cell division located at the periphery of the follicle (which is now known as a secondary follicle). The GC grows in size as the immune response proceeds and two distinct areas emerge, the light zone and dark zone; B cells divide and mutate in the dark zone and then move to the light zone, which is rich in follicular dendritic cells (FDCs), where they undergo selection based upon their continued ability to bind to their antigen (Allen et al., 2007). The enzyme activation–induced cytidine deaminase (AID), a member of the cytidine deaminase family, is the initiator of SHM (as well as class switching and gene conversion) in the GC (Muramatsu *et al.*, 2000; Arakawa *et al.*, 2002). Expressed only in B cells, AID is essential for secondary antibody diversification and functions by removing the amino group from cytidine bases located within *hot-spot* motifs in the immunoglobulin gene, converting them to uracil. Repair of AID-induced lesions within the V region by various pathways results in changes to the coding sequence: (1) the lesion can be replicated creating a transition (C > T) mutation; (2) the uracil can be identified by uracil-DNA glycosylase (UNG), a base-excision repair enzyme, and removed. The abasic site can then be randomly filled with any nucleotide or cleaved by the endonuclease APE1 creating a single-strand break in the DNA. If two such breaks occur in proximity, forming a double-strand break (DSB) within the V region, it can induce repair through gene conversion; (3) the lesion can be recognized by the DNA mismatch repair machinery which, through its exonucleolytic activity, exposes a single-strand region of DNA, which is then back-filled by an error-prone DNA polymerase introducing multiple mutations across the gap (Maul and Gearhart, 2010).

In mammals, the rate of mutation is such that only every other B cell will acquire a mutation in its V region per cell division. However, these mutations accumulate in a stepwise manner as descendants of an individual B cell proliferate in the GC; in this way, SHM builds a series of related clones that differ subtly in their affinity for antigen. Interestingly, the cartilaginous fishes differ slightly from this model, showing a very high level of mutation for some of their Ig isotypes (at times surpassing those reported for mammalian V regions [Diaz *et al.*, 1999]) and a high frequency of tandem mutations (Diaz *et al.*, 1998; Lee *et al.*, 2002), perhaps suggesting a reliance upon the DNA mismatch repair pathway (pathway #3 earlier) or possibly even the presence of a novel mutational repair pathway in this lineage.

Most of the mutations introduced to a V region have a negative impact upon antigen binding meaning the cells carrying these receptors cannot compete with their siblings for the required survival signals and so are deleted by apoptosis. More rarely, a mutation will improve the affinity of antigen binding leading to positive selection and expansion of that particular B cell. Both positive and negative selections leave traces on V region sequences of mutated antibodies; very few mutations are found in the framework residues; rather, mutations tend to accumulate in the hypervariable loops (also called complementarity determining regions or CDRs) and are usually located at the periphery of the binding site. This process, in which antibody specificity and affinity are continually refined during an immune response, is called affinity maturation. It should, however, be noted that although all vertebrates have compartmentalized secondary lymphoid tissues containing accumulations of B cells, only birds and mammals can form bona fide GCs; in those species lacking GCs, affinity maturation is not as apparent as it is in mammals. Nevertheless, at least some of the species without GCs show high levels of SHM in their V regions, with obvious evidence of mutation selection (Hsu, 1998), and so questions have been raised regarding the exact role of GCs or if high-affinity antibodies are actually essential for immune protection.

The first antibody isotype to be produced during a mammalian immune response is IgM, which is particularly effective at activating complement through the classical pathway (see earlier). However, due to the configuration of their Ig genes, mammalian B cells can switch their antigen-specific V region onto the C region exons of other isotypes in response to signals received through cytokines. As mentioned earlier, the process of class switching is also catalyzed by AID; in this instance, AID introduces lesions into the highly repetitive *switch regions*, located between the constant regions of the different isotypes, and leads to the deletion of the μ C regions and their replacement with those of a different heavy chain isotype (Muramatsu *et al.*, 2000). The isotype targeted during the switch is governed by the cytokines produced during the ongoing immune response and allows antibody effector function(s) to be optimized for pathogen clearance. Although true class switching is not observed in any lineage below amphibians, it has been shown that AID from bony fishes is able to catalyze class switch recombination when transfected into mammalian B cells (Barreto *et al.*, 2005; Wakae *et al.*, 2006). More recently, a somatic recombination process that appears to be the precursor to tetrapod class switching has been identified in cartilaginous fishes (Zhu *et al.*, 2012); as mentioned earlier, rearrangement occurs only within a cluster in sharks, but it is now apparent that postrearrangement IgM V regions generated can be *switched* onto the constant regions of a different cluster and that the target constant regions may be of the same or a different isotype. Although this process is almost certainly AID mediated (barely detectable in shark pups that express low levels of AID and highest in deliberately immunized adult animals), it differs from the tetrapod class switch in that it occurs concomitant with SHM and may well involve breakage and joining between nonhomologous chromosomes (Zhu *et al.*, 2012). Thus, it seems AID had the potential to catalyze the switching reaction long before the divergence of amphibians but required some other factor, perhaps the emergence of true switch regions or temporal/spatial separation from SHM, to optimize the process.

Eventually, successfully mutated B cells exit the GC either as long-lived plasma cells that secrete large quantities of high-affinity antibody of different isotypes or as memory B cells that can be reactivated quickly in case of subsequent encounter of the same antigen.

Many aspects of this process can be identified in nonmammalian species and have been studied to better our understanding of when the various components emerged. Cartilaginous fishes have three heavy chain isotypes (IgM, the lineage-specific isotype IgNAR and IgW, the shark IgD orthologue), and these are used in concert to produce a highly complex, multilayered humoral response showing affinity maturation and immunological memory. The pentameric form of IgM is used as the *first line of defense*, binding antigen through high-avidity, low-affinity interactions. In contrast, the monomeric form of IgM, along with IgNAR, acts as the functional equivalents of mammalian IgG; after a lag phase, these isotypes show a highly antigen-specific response which affinity matures over the course of the response. It is these isotypes that are also responsible for the memory response (Dooley and Flajnik, 2006). Although the kinetics of the primary response in sharks is much slower than that of mammals (~4–6 months in shark compared to ~5–7 days for mammals), it has all the same hallmarks of T dependency.

Bony fishes also have multiple antibody heavy chain isotypes, all species studied so far have both IgM and IgD, and some, but not all, species also have a third isotype called IgT (Hansen *et al.*, 2005) or IgZ (Danilova *et al.*, 2005) dependent upon the fish species. IgT is found at low levels as a monomer in serum, but is expressed at much higher levels in the gut where it is found as a tetramer. Following the infection of fish with an intestinal parasite, an antigen-specific IgT response is produced in the GALT, and most intestinal bacteria are found to be coated with IgT. Thus, bony fishes are the most ancient phylogenetic group proven to have an antibody isotype functionally analogous to mammalian IgA and dedicated to mucosal protection (Zhang *et al.*, 2010). Amphibians have four heavy chain isotypes (IgM, IgD, IgY, and the mucosal isotype IgX) and as mentioned earlier are the most evolutionarily ancient vertebrates to isotype class switch. Frog IgY is functionally analogous to mammalian IgG and is thought to have shared a common ancestor with both IgG and IgE (Warr *et al.*, 1995). Although amphibians do not generate GCs, SHM is not lacking in amphibians. However, they produce antibodies of low heterogeneity that show little increase in binding affinity with time after immunization, perhaps indicating that the mutations introduced into the V regions do not undergo effective selection (Hsu, 1998). Finally, as mentioned earlier, true GCs have been found only in birds and mammals, and although affinity maturation has been observed in some non-GC species such as the cartilaginous fishes (Dooley and Flajnik, 2005), the highly organized structure of the GC obviously facilitates the efficient selection of generated mutants.

CONCLUDING REMARKS

As we have detailed herein, the IIS had already reached a high level of sophistication in some nonvertebrates; however, an AIS, able to generate a highly diverse repertoire of receptors by somatic rearrangement and demonstrating immunological memory, is found only in the vertebrates. Further, although the jawless fishes have an alternate form of AIS (based upon VLRs that are somatically *converted* by members of the cytidine deaminase family of enzymes), the AIS, as we know it from mammals, is found only in the jawed vertebrates. Thus, the two sequential GWD events, the first hypothesized to have occurred in an ancestor of all vertebrates, the second in an ancestor of the jawed vertebrates, and the invasion of the RAG transposon seem to have provided both the genetic raw material and evolutionary opportunity to facilitate this *immunological revolution*. The cartilaginous fishes, being the most ancient jawed vertebrate lineage, are poised right at this pivotal point in evolution. Study of this lineage has previously—and no doubt will continue to—offer great insight into the emergence, and subsequent fine-tuning, of the jawed vertebrate AIS. Perhaps if we show the wisdom and skill exemplified by Athena in the study of this key phylogenetic group, we will succeed in our odyssey to unravel the remaining secrets of the vertebrate immune system.

ACKNOWLEDGMENT

I thank Dr. Kathryn Crouch for critical reading and discussion of this work. The author has no conflicts of interest to declare.

REFERENCES

Allen, C.D., Okada, T., and Cyster, J.G. Germinal-center organization and cellular dynamics. *Immunity* 2007; 27 (2), 190–202.

Anderson, M.S., Venanzi, E.S., Chen, Z., Berzins, S.P., Benoist, C., and Mathis, D. The cellular mechanism of Aire control of T cell tolerance. *Immunity* 2005; 23 (2), 227–239.

Arakawa, H., Hauschild, J., and Buerstedde, J.M. Requirement of the activation-induced deaminase (AID) gene for immunoglobulin gene conversion. *Science* 2002; 295 (5558), 1301–1306.

Bajoghli, B., Guo, P., Aghaallaei, N., Hirano, M., Strohmeier, C., McCurley, N., Bockman, D.E., Schorpp, M., Cooper, M.D., and Boehm, T. A thymus candidate in lampreys. *Nature* 2011; 470 (7332), 90–94.

Barreto, V.M., Pan-Hammarstrom, Q., Zhao, Y., Hammarstrom, L., Misulovin, Z., and Nussenzweig, M.C. AID from bony fish catalyzes class switch recombination. *J. Exp. Med.* 2005; 202 (6), 733–738.

Bartl, S. What sharks can tell us about the evolution of MHC genes. *Immunol. Rev.* 1998; 166: 317–331.

Bengten, E., Clem, L.W., Miller, N.W., Warr, G.W., and Wilson, M. Channel catfish immunoglobulins: Repertoire and expression. *Dev. Comp. Immunol.* 2006; 30 (1/2), 77–92.

Bernstein, R.M., Schluter, S.F., Bernstein, H., and Marchalonis, J.J. Primordial emergence of the recombination activating gene 1 (RAG1): Sequence of the complete shark gene indicates homology to microbial integrases. *Proc. Natl. Acad. Sci. USA* 1996; 93 (18), 9454–9459.

Blair, J.E. and Hedges, S.B. Molecular phylogeny and divergence times of deuterostome animals. *Mol. Biol. Evol.* 2005; 22 (11), 2275–2284.

Boehm, T. Design principles of adaptive immune systems. *Nat. Rev. Immunol.* 2011; 11 (5), 307–317.

Champagne, E. Gamma-delta T cell receptor ligands and modes of antigen recognition. *Arch. Immunol. Ther. Exp. (Warsz.)* 2011; 59 (2), 117–137.

Chen, H., Bernstein, H., Ranganathan, P., and Schluter, S.F. Somatic hypermutation of TCR gamma V genes in the sandbar shark. *Dev. Comp. Immunol.* 2012; 37 (1), 176–183.

Chen, H., Kshirsagar, S., Jensen, I., Lau, K., Covarrubias, R., Schluter, S.F., and Marchalonis, J.J. Characterization of arrangement and expression of the T cell receptor gamma locus in the sandbar shark. *Proc. Natl. Acad. Sci. USA* 2009; 106 (21), 8591–8596.

Chen, K. and Cerutti, A. The function and regulation of immunoglobulin D. *Curr. Opin. Immunol.* 2011; 23 (3), 345–352.

Cooper, M.D., Peterson, R.D., and Good, R.A. Deliniation of the thymic and bursal lymphoid systems in the chicken. *Nature* 1965; 205: 143–146.

Criscitiello, M.F. and Flajnik, M.F. Four primordial immunoglobulin light chain isotypes, including lambda and kappa, identified in the most primitive living jawed vertebrates. *Eur. J. Immunol.* 2007; 37 (10), 2683–2694.

Criscitiello, M.F., Ohta, Y., Saltis, M., McKinney, E.C., and Flajnik, M.F. Evolutionarily conserved TCR binding sites, identification of T cells in primary lymphoid tissues, and surprising trans-rearrangements in nurse shark. *J. Immunol.* 2010; 184 (12), 6950–6960.

Criscitiello, M.F., Saltis, M., and Flajnik, M.F. An evolutionarily mobile antigen receptor variable region gene: Doubly rearranging NAR-TcR genes in sharks. *Proc. Natl. Acad. Sci. USA* 2006; 103 (13), 5036–5041.

Danilova, N., Bussmann, J., Jekosch, K., and Steiner, L.A. The immunoglobulin heavy-chain locus in zebrafish: Identification and expression of a previously unknown isotype, immunoglobulin Z. *Nat. Immunol.* 2005; 6 (3), 295–302.

Deng, L., Velikovsky, C.A., Xu, G., Iyer, L.M., Tasumi, S., Kerzic, M.C., Flajnik, M.F., Aravind, L., Pancer, Z., and Mariuzza, R.A. A structural basis for antigen recognition by the T cell-like lymphocytes of sea lamprey. *Proc. Natl. Acad. Sci. USA* 2010; 107 (30), 13408–13413.

Desjardins, M., Houde, M., and Gagnon, E. Phagocytosis: The convoluted way from nutrition to adaptive immunity. *Immunol. Rev.* 2005; 207: 158–165.

Diaz, M., Greenberg, A.S., and Flajnik, M.F. Somatic hypermutation of the new antigen receptor gene (NAR) in the nurse shark does not generate the repertoire: Possible role in antigen-driven reactions in the absence of germinal centers. *Proc. Natl. Acad. Sci. USA* 1998; 95 (24), 14343–14348.

Diaz, M., Velez, J., Singh, M., Cerny, J., and Flajnik, M.F. Mutational pattern of the nurse shark antigen receptor gene (NAR) is similar to that of mammalian Ig genes and to spontaneous mutations in evolution: The translesion synthesis model of somatic hypermutation. *Int. Immunol.* 1999; 11 (5), 825–833.

Dooley, H. and Flajnik, M.F. Shark immunity bites back: Affinity maturation and memory response in the nurse shark, Ginglymostoma cirratum. *Eur. J. Immunol.* 2005; 35 (3), 936–945.

Dooley, H. and Flajnik, M.F. Antibody repertoire development in cartilaginous fish. *Dev. Comp. Immunol.* 2006; 30 (1/2), 43–56.

Finstad, J. and Good, R.A. The evolution of the immune response. III. Immunologic responses in the lamprey. *J. Exp. Med.* 1964; 120: 1151–1168.

Flajnik, M.F. Comparative analyses of immunoglobulin genes: Surprises and portents. *Nat. Rev. Immunol.* 2002; 2 (9), 688–698.

Flajnik, M.F. and Kasahara, M. Comparative genomics of the MHC: Glimpses into the evolution of the adaptive immune system. *Immunity* 2001; 15 (3), 351–362.

Flajnik, M.F. and Kasahara, M. Origin and evolution of the adaptive immune system: Genetic events and selective pressures. *Nat. Rev. Genet.* 2010; 11 (1), 47–59.

Fugmann, S.D., Messier, C., Novack, L.A., Cameron, R.A., and Rast, J.P. An ancient evolutionary origin of the Rag1/2 gene locus. *Proc. Natl. Acad. Sci. USA* 2006; 103 (10), 3728–3733.

Graham, M., Shin, D.H., and Smith, S.L. Molecular and expression analysis of complement component C5 in the nurse shark (*Ginglymostoma cirratum*) and its predicted functional role. *Fish Shellfish Immunol.* 2009; 27 (1), 40–49.

Guo, P., Hirano, M., Herrin, B.R., Li, J., Yu, C., Sadlonova, A., and Cooper, M.D. Dual nature of the adaptive immune system in lampreys. *Nature* 2009; 459 (7248), 796–801.

Hansen, J.D. Characterization of rainbow trout terminal deoxynucleotidyl transferase structure and expression. TdT and RAG1 co-expression define the trout primary lymphoid tissues. *Immunogenetics* 1997; 46 (5), 367–375.

Hansen, J.D., Landis, E.D., and Phillips, R.B. Discovery of a unique Ig heavy-chain isotype (IgT) in rainbow trout: Implications for a distinctive B cell developmental pathway in teleost fish. *Proc. Natl. Acad. Sci. USA* 2005; 102 (19), 6919–6924.

Herrin, B.R., Alder, M.N., Roux, K.H., Sina, C., Ehrhardt, G.R., Boydston, J.A., Turnbough, C.L., Jr., and Cooper, M.D. Structure and specificity of lamprey monoclonal antibodies. *Proc. Natl. Acad. Sci. USA* 2008; 105 (6), 2040–2045.

Hirano, M., Guo, P., McCurley, N., Schorpp, M., Das, S., Boehm, T., and Cooper, M.D. Evolutionary implications of a third lymphocyte lineage in lampreys. *Nature* 2013; 501 (7467), 435–438.

Hoffmann, J.A. The immune response of Drosophila. *Nature* 2003; 426 (6962), 33–38.

Hsu, E. Mutation, selection, and memory in B lymphocytes of exothermic vertebrates. *Immunol. Rev.* 1998; 162: 25–36.

Jensen, J.A., Festa, E., Smith, D.S., and Cayer, M. The complement system of the nurse shark: Hemolytic and comparative characteristics. *Science*, 1981; 214 (4520), 566–569.

Kasahara, M. The 2R hypothesis: An update. *Curr. Opin. Immunol.* 2007; 19 (5), 547–552.

Kasamatsu, J., Oshiumi, H., Matsumoto, M., Kasahara, M., and Seya, T. Phylogenetic and expression analysis of lamprey toll-like receptors. *Dev. Comp. Immunol.* 2010; 34 (8), 855–865.

Kaufman, J., Milne, S., Gobel, T.W., Walker, B.A., Jacob, J.P., Auffray, C., Zoorob, R., and Beck, S. The chicken B locus is a minimal essential major histocompatibility complex. *Nature* 1999; 401 (6756), 923–925.

Kawai, T. and Akira, S. Toll-like receptors and their crosstalk with other innate receptors in infection and immunity. *Immunity* 2011; 34 (5), 637–650.

Kimura, A., Ikeo, K., and Nonaka, M. Evolutionary origin of the vertebrate blood complement and coagulation systems inferred from liver EST analysis of lamprey. *Dev. Comp. Immunol.* 2008; 33 (1), 77–87.

Kimura, A. and Nonaka, M. Molecular cloning of the terminal complement components C6 and C8beta of cartilaginous fish. *Fish Shellfish Immunol.* 2009; 27 (6), 768–772.

Kimura, A., Sakaguchi, E., and Nonaka, M. Multi-component complement system of Cnidaria: C3, Bf, and MASP genes expressed in the endodermal tissues of a sea anemone, *Nematostella vectensis*. *Immunobiology* 2009; 214 (3), 165–178.

Kokubu, F., Litman, R., Shamblott, M.J., Hinds, K., and Litman, G.W. Diverse organization of immunoglobulin VH gene loci in a primitive vertebrate. *EMBO J.* 1988; 7 (11), 3413–3422.

Lanier, L.L. NK cell recognition. *Annu. Rev. Immunol.* 2005; 23: 225–274.

Lee, S.S., Fitch, D., Flajnik, M.F., and Hsu, E. Rearrangement of immunoglobulin genes in shark germ cells. *J. Exp. Med.* 2000; 191 (10), 1637–1648.

Lee, S.S., Tranchina, D., Ohta, Y., Flajnik, M.F., and Hsu, E. Hypermutation in shark immunoglobulin light chain genes results in contiguous substitutions. *Immunity* 2002; 16 (4), 571–582.

Li, J., Barreda, D.R., Zhang, Y.A., Boshara, H., Gelman, A.E., Lapatra, S., Tort, L., and Sunyer, J.O. B lymphocytes from early vertebrates have potent phagocytic and microbicidal abilities. *Nat. Immunol.* 2006; 7: 1116–1124.

Litman, G.W. and Cooper, M.D. Why study the evolution of immunity? *Nat. Immunol.* 2007; 8 (6), 547–548.

Malecek, K., Lee, V., Feng, W., Huang, J.L., Flajnik, M.F., Ohta, Y., and Hsu, E. Immunoglobulin heavy chain exclusion in the shark. *PLoS Biol.* 2008; 6 (6), e157.

Maul, R.W. and Gearhart, P.J. AID and somatic hypermutation. *Adv. Immunol.* 2010; 105: 159–191.

Mayer, W.E., Uinuk-Ool, T., Tichy, H., Gartland, L.A., Klein, J., and Cooper, M.D. Isolation and characterization of lymphocyte-like cells from a lamprey. *Proc. Natl. Acad. Sci. USA* 2002; 99 (22), 14350–14355.

McKitrick, T.R. and De Tomaso, A.W. Molecular mechanisms of allorecognition in a basal chordate. *Semin. Immunol.* 2010; 22 (1), 34–38.

Miracle, A.L., Anderson, M.K., Litman, R.T., Walsh, C.J., Luer, C.A., Rothenberg, E.V., and Litman, G.W. Complex expression patterns of lymphocyte-specific genes during the development of cartilaginous fish implicate unique lymphoid tissues in generating an immune repertoire. *Int. Immunol.* 2001; 13 (4), 567–580.

Muramatsu, M., Kinoshita, K., Fagarasan, S., Yamada, S., Shinkai, Y., and Honjo, T. Class switch recombination and hypermutation require activation-induced cytidine deaminase (AID), a potential RNA editing enzyme. *Cell* 2000; 102 (5), 553–563.

Nagawa, F., Kishishita, N., Shimizu, K., Hirose, S., Miyoshi, M., Nezu, J., Nishimura, T. *et al.* Antigen-receptor genes of the agnathan lamprey are assembled by a process involving copy choice. *Nat. Immunol.* 2007; 8 (2), 206–213.

Oettinger, M.A., Schatz, D.G., Gorka, C., and Baltimore, D. RAG-1 and RAG-2, adjacent genes that synergistically activate V(D)J recombination. *Science* 1990; 248 (4962), 1517–1523.

Ohno, S. Gene duplication and the uniqueness of vertebrate genomes circa 1970–1999. *Semin. Cell Dev. Biol.* 1999; 10 (5), 517–522.

Ohta, Y. and Flajnik, M. IgD, like IgM, is a primordial immunoglobulin class perpetuated in most jawed vertebrates. *Proc. Natl. Acad. Sci. USA* 2006; 103 (28), 10723–10728.

Ohta, Y., McKinney, E.C., Criscitiello, M.F., and Flajnik, M.F. Proteasome, transporter associated with antigen processing, and class I genes in the nurse shark Ginglymostoma cirratum: Evidence for a stable class I region and MHC haplotype lineages. *J. Immunol.* 2002; 168 (2), 771–781.

Ohta, Y., Okamura, K., McKinney, E.C., Bartl, S., Hashimoto, K., and Flajnik, M.F. Primitive synteny of vertebrate major histocompatibility complex class I and class II genes. *Proc. Natl. Acad. Sci. USA* 2000; 97 (9), 4712–4717.

Palti, Y. Toll-like receptors in bony fish: From genomics to function. *Dev. Comp. Immunol.* 2011; 35 (12), 1263–1272.

Parra, D., Rieger, A.M., Li, J., Zhang, Y.A., Randall, L.M., Hunter, C.A., Barreda, D.R., and Sunyer, J.O. Pivotal advance: Peritoneal cavity B-1 B cells have phagocytic and microbicidal capacities and present phagocytosed antigen to CD4+ T cells. *J. Leukoc. Biol.* 2012; 91 (4), 525–536.

Parra, Z.E., Baker, M.L., Schwarz, R.S., Deakin, J.E., Lindblad-Toh, K., and Miller, R.D. A unique T cell receptor discovered in marsupials. *Proc. Natl. Acad. Sci. USA* 2007; 104 (23), 9776–9781.

Rast, J.P., Anderson, M.K., Ota, T., Litman, R.T., Margittai, M., Shamblott, M.J., and Litman, G.W. Immunoglobulin light chain class multiplicity and alternative organizational forms in early vertebrate phylogeny. *Immunogenetics* 1994; 40 (2), 83–99.

Rast, J.P., Anderson, M.K., Strong, S.J., Luer, C., Litman, R.T., and Litman, G.W. Alpha, beta, gamma, and delta T cell antigen receptor genes arose early in vertebrate phylogeny. *Immunity* 1997; 6 (1), 1–11.

Rast, J.P. and Messier-Solek, C. Marine invertebrate genome sequences and our evolving understanding of animal immunity. *Biol. Bull.* 2008; 214 (3), 274–283.

Reynaud, C.A., Dahan, A., Anquez, V., and Weill, J.C. Somatic hyperconversion diversifies the single Vh gene of the chicken with a high incidence in the D region. *Cell* 1989; 59 (1), 171–183.

Reynaud, C.A., Mackay, C.R., Muller, R.G., and Weill, J.C. Somatic generation of diversity in a mammalian primary lymphoid organ: The sheep ileal Peyer's patches. *Cell* 1991; 64 (5), 995–1005.

Rogozin, I.B., Iyer, L.M., Liang, L., Glazko, G.V., Liston, V.G., Pavlov, Y.I., Aravind, L., and Pancer, Z. Evolution and diversification of lamprey antigen receptors: Evidence for involvement of an AID-APOBEC family cytosine deaminase. *Nat. Immunol.* 2007; 8 (6), 647–656.

Roux, K.H., Greenberg, A.S., Greene, L., Strelets, L., Avila, D., McKinney, E.C., and Flajnik, M.F. Structural analysis of the nurse shark (new) antigen receptor (NAR): Molecular convergence of NAR and unusual mammalian immunoglobulins. *Proc. Natl. Acad. Sci. USA* 1998; 95 (20), 11804–11809.

Rumfelt, L.L., Avila, D., Diaz, M., Bartl, S., McKinney, E.C., and Flajnik, M.F. A shark antibody heavy chain encoded by a nonsomatically rearranged VDJ is preferentially expressed in early development and is convergent with mammalian IgG. *Proc. Natl. Acad. Sci. USA* 2001; 98 (4), 1775–1780.

Saha, N.R., Ota, T., Litman, G.W., Hansen, J., Parra, Z., Hsu, E., Buonocore, F., Canapa, A., Cheng, J.F., and Amemiya, C.T. Genome complexity in the coelacanth is reflected in its adaptive immune system. *J. Exp. Zool. B. (Mol. Dev. Evol.)* 2014; 9999B: 1–26.

Sakano, H., Huppi, K., Heinrich, G., and Tonegawa, S. Sequences at the somatic recombination sites of immunoglobulin light-chain genes. *Nature* 1979; 280 (5720), 288–294.

Shiina, T., Shimizu, S., Hosomichi, K., Kohara, S., Watanabe, S., Hanzawa, K., Beck, S., Kulski, J.K., and Inoko, H. Comparative genomic analysis of two avian (quail and chicken) MHC regions. *J. Immunol.* 2004; 172 (11), 6751–6763.

Smith, L.C., Clow, L.A., and Terwilliger, D.P. The ancestral complement system in sea urchins. *Immunol. Rev.* 2001; 180: 16–34.

Sun, J.C. and Lanier, L.L. The natural selection of herpesviruses and virus-specific NK cell receptors. *Viruses* 2009; 1 (3), 362.

Tasumi, S., Velikovsky, C.A., Xu, G., Gai, S.A., Wittrup, K.D., Flajnik, M.F., Mariuzza, R.A., and Pancer, Z. High-affinity lamprey VLRA and VLRB monoclonal antibodies. *Proc. Natl. Acad. Sci. USA* 2009; 106 (31), 12891–12896.

The MHC sequencing consortium. Complete sequence and gene map of a human major histocompatibility complex. *Nature* 1999; 401 (6756), 921–923.

Uinuk-Ool, T., Mayer, W.E., Sato, A., Dongak, R., Cooper, M.D., and Klein, J. Lamprey lymphocyte-like cells express homologs of genes involved in immunologically relevant activities of mammalian lymphocytes. *Proc. Natl. Acad. Sci. USA* 2002; 99 (22), 14356–14361.

Valanne, S., Wang, J.H., and Ramet, M. The Drosophila Toll signaling pathway. *J. Immunol.* 2011; 186 (2), 649–656.

Venkatesh, B., Lee, A.P., Ravi, V., Maurya, A.K., Lian, M.M., Swann, J.B., Ohta, Y. *et al.* Elephant shark genome provides unique insights into gnathostome evolution. *Nature* 2014; 505 (7482), 174–179.

Vivier, E., Raulet, D.H., Moretta, A., Caligiuri, M.A., Zitvogel, L., Lanier, L.L., Yokoyama, W.M., and Ugolini, S. Innate or adaptive immunity? The example of natural killer cells. *Science* 2011; 331 (6013), 44–49.

Wakae, K., Magor, B.G., Saunders, H., Nagaoka, H., Kawamura, A., Kinoshita, K., Honjo, T., and Muramatsu, M. Evolution of class switch recombination function in fish activation-induced cytidine deaminase, AID. *Int. Immunol.* 2006; 18 (1), 41–47.

Warr, G.W., Magor, K.E., and Higgins, D.A. IgY: Clues to the origins of modern antibodies. *Immunol. Today* 1995; 16 (8), 392–398.

Zapata, A. and Amemiya, C.T. Phylogeny of lower vertebrates and their immunological structures. *Curr. Top. Microbiol. Immunol.* 2000; 248: 67–107.

Zhang, Y.A., Salinas, I., Li, J., Parra, D., Bjork, S., Xu, Z., LaPatra, S.E., Bartholomew, J., and Sunyer, J.O. IgT, a primitive immunoglobulin class specialized in mucosal immunity. *Nat. Immunol.* 2010; 11 (9), 827–835.

Zhu, C., Lee, V., Finn, A., Senger, K., Zarrin, A.A., Du, P.L., and Hsu, E. Origin of immunoglobulin isotype switching. *Curr. Biol.* 2012; 22 (10), 872–880.

CHAPTER 3

Shark Reproduction, Immune System Development and Maturation
A Review

Lynn L. Rumfelt

CONTENTS

SUMMARY

Elasmobranchs (sharks, skates, and rays) have innate and adaptive immunity on par with mammalian animals and demonstrate a strong conservation of genes, cells, tissues, tissue organization, and primary and secondary lymphoid organs needed to function to protect the individual organism. Similar to mammals, data suggest cartilaginous fish have a developmental program for expression of their immunoglobulin (Ig) genes that progress from an innate-like restricted repertoire to a complex diverse repertoire along with maturation of both primary and secondary lymphoid tissues and expression of polymorphic major histocompatibility complex (MHC) class II genes. This chapter reviews the current knowledge on shark phylogeny, reproduction, embryonic and postnatal development of innate and adaptive immunity in these intriguing and evolutionarily successful animals. The focus of this review is the nurse shark, *Ginglymostoma cirratum*; however, research on other cartilaginous and bony fish is included.

SHARK CLASSIFICATION AND ANCESTRAL ORIGINS

Sharks are members of the group of vertebrate-jawed animals, whose internal skeleton is composed of cartilage rather than bone. Based on their cartilaginous skeleton and 5–7 gill slits, these animals are members of the class Chondrichthyes. Chondrichthyans are comprised of two main groups or subclasses: Elasmobranchs, which includes sharks, skates, and rays, and Holocephali, which includes ratfish or chimeras (Carroll, 1988). Together Chondrichthyans comprise 900–1200 known species (Compagno, 1999; Ratnasingham and Hebert, 2013; Weisz, 2012) and are distantly related to the large and diverse bony fish group in the class of Osteichthyes by divergent evolution from a common ancestral gene pool approximately 450 million years ago (MYA) (Figure 3.1). Continuation of these cartilaginous animals in modern times demonstrates their success in survival and reproductive strategies (see Chapter 1 for additional details).

LIVING LA VIDA TIBURON (LIVING THE LIFE OF A SHARK)

Where Do They Live?

Nurse sharks are commonly found in the shallow warm waters of the surf zone of the eastern United States and Gulf of Mexico (Guarracino, 2013). Free-ranging nurse sharks examined in the waters in northeast Brazil showed a water temperature preference of 25°C–30°C and salinity levels between 3.4% and 3.7% (average ocean water salinity is 3.5%) (Ferreira, 2013).

What Size Do They Grow To?

A nine-year study between 1955 and 1963 by Dr. Eugenie Clark and colleague identified that nurse sharks in the area of Central Gulf Coast of Florida ranged in length 210–270 cm and mass 80–100 kg (Clark and von Schmidt, 1965). In a more recent study, adult nurse sharks did not exceed 265 cm total length and 115 kg mass (Castro, 2000). A growth rate study (Carrier and Luer, 1990) using tag and recapture of free-ranging adult nurse sharks showed yearly growth rates averaged 13.1 ± 9.5 cm

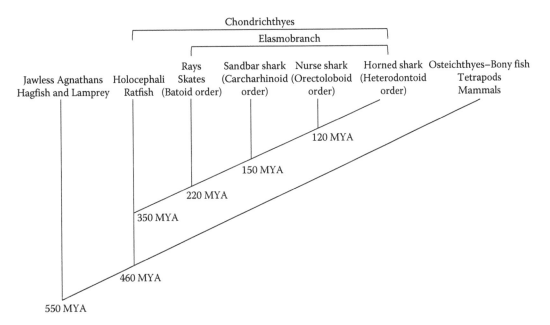

Figure 3.1 Phylogeny of cartilaginous fish. Cartilaginous fish (Chondrichthyes) comprise the organismal groups of sharks, skates, and rays (Elasmobranchs) and chimaera/ratfish (Holocephali) based on morphology, the fossil record, and molecular biology research. These animals are distantly related to bony fish and have many genetic similarities to more recent organisms in the tetrapod and mammalian groups. (Adapted from Flajnik, M.F., and Rumfelt, L.L., *Curr. Top. Microbiol. Immunol.* 248, 249–270, 2000.)

and 2.3 ± 1.3 kg. This study also demonstrated that captive nurse sharks had a faster rate of growth 19.1 ± 4.9 cm/year and 4.0 ± 1.7 kg/year, which slowed down as the study group aged, although the sample size was small (3 captive sharks compared to 44 free-ranging sharks). The difference in growth rate between captive and free-ranging sharks is not surprising because the sharks would be fed regular meals in captivity rather than having to hunt for their food in the wild. The nurse shark lives well in captivity and has been kept alive in public aquariums for over 25 years (Clark, 1963).

What and How Do They Eat?

Nurse sharks eat mainly shellfish, octopus, squid, sea urchins, and fishes (Robinson and Motta, 2002). The nurse shark throughout ontogeny, the horn shark (Edmonds *et al.*, 2001), and at least eight different elasmobranch families (Motta *et al.*, 2002) share the same manner of eating as bony fish, called inertial suction feeding. They eat their prey by inertial suction through creation of a negative pressure in their oral cavity by rapidly opening their jaw and controlling the size of mouth opening, which draws prey into the mouth, rather than biting and chewing of prey. Nurse sharks have specializations for suction feeding by having small teeth without overlap, small mouth, large muscles for expanding the mouth and gill cavities (orobranchial chamber), and the ability to open their mouths very fast (Robinson and Motta, 2002). This rapid suctioning helps sharks to capture their prey and eat them whole or in large pieces, rather than escape to live another day (a video of nurse shark feeding—http://www.discovery.com/tv-shows/shark-week/videos/shark-diet.htm).

How Do They Make Babies?

Animals in the class Chondrichthyes reproduce by internal fertilization. The ability to sexually reproduce to produce shark pups occurs in the nurse shark when males and females are 83% and 86%

(females 223–231 cm) of their maximum adult size, respectively (Castro, 2000). Internal fertilization of a female shark by a male occurs through insertion of the male genital appendage called claspers into the cloaca of the female (Pratt and Carrier, 2001). Free-living nurse shark mating is seasonal, occurring in June in the Florida Keys, during the day and night (Carrier *et al.*, 1994). The events that lead to selection, courtship, and mating are not well understood in the shark, nor is it understood how these animals choose specific locations as their mating grounds, which are returned to annually (Carrier *et al.*, 2002, 2003). That said mating behavior in the nurse shark is easy to detect at the water's surface, because these are shallow water-living creatures. A mating event is thought to occur when tail and pectoral fins thrashing is observed, and the shark's light-colored ventral (belly) side is visible (Carrier *et al.*, 1994). Mating in nurse sharks has been observed to be cooperative with multiple males working together to immobilize a female and inject sperm (Carrier *et al.*, 1994). The study of reproductive behaviors of nurse sharks by Carrier and colleagues directly observed 62.5% of the matings (10 out of 16 matings) involved multiple males mating with one female (polyandry), whereas 37.5% of the matings involved a solitary pair of nurse sharks. Confirmation of polyandry has been shown molecularly by restriction fragment length polymorphism analysis of MHC class Iα, MHC class IIα, and MHC class IIβ genes in two nurse shark families that detected at least three fathers in one family of 17 pups and at least four fathers in another family of 39 pups (Ohta *et al.*, 2000). More recently, a different genotyping method from Heist and collaborators analyzed polyandry in 29–39 pups per litter in three nurse shark litters (Heist *et al.*, 2011). They found each litter resulted from 5 to 7 males fathering from 1 to 17 pups (Heist *et al.*, 2011). Heist *et al.* suggested that the nurse shark might require multiple inseminations to fertilize an entire litter because these animals do not store sperm in the oviductal gland, a gland present in the nurse shark, which produces mature oocytes, unlike other elasmobranchs such as the dogfish (Moura *et al.*, 2011). Further work is needed to confirm this hypothesis. Polyandry has also been observed in the reproductive behaviors of the sandbar shark, *Carcharhinus plumbeus*, and the examination of polymorphic microsatellite loci in a commercially unexploited population in Hawaii determined 8 of 20 litters (40%) resulted from multiple males siring the sandbar shark pups (Daly-Engel *et al.*, 2007). Thus, Fitzpatrick and colleagues have concluded that polyandry evolved early in elasmobranch evolution approximately 350 MYA, because it has been identified in five of the eight existing orders of sharks (Fitzpatrick *et al.*, 2012). It is apparent that polyandry increases genetic diversity and potentially enhances immune responses to assist in animal survival and reproduction, thus may be one of several reasons for the evolutionary success of these creatures.

How Does a Mother Nurse Shark Feed Her Embryonic Pups?

Once the mature nurse shark oocytes are fertilized by sperm, contributed by one or more males, embryonic development ranges between 131 and 204 days (5–6 months) (Castro, 2000). Nurse shark embryos develop within egg cases carried by the mother in the paired uteri (Figure 3.2; Wourms and Lombardi, 1992). The embryos hatch from their egg cases *in utero*, prior to delivery, and are born live at a later and appropriate stage of development. It is hypothesized that live-bearing young evolved as a means to shield offspring from predation, temperature extremes, anoxia, and osmotic stress (Goodwin *et al.*, 2002).

Nutrition to the developing nurse shark embryos is provided by the yolk from a structure called the yolk sac, which is attached directly to their intestines; thus, these animals are classified as lecithotrophic (yolk feeders) (Heming and Buddington, 1988). Yolk contains nutrients, which include lipoglycophosphoproteins, and oil globules (Heming and Buddington, 1988). Nurse shark embryos feeding on the nutritious yolk within the yolk sac, rather than directly from the mother through a placental structure, are classified as aplacental viviparous or ovoviviparous (Carrier *et al.*, 2003; Castro, 2000). Most sharks and all rays are viviparous when fed by placental yolk sac or aplacental development from egg case within the mother, for example, the nurse shark (Wourms, 1977). Other

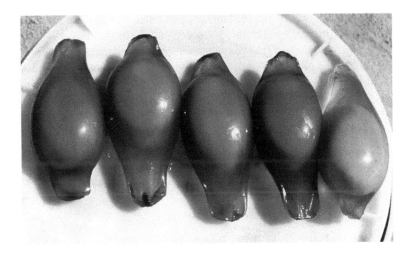

Figure 3.2   Nurse sharks are ovoviviparous and develop within an egg case over 5- to 6-month period within the uterus of the female nurse shark, and then hatch from these cases and are born live over a period of several days. Egg cases shown were collected from a nurse shark in Mayport, Florida. Note the abundant yolk visible within the egg cases. (Image supplied by George H. Burgess, Florida Museum of Natural History, Gainesville, FL, with permission.)

types of embryonic development are oviparous, where the embryos develop from a fertilized egg within an egg case, which is located outside of the body of the mother shark, for example, the horn shark, *Heterodontus francisci*, and is similar to bony fish and amphibians.

A nurse shark study by Castro indicated newborn pups are 28–30 cm in length, with a positive correlation between *in utero* development time and total length of the embryo (Castro, 2000). Another study, using invasive imaging to study gestation in the nurse shark, found that newborn nurse shark pups length range was somewhat smaller, between 20 and 25 cm (average 21.7 cm) (Carrier *et al.*, 2003). Live birth of the pups may occur over several days, and at the time of birth, the nurse shark pups have completely absorbed their yolk sac, and may still have an open yolk sac scar or a closed scar (Castro, 2000). The nurse shark embryos do not show evidence of intrauterine embryonic cannibalism, identified in other species such as the sand tiger shark, *Odontaspis taurus* (Gilmore *et al.*, 1983).

We observed that nurse shark pups delivered near term by Caesarian birth always had a yolk sac and a larger yolk sac correlated with shorter pup length (less mature) (Ohta *et al.*, 2000; Rumfelt, 2000). Newborn nurse shark pups averaged 20–24 cm in length and weighed 120–150 g. It is predicted that these litters were the result of polyandry, as discussed earlier; however, this was not explicitly investigated. When the pups were delivered in this unnatural way, they were able to continue development *ex utero* through continued feeding from their associated yolk sac, which became internalized as they matured and transitioned to a normal adult diet.

How Are the Embryonic Pups Immunologically Protected *In Utero*?

One means of protection for the developing embryo is through passive immunity, provided by the maternal transfer of antibodies to the developing embryo, either through placental transfer *in utero* or by transfer into the embryo's or newborn's food supply of yolk, milk, or colostrum (Grindstaff *et al.*, 2003). Presumably, these antibodies would have been produced primarily by an adaptive immune response to pathogens or antigens seen in the mother's environment. It has been well documented that female fish transfer maternal antibodies into their eggs (Grindstaff *et al.*, 2003). This has been confirmed for nurse sharks by Haines and colleagues, who examined antibodies present in the yolk of developing embryos and found that two classes of antibodies were present, likely to protect the embryos from waterborne pathogens (Haines *et al.*, 2005). The antibody classes present in yolk sac are IgM in the monomeric form (two heavy chains each disulfide bonded to a light chain and disulfide-bonded to each other, 7S) and immunoglobulin new antigen receptor (IgNAR) (two heavy chains disulfide bonded together with no light chain association; Figure 3.3). IgNAR and 7S IgM in adult sharks have been shown to be adaptive responses that undergo affinity maturation (Diaz *et al.*, 1998, 1999; Dooley and Flajnik, 2005; Dooley *et al.*, 2006; Greenberg *et al.*, 1995). Based on a long-term immunization study using hen egg-white lysozyme

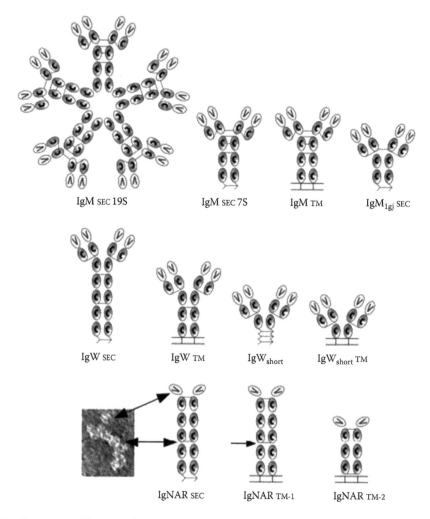

IgM SEC 19S   IgM SEC 7S   IgM TM   IgM₁gj SEC

IgW SEC   IgW TM   IgW short   IgW short TM

IgNAR SEC   IgNAR TM-1   IgNAR TM-2

Figure 3.3 Secretory and transmembrane immunoglobulins IgM, IgW/D, and IgNAR. Cartilaginous fish have three isotypes of immunogobulins (Ig): IgM, IgW, and IgNAR. IgM forms a pentamer (19S) and monomer (7S). IgM₁gj is a subclass of IgM that is expressed preferentially at birth and early development with a germline-joined heavy chain variable region and thus has a restricted specificity repertoire. IgW, now classified as IgD, has multiple transmembrane (TM) and secretory (SEC) forms and expressed differently between individuals of the same species. IgNAR is a monomer composed of two heavy chains without associated light chains, and its variable region exhibits considerable rotation. All Ig isotypes, except IgM₁gj, have been identified in elasmobranchs examined thus far. (Electron micrograph image from Roux, K.H. *et al.*, *Proc. Natl. Acad. Sci. USA.* 95, 11804–11809, 1998; diagram of Ig classes from Rumfelt *et al.*, 2004b.)

(HEL) in adult nurse sharks, Dooley and Flajnik have shown IgM monomeric 7S and IgNAR are the isotypes that will provide an antigen-specific response with affinity maturation and memory, under appropriate stimulation conditions (Dooley and Flajnik, 2005). These data suggest that the transfer of these antibody classes into the unfertilized ovum (transfer is assumed to occur prior to fertilization, similar to chicken; Hamal *et al.*, 2006) will provide protection to developing embryos. In addition, adult nurse shark IgM heavy and light chains are junctionally diverse in the variable domains, and these domains undergo an increase in somatic hypermutation as the animal ages (Lee *et al.*, 2002; Malecek *et al.*, 2005). Together these data suggest that passive transfer of antibodies from mother to fetus through yolk sac would be diverse in their specificities and protective through affinity maturation against various pathogens present in the uterine environment.

In addition to immunity from passive transfer of antibodies and physical sequestration by developing *in utero*, the expression of inflammatory cytokines may assist in immunoprotection during embryonic development. Analysis of a smoothhound shark, *Mustelis canis*, near-term gestation found the interleukin (IL)-1 α, IL-1 β, and the functional IL-1 receptor type I expressed in cells in the mother's uterus, fetus' egg envelope, and placental tissues (Cateni *et al.*, 2003). The smooth-hound shark, aplacental viviparous shark, develops *in utero* from a yolk sac that changes into a yolk sac placenta after depletion of the yolk sac nutrients. IL-1 cytokine functions to activate the innate immunity system, and thus, its expression may enhance immune responses to microbial products (Cateni *et al.*, 2003). The IL-1β gene identified in the small-spotted catshark, *Scyliorhinus canicula*, is expressed in the spleen and testes, and its expression increased when splenocytes were incubated with the Gram-negative bacterial outer membrane component, lipopolysaccharide (LPS) (Bird *et al.*, 2002). Thus, IL-1 in the shark appears to function, similar to mammal, as an inflammatory cytokine and likely will be found protective for developing nurse shark embryos (for further details refer to Secombes *et al.*, Chapter 7 in this book).

What Is the *In Utero* Environment Like for Shark Embryos?

The nurse shark embryos develop within egg capsules in the enlarged uteri for 12–14 weeks of gestation and then hatch at approximate length of 218–233 mm and continue development within the uteri (Castro, 2000). This makes for a crowded environment in the mother's uteri for litters of 25–40 pups on average (Castro, 2000; Heist *et al.*, 2011), which may result in hypoxia or osmotic imbalances, so the mother or the pups must control these factors in the uteri at a level sufficient for embryonic life to be sustained. How does this occur? Living in the salty marine environment requires elasmobranchs to regulate osmotic pressures, so excess salts are excreted and water maintained at optimal levels. Elasmobranchs blood osmolarity is similar to seawater by excretion of excess Na^+ and Cl^- in urine from kidneys, through a unique tissue called the rectal gland, and in some species the gills; thus, plasma concentrations of these ions are lower than in seawater (Pang *et al.*, 1977). High osmolarity of plasma fluid occurs by retaining high levels of organic nitrogenous compounds: urea and trimethylamine oxide (TMAO) (Hammerschlag, 2006). TMAO is thought to counteract the detrimental effects of urea on proteins and thus is an important solute in osmolarity maintenance (Seibel and Walsh, 2002). These data are based on the analysis of adult elasmobranchs, so how do shark pups undergo osmoregulation as they develop *in utero*? Help may be provided by the mother providing urea in the sealed egg capsule as well as bathing the early ovoviviparous embryos (*Squalus acanthias* and *M. canis*) in levels of urea equivalent to levels in the maternal blood (Price and Daiber, 1967). Oxygen and water for the embryos are provided by the uterine fluid of the mother, which is close in salinity to seawater, and thus, it has been assumed that the gestating mother regularly flushes the uterine environment by opening of the cloacal orifice (Price and Daiber, 1967).

Exposure to the marine environment during embryonic development due to proposed flushing of the uterine horns would remove metabolic waste products yet introduce seaborne microbes, viruses, worms, and other pathogens. This suggests immune defense is necessary and important for survival in embryonic development to birth and beyond. Thus, transfer of maternal adaptive immunity type of antibodies, 7S IgM and IgNAR, into yolk sac and the presence of inflammatory cytokines, such as IL-1, may be important to the developing animal as it starts hematopoiesis to produce embryonic B and T cells.

ONTOGENY OF LYMPHOID TISSUE/ORGANS IN CARTILAGINOUS FISH

The Innate and Adaptive Immune Systems in Sharks

Similar to other jawed and jawless vertebrate animals, sharks and their offspring require protection against pathogens through the ability to recognize and defend against organisms and their molecules, which do not arise within the host, but are of nonself origins (Beutler, 2004; Jerne, 1955, 1974). The innate immune system works to detect and respond immediately to pathogenic invaders in a general, rapid, and nonspecific manner, using a specific and limited set of cells and molecules (Beutler, 2004; Janeway et al., 2001). These innate immune cells are able to detect and destroy nonself cells by recognition of invariant patterns in molecules associated with pathogens, resulting in an activation cascade that induces a cellular and molecular response to remove and/or inactivate the pathogen (Medzhitov and Janeway, 1997). Janeway and Medzhitov's review on innate immunity recognition states, "Innate immunity is an evolutionarily ancient part of the host defense mechanisms: the same molecular modules are found in plants and animals, meaning it arose before the split into these two kingdoms," which occurred approximately 1.6 billion years ago based on molecular analyses (Hoffmann et al., 1999; Janeway and Medzhitov, 2002; Wang et al., 1999). Thus, innate immunity is thought to be the first type of immune system to develop in the metazoan tree of life, and it has evolved to function in conjunction with the adaptive immune system in vertebrates (Beutler, 2004). Unlike innate immunity, the adaptive immune response is marked by its specificity, its diversity in receptors to recognize nonself molecules, and its induction when the host organism is exposed to a specific pathogen. Adaptive immunity is slower in response time and *remembers* the specific pathogen, so that host reexposure to the same pathogen induces a stronger and more rapid immune response (Janeway et al., 2001).

Innate Immunity in Sharks and Other Fish

The cells of the innate immune system are thought to develop mainly from the myeloid lineage during hematopoiesis, and in mammals, these cells consist of monocytes/macrophages, dendritic cells (DCs), and granulocytes: neutrophils, eosinophils, and basophils. These cells function to remove pathogens by phagocytosis, activation of the complement pathway, and activation of the adaptive immune system naïve mature or memory B and T cells to become effector cells. The humoral component of innate immunity is the complement system of plasma proteins, which become activated by antibodies binding pathogen surfaces (classical pathway of complement activation), or by antibody-independent mechanisms, the lectin or alternative pathways (Janeway et al., 2001) (refer to Smith and Nonaka Chapter 8 in this book for further details). Further examples of humoral components of innate immunity are the enzyme lysozyme, which degrades bacterial cell walls, transferrin, an iron chelator and microbial growth inhibitor, and antimicrobial molecules. Together these secreted molecules and cells of the innate immune system function to help kill invading pathogens and promote survival of the host.

Based on the sharing of innate immune system molecules since ancient times, one would expect a similar repertoire of cells and molecules in the shark, compared to mammals. Macrophages, monocytes, interdigitating cells (dendritic-like cells), and four types of granulocytes, including neutrophils and eosinophils, have been identified by light and electron microscopy and immunohistochemistry in tissues including blood (Fänge and Pulsford, 1983; Hyder *et al.*, 1983; Morrow and Pulsford, 1980; Rumfelt *et al.*, 2002; Zapata *et al.*, 1996b). Cellular analysis of blood in the adult nurse shark is approximately 73% lymphocytes, 25% granulocytes, and 2% monocytes, similar to the adult blacktip shark, *Carcharhinus limbatus* (Walsh and Luer, 2004).

Shark monocytes and granulocytes function similarly to mammalian cells. At least some nurse shark leukocytes have receptors for the constant region of IgM heavy chain (μ), fragment crystallizable receptor (FcμR), and are granulocytes: eosinophils or neutrophils (Haynes *et al.*, 1988; McKinney and Flajnik, 1997). Further analysis found that spontaneous cytotoxicity mediated by macrophage cells was antibody independent (Hyder *et al.*, 1983; McKinney and Flajnik, 1997; Morrow and Pulsford, 1980). Shark neutrophils are phagocytic and cytotoxic, based on antibody-dependent FcμR binding to pentameric IgM (19S) or monomeric IgM (7S) (Haynes *et al.*, 1988; McKinney and Flajnik, 1997). Similar to mammalian eosinophils, shark eosinophils appear not to be actively phagocytic (Hyder *et al.*, 1983; Weiss, 1983). Granulocytes in elasmobranchs contain hydrolytic enzymes, such as acid phosphatase, glycosidase, and chitinase, and neutrophils contain peroxidase similar to mammals (Johansson, 1973; Zapata *et al.*, 1996b). Macrophages cluster with lymphocytes in the elasmobranch brain and spleen, presumably to activate an adaptive immune response (Pulsford and Zapata, 1989; Torroba *et al.*, 1995) (refer to Haines and Arnold, Chapter 5 and Walsh and Luer, Chapter 6 for further review of these topics).

Humoral components of innate immunity present in elasmobranchs: lysozyme, chitinase, which degrades chitin, and transferrin are present in lymphomyeloid/hematopoietic tissues (Clem and Small, 1967; Fänge *et al.*, 1976; Got *et al.*, 1967). Antimicrobial molecules, such as squalamine, have been found in the stomach of dogfish and shown to kill in vitro Gram-positive and Gram-negative bacteria, and fungi (Moore *et al.*, 1993). Classical and alternative complement pathways and components of the ancient lectin complement pathway are present in the nurse shark (Smith, 1998). For further details on antimicrobial molecules and complement pathways refer to Chapters 8 and 15.

Innate Antibodies

In mice and humans, there exists a population of B cells called B1 B that form during fetal and newborn development and reside in several tissues, which includes the peritoneal and pleural cavities, spleen, and intestines (Kantor *et al.*, 1992; reviewed by Dorshkind and Montecino-Rodriguez, 2007). B-1 B cell variable regions are considered to be innate-like due to their lack of diversity compared to the adult (Feeney, 1990). B1 B cells have been further subdivided into B1a and B1b cells due to functional differences. B1a cells spontaneously (in the absence of antigen) secrete pentameric IgM that is innate-like due to a nondiverse repertoire and polyreactivity to T-cell-independent antigens, such as carbohydrates. This early-secreted IgM from B1a B cells has been shown to be protective against encapsulated bacteria, such as *Streptococcus pneumonia*, and is thought to be a first line of defense, whereas the B1b response is induced, thus antigen-responsive, and important for removal of the pathogen and long-term protection (Haas *et al.*, 2005).

Do Cartilaginous Fish Have Early Innate-Like B Cells during Development of the Adaptive Immune System and Lymphatics?

Sharks may have a population of B cells that function similar to mouse innate B-1 cells. Neonatal nurse shark pups preferentially express an innate-like Ig identified as a subclass of immunoglobulin M (IgM), IgM$_{1gj}$ (Castro *et al.*, 2013; Rumfelt *et al.*, 2001). This unusual Ig associates with light chains and has restricted specificity in its heavy chain variable region (V$_{H1gj}$) consisting one variable (V) gene, one diversity (D) gene, and one joining (J) gene (V-D-J), which must have been rearranged in germ cells (*germline-joined*) and thus is a nonsomatically rearranged VDJ. Germline-joined V_{HIgj} is thought to be RAG-mediated due to potential P nucleotide addition of TC dinucleotide in its V-D junction (Rumfelt *et al.*, 2001). V_{HIgj} region complementarity determining region 3 (CDR3) is short, in part because there is only one D gene and because there does not appear to have been much trimming or addition of nucleotides (N-region addition) during the processing of RAG-cleaved coding ends in its VDJ rearrangement. Lack of N-region addition in V_{HIgj} CDR3 is not due to lack of expression of terminal deoxynucleotidyl transferase (TdT), the enzyme responsible for template-independent deoxynucleotide addition at 3'OH terminus of cleaved DNA regions during somatic gene recombination, because we have shown nurse sharks express this enzyme throughout ontogeny in the epigonal organ (Rumfelt *et al.*, 2001). IgM_{1gj} is a smaller-sized antibody, similar to mammalian IgG, because it lacks a CH_2 domain compared to classical μ heavy chain. IgM_{1gj} associates with joining chain (J chain), a small protein important for multimerization of Igs, and is secreted in the newborn shark tissues of spleen, epigonal organ, and plasma, and by circulating young and adult peripheral blood leukocytes (PBL) (Castro *et al.*, 2013; Hohman *et al.*, 2003; Rumfelt *et al.*, 2001). Castro and colleagues have shown that V_{HIgj} associates with κ light chains, of which some are germline joined, producing at least a proportion of IgM_{1gj} molecules, which contain both invariant heavy and light chain variable regions of a single specificity (Castro, 2013). Castro has shown that IgM_{1gj} specifically binds to laminin, a glycoprotein present in basement membranes and important for cell differentiation, migration, and adhesion (Castro, 2013; Timpl *et al.*, 1979). Possibly innate antibodies, such as IgM_{1gj}, in association with laminin, may assist in the movement of lipidated ligands, such as Sonic Hedgehog, to activate signal transduction pathways, such as Hedgehog, to induce cell-fate decisions and development of tissues in the shark (Pires-daSilva and Sommer, 2003).

The lymphatic system is important to both innate and adaptive immunity. It serves as a secondary vascular system in mammalian vertebrates, formed by endothelial blind-ended capillaries present in most tissues that function to drain and collect extracellular fluids and extravasated cells and return these to circulating blood (Butler *et al.*, 2009). The lymphatic circulatory system transports antigens/pathogens and lymphocytes from a site of infection in distant tissues to the secondary lymphoid organs, where adaptive immune responses may be activated.

Cartilaginous fish do have a secondary lymphatic-like vascular system; however, it contains erythrocytes, thus this system is part of the circulatory system rather than a true and separate lymphatic system (Butler *et al.*, 2009; Kampmeier, 1969). Because red blood cells do not extravasate from blood vessels, lymph fluid in true lymphatic system does not contain red blood cells under normal circumstances. Another ancient vertebrate group, the bony fish, along with amphibians, reptiles, and birds, do have true lymphatic vascular systems (Butler *et al.*, 2009).

Adaptive Immunity in Sharks and Other Fish

A model of cartilaginous fish with central lymphoid organs, consisting the epigonal organ, Leydig organ, thymus, meninges of the brain, and eye orbit; peripheral lymphoid organs, consisting the spleen; and gut-associated lymphoid tissue (GALT) are shown in Figure 3.4 (Chiba *et al.*, 1988; Rumfelt *et al.*, 2002; Tomonaga *et al.*, 1986; Torroba, 1995).

The Immunoglobulins

IgM and IgM$_{1gj}$

IgM was the first Ig class identified in cartilaginous fish: dogfish shark by Edelman's group and lemon and nurse sharks by Clem's group, and is orthologous to mammalian IgM (refer to Figure 3.3 for immunoglobulin classes expressed in cartilaginous fish; Clem and Small, 1967; Clem *et al.*, 1967; Marchalonis and Edelman, 1965, 1966). Cartilaginous fish IgM exists as monomers (7S) or

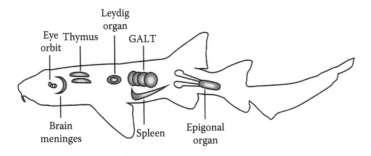

Figure 3.4   Putative immune-developing (primary) tissues: thymus, epigonal organ, Leydig organ, meninges of the brain, and eye orbit, and immune-responding (secondary) tissues: spleen and gut-associated lymphoid tissue (GALT), in Elasmobranch sharks, skates, and rays. (From Rumfelt, L.L. *et al.*, *Scand. J. Immunol.* 56, 130–148, 2002.)

pentamers (19S), and monomeric IgM is not a precursor or degradation product of pentameric IgM (Clem *et al.*, 1967; Small *et al.*, 1970). Similar to mammalian IgM, shark pentameric IgM associates with J-chain to form multimers and is restricted to the blood, and comparable to mammalian IgG; shark monomeric IgM is present in both tissues and blood and does not associate with J-chain proteins (Castro *et al.*, 2013; Clem *et al.*, 1967; Small *et al.*, 1970). In adult nurse sharks, about 50% of plasma protein is composed of IgM in roughly equal concentrations of pentameric and monomeric IgM (Clem *et al.*, 1967). Much evidence suggests that pentameric IgM is present prior to immunization and polyreactive and does not undergo affinity maturation postimmunization, whereas monomeric IgM is an adaptive response that increases in concentration over time, undergoes affinity maturation, and shows a memory response (Clem *et al.*, 1967; Dooley and Flajnik, 2005). In support of the ontogeny of IgM in the nurse shark, Castro and colleagues have shown by *in situ* hybridization analysis that all secretory plasma cells are J-chain[+] in neonatal pups (Castro *et al.*, 2013). In addition, splenic secretory pentameric IgM[+] cells in newborn pups are J-chain[+], and one to two months of maturation is required before splenic monomeric IgM-secreting J-chain[−] cells are detected. This work confirms previous work showing a delay in the detection of monomeric IgM in blood and PBLs (Clem *et al.*, 1967; Rumfelt *et al.*, 2001).

Nurse shark pups *in utero* do not receive Igs from their mother transplacentally; however, their yolk sacs contain adaptive immunity-type Igs: monomeric IgM and IgNAR, likely to provide another layer of immune protection for the embryos (Haines *et al.*, 2005). Newborn nurse sharks have low serum levels of IgM at birth that is 19S pentameric IgM (Fidler *et al.*, 1969). Monomeric 7S IgM appears later in development, whereas both 19S and 7S IgM increase in concentration during the first month of life. At this early stage of development, IgM concentration is less than 50% of adult nurse sharks levels (Fidler *et al.*, 1969). Examination of IgM *VH* repertoire in newborn nurse sharks is similar to the ratfish, sandbar shark, and horn shark, and most IgM *VH* genes are from one *VH* family with multiple, heterogeneous loci (Rumfelt *et al.*, 2004a). The Ig *VH* CDR3 repertoire is extensively diversified by the addition of nontemplated nucleotides, yet is shorter than adult VDJ junctions, suggesting one means of postnatal repertoire diversification is to select for longer CDR3 junction length. Analysis of splenic surface IgM[+] single B cells from an unimmunized nurse shark pup with an immature immune system (~2 months old) showed IgM *VH* is essentially not hypermutated (Zhu *et al.*, 2011). In contrast, an immunized seven-year-old adult showed a 62% hypermutation frequency in surface IgM[+] splenic B cells (Zhu *et al.*, 2011). Together these data suggest that mounting an antigen-driven adaptive immune response requires a maturation program during postnatal development to enable selection of IgM-expressing B cells and *VH* mutation to improve antibody affinity for its epitope. Although 45 years ago we thought nurse sharks pups had very little immunoglobulin in their serum, we now know otherwise. IgM_{1gj} is present in abundant amounts in serum of every newborn nurse shark pup examined (16 pups from three different families) (Rumfelt et al., 2001). These pups were delivered near or at term by Caesarian section yet they already secreted this Ig class; thus, IgM_{1gj} is expressed early in the shark Ig repertoire. IgM_{1gj} VH is most similar to horn shark VH (65% amino acid homology) and is a new nurse shark V family most related to conventional IgM VH (Rumfelt et al., 2001). As discussed, IgM_{1gj} is restricted in its repertoire due to the germline joined VH_{1gj}, which associates with germline joined VL in some IgM_{1gj} expressing B cells (Castro, 2013; Rumfelt et al., 2001). Castro's work identified that IgM_{1gj} binds most specifically to laminin, which was selected for testing, because laminin is an antigen recognized by mouse B1 B-cell antibodies made as part of the innate repertoire before the animal has a fully developed adaptive immune system (Castro, 2013). Laminin binding suggests that IgM_{1gj} may function as a defense against an invariant pathogen or may have a nonimmunological function, such as assistance in embryogenesis. At birth, IgM_{1gj} is most highly expressed in both epigonal organ and spleen tissues yet becomes restricted to only the epigonal organ in the adult. This further evidence supports the hypothesis that the epigonal organ is the site of hematopoiesis/lymphopoiesis and a reservoir for secretory B cells, similar to mammalian bone marrow (Rumfelt et al., 2001). As the nurse shark pup matures, large amounts of IgM are produced that dilute IgM_{1gj} serum levels, which makes the IgM_{1gj} concentration in plasma appear to decrease. ELISPOT data have now shown PBLs enriched for plasma cells in both young (six months old) and adult (nine years old) sharks secrete IgM_{1gj} at detectable levels with the young shark having more than twofold the number of IgM_{1gj}-expressing plasma cells compared to the adult (Castro et al., 2013). Based on this evidence, it appears IgM_{1gj}^+ plasma cells traffic through the blood from the epigonal organ because they do not reside in adult spleen (Castro et al., 2013; Rumfelt et al., 2001, 2002).

Clearly, IgM_{1gj} is an IgM-related molecule expressed early in development as a multimer associating with J-chain. At this early stage of shark development, IgM_{1gj} is likely a spontaneous Ig (i.e., not in response to antigen) and may function in the first line of defense against a pathogen or during tissue remodeling. Thus, IgM_{1gj}-expressing B cells have several characteristics of fetal and newborn mouse and human B1a B cells and may be the shark version of a natural antibody.

IgW/IgD

Another class of immunoglobulin was identified in several elasmobranch species and given many names: IgX_{short} (clearnose skate, Harding et al., 1990b), IgX_{long} (clearnose skate, Anderson et al., 1999; Harding et al., 1990b), IgNARC (nurse shark, Greenberg et al., 1996), and now IgW (sandbar shark, Bernstein et al., 1996b). For a period of time, it was thought that IgW existed only in cartilaginous fish, then a homologue was identified recently in the lungfish (Ota et al., 2003), and thereafter, it was found in the amphibian Xenopus tropicalis (Ohta and Flajnik, 2006) as an orthologue to mammalian and bony fish IgD (Wilson et al., 1997). Thus, the missing link of the origins of IgD has been solved, and this Ig class stretches back in vertebrate history 450 MYA as IgW in cartilaginous fish.

IgW is the most heterogeneous class discovered to date because multiple forms of transmembrane and secretory IgW have been identified: secreted long and short forms, and transmembrane long and short forms. The secreted long form, IgW_{long} heavy chain composed of seven domains: one variable and six constant domains, was identified in shark (Figure 3.3; Greenberg et al., 1996), and the secreted short form, IgW_{short} heavy chain composed of three domains: one variable and two constant domains, was identified in skate (Harding, 1990). Both long and short IgW secretory forms are present in the nurse and horn sharks (Rumfelt et al., 2004a). The transmembrane IgW_{long} heavy chain is alternatively spliced to produce a shorter heavy chain with five domains: one variable and four constant domains, or an even shorter three-domain heavy chain: one variable and two constant domains, in the horn and nurse sharks (Rumfelt et al., 2004a).

The number of IgW genes varies between individual nurse sharks as analyzed by Southern blot under high-stringency conditions and IgW $CH1$ probe (Greenberg *et al.*, 1996). Zhang and colleagues examined IgW clusters expressed in the adult nurse shark and found transmembrane IgW_{short} similar to the previous findings with two CH domains (Rumfelt *et al.*, 2004b; Zhang *et al.*, 2013). Zhang describes secretory IgW_{short} containing two CH domains but it is unclear if this form associates with a noncanonical secretory tail with seven cysteines similar to the previous findings for secretory IgW_{short} (Rumfelt *et al.*, 2004b). Their analysis of nurse shark IgW gene clusters W1–W5 found alternative-spliced products, not been previously described, that included secretory IgW with four CH domains associated with the noncanonical secretory tail, and an unusual transcript with eight CH domains due to duplication of $CH3$ and $CH4$, but missing its secretory tail (Zhang *et al.*, 2013). These unusual forms of secretory IgW, IgW with nine domains (1 VH and 8 CH) and IgW with five domains (1 VH and 4 CH), have not been identified as proteins expressed in primary or secondary tissues in adult/newborn nurse sharks and skates (Greenberg *et al.*, 1996; Rumfelt *et al.*, 2004b). A recent study of genes in the bony fish coelacanth, in phylogeny more closely related to tetrapods than bony fish, found two IgW loci containing multiple tandem repeats of constant region exons supporting the findings in the nurse shark (Saha *et al.*, 2014; Zhang *et al.*, 2013).

Northern analysis of the IgW secretory forms showed that IgW_{long} expressed at approximately the same levels in the spleen of adult nurse shark individuals, and IgW_{short} expression was variable between individuals, from very little to strong expression (Rumfelt *et al.*, 2004a). It is unclear the reasons for this variation of IgW_{short} expression between individuals. Adult nurse shark IgW_{long} does associate with the J-chain in spleen and epigonal organ cells using *in situ* hybridization (Castro *et al.*, 2013).

Newborn nurse shark IgW cDNA clonal analysis revealed that the ωVH repertoire is highly diverse with unique CDR3s by N-region addition, VDJ rearrangement, and the CDR3s were significantly shorter than the adult ωVH, similar to IgM VH (Rumfelt *et al.*, 2004b). IgW VH has diverged sufficiently to form three families expressed at birth, differing substantially from the single VH family for IgM. Northern analysis showed all forms of IgW: long and short are expressed early in ontogeny (Rumfelt *et al.*, 2004b). Northern analysis of IgW expression using ωVH probe among six adults and five newborn pup individuals found highest IgW VH expression in the spleen and very little expressed in the epigonal organ. Low to negligible amounts of IgW_{long} are secreted by epigonal organ and splenic B cells in the newborn nurse shark, which contrasts with readily detectable amounts of IgW_{long} in the adult (Rumfelt *et al.*, 2001). Together these data suggest that, at least for IgW_{long}, there is no innate-like version expressed early in postnatal development, and maturation of adaptive immune tissues may be required prior to secretion of this IgW form.

IgNAR

IgNAR, first identified in the serum of nurse sharks, is an Ig class restricted to elasmobranchs thus far (Greenberg *et al.*, 1995). It is unusual in its shape in that it is composed of two heavy chains disulfide bonded together, which do not associate with light chains (Figure 3.3), unlike most other vertebrate Igs, with the exception of the camelid IgG (Hamers-Casterman *et al.*, 1993). IgNAR heavy chains have one variable and five constant domains as secretory and transmembrane forms and a shorter transmembrane form, $IgNAR_{TM-2}$, with one variable and three constant domains (Greenberg *et al.*, 1995; Rumfelt *et al.*, 2004b). Interestingly, the variable domain on each heavy chain moves freely and is flexible due to hinge-like regions (Roux *et al.*, 1998). Its variable region is generated by four rearrangement events between one V gene, three D genes, and one J gene resulting in long CDR3s containing cysteine residues. Evidence suggests that IgNAR is diverse in its primary repertoire of antigen receptors by somatic cell rearrangement events mediated by *RAG* and then further diversifies by an antigen-reactive process of somatic hypermutation, which occurs in response to recognition of antigen bound to its CDR loops (Diaz *et al.*, 1998; Dooley and Flajnik,

2005; Dooley *et al.*, 2006; Greenberg *et al.*, 1995). Analysis of IgNAR mutation frequency is very high and selected for in the secretory form rather than transmembrane form (Diaz *et al.*, 1998, 2002). Analysis of the binding capacity of IgNAR V domain using phage-displayed libraries found this unusual V domain binds antigen with high affinity as single V domain (Dooley *et al.*, 2003). The crystal structure of type I IgNAR V domain complexed with HEL antigen, showed a minimal antigen-binding domain that contains only two CDRs (rather than three) but still binds antigen with nanomolar affinity, similar to conventional antibodies associating with light chains, such as IgM and IgW (Stanfield *et al.*, 2004). IgNAR variable domain binds at increased affinity by increasing the number of contacts of the two CDR loops (Stanfield *et al.*, 2007).

Much evidence strongly suggests that IgNAR is an immunoglobulin response to antigen that occurs after maturation of the immune system by antigenic stimulation. Examination of newborn nurse shark IgNAR mRNA and protein compared to the adult supports this model of requirement for maturation. In newborn nurse sharks, secreted IgNAR was negligible in plasma, epigonal organ, and splenic B cells using three different IgNAR-specific monoclonal antibodies (mAbs) (Rumfelt, 2000; Rumfelt *et al.*, 2001). This contrasts with the adult shark, which has abundant secreted IgNAR present in plasma, epigonal organ, and spleen, in approximately 10-fold lower quantities than pentameric/monomeric IgM using the same IgNAR-specific mAbs. Maturation of the pup's adaptive immune system occurs at approximately six months of age, because IgNAR was present in plasma and expressed by spleen and epigonal organ cells at levels equivalent to the adult shark (Rumfelt, 2000; Rumfelt *et al.*, 2001). A nurse shark ontogeny study by Diaz and colleagues found a novel IgNAR (type 3), preferentially expressed in newborn and young sharks with innate-like characteristics: infrequently mutated *V* regions not targeted to the CDRs, smaller CDR3 by use of two rather than three *D* genes, which are germline joined (!), CDR3s have highly similar amino acid sequence restricted to one length of 16 amino acids and very little N-region addition (Diaz *et al.*, 2002). It was unexpected that this adaptive response-type Ig would be expressed in an innate-like form at birth that is gradually superseded by the other two types of IgNAR, which do undergo high levels of somatic mutation with increased CDR3 length and variability as the animals mature (Diaz *et al.*, 2002; Rumfelt *et al.*, 2004b). Similar to IgM_{1gj}, IgNAR type 3 maintains its expression in adults only in the epigonal organ (Diaz *et al.*, 2002; Rumfelt *et al.*, 2001).

The Tissues

Give Me a T for Thymus and T-Cell Development

All cartilaginous-jawed fish species examined thus far have thymus and spleen organs (Good *et al.*, 1966; Zapata *et al.*, 1996b). Twenty-two species of elasmobranch fish (sharks, skates, and rays) were examined for thymic architecture, cellular components, anatomical location, thymocyte function, and morphology, and each possessed a thymus with separate medullary and cortical regions, similar to mammalian thymus (Luer *et al.*, 1995). In some species, *Scyliorhinus* and *G. cirratum*, the thymus involutes with age, similar to mammals; however, in *Heterodontus* and some rays, the thymus remains large throughout ontogeny (Fänge, 1987). Morphological studies have shown the thymus is well-vascularized, innervated, multilobed bilateral organ located near the gills, surrounded by a connective tissue capsule containing macrophages, fibroblasts, and collagenous fibers (Fänge, 1982). The thymic cortex contains lymphoblasts undergoing cell division and thymocytes of various sizes closely packed among a network of epithelial cell processes, whereas the medulla contains mainly epithelial cells. Macrophages are frequently found in the cortex and medulla with occasional macrophage-lymphocyte clusters observed. Interdigitating cells (likely DCs) appear predominantly at the corticomedullary border, and the medullary organization is similar to mammalian thymus (Zapata *et al.*, 1996a). A marker for immature thymocytes in mammals, peanut agglutinin receptor (C-type lectin receptor), is present on the shark cortical thymocytes, similar to other vertebrate studies (Luer *et al.*, 1995). Analogous to mammals, T-cell receptor (TCR) α, β, γ, and δ genes are present and expressed along with the genes involved in TCR V(D)J gene rearrangement and TCR sequence diversity: recombination-activating gene 1 (*RAG1*) and terminal deoxynucleotidyl transferase (TdT) in thymi of the horn shark, *H. francisci*, clearnose skate, *Raja eglanteria*, and nurse shark, *G. cirratum* (Bartl *et al.*, 2003; Bernstein *et al.*, 1996a; Criscitiello *et al.*, 2010; Miracle *et al.*, 2001; Rast and Litman, 1994; Rast *et al.*, 1995, 1997; Rumfelt *et al.*, 2002). *RAG1* is expressed throughout ontogeny in the thymus by northern blot analysis and *in situ* hybridization (Rumfelt *et al.*, 2001, 2002). More recently, *in situ* hybridization analysis found *RAG1* and *TdT* most highly expressed in the thymic cortical subcapsular region, where TCR rearrangement events occur in mammals (Criscitiello *et al.*, 2010). Numerous cells positive for Ig (possibly mature plasma cells) have been detected by immunofluorescence in skate thymus from *Raja naevus* (Ellis and Parkhouse, 1975), IgNAR transcripts isolated from young (six-month old) thymus in the nurse shark (Diaz *et al.*, 2002), and IgW$_{long}$ transcripts isolated from adult thymus in the sandbar shark, *C. plumbeus* (Bernstein *et al.*, 1996b), likely due to circulating plasma cells homing back to epigonal organ or perhaps important for negative selection of thymo-cyte T-cell receptor (TCR).

γδ T cells develop before birth and have been described as innate lymphocytes (Prinz *et al.*, 2013). γδ T cells may develop from several subsets of double negative (DN) (CD4−CD8−surface TCR−) thymocytes: DN2a, DN2b, and DN3a (Rothenberg *et al.*, 2010). Criscitiello has shown that adult nurse shark TCR γδ is most highly expressed in the outer half of the cortex (Zone 2 in Petrie and Zúñiga-Pflücker, 2007) and subcapsular region (Zone 3 in Petrie and Zúñiga-Pflücker, 2007), where commitment to the T-cell lineage and divergence between TCRα/β versus TCRγ/δ lineages occur in mammalian adult thymus (Criscitiello *et al.*, 2010; Petrie and Zúñiga-Pflücker, 2007). Whether this pattern holds true for fetal and neonatal thymic environment is not known in mammals (Petrie and Zúñiga-Pflücker, 2007), and ontogeny of expression of TCRα/β versus TCRγ/δ is still to be determined in cartilaginous fish.

In the more primitive jawless fish group that includes hagfish, *Eptatretus stoutii*, and lamprey, *Petromyzon marinus*, neither spleen nor thymus has been identified (Good and Finstad, 1968). Recently, lymphoepithelial structures called thymoids in the lamprey larvae gill basket showed molecular expression of the orthologous forkhead box N1 (FOXN1), a marker of thymopoiesis in jawed vertebrates, and cytosine deaminase 1 (CDA-1) thought to be involved in gene conversion to produce type A variable lymphocyte receptors (VLRAs), the lamprey equivalent of T-cell receptors (Bajoghli *et al.*, 2011). These data suggest that the thymoid structure functions as a site of lamprey T-cell-like development (Bajoghli *et al.*, 2011).

In mammalian T-cell development, negative selection of self-reactive T cells is dependent on autoimmune regulator (*Aire*) gene expression in a subset of medullary thymic epithelial cells, which induce expression of tissue-specific self-antigens (e.g., insulin protein expressed in thymus rather than pancreatic islet of Langerhans β cells) (Anderson and Su, 2011). T-cell development in carti-laginous fish will likely utilize this enzyme similar to mammals. Although Saltis and colleagues identified *Aire* in bony fish, amphibian, and chicken, it has not yet been isolated in cartilaginous fish (Saltis *et al.*, 2008) (refer to Criscitiello Chapter 12 for detailed discussion on shark T-cell develop-ment and T-cell receptors).

Epigonal Organ and Leydig Organ–Primary Lymphoid Tissue and Bone-Marrow Equivalent in Shark

The elasmobranch endoskeleton is composed of cartilage rather than bone; therefore, these ani-mals do not possess bone marrow and are unable to undergo hematopoiesis in the same location as amphibians (some), birds, and mammals (Zon, 1995). So which cartilaginous fish tissues are the site(s) of hematopoiesis?

Two such organs exist in these animals. The first described was the Leydig organ, a gland-like structure associated with the esophagus described in rays by the Danish anatomist, Nicolaus Steno, in 1685 (Mattisson and Fänge, 1982); the second described was the epigonal organ, physically attached to gonads and similar in structure and organization to the Leydig organ (Leydig, 1857; Matthews, 1950; Monro, 1785). The Leydig organ is composed of a reticulum packed with large numbers of leukocytes, penetrated by few arteries and capillaries (Leydig, 1857; Mattisson and Fänge, 1982). The leukocytes include neutrophils, eosinophils, and other granulocytes, which are less mature in the Leydig organ than in blood. Putative lymphocytes of various sizes are abundant and form loose follicle-like aggregates with scattered plasma cells (Fänge, 1987). The epigonal organ is similar in structure and organization to the Leydig organ, except it is physically connected with gonadal tissue. In the ray species *Rhinobatos rhinobatos*, the epigonal organ is highly vascularized, completely encloses the gonads, and contains undifferentiated male or female germ cells (spermatogonia and oogonia) (Bircan-Yildirim *et al.*, 2011). Either the Leydig organ or epigonal organ, or both, are present in all cartilaginous fish examined; the nurse shark has only an epigonal organ (Fänge, 1977, 1984; Fänge and Mattisson, 1981; Pratt, 1988).

The newborn nurse shark epigonal organ cells are mainly nongranular by Wright–Giemsa granulocyte staining, and as the animals mature, the frequency of granular cells and level of endogenous peroxidase increases, suggesting myelopoiesis occurs in the epigonal organ (Rumfelt, 2000; Rumfelt *et al.*, 2002).

RAG1 and *TdT* are expressed in newborn and adult epigonal organ cells, not in spleen, again suggesting epigonal organ is a site of V(D)J gene rearrangement and B-cell development (Criscitiello *et al.*, 2010; Rumfelt *et al.*, 2001, 2002). We discovered that newborn nurse shark epigonal organ shows small distinct clusters containing 20–50 cells/cluster surface IgM (sIgM+), whereas separate clusters secrete IgM_{1gj} (Rumfelt *et al.*, 2002). Many newborn epigonal organ cells are MHC class II+ and surround clusters of MHC class II[low/negative]-expressing cells. Sequential sections demonstrated that the MHC class II[low/negative] cell clusters contained some sIgM+ B cells. Low/null MHC class II expression on sIgM+ B cells is similar to the observations of delayed MHC class II expression in developing B cells in the murine fetal liver and sIgM+ splenic B cells in perinatal mouse (Hayakawa *et al.*, 1994; Rumfelt *et al.*, 2002). Some cells within sIgM+ clusters were sIgM− cells and may be undergoing V(D)J gene rearrangement. As the shark pups matured by five months of age, cells in the epigonal organ were MHC class II+ in small clusters rather than on many cells throughout the tissue as seen in neonates (Rumfelt *et al.*, 2002). Five-month-old epigonal organ cells showed clusters of MHC class II+ and IgM+ secretory cells, suggesting they had matured into plasma cells from a naïve type of B cell present at birth. ELISPOT analysis of immunoglobulin secretion in a more mature one-year-old shark (shark EN) found significantly more IgM_{1gj}-secreting plasma cells in epigonal organ than in spleen (Castro *et al.*, 2013).

Expression of the important transcription factor, PU.1, for mouse and human B-cell, T-cell, and myeloid-cell development is limited to the epigonal and Leydig organs in the skate (these animals have both) (Anderson *et al.*, 2001; Scott *et al.*, 1994). Further work by Anderson and colleagues has shown the hematopoietic transcription factors important for B- and T-cell development: GATA-3, Ikaros, Aiolos, Helios, EBF-1, Spi-C, Spi-D, Runx-2, and Pax-5 are expressed in the Leydig and epigonal organs of embryonic clearnose skate (Anderson *et al.*, 2004).

Functional studies have shown that very little antigen is trapped in the epigonal/Leydig organs when compared to spleen and liver, further supporting the idea that lymphopoiesis and myelopoiesis are the main purpose of these organs (Rowley *et al.*, 1988). There may be differences between various cartilaginous fish because *RAG1* is expressed in the Leydig organ, epigonal organ, and thymus of the adult clearnose skate, *R. eglanteria*. This is similar to the results of the nurse shark, yet the clearnose skate also expressed *RAG1* in spleen and nonlymphoid tissues: muscle, intestine, rectal gland, and liver, which has not been found in the nurse shark (Miracle *et al.*, 2001; Rumfelt *et al.*, 2001). Together these morphological, molecular, and functional studies point to the Leydig and epigonal organs as the bone-marrow equivalent in cartilaginous fish and site of myelopoiesis and lymphopoiesis.

Spleen-Secondary Lymphoid Tissue Needs to Mature before Functioning in Adaptive Immune Responses

As in other vertebrates, the elasmobranch spleen is a highly differentiated and vascularized structure that functions as the major site for antigen stimulation leading to antibody synthesis (Morrow *et al.*, 1982; Tomonaga *et al.*, 1984, 1992; Zapata *et al.*, 1996b). Splenic white pulp (splenic WP) is well defined, consisting dense aggregates of various-sized lymphocytes, including mature and developing plasma cells, and missing mammalian-like marginal zones and periarterial lymphoid sheath (PALS) (Fänge and Nilsson, 1985; Zapata *et al.*, 1996b). The absence of PALS regions in the spleen makes sense in light of the fact that cartilaginous fish do not have a separate lymphatic system dedicated to white blood cell movement, unlike bony fish, and more recently evolved vertebrates.

Close association between macrophages, lymphocytes, and plasma cells in splenic WP has been observed; however, germinal centers similar to mammals have not been found in any ectothermic vertebrate so far (Pulsford and Zapata, 1989; Pulsford *et al.*, 1982; Zapata *et al.*, 1996a). Melanomacrophage centers in bony fish hematopoietic tissue of the kidney contain large aggregates of melanin-containing macrophages, which are proposed as germinal center precursors in phylogeny (Lamers, 1985). In cartilaginous fish, these melanin-containing macrophages occur as isolated cells in spleen and liver and do not form large aggregates as seen in bony fish (Agius, 1980).

Polymorphic molecules of *MHC* class I, *MHC* class II, and *MHC* class II invariant chain Ii are present and expressed in elasmobranchs (Bartl and Weissman, 1994; Criscitiello *et al.*, 2012; Hashimoto *et al.*, 1992; Kasahara *et al.*, 1992). Ig-expressing (not secreting) B cells in adult nurse shark spleen, PBL, and dendritic-like cells are MHC class II[+], similar to mammals (Rumfelt *et al.*, 2002).

The ontogeny and expression of IgM, IgM_{1gj}, and IgNAR-expressing B cells in the secondary lymphoid tissue of the spleen was examined using immunohistochemistry, a panel of mAbs, and MHC class II antisera to compare tissue organization and structure of the adult to newborn through five-month-old nurse shark (Rumfelt *et al.*, 2002). In the adult shark, splenic WP is well organized into B-cell zones comprised of mainly *naïve* surface immunoglobulin positive (sIg[+]) for IgM or IgNAR and surface MHC class II[+] (sMHC II[+]) B cells (Figure 3.5). The frequency of sIgM[+] B cells is approximately 10-fold greater than sIgNAR[+] B cells, on par with serum levels of IgM and IgNAR (Greenberg *et al.*, 1995; Rumfelt *et al.*, 2002). B-cell zones surround centralized *putative T-cell* zones, based on lack of B-cell markers: sIgM[−], secreted IgM[−], secreted IgM_{1gj}[−], and lack of sMHC class II expression on cells with lymphocyte phenotype by morphology and high nuclear/cytoplasmic ratio. Within these *T-cell-like zones* reside a population of cells, which are large and express much higher levels of sMHC class II[+] yet are negative for surface or secreted Igs and contain putative dendritic processes. Based on location, morphology, and phenotype, these *T-cell-like* zone large cells are likely DCs and appear similar to murine splenic DCs observed in T-cell zones (Steinman *et al.*, 1997). A three-dimensional model was developed based on the analysis of serial sectioning and staining of adult splenic tissue shown in Figure 3.5i. The compartmentalization of B cells, putative T cells, and DCs suggests that mammalian-like antigen activation of B cells by T cells and of T cells by MHC class II[+] DCs may occur in the splenic secondary lymphoid tissue and requires further analysis (Rumfelt *et al.*, 2002).

Defined mammalian-like germinal centers where affinity maturation occurs may not exist in cartilaginous fish (Zapata and Amemiya, 2000). Tracking of immunized antigen in adult nurse sharks showed movement of the antigen from the subcutaneous injection site in fin muscle to the spleen after several weeks (Rumfelt, 2000). The antigen presented on putative DCs in the WP *T-cell-like* zone and presence of IgNAR[+] secretory cells formed in small clusters of 5–20 cells within this same antigen-presenting T-cell zone suggests these IgNAR[+] clusters may be analogous to mammalian germinal

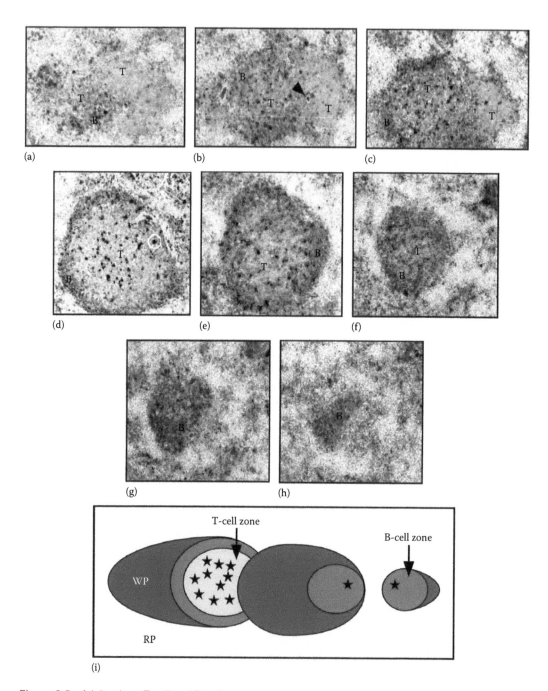

Figure 3.5 Adult spleen T-cell and B-cell zones are components of single WP. Adult nurse shark spleen
was sequentially sectioned, stained with MHC class IIβ antiserum and counterstained with
methyl green. Parts (a–h) show representative sections from a 750-μm block of tissue. T cells
(T) and putative DCs white circle (d) are located within T-cell zones within the WP, and B cells
form the outer edge. Part (i) shows a proposed model of the adult B-cell and T-cell zone organization within splenic WP. Sectioning near the end of the WP displays a B-cell zone (purple),
and sectioning WP more centrally shows a region enriched in putative DCs (stars) and T cells
(green). (From Rumfelt et al., *Scand. J. Immunol.* 2002; 56: 130–148.)

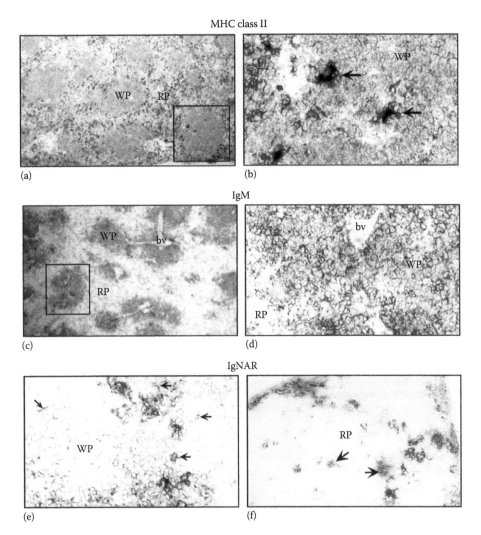

Figure 3.6 Newborn spleen consists of WPs comprising exclusively of B-cell zones with MHC class II[low/+] panels (a–b), surface IgM[+] B cells panels (c–d), and red pulp zones that are strongly MHC class II[+] panels (a–b). IgNAR-expressing cells are very infrequent in the newborn splenic WP as surface IgNAR[+] and red pulp as secretory IgNAR[+] B cells panels (e–f).

centers (Rumfelt, 2000). Dooley and colleagues have shown IgNAR binds antigen with high affinity and mapped differential mutations from the same ancestral clone during *in vivo* expansion (Dooley and Flajnik, 2005; Dooley *et al.*, 2006). They found most clones derived from the ancestral clone had further increased affinity for antigen by about 10-fold, demonstrating affinity maturation does occur.

In newborn nurse sharks, splenic WPs are comprised exclusively of B-cells zones containing sIgM[+] MHC class II[low] B cells and of T-cell-like zones containing DCs as well as sIgNAR-expressing B cells are essentially absent at birth (Figure 3.6; Rumfelt *et al.*, 2002). Further analysis using differential *in situ* hybridization has shown pentameric secretory IgM[+] and IgM_{1gj}^+ in newborn splenic red pulp is J-chain[+], and these are the predominant cell types in newborn spleen (Castro *et al.*, 2013). Because newborn serum has been shown to contain largely pentameric IgM, mono-meric/dimeric IgM_{1gj}, and very little monomeric IgM and IgNAR, the *in situ* results fit well with the previous data (Rumfelt *et al.*, 2001). Secretion of newborn pentameric IgM and IgM_{1gj} is not depen-dent on B-lymphocyte-induced maturation protein 1 (Blimp-1), the master regulator of mammalian

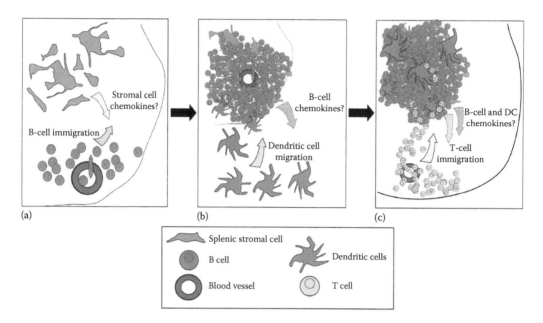

Figure 3.7   Developmental model of spleen organogenesis. (a) Stromal/reticuloendothelial cells form the embryonic spleen structure and potentially secrete chemokines/cytokines such as CCL19 and CXCL13 to migrate into splenic tissue. (b) B cells aggregate, forming immature WPs within the stromal cell structure, which may secrete chemokines causing DCs migration into the WP and formation of B-cell-rich areas without the presence of T cells. (c) Collectively, stromal cells, B cells, and/or DCs may secrete chemoattractants such as CCL19/CCL21 to induce T cells to migrate into WP within the DC network and displace B cells to WP periphery. (Figure from Rumfelt, *et al.*, *Scand. J. Immunol.* 2002; 56: 130–148.)

plasma cell development (Castro *et al.*, 2013). This requirement changes during ontogeny because adult monomeric IgM (7S) secretors are also positive for Blimp-1. These results suggest that pentameric IgM and IgM$_{1gj}$ are programmed to become plasma cells in newborns, prior to maturation of the spleen. Once the splenic tissue has matured sufficiently to support an adaptive-type Ig response, then the secretion of monomeric IgM requires Blimp-1 to undergo terminal differentiation into plasma cell secretor of monomeric IgM (Castro *et al.*, 2013).

The nurse shark pups required about five months of postnatal development to undergo maturation of the splenic compartment to adult-like features: significant numbers of sIgNAR$^+$ B cells in splenic WP, *T-cell-like* zones inhabited by MHC class II-rich *dendritic cells*, and many secretory IgNAR plasma cells present in the splenic red pulp. This delay in splenic tissue development in the shark is similar to murine studies that show T cells populating the spleen four days after birth and displacing B cells by occupying the perivascular area and establishing adult-like WP organization (Friedberg and Weissman, 1974) (refer to Figure 3.7 for our model of spleen organogenesis).

FUTURE WORK

More research is needed to better understand the development of adaptive immunity in fetal and newborn sharks. Is T-cell development similar in the adult shark compared to the newborn pup? An ontogeny study of thymic T-cell development and localization of TCRα/β versus TCRγ/δ in these animals would go far in helping us understand the differences. Are natural killer cells important for defense in the newborn to assist in broadening the newborn limited repertoire? Little is known about natural killer cells and their development, function, and localization in cartilaginous fish, so much work is needed in this area. IgW has multiple transmembrane and secretory forms present in the shark. How is IgW engaged in an immune response and how are the production of IgW long and short forms regulated? Finally, cytokines and chemokines are important for migration and differentiation of progenitors to become naïve B and T cells in mammals. Will this hold true for cartilaginous fish? We shall see as we progress in our discoveries of the ancient and successful pathway of adaptive immunity in these fascinating animals.

ACKNOWLEDGMENTS

Thanks to George H. Burgess for use of the egg case image in Figure 3.2; Sylvia Smith for extreme patience, persistence, and kindness in helping me complete the chapter; and Martin Flajnik for reviewing and suggesting changes to improve the writing.

REFERENCES

Agius, C. Phylogenetic development of melano–macrophage centres in fish. *Journal of Zoology* 1980; 191:11–31.

Anderson, M.K., Pant, R., Miracle, A.L., Sun, X., Luer, C.A., Walsh, C.J., Telfer, J.C., Litman, G.W., and Rothenberg, E.V. Evolutionary origins of lymphocytes: Ensembles of T cell and B cell transcriptional regulators in a cartilaginous fish. *Journal of Immunology* 2004; 172:5851–5860.

Anderson, M.K., Strong, S.J., Litman, R.T., Luer, C.A., Amemiya, C.T., Rast, J.P., and Litman, G.W. A long form of the skate IgX gene exhibits a striking resemblance to the new shark IgW and IgNARC genes. *Immunogenetics* 1999; 49:56–67.

Anderson, M.K., Sun, X., Miracle, A., Litman, G.W., and Rothenberg, E. Evolution of hematopoiesis: Three members of the PU.1 transcription factor family in a cartilaginous fish, *Raja eglanteria. Proceedings of the National Academy of Sciences USA* 2001; 98:553–558.

Anderson, M.S., and Su, M.A. Aire and T cell development. *Current Opinion in Immunology* 2011; 23(2):198–206.

Bajoghli, B., Guo, P., Aghaallaei, N., Hirano, M., Strohmeier, C., McCurley, N., Bockman, D.E., Schorpp, M., Cooper, M.D., and Boehm, T. A thymus candidate in lampreys. *Nature* 2011; 470:90–94.

Bartl, S., Miracle, A.L., Rumfelt, L.L., Kepler, T.B., Mochon, E., Litman, G.W., and Flajnik, M.F. Terminal deoxynucleotidyltransferases from elasmobranchs reveal structural conservation within vertebrates. *Immunogenetics* 2003; 55:594–604.

Bartl, S., and Weissman, I.L. Isolation and characterization of major histocompatibility complex class IIB genes from the nurse shark. *Proceedings of the National Academy of Sciences USA* 1994; 91:262–266.

Bernstein, R.M., Schluter, S.F., Bernstein, H., and Marchalonis, J.J. Primordial emergence of the recombination activating gene 1 (*RAG1*): Sequence of the complete shark gene indicates homology to microbial integrases. *Proceedings of the National Academy of Sciences USA* 1996a; 93:9454–9459.

Bernstein, R.M., Schluter, S.F., Shen, S., and Marchalonis, J.J. A new high molecular weight immunoglobulin class from the carcharhine shark: implications for the properties of the primordial immunoglobulin. *Proceedings of the National Academy of Sciences USA* 1996b; 93:3289–3293.

Beutler, B. Innate immunity: An overview. *Molecular Immunology* 2004; 40:845–859.

Bircan-Yildirim, Y., Çek, Ş., Başusta, N., and Atik, E. Histology and morphology of the epigonal organ with special reference to the Lymphomyeloid system in *Rhinobatos rhinobatos. Turkish Journal of Fisheries and Aquatic Sciences* 2011; 11:351–358.

Bird, S., Wang, T., Zou, J., Cunningham, C., and Secombes, C.J. The first cytokine sequence within cartilaginous fish: IL-1 in the small spotted Catshark (*Scyliorhinus canicula*). *Journal of Immunology* 2002; 168:3329–3340.

Burgess, G. NOAA Fisheries, 1999. http://www.flmnh.ufl.edu/fish/sharks/ISAF/ISAF.htm. Accessed May 27, 2013.

Butler, M.G., Isogai, S., and Weinstein, B.M. Lymphatic development. *Birth Defects Research. Part C, Embryo Today: Reviews* 2009; 87:222–231.

Carrier, J.C., and Luer, C. Growth rates in the nurse shark, *Ginglymostoma cirratum. Copeia* 1990; 3:686–692.

Carrier, J.C., Murru, F.L., Walsh, M.T., and Pratt, H.L. Assessing reproductive potential and gestation in nurse sharks (*Ginglymostoma cirratum*) using ultrasonography and endoscopy: An example of bridging the gap between field research and captive studies. *Zoo Biology* 2003; 22:179–187.

Carrier, J.C., Pratt, H.L., and Martin, L.K. Group reproductive behaviors in free living nurse sharks, *Ginglymostoma cirratum. Copeia* 1994; 3:646–656.

Carroll, R.L. *Vertebrate Paleontology and Evolution.* New York: W.H. Freeman and Company, 1988.

Castro, C.D. Characterization of Plasma Cell Lineages in the Nurse Shark. Ph.D. Dissertation, University of Maryland at Baltimore Digital Archive, 2013. http://hdl.handle.net/10713/3638. Accessed March 29, 2014.

Castro, C.D., Ohta, Y., Dooley, H., and Flajnik. M.F. Noncoordinate expression of J-chain and Blimp-1 define nurse shark plasma cell populations during ontogeny. *European Journal of Immunology* 2013; 43:3061–3075.

Castro, J.I. The biology of the Nurse Shark, *Ginglymostoma cirratum,* off the Florida East Coast and the Bahama Islands. *Environmental Biology of Fishes* 2000; 58:1–22.

Cateni, C., Paulesu, L., Bigliardi, E., and Hamlett, W.C. The interleukin 1 (IL-1) system in the uteroplacental complex of a cartilaginous fish, the smoothhound shark, *Mustelu scanis. Reproductive Biology Endocrinology* 2003; 1:25–34.

Clark, E. Maintenance of sharks in captivity with a report on their instrumental conditioning. In: Gilbert, P.W., editor. *Sharks and Survival.* Boston, MA: Heath & Co: 1963, Section II, Chapter 4, pp. 115–149.

Clark, E., and von Schmidt, K. Sharks of the Central Gulf Coast of Florida. *Bulletin of Marine Science* 1965; 15:13–83.

Clem, L.W., De Boutaud, F., and Sigel, M.M. Phylogeny of immunoglobulin structure and function. II. Immunoglobulins of the nurse shark. *Journal of Immunology* 1967; 99:1226–1235.

Clem, L.W., and Small, P.A. Phylogeny of immunoglobulin structure and function. I. Immunoglobulins of the lemon shark. *Journal of Experimental Medicine* 1967;125:893–920.

Compagno, L.J.V. Systematics and body form. In: Hamlett, W.C., editor. *Sharks, Skates, and Rays: The Biology of Elasmobranch Fishes.* Baltimore, MD: Johns Hopkins University Press, 1999, pp. 1–42.

Criscitiello, M.F., and de Figueiredo, P. Fifty shade of immune defense. *PLoS Pathogens* 2013; 9:e1003110.

Criscitiello, M.F., Ohta, Y., Graham, M.D., Eubanks, J.O., Chen, P.L., and Flajnik, M.F. Shark class II invariant chain reveals ancient conserved relationships with cathepsins and MHC class II. *Developmental and Comparative Immunology* 2012; 36:521–533.

Criscitiello, M.F., Ohta, Y., Saltis, M., McKinney, E.C., and Flajnik, M.F. Evolutionarily conserved TCR binding sites, identification of T cells in primary lymphoid tissues, and surprising trans-rearrangements in nurse shark. *Journal of Immunology* 2010; 184:6950–6960.

Daly-Engel, T.S., Grubbs, R.D., Bowen, B.W., and Toonen, R.J. Frequency of multiple paternity in an unexploited tropical population of sandbar sharks (*Carcharhinus plumbeus*). *Canadian Journal of Fisheries and Aquatic Sciences* 2007; 64:198–204.

Diaz, M., Greenberg, A.S., and Flajnik, M.F. Somatic hypermutation of the new antigen receptor gene (NAR) in the nurse shark does not generate the repertoire: Possible role in antigen-driven reactions in the absence of germinal centers. *Proceedings of the National Academy of Sciences USA* 1998; 95:14343–14348.

Diaz, M., Stanfield, R.L., Greenberg, A.S., and Flajnik, M.F. Structural analysis, selection, and ontogeny of the shark new antigen receptor (IgNAR): Identification of a new locus preferentially expressed in early development. *Immunogenetics* 2002; 54:501–512.

Diaz, M., Velez, J., Singh, M., Cerny, J., and Flajnik, M.F. Mutational pattern of the nurse shark antigen receptor gene (NAR) is similar to that of mammalian Ig genes and to spontaneous mutations in evolution: The translesion synthesis model of somatic hypermutation. *International Immunology* 1999; 11:825–833.

Dooley, H., and Flajnik, M.F. Shark immunity bites back: Affinity maturation and memory response in the nurse shark, *Ginglymostoma cirratum. European Journal of Immunology* 2005; 35:936–945.

Dooley, H., Flajnik, M.F., and Porter, A.J. Selection and characterization of naturally occurring single-domain (IgNAR) antibody fragments from immunized sharks by phage display. *Molecular Immunology* 2003; 40:25–33.

Dooley, H., Stanfield, R.L., Brady, R.A., and Flajnik, M.F. First molecular and biochemical analysis of in vivo affinity maturation in an ectothermic vertebrate. *Proceedings of the National Academy of Sciences USA* 2006; 103:1846–1851.

Dorshkind, K., and Montecino-Rodriguez, E. Fetal B-cell lymphopoiesis and the emergence of B-1-cell potential. *Nature Reviews Immunology* 2007; 7:213–219.

Edmonds, M.A., Motta, P.J., and Hueten, R.E. Food capture kinematics of the suction feeding horn shark, *Heterodontus francisci*. *Environmental Biology of Fishes* 2001; 62:415–427.

Ellis, A.E., and Parkhouse, R.M.E. Surface immunoglobulins on the lymphocytes of the skate, *Raja naevus*. *European Journal of Immunology* 1975; 5:726–728.

Fänge, R. Size relations of Lymphomyeloid organs in some Cartilaginous Fish. *Acta Zoologica* 1977; 58:125–128.

Fänge, R. Lymphomyeloid tissues in fish. *Vidensk Meddrdansknaturh Foren* 1984; 145:125–128.

Fänge, R. Lymphomyeloid system and blood cell morphology in elasmobranchs. *Archives Biology (Brussels)* 1987; 98:187–208.

Fänge, R., Lundblad, G., and Lind, J. Lysozyme and chitinase in blood and lymphomyeloid tissues of marine fish. *Marine Biology* 1976; 36:277–282.

Fänge, R., and Mattisson, A. The lymphomyeloid (hemopoietic) system of the Atlantic Nurse Shark, *Ginglymostoma cirratum*. *Biological Bulletin* 1981; 160:240–249.

Fänge, R., and Nilsson, S. The fish spleen: Structure and function. *Experentia* 1985; 41:152–158.

Fänge, R., and Pulsford, A. Structural studies on lymphomyeloid tissues of the dogfish, *Scyliorhinus canicula*. *Cell Tissue Research* 1983; 230:337–351.

Feeney, A.J. Lack of N-regions in fetal and neonatal mouse immunoglobulin V-D-J junctional sequences. *Journal Experimental Medicine* 1990; 172:1377–1390.

Ferreira, L.C., Afonso, A.S., and Castilho, P.C. Habitat use of the nurse shark, *Ginglymostoma cirratum*, off Recife, Northeast Brazil: A combined survey with longline and acoustic telemetry. *Environmental Biology of Fishes* 2013; 96:735–745.

Fidler, J.E., Clem, L.W., and Small, P.A. Immunoglobulin synthesis in neonatal nurse sharks (*Ginglymostoma cirratum*). *Comparative Biochemistry and Physiology* 1969; 31:365–371.

Fitzpatrick, J.L., Kempster, R.M., Daly-Engel, T.S., Collins, S.P., and Evans, J.P. Assessing the potential for post-copulatory sexual selection in elasmobranchs. *Journal of Fish Biology* 2012; 80:1141–1158.

Flajnik, M.F., and Rumfelt, L.L. The immune system of cartilaginous fish. *Current Topics in Microbiology and Immunology* 2000; 248:249–270.

Friedberg, S.H., and Weissman, I.L. Lymphoid tissue architecture. II. Ontogeny of peripheral T and B cells in mice: Evidence against Peyer's patches as the site of generation of B cells. *Journal of Immunology* 1974; 113:1477–1492.

Gilmore, R.G., Dodrill, J.W., and Linley, P.A. Reproduction and embryonic development of the sand tiger shark, *Odontaspis taurus* (Rafinesque). *Fishery Bulletin* 1983; 81:201–225.

Good, R.A., and Finstad, J. The development and involution of the Lymphoid system and immunologic capacity. *Transactions of the American Clinical and Climatological Association* 1968; 79:69–107.

Good, R.A., Finstad, J., Pollara, B., and Gabrielsen, A.E. Morphologic studies on the lymphoid tissues among the lower vertebrates. In: Smith, R.T., Miescher, P.A., and Good, R.A., editors. *Phylogeny of Immunity*. Gainesville, FL: University of Florida Press, 1966.

Goodwin, N.B., Dulvy, N.K., and Reynolds, J.D. Life-history correlates of the evolution of live bearing in fishes. *Philosophical Transactions of the Royal Society Biological Sciences* 2002; 357:259–267.

Got, R., Font, J., and Goussault, Y. Comparative study of the physicochemical characteristics of human transferrin and the transferrin of a selachian (*Scylliumstellare*). In: Iwama, G.K., and Nakanishi, T., editors. *The Fish Immune System: Organism, Pathogen, and Environment*. San Diego, CA: Academic Press, 1996, pp. 127–128.

Greenberg, A.S., Avila, D., Hughes, M., Hughes, A., McKinney, E.C., and Flajnik, M.F. A new antigen receptor gene family that undergoes rearrangement and extensive somatic diversification in sharks. *Nature* 1995; 374:168–173.

Greenberg, A.S., Hughes, A.L., Guo, J., Avila, D., McKinney, E.C., and Flajnik, M.F. A novel "chimeric" antibody class in cartilaginous fish: IgM may not be the primordial immunoglobulin. *European Journal of Immunology* 1996; 26:1123–1129.

Grindstaff, J.L., Brodie, E.D., and Ketterson, E.D. Immune function across generations: Integrating mechanism and evolutionary process in maternal antibody transmission. *Proceedings of the Royal Society B: Biological Sciences* 2003; 270:2309–2319.

Guarracino, M. Education: Biological profile on the Nurse Shark. In: *Florida Museum of Natural History*. https://www.flmnh.ufl.edu/fish/Gallery/descript/nurseshark/nurseshark.htm

Haas, K.M., Poe, J.C., Steeber, D.A., and Tedder, T.F. B-1a and B-1b cells exhibit distinct developmental requirements and have unique functional roles in innate and adaptive immunity to *S. pneumoniae*. *Immunity* 2005; 23:7–18.

Haines, A., Flajnik, M.F., Rumfelt, L.L., and Wourms, J.P. Immunoglobulins in the eggs of the nurse shark, *Ginglymostoma cirratum*. *Developmental and Comparative Immunology* 2005; 29:417–430.

Hamal, K.R., Burgess, S.C., Pevzner, I.Y., and Erf, G.F. Maternal antibody transfer from dams to their egg yolks, egg whites, and chicks in meat lines of chickens. *Poultry Science* 2006; 85(8):1364–1372.

Hamers-Casterman, C., Atarhouch, T., Muyldermans, S., Robinson, G., Hammers, C., BajyanaSonga, E., Bendahman, N., and Hammers, R. Naturally occurring antibodies devoid of light chains. *Nature* 1993; 363:446–448.

Hammerschlag, N. Osmoregulation in elasmobranchs: A review for fish biologists, behaviourists and ecologists. *Marine and Freshwater Behaviour and Physiology* 2006; 39:209–228.

Harding, F.A., Amemiya, C.T., Litman, R.T., Cohen, N., and Litman, G.W. Two distinct immunoglobulin heavy chain isotypes in a primitive, cartilaginous fish, *Raja erinacea*. *Nucleic Acids Research* 1990b; 18:6369–6376.

Hashimoto, K., Nakanishi, T., and Kurosawa, Y. Identification of a shark sequence resembling the major histocompatibility complex class I alpha 3 domain. *Proceedings of the National Academy of Sciences USA* 1992; 89:2209–2212.

Hayakawa, K., Tarlinton, D., and Hardy, R.R. Absence of MHC class II expression distinguishes fetal from adult B lymphopoiesis in mice. *The Journal of Immunology* 1994; 152:4801–4807.

Haynes, L., Fuller, L., and McKinney, E.C. Fc receptor for shark IgM. *Developmental and Comparative Immunology* 1988; 12:561–571.

Heist, E.J., Carrier, J.C., Pratt, H.L., and Pratt, T.C. Exact enumeration of sires in the polyandrous nurse shark (*Ginglymostoma cirratum*). *Copeia* 2011; 4:539–544.

Heming, T.A., and Buddington, R.K. Yolk absorption in embryonic and larval fishes. In: Hoar, W.S., and Randall, D.J., editors. *Fish Physiology*. London: Academic, 1988, Vol. 11, Part A, pp. 407–446.

Herzenberg, L.A., Kantor, A.B., and Herzenberg, L.A. Layered evolution in the immune system. A model for the ontogeny and development of multiple lymphocyte lineages. *Annals of the New York Academy of Sciences* 1992; 651:1–9.

Hoffmann, J.A., Kafatos, F.C., Janeway, C.A., and Ezekowitz, R.A. Phylogenetic perspectives in innate immunity. *Science* 1999; 284:1313–1318.

Hohman, V.S., Stewart, S.E., Rumfelt, L.L., Greenberg, A.S., Avila, D.W., Flajnik, M.F., and Steiner L.A. J Chain in the Nurse Shark: Implications for function in a lower vertebrate. *Journal of Immunology* 2003; 170:6016–6023.

Hyder, S.L., Cayer, M.L., and Pettey, C.L. Cell types in peripheral blood of the nurse shark: An approach to structure and function. *Tissue and Cell* 1983; 15:437–455.

Isogai, S., Hitomi, J., Yaniv, K., and Weinstein, B. Zebrafish as a new animal model to study lymphangiogenesis. *Anatomical Science International* 2009; 84:102–111.

Itohara, S., Nakanishi, N., Kanagawa,O., Kubo, R., and Tonegawa, S. Monoclonal antibodies specific to native murine T-cell receptor gamma delta: Analysis of gamma delta T cells during thymic ontogeny and in peripheral lymphoid organs. *Proceedings of the National Academy of Sciences USA* 1989; 86:5094–5098.

Janeway, C.A. Innate Immunity. In: Janeway, C.A., Jr., Travers, P., Walport, M., and Shlomchik, M.J., editors. *Immunobiology: The Immune System in Health and Disease*. New York: Garland Publishing, 2001, pp. 16–44.

Jerne, N. The natural selection theory of antibody formation. *Proceedings of the National Academy of Sciences USA* 1955; 41:849–857.

Jerne, N.K. Towards a network theory of the immune system. *Annals of Immunology* 1974; 125C:373–389.

Johansson, M.L. Peroxidase in blood cells of fishes and cyclostomes. *Acta Regiae Societatis Scientarumet Litterarum Gothoburgensis Zoologica* 1973; 8:53–56.

Kampmeier O.F. Lymphatic system of the cartilaginous fishes. In: Thomas, Charles C., editor. *Evolution and Comparative Morphology of the Lymphatic System*. Springfield, IL: Thomas, 1969, pp. 211–231.

Kasahara, M., Vazquez, M., Sato, K., McKinney, E.C., and Flajnik, M.F. Evolution of the major histocompatibility complex: Isolation of class II A cDNA clones from the cartilaginous fish. *Proceedings of the National Academy of Sciences USA* 1992; 89:6688–6692.

Lamars, C.H.J. The Reaction of the Immune System of Fish to Vaccination. Ph.D. Thesis, Agricultural University, Wageningen, the Netherlands, 1985, 256 pp.

Lee, S.S., Tranchina, D., Ohta, Y., Flajnik, M.F., and Hsu, E. Hypermutation in Shark immunoglobulin light chain genes results in contiguous substitutions. *Immunity* 2002; 16:571–582.

Leydig, F. *Lehrbuch der Histologie.* Frankfurt A.M., Germany: Meidinger, Sohn, and Comp., 1857.

Luer, C.A., Walsh, C.J., Bodine, A.B., Wyffels, J.T., and Scott, T.R. The elasmobranch thymus: Anatomical, histological, and preliminary functional characterization. *Journal of Experimental Zoology* 1995; 273:342–354.

Malecek, K., Brandman, J., Brodsky, J.E., Ohta, Y., Flajnik, M.F., and Hsu, E. Somatic hypermutation and junctional diversification at Ig heavy chain Loci in the Nurse Shark. *Journal of Immunology* 2005; 175:8105–8115.

Marchalonis, J.J., and Edelman, G.M. Phylogenetic origins of antibody structure. I. Multichain structure of immunoglobulins in the smooth dogfish (*Mustelus canis*). *Journal of Experimental Medicine* 1965; 122:601–618.

Marchalonis, J.J., and Edelman, G.M. Polypeptide chains of immunoglobulins from the smooth dogfish (*Mustelus canis*). *Science* 1966; 154:1567–1568.

Matthews, L.H. Reproduction in the Basking Shark, *Cetorhinus maximus* (Gunner). *Philosophical Transactions of the Royal Society of London. Series B, Biological Sciences* 1950; 234:247–316.

Mattisson, A., and Fänge, R. The cellular structure of the leydig organ in the shark, *Etmopterus spinax* (L.). *Biological Bulletin* 1982; 162:182–194.

McKinney, E.C., and Flajnik, M.F. IgM-mediated opsonization and cytotoxicity in the shark. *Journal of Leukocyte Biology* 1997; 61:141–146.

Medzhitov, R., and Janeway, C.A. Innate immunity: The virtues of a nonclonal system of recognition. *Cell* 1997; 91:295–298.

Miracle, A.L., Anderson, M.K., Litman, R.T., Walsh, C.J., Luer, C.A., Rothenberg, E.V., and Litman, G.W. Complex expression patterns of lymphocyte-specific genes during the development of cartilaginous fish implicate unique lymphoid tissues in generating an immune repertoire. *International Immunology* 2001; 13:567–580.

Monro, A. *The Structure and Physiology of Fishes Explained and Compared with Those of Man and Other Animals.* Edinburgh, Scotland: Charles Elliot, 1785.

Montecino-Rodriguez, E., Leathers, H., and Dorshkind, K. Identification of a B-1 B cell–specified progenitor. *Nature Immunology* 2006; 7(3):293–301.

Moore, K.S., Wehrli, S., Roder, H., Rogers, M., Forrest, J.N., McCrimmon, D., and Zasloff, M. Squalamine: An aminosterol antibiotic from the shark. *Proceedings of the National Academy of Sciences USA* 1993; 90:1354–1358.

Morrow, W.J.W., Harris, J.E., and Pulsford, A. Immunological responses of the dogfish (*Scyliorhinus canicula L.*) to cellular antigens. *Acta Zoologica (Stockholm)* 1982; 63:153–159.

Morrow, W.J.W., and Pulsford, A. Identification of peripheral blood leucocytes of the dogfish (*Scyliorhinus canicula L.*) by electron microscopy. *Journal of Fish Biology* 1980; 17:461–475.

Motta, P.J., Hueter, R.E., Tricas, T.C., and Summers, A.P. Kinematic analysis of suction feeding in the Nurse Shark, *Ginglymostoma cirratum* (Orectolobiformes, Ginglymostomatidae). *Copeia* 2002; 1:24–38.

Moura, T., Serra-Pereira, B., Gordo, L.S., and Figueiredo, I. Sperm storage in males and females of the deepwater shark Portuguese dogfish with notes on oviducal gland microscopic organization. *Journal of Zoology* 2011; 283:210–219.

Ohta, Y., and Flajnik, M.F. IgD, like IgM, is a primordial immunoglobulin class perpetuated in most jawed vertebrates. *Proceedings of the National Academy of Sciences USA* 2006; 103:10723–10728.

Ohta, Y., Okamura, K., McKinney, E.C., Bartl, S., Hashimoto, K., and Flajnik, M.F. Primitive synteny of vertebrate major histocompatibility complex class I and class II genes. *Proceedings of the National Academy of Sciences USA* 2000; 97:4712–4717.

Ota, T., Rast, J.P., Litman, G.W., and Amemiya, C.T. Lineage-restricted retention of a primitive immunoglobulin heavy chain isotype within the Dipnoi reveals an evolutionary paradox. *Proceedings of the National Academy of Sciences USA* 2003; 100:2501–2506.

Pang, P.K.T., Griffith, R.W., and Atz, J.W. Osmoregulation in elasmobranchs. *American Zoologist* 1977; 17:365–377.

Petrie, H.T., and Zúñiga-Pflücker, J.C. Zoned out: Functional mapping of stromal signaling microenvironments in the thymus. *Annual Review of Immunology* 2007; 25:649–679.

Pires-daSilva, A., and Sommer, R. The evolution of signalling pathways in animal development. *Nature Reviews Genetics* 2003; 4:39–49.

Pratt, H.L., Jr. Elasmobranch Gonad Structure: A Description and Survey. *Copeia*, 1988; 3:719–729.

Pratt, H.L., Jr., and Carrier, J.C. A review of Elasmobranch reproductive behavior with a case study on the Nurse Shark, *Ginglymostoma cirratum*. *Environmental Biology of Fishes* 2001; 60:157–188.

Price, K.S., and Daiber, F.C. Osmotic environments during fetal development of Dogfish, *Mustelus canis* (Mitchill) and *Squalus acanthias* Linnaeus, and some comparisons with Skates and Rays. *Physiological Zoology* 1967; 40:248–260.

Prinz, I., Silva-Santos, B., and Pennington, D.J. Functional development of γδ T cells. *European Journal of Immunology* 2013; 43:1988–1994.

Pulsford, A., Fänge, R., and Morrow, W.J.W. Cell types and interactions in the spleen of the dogfish, *Scyliorhinus canicula* L. An electron microscope study. *Journal of Fish Biology* 1982; 21:649–662.

Pulsford, A., and Zapata, A.G. Macrophages and reticulum cells in the spleen of the dogfish, *Scyliorhinus canicula*. *Acta Zoologica* 1989; 70:221–227.

Rast, J.P., Anderson, M.K., Strong, S.J., Luer, C., Litman, R.T., and Litman, G.W. Alpha, beta, gamma, and delta T cell antigen receptor genes arose early in vertebrate phylogeny. *Immunity* 1997; 6:1–11.

Rast, J.P., Haire, R.N., Litman, R.T., Pross, S., and Litman, G.W. Identification and characterization of T-cell antigen receptor-related genes in phylogenetically diverse vertebrate species. *Immunogenetics* 1995; 42:204–212.

Rast, J.P., and Litman, G.W. T-cell receptor gene homologs are present in the most primitive jawed vertebrates. *Proceedings of the National Academy of Sciences USA* 1994; 91:9248–9252.

Ratnasingham, S., and Hebert, P.D.N. Barcode of Life Data Systems: Encyclopedia of Life, 2013. http://eol.org/data_objects/23655087. Accessed June 25, 2013.

Robinson, M.P., and Motta, P.J. Patterns of growth and the effects of scale on the feeding kinematics of the nurse shark (*Ginglymostoma cirratum*). *Journal of Zoology (London)* 2002; 256:449–462.

Rothenberg, E., Zhang, J., and Li, L. Multilayered specification of the T-cell lineage fate. *Immunological Reviews* 2010; 238:150–168.

Roux, K.H., Greenberg, A.S., Greene, L., Strelets, L., Avila, D., McKinney, E.C., and Flajnik, M.F. Structural analysis of the nurse shark (new) antigen receptor (NAR): Molecular convergence of NAR and unusual mammalian immunoglobulins. *Proceedings of the National Academy of Sciences USA* 1998; 95:11804–11809.

Rowley, A.F., Hunt, T.C., Page, M., and Mainwaring, G. Fish. In: Rowley, A.F., and Ratcliffe, N.A., editors. *Vertebrate Blood Cells*. Cambridge, MA: Cambridge University Press, 1988, pp. 19–127.

Rumfelt, L.L. Studies of Immunoglobulin Expression and Lymphoid Tissue Development during Ontogeny in the Nurse Shark, *Ginglymostoma cirratum*. PhD Dissertation, Scholarly Repository at University of Miami Library, Coral Gables, FL: University of Miami. Paper3826.2000. http://scholarlyrepository.miami.edu/dissertations/3826. Accessed July 2, 2013.

Rumfelt, L.L, Avila, D., Diaz, M., Bartl, S., McKinney, E., and Flajnik, M. A shark antibody heavy chain encoded by a non-somatically rearranged VDJ is preferentially expressed in early development and is convergent with mammalian IgG. *Proceedings of the National Academy of Sciences USA* 2001; 98:1775–1780.

Rumfelt, L.L., Diaz, M., Lohr, R.L., Mochon, E., and Flajnik, M.F. Unprecedented multiplicity of Ig transmembrane and secretory mRNA forms in the Cartilaginous Fish. *Journal of Immunology* 2004a; 173:1129–1139.

Rumfelt, L.L., Lohr, R.L., Dooley, H., and Flajnik, M.F. Diversity and repertoire of IgW and IgM VH families in the newborn nurse shark. *BMC Immunology* 2004b; 5:8.

Rumfelt, L.L., McKinney, E.C., Taylor, E., and Flajnik, M.F. The development of primary and secondary lymphoid tissues in the nurse shark *Ginglymostoma cirratum*: B-cell zones precede dendritic cell immigration and T-cell zone formation during ontogeny of the spleen. *Scandinavian Journal of Immunology* 2002; 56:130–148.

Saha, N.R., Ota, T., Litman, G.W., Hansen, J., Parra, Z., Hsu, E., Buonocore, F., Canapa, A., Cheng, J.F., and Amemiya, C.T. Genome complexity in the coelacanth is reflected in its adaptive immune system. *Journal of Experimental Zoology Part B: Molecular and Developmental Evolution* 2014; 9999B:1–26.

Saltis, M., Criscitiello, M.F., Ohta, Y., Keefe, M., Trede, N.S., Goitsuka, R., and Flajnik, M.F. Evolutionarily conserved and divergent regions of the autoimmune regulator (Aire) gene: A comparative analysis. *Immunogenetics* 2008; 60(2):105–114.

Saville, K.J., Lindley, A.M., Maries, E.G., Carrier, J.C., and Pratt, H.L. Multiple paternity in the nurse shark, *Ginglymostoma cirratum*. *Environmental Biology of Fishes* 2002; 63:347–351.

Scott, E., Simon, M., Anastasi, J., and Singh, H. Requirement of transcription factor PU.1 in the development of multiple hematopoietic lineages. *Science* 1994; 265:1573–1577.

Seibel, B.A., and Walsh, P.J. Trimethylamine oxide accumulation in marine animals: Relationship to acylglycerol storage. *Journal of Experimental Biology* 2002; 205:297–306.

Sigel, M.M., and Clem, L.W. Antibody response of fish to viral antigens. *Annals of the New York Academy of Sciences* 1965; 126:662–677.

Small, P.A., Jr., Klapper, D.G., and Clem, L.W. Half-lives, body distribution and lack of interconversion of serum 19S and 7S IgM of sharks. *Journal of Immunology* 1970; 105:29–37.

Smith, S.L. Shark complement: An assessment. *Immunological Reviews* 1998; 166:67–78.

Stanfield, R.L., Dooley, H., Flajnik, M.F., and Wilson, I.A. Crystal structure of a shark single-domain antibody V region in complex with lysozyme. *Science* 2004; 305:1770–1773.

Stanfield, R.L., Dooley, H., Verdino, P., Flajnik, M.F., and Wilson, I.A. Maturation of shark single-domain (IgNAR) antibodies: Evidence for induced-fit binding. *Journal of Molecular Biology* 2007; 307:358–372.

Steinman, R., Pack, M., and Inaba, K. Dendritic cells in the T-cell areas of lymphoid organs. *Immunological Reviews* 1997; 156:25–37.

Timpl, R., Rohde, H., Robey, P.G., Rennard, S.I., Foidart, J.M., and Martin, G.R. Laminin—A glycoprotein from basement membranes. *Journal of Biological Chemistry* 1979; 254:9933–9937.

Tomonaga, S., Kobayashi, K., Hagiwara, K., Yamaguchi, K., and Awaya, K. Gut-associated lymphoid tissue in elasmobranchs. *Zoological Science* 1986; 3:453–458.

Tomonaga, S., Kobayashi, K., Kajii, T., and Awaya, K. Two populations of immunoglobulin-forming cells in the skate, *Raja kenojei*: Their distribution and characterization. *Developmental and Comparative Immunology* 1984; 8:803–812.

Tomonaga, S., Zhang, H., Kobayashi, K., Fujii, R., and Teshima, K. Plasma cells in the spleen of the Aleutian skate, *Bathyraja aleutica*. *Archives of Histology and Cytology* 1992; 55:287–294.

Torroba, M., Chiba, A., Vicente, A., Varas, A., Sacedón, R., Jimenez, E., Honma, Y., and Zapata, A.G. Macrophage-lymphocyte cell clusters in the hypothalamic ventricle of some elasmobranch fish: Ultrastructural analysis and possible functional significance. *Anatomical Record* 1995; 242:400–410.

Walsh, C.J., and Luer, C.A. Elasmobranch Hematology. In: Smith, M., Warmolts, D., Thoney, D., and Hueter, R., editors. *The Elasmobranch Husbandry Manual: Captive Care of Sharks, Rays and Their Relatives.* Columbus, OH: Ohio Biological Survey, Inc., 2004, Chapter 23, pp. 307–323.

Wang, D.Y., Kumar, S., and Hedges, S.B. Divergence time estimates for the early history of animal phyla and the origin of plants, animals and fungi. *Proceedings of the Royal Society B: Biological Sciences* 1999; 266:163–171.

Weiss, L., editor. *Cell and Tissue Biology.* Baltimore, MD: Urban & Schwarzenberg, Inc., 1983, pp. 425–442.

Weisz, N. Introduction: Elasmobranchii: Sharks, Skates, and Rays. Encyclopedia of Life, 2012. http://eol.org/pages/1857/overview.

Wilson, M., Bengtén, E., Miller, N.W., Clem, L.W., Du Pasquier, L., and Warr, G.W. A novel chimeric Ig heavy chain from a teleost fish shares similarities to IgD. *Proceedings of the National Academy of Sciences USA* 1997; 94:4593–4597.

Wourms, J.P. Reproduction and development in Chondrichthyan fishes. *American Zoologist* 1977; 17:379–410.

Wourms, J.P., and Lombardi, J. Reflections on the evolution of piscine viviparity. *American Zoologist* 1992; 32:276–293.

Zapata, A., and Amemiya, C.T. Phylogeny of lower vertebrates and their immunological structures. *Current Topics in Microbiology and Immunology* 2000; 248:67–107.

Zapata, A., Chibá, A., and Varas, A. Cells and tissues of the immune system of fish. In: Iwama, G., Nakanishi, T., editors. *The Fish Immune System: Organism, Pathogen, and Environment.* San Diego, CA: Academic Press; 1996b. pp. 1–62.

Zapata, A.G., Torroba, M., Sacedón, R., Varas, A., and Vicente, A. Structure of the lymphoid organs of elasmobranchs. *Journal of Experimental Zoology* 1996b; 275:125–143.

Zhang, C., DuPasquier, L., and Hsu, E. Shark IgW C region diversification through RNA processing and isotype switching. *Journal of Immunology* 2013; 191:3410–3418.

Zhu, C., Feng, W., Weedon, J., Hua, P., Stefanov, D., Ohta, Y., Flajnik, M.F., and Hsu, E. The multiple shark immunoglobulin heavy chain genes rearrange and hypermutate autonomously. *Journal of Immunology* 2011; 5:2492–2501.

Zon, L. Developmental biology of hematopoiesis. *Blood* 1995; 86:2876–2891.

CHAPTER 4

Sites of Immune Cell Production in Elasmobranch Fishes
Lymphomyeloid Tissues and Organs

Carl A. Luer, Catherine J. Walsh, and A.B. Bodine

CONTENTS

SUMMARY

Immune cells in elasmobranch fishes are produced in a variety of tissue and organ sites, some of which are common to higher vertebrate immune systems, whereas others are sites not found in any other vertebrate animal group. Thymus and spleen participate in elasmobranch lymphopoiesis in much the same way as they do in higher vertebrates, but in the absence of bony skeletons and lymphatic systems, sharks and their batoid relatives possess unique *bone-marrow equivalent* tissues that effectively compensate for the lack of marrow and lymphatic tissues. These unique lymphomyeloid tissues are the epigonal and Leydig organs, associated with the gonads and esophagus, respectively. These four tissues/organs constitute the primary sites for leukocyte production in the elasmobranch immune system.

INTRODUCTION

Tissues and organs that provide the environments for immune cell production in representatives of the class Chondrichthyes (characterized by members of subclass Elasmobranchii, commonly called elasmobranch fishes, or simply elasmobranchs) include a variety of sites, some of which are common to other vertebrate animal immune systems and some that are found only in sharks, skates, and rays (Luer *et al.*, 2004). Of the four primary sites of immune cell production characteristic of vertebrate immune systems (bone marrow, thymus, spleen, and lymph nodes), only the thymus and spleen are found in elasmobranchs. Not only are the thymus and spleen present in elasmobranchs, but these lymphoid tissues have their earliest phylogenetic appearances in this subclass of fishes.

Because elasmobranchs possess cartilaginous skeletons and lack a lymphatic system with lymphatic vessels and ducts, bone marrow and lymph nodes do not exist. In their place, unique lymphomyeloid tissues with immune functions remarkably similar to those of bone marrow and lymph nodes have evolved in sharks, skates, and rays. Often referred to as *bone-marrow equivalent* tissues, these include the epigonal organ, occurring in all elasmobranchs, and the Leydig organ, occurring only in certain elasmobranch species.

The purpose of this chapter is to describe the anatomy and histology of these primary lymphoid and lymphomyeloid sites, namely, the thymus, spleen, epigonal organ, and Leydig organ, as they occur in the elasmobranch fishes. To the extent that information or speculation exists, lymphoid aggregates, as opposed to isolated lymphoid organs, that could potentially serve as minor sites of immune cell production are described.

THYMUS

As in embryos of higher vertebrates, the thymus is the first lymphoid structure to develop in embryonic elasmobranchs, originating from epithelial outgrowths of the third pharyngeal pouches. Unlike higher vertebrates, however, in which thymus tissue migrates ventrally to form a single midline organ, developing thymus tissue in sharks, skates, and rays remains paired and migrates dorsally to form a pair of glands situated at dorsal and medial to both sets of gill arches (Luer *et al.*, 1995).

In embryonic and early-stage juvenile sharks, the thymus lies in a slight depression lateral to the epaxial musculature and just beneath a thin layer of muscle composed of the medial margins of the levator hyomandibulae, dorsal hyoid constrictor, and first three superficial brachial constrictors (Figure 4.1). During these early stages, the thymus extends anteriorly to the chondrocranium and posteriorly across the epibranchial portions of gill arches three through five (Figure 4.2). As the surrounding musculature changes with somatic growth and aging, the thymus becomes increasingly difficult to identify. In subadults and sexually mature sharks, the thymus is no longer superficial but rather deep in a groove between the constrictors and the epaxial musculature, *buried* as a result of growth and development of the muscles in the gill arch region. Although its location in mature animals is considerably deeper than in embryonic and immature specimens, its position relative to the posterior edge of the chondrocranium and the epibranchial portions of the gill arches is retained.

The elasmobranch thymus is encased in a connective tissue capsule and arranged in distinct lobes separated by trabeculae that are continuous with the connective tissue encapsulating the

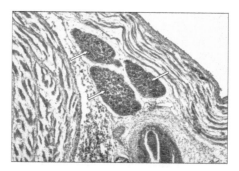

Figure 4.1 Paraffin-embedded 6-μ section through gill region of newborn bonnethead shark, *Sphyrna tiburo*, showing anatomical location of lobed thymus (arrows) beneath dorsal musculature and dorsomedial to gills (stain: hematoxylin and eosin; original magnification: 100X).

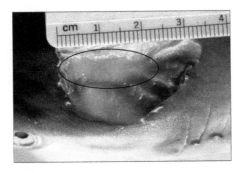

Figure 4.2 Dorsal view of the dissection of a near-term fetal blacknose shark, *Carcharhinus acronotus*, showing the *in situ* location of the thymus (circled region). (From Luer, C.A. *et al.*, *Biology of Sharks and Their Relatives*, CRC Press, Boca Raton, FL, 2004, pp. 369–395. Republished with permission.)

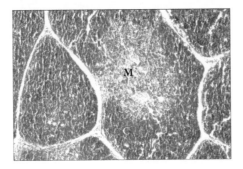

Figure 4.3 Paraffin-embedded 10-μ section of thymus from a near-term fetal sandbar shark, *C. plumbeus*, showing characteristic lobular architecture composed of outer cortical regions (C) of tightly packed thymocytes and centrally located medullary regions (M) of less densely populated thymocytes (stain: hematoxylin and eosin; original magnification: 100X).

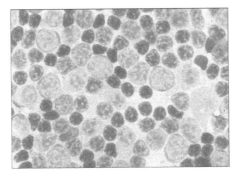

Figure 4.4 Tissue imprint of thymus from a juvenile nurse shark, *Ginglymostoma cirratum*, showing small, darkly staining mature thymocytes and larger immature thymocytes of varying sizes (stain: methylene blue; original magnification: 1000X).

organ (Zapata, 1980). As in higher vertebrates, each lobe consists of an outer cortex and an inner medulla (Figure 4.3). The cortex is composed of tightly packed thymocytes and occupies most of the space in a given lobule. Medullary regions contain less densely populated thymocytes and tend to be more centrally located within the lobule. Thymic tissue imprints reveal that the cortex and medulla contain thymocytes at varying sizes, corresponding to thymocytes at different stages of maturation (Figure 4.4), although only a small percentage will complete their maturation in the thymus prior to release into the peripheral circulation as thymus-derived lymphocytes (T-lymphocytes or

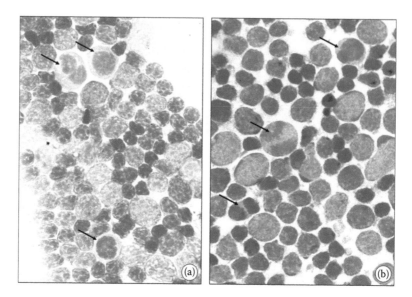

Figure 4.5   Tissue imprints of thymi from a juvenile nurse shark, *G. cirratum*, (a) and a clearnose skate, *Raja eglanteria*, (b) showing thymocytes in various stages of mitotic activity (arrows) (stain: methylene blue; original magnification: 1000X).

Table 4.1   **Cell Flow Cytometric Analysis of Cell Cycle Phases of Thymocytes from Juvenile Nurse Shark, *G. cirratum* ($n = 3$)**

Phase of Cell Cycle	Mean Number of Cells (%)	Range (%)
G_0/G_1	81	73–86
S	15	11–22
G_2/M	4	3–5

G_0 (Gap 0), resting phase; G_1 (Gap 1), phase of cell growth prior to DNA synthesis; S (Synthesis), DNA replication; G_2 (Gap 2), gap between DNA synthesis and mitosis; M (Mitosis), cell division into two daughter cells.

T-cells). Thymic cells in the thymocyte lineage range from large blast-like cells to maturing cells of intermediate and smaller sizes, whose increase in staining intensity reflects the transition from loosely to densely packed chromatin. Cells in prophase/metaphase are regularly observed in thymus tissue imprints (Figure 4.5a), and cells in more advanced stages of mitosis are not uncommon (Figure 4.5b). Consistent with these observations of mitotic activity, flow cytometric cell cycle analysis of freshly isolated, unstimulated elasmobranch thymocytes (Table 4.1) demonstrates that approximately 15% of these cells are actively synthesizing DNA (S phase), whereas nearly 4% are in the process of mitotic division (G2/M phase) (Luer *et al.*, 1995).

Frequently observed in elasmobranch thymic imprints are large cells several times the size of thymocytes. Often referred to as thymic nurse cells (De Waal Malefijt *et al.*, 1986; Wekerle *et al.*, 1980), they are more accurately a thymic lymphoepithelial cell complex, consisting a thymic epithelial cell and variable numbers of internalized thymocytes (Webb *et al.*, 2004). In mammalian and avian species, these specialized microenvironments are not only considered to be sites for the positive selection of thymocytes destined to become functional T-lymphocytes but also thought to participate in the clearance of nonfunctional, negatively selected, apoptotic thymocytes (Aguilar *et al.*, 1994).

As in teleost fishes (Flaño *et al.*, 1996), elasmobranch thymic nurse cells typically are round with a large, nonsegmented and eccentric nucleus (Luer *et al.*, 1995). Their cytoplasm appears slightly granular, but granules do not stain with methylene blue and are only faintly eosinophilic

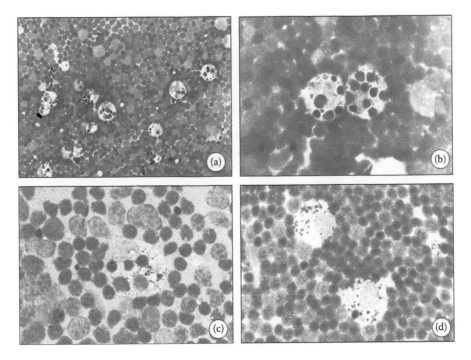

Figure 4.6   Imprints of thymic tissue from a bonnethead shark, *S. tiburo*, (a, b, and c) and nurse shark, *G. cirratum*, (d) showing thymic nurse cells (TNC). (a) shows the prevalence of TNC in the thymus, (b) shows TNC with internalized thymocytes, (c) shows a TNC with internalized thymocytes plus cellular debris, and (d) shows TNC with granular cell debris but no internalized thymocytes (stain: methylene blue; original magnifications: a, 40X; b, c, and d, 1000X).

with Wright's stain (Figure 4.6a). Thymic nurse cells often contain as many as 10 thymocytes in their cytoplasm (Figure 4.6b). In place of or in addition to intact cells, the cytoplasm occasionally retains cellular or nuclear debris from the destruction of negatively selected thymocytes (Figure 4.6c and d).

SPLEEN

The spleen is readily distinguished among elasmobranch visceral organs by its rich dark red to purplish color. In sharks, the spleen is elongate and positioned along the outer margin of the cardiac and pyloric regions of the stomach (Figure 4.7a). In batoids, however, with their peritoneal cavity being relatively compressed compared to that of sharks, the spleen is more compact and situated along the inner margin of the stomach (Figure 4.7b).

Histologically, the structural organization of the elasmobranch spleen is remarkably similar to the spleens of higher vertebrates (Fänge and Nilsson, 1985), being composed of easily recognizable regions of red and white pulp (Figure 4.8a and b). The scattered regions of white pulp are dense accumulations of small lymphocytes with asymmetrically placed central arteries. Areas of white pulp are surrounded by less dense areas of red pulp containing venous sinuses (Zapata, 1980). Instead of being filled with lymph, as in mammals, these sinuses are filled primarily with erythrocytes and to a lesser extent with lymphocytes (Andrew and Hickman, 1974).

The presence of mature, immature, and dividing lymphocytes and erythrocytes in splenic imprints supports the notion that this organ is a primary site for lymphopoiesis and erythropoiesis (Figure 4.9a). Although not common, immature heterophils and eosinophils occasionally appear in splenic imprints, suggesting that granulocytes may also be produced in the spleen (Figure 4.9b).

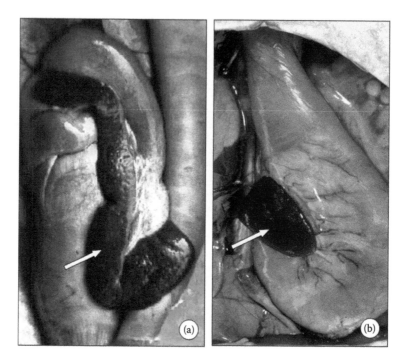

Figure 4.7 Ventral views of internal organs of a nurse shark, *G. cirratum*, (a) and a clearnose skate, *R. eglanteria*, (b) showing the location of the spleen, typically situated along the outer curvature of the cardiac and pyloric regions of the stomach in sharks, and on the inner curvature of the stomach in batoids. (From Luer, C.A. *et al.*, *Biology of Sharks and Their Relatives*, CRC Press, Boca Raton, FL. Republished with permission.)

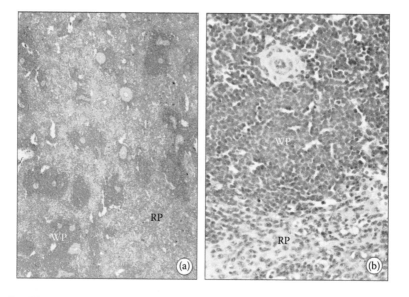

Figure 4.8 Paraffin-embedded 10-μ sections of spleen from a nurse shark, *G. cirratum*, showing characteristic red pulp (RP) composed of venous sinuses filled with red blood cells, and white pulp (WP) composed of dense accumulations of leukocytes (stain: hematoxylin and eosin; original magnifications: a, 40X; b, 200X).

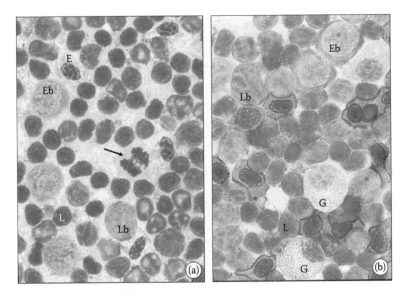

Figure 4.9 Tissue imprints of spleen from a horn shark, *Heterodontus francisci*, showing lymphocytes (L), lymphoblasts (Lb), granulocytes (G), erythrocytes (E), and erythroblasts (Eb). An erythrocyte in anaphase can be seen in (a) (arrow). (stains: [a], methylene blue; [b], Wright-Giemsa; original magnification: 1000X).

EPIGONAL ORGAN

The most conspicuous of the bone-marrow equivalent tissues is the epigonal organ. This organ, described by comparative anatomists long before its function was realized, is unique to the elasmobranch fishes (Fänge and Mattisson, 1981; Honma *et al.*, 1984). The epigonal organ continues caudally from the posterior margin of the gonads in all shark and batoid species (Figure 4.10). Histologically, the epigonal organ is composed of sinuses reminiscent of mammalian bone marrow (Figure 4.11a), except for the absence of adipose cells (fat cells). Its size and shape vary dramatically depending upon the species, but also as a function of gonadal cycles. Also, because the postmortem deterioration of this tissue is extremely rapid, it is often not recognizable if an animal has been dead too long before examination. These observations could explain the mistaken conclusion that an epigonal organ might not exist in a particular specimen. Tissue imprints demonstrate that epigonal organs are truly lymphomyeloid, with both granulocytes and lymphocytes regularly present during mitosis and throughout all stages of maturation. Although most of the cells are granulocytes, lymphocytes are present in a significant but lesser degree (Figure 4.11b).

LEYDIG ORGAN

Like the epigonal organ, anatomical descriptions of Leydig organs have been in the literature for many years (Fänge, 1968; Mattisson and Fänge, 1982; Oguri, 1983). Unlike the epigonal organ, the Leydig organ is not ubiquitous among elasmobranch species. Anecdotal observations support the notion that species possessing Leydig organs tend to have smaller epigonal organs, fueling speculation that Leydig tissue may compensate for the lack of lymphomyeloid tissue when epigonal tissue is limited (Fänge, 1977). When present, Leydig organs can be visualized as whitish masses beneath the epithelium on both dorsal and ventral sides of the esophagus (Figure 4.12a). Leydig organ histology is similar to that of the epigonal organ (Honma *et al.*, 1984), composed of sinuses that again are

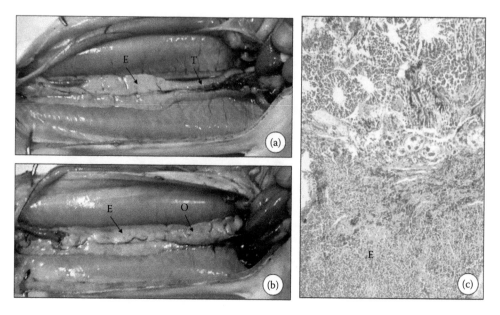

Figure 4.10   Dissections of male (a) and female (b) bonnethead sharks, *S. tiburo*, showing the anatomical
               location of the epigonal organ (E), relative to the testis (T) or ovary (O). (From Luer, C.A. *et al.*,
               Biology of Sharks and Their Relatives, CRC Press, Boca Raton, FL. Republished with permission.)
               Paraffin-embedded 10-μ section of tissue from a nurse shark, *G. cirratum*, (c) at the interface of
               the testis and epigonal organ (stain: hematoxylin and eosin; original magnification: 40X).

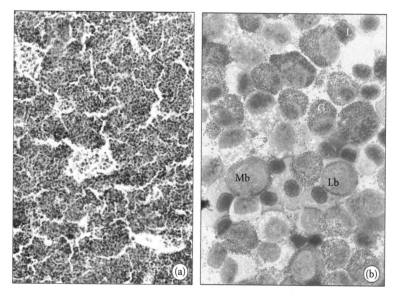

Figure 4.11   Paraffin-embedded 10-μ section of epigonal organ from a nurse shark, *G. cirratum*, (a) show-
               ing leukocyte-filled sinuses reminiscent of mammalian bone marrow (stain: hematoxylin and
               eosin; original magnification: 200X). Tissue imprint of epigonal organ from a blacknose shark,
               C. acronotus, (b) showing the presence of granulocytes (G), myeloblasts (Mb), lymphocytes (L),
               and lymphoblasts (Lb) (stain: Wright-Giemsa; original magnification: 1000X).

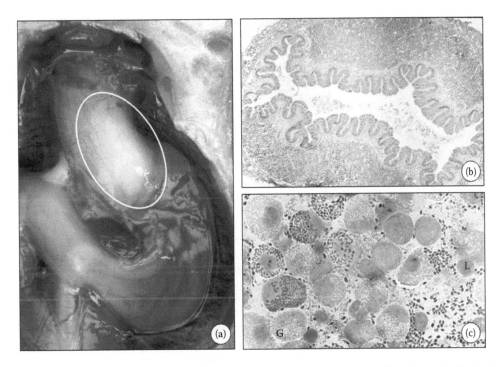

Figure 4.12 Ventral view of the esophagus and stomach of a clearnose skate, *R. eglanteria*, (a) showing the anatomical location of the Leydig organ (circled area). Paraffin-embedded 6-μ section through the esophagus (b) reveals dorsal and ventral lobes of Leydig organ on the respective surfaces of the esophagus, consisting of leukocyte-filled sinuses much like the epigonal organ (stain: hematoxylin and eosin; original magnifications: 40X). Tissue imprint of Leydig organ from a clearnose skate, *R. eglanteria* (c) showing the presence of granulocytes (G) and lymphocytes (L). Granulocytes with granules of different sizes, shapes, and staining intensities are visible in this species (stain: Wright-Giemsa; original magnification: 1000X).

reminiscent of mammalian bone marrow (Figure 4.12b). Tissue imprints are also similar to those of epigonal tissue, indicating leukocytes at various stages of maturation (Figure 4.12c). Again, cells are primarily granulocytes, although lymphocytes are also present.

MISCELLANEOUS SITES

In addition to the well-defined, encapsulated lymphomyeloid tissues described previously, pockets or aggregations of leukocytes can be found in various locations ranging from the intestinal mucosa to the meninges of the brain (Chiba *et al.*, 1988; Zapata *et al.*, 1996) and occasionally in the rectal gland (Walsh and Luer unpublished). Intestinal aggregations known as gut-associated lymphoid tissue, or GALT, can often be substantial (Hart *et al.*, 1988; Tomonaga *et al.*, 1986). Although it is tempting to speculate that these leukocyte aggregations may serve lymphomyeloid functions, they appear to be sites where immune cells accumulate rather than sites of immune cell production. The only site outside of the encapsulated lymphoid organs where cycling of leukocytes does appear to take place is the peripheral circulation. Although peripheral replication of leukocytes is not observed in higher vertebrates, flow cytometric analysis of elasmobranch peripheral blood leukocytes (Table 4.2) reveals as many as 20%–23% of the circulating leukocytes can be actively synthesizing DNA (S phase) and 2%–7% can be in some stage of mitosis (G_2/M phase) (Bodine and Luer unpublished).

Table 4.2   **Cell Cycle Analysis of Nurse Shark (G. cirratum)**
Peripheral Blood Leukocytes (Three Sharks
Sampled Weekly for Eight Weeks)

Phase of Cell Cycle	Mean Number of Cells (%)	Range (%)
G_0/G_1	73	70–76
S	22	20–23
G_2/M	5	2–7

REFERENCES

Aguilar, L.K., Aguilar-Cordova, E., Cartwright, J., Jr., and Belmont, J.W. Thymic nurse cells are sites of thymocyte apoptosis. *Journal of Immunology* 1994; 152:2645–2651.

Andrew, W., and Hickman, C.P. Circulatory systems. Ch. 8. In: *Histology of the Vertebrates—A Comparative Text*. St. Louis, MO: The C.V. Mosby Company, 1974, pp. 133–165.

Chiba, A., Torroba, M., Honma, Y., and Zapata, A.G. Occurrence of lymphohaemopoietic tissue in the meninges of the stingray *Dasyatis akajei* (Elasmobranchii, chondrichthyes). *American Journal of Anatomy* 1988; 183:268–276.

De Waal Malefijt, R., Leene, W., Roholl, P.J.M., Wormmeester, J., and Hoeben, K.A. T cell differentiation within thymic nurse cells. *Laboratory Investigation* 1986; 55:25–34.

Fänge, R. The formation of eosinophilic granulocytes in the oesophageal lymphomyeloid tissue of the elasmobranchs. *Acta Zoologica* 1968; 49:155–161.

Fänge, R. Size relations of lymphomyeloid organs in some cartilaginous fish. *Acta Zoologica* 1977; 58:125–128.

Fänge, R., and Mattisson, A. The lymphomyeloid (hemopoietic) system of the Atlantic nurse shark, *Ginglymostoma cirratum. Biological Bulletin* 1981; 160:240–249.

Fänge, R., and Nilsson, S. The fish spleen: Structure and function. *Cellular and Molecular Life Sciences* 1985; 41:152–158.

Flaño, E., Álvarez, F., López-Fierro, P., Razquin, B., Villena, A., and Zapata, A. In vitro and in situ characterization of fish thymic nurse cells. *Developmental Immunology* 1996; 5:17–24.

Hart, S., Wrathmell, A.B., Harris, J.E., and Grayson, T.H. Gut immunology in fish: A review. *Developmental and Comparative Immunology* 1988; 12:453–480.

Honma, Y., Okabe, K., and Chiba, A. Comparative histology of the Leydig and epigonal organs in some elasmobranchs. *Japanese Journal of Ichthyology* 1984; 31:47–54.

Luer, C.A., Walsh, C.J., and Bodine, A.B. The immune system of sharks, skates, and rays. In: *Biology of Sharks and Their Relatives*. Carrier, J., Musick, J., and Heithaus, M., editors. Boca Raton, FL: CRC Press, 2004, pp. 369–395.

Luer, C.A., Walsh, C.J., Bodine, A.B., Wyffels, J.T., and Scott, T.R. The elasmobranch thymus: Anatomical, histological, and preliminary functional characterization. *Journal of Experimental Zoology* 1995; 273:342–354.

Mattisson, A., and Fänge, R. The cellular structure of the Leydig organ in the shark, *Etmopterus spinax* (L.). *Biological Bulletin* 1982; 162:182–194.

Oguri, M. On the Leydig organ in the esophagus of some elasmobranchs. *Bulletin of the Japanese Society of Scientific Fisheries* 1983; 49:989–991.

Tomonaga, S., Kobayashi, K., Hagiwara, K., Yamaguchi, K., and Awaya, K. Gut-associated lymphoid tissue in the elasmobranchs. *Zoological Science* 1986; 3:453–458.

Webb, O., Kelly, F., Benitez, J., Juncheng, L., Parker, M., Martinez, M., Samms, M., Blake, A., Pezzano, M., and Guyden, J. The identification of thymic nurse cells in vivo and the role of cytoskeletal proteins in thymocyte internalization. *Cellular Immunology* 2004; 228:119–129.

Wekerle, H., Ketelsen, U.-E., and Ernst, M. Thymic nurse cells. Lymphoepithelial cell complexes in murine thymuses: Morphological and serological characterization. *Journal of Experimental Medicine* 1980; 151:925–944.

Zapata, A. Ultrastructure of elasmobranch lymphoid tissue. I. Thymus and spleen. *Developmental and Comparative Immunology* 1980; 4:459–472.

Zapata, A.G., Torroba, M., Sacedon, R., Varas, A., and Vicente, A. Structure of the lymphoid organs of elasmobranchs. *Journal of Experimental Zoology* 1996; 275:125–143.

CHAPTER 5

Elasmobranch Blood Cells

Ashley N. Haines and Jill E. Arnold

CONTENTS

SUMMARY

This chapter addresses the form and function of elasmobranch peripheral blood cells. It provides an overview, including images, of the cell types found across several families of sharks and compares them to those of skates and rays. The selected species in this chapter represent those commonly found in public aquaria and/or research laboratories. Although blood cells have been studied previously by multiple authors, the result has been a variety of naming schemes that can hinder scientific communication regarding the functions of these cells. The primary goal of this chapter is to provide a standardized nomenclature for elasmobranch blood cells that is supported by color micrographs of each cell type, with emphasis on the eosinophilic granulocytes, which are the most common source of confusion. As such, the information in this chapter should provide sufficient details to identify cell types in many species. We hope it will serve as a guide for interpreting previous reports on elasmobranch hematology and provide a single nomenclature strategy for future publications.

INTRODUCTION

Elasmobranchs comprise a large and diverse group of animals, representing 60 families, 185 genera, and approximately 1200 species; 96% are sharks and rays and 4% represent chimaeras (Compango, 1999). As in other animals, immune cells are an important component of the blood and can be collected readily for use in the study of immune function. However, such studies must be predicated upon an understanding of the cells found in peripheral blood. Detailed descriptions of peripheral blood cells observed in select elasmobranch species have been reported, using a variety of nomenclature strategies based upon ultrastructure, cytochemistry, and light microscopy; however, the difficulty in assigning a single nomenclature strategy continues today. A common source of confusion is the heterogeneity of the granulocytes, especially the acidophilic cells, among non-mammalian vertebrate phylogenetic groups, including birds, reptiles, amphibians, primitive fishes, and teleosts.

Literature shows that interest in understanding the blood cells of fish dates to the mid-1800s. In 1857, Leydig described an esophogeal gland-like structure, which he interpreted as a lymph node, in several elasmobranchs as a source of granulocytopoiesis; this tissue is now often referred to as the organ of Leydig (Mattison and Fänge, 1982). Fänge and colleagues further studied hematopoiesis in a wide variety of elasmobranch species and described the following tissues: the organ of Leydig, a white mass located in the dorsal and ventral wall of the esophagus (abundant granulocytes and lymphocytes) (Fänge, 1968); the epigonal organ, associated with the gonads (abundant granulocytes and undistinguished blast-type cells); the spleen (white pulp primarily lymphocytes and red pulp primarily erythrocytes); and the thymus (lymphoid only) (Fänge, 1977; Fänge and Mattisson, 1981). In the more primitive holocephalans, the site of granulocytopoiesis is found in the tissues within the cranium (Mattisson and Fänge, 1986). The major organ of hematopoiesis in teleosts, by comparison, is the anterior portion of the kidney, referred to as the head kidney, with minor sites including the spleen, liver, and thymus (Ellis, 1977; Zapata and Amemiya, 2000).

Review articles by Fänge (1992) and Hine (1992) summarize the work of earlier authors who studied the ultrastructure, cytochemistry, and morphology of elasmobranch cells from Romanowsky stained smears. Mainwaring and Rowley's (1985) commonly cited nomenclature for dogfish granulocytes uses a numerical series (G1–3) and described a fourth type (G4) that best resembles a thrombocyte containing abundant azurophilic granules (Figures 5.2 and 5.3; GT). Hine *et al.*'s (1987) extensive study of the cytochemistry of fish leukocytes concluded that identification should be based upon cell morphology, granule size and morphology, granule appearance in periodic acid–Schiff (PAS) stained films, appearance of granule degradation and degranulation, tinctorial properties in Wright's stained films, and enzyme cytochemistry profile, in that order of priority. Based on their findings from cytochemical stains of blood from >100 fish species, they proposed three separate granulocyte nomenclature strategies: agnathans (granulocyte), holocephalans (fine and coarse granulocytes), and elasmobranchs (fine and coarse eosinophilic granulocytes and neutrophilic granulocytes), and there is no mention of a cell matching Mainwaring and Rowley's G4.

A number of authors agree that the terminology used to describe avian and reptilian granulated cells (heterophil, eosinophil, and basophil) is the most appropriate choice for elasmobranch species (Campbell, 2007; Clauss *et al.*, 2008; Dove *et al.*, 2010; Saunders, 1966; Stoskopf, 1993; Van Rijn and Reina, 2010; Walsh and Luer, 2004). Lucas and Jamroz (1961) discussed the evolution of the term *heterophil*, considered the avian functional equivalent of the mammalian neutrophil, and provided a full array of color plates to illustrate the developmental stages of poultry leukocytes. They emphasize that, while not all will agree with the cell names they assigned, "it is important that the mental image of a particular cell be the same in the minds of all who discuss it." Therefore, despite the variation in morphology between species (Figures 5.1 through 5.5), it is generally possible to assign observed granulocytes to the category of basophil, eosinophil, heterophil, or neutrophil based upon morphological and tinctorial properties. As this strategy has been recommended by

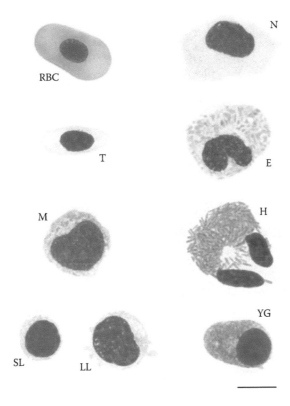

Figure 5.1   Romanowsky stained peripheral blood cells typical for the nurse shark, *G. cirratum*. Cell preparation by cytocentrifugation. RBC, red blood cell; T, thrombocyte; M, monocyte; SL, small lymphocyte; LL, large lymphocyte; N, neutrophil/G2 (Mainwaring); E, eosinophil/coarse eosinophilic
granulocyte (Hine *et al.*, 1987)/G1 (Mainwaring); H, heterophil/fine eosinophilic granulocyte (Hine
et al., 1987)/G3 (Mainwaring); and YG, young granulocyte. Mark = 10 μm.

multiple authors, it will be followed in this chapter with annotations of corresponding names found
in literature.

The purpose of this chapter is to provide a comprehensive list, with micrographs, of the cell
types commonly observed in elasmobranchs by light microscopy of Romanowsky stained blood
films. This list is accompanied by brief descriptions of cell morphology at the electron microscope
level, as well as a brief description of cell functions, which are largely covered elsewhere in this
text. It should be noted that tinctorial properties will vary with different stains (Wright's, Wright-
Giemsa, three-dip stains, and so on), and it is advisable to prepare multiple blood smears for comparison. With the exception of the cells shown in Figure 5.1 (nurse shark, *Ginglymostoma cirratum*),
prepared by cytocentrifugation, the images shown here were collected from smears prepared by the
two-slide method. Cytocentrifuge preparations are ideal for observing detailed cell morphology but
may be impractical for some laboratories and for routine hematology specimens. The cell nomenclature used in this chapter is as follows: erythrocyte or red blood cell (RBC), immature erythrocyte
or red blood cell (iRBC), thrombocyte (T), granulated thrombocyte (GT), monocyte (M), small
lymphocyte (up to 10 μm) (SL), large lymphocyte (>10 μm) (LL), neutrophil (N), heterophil (H),
eosinophil (E), and basophil (B). To aid in comparison of published reports, the nomenclature strategies of Mainwaring and Rowley's (1985) G1–4 series and of Hine *et al.*'s (1987) fine eosinophilic
granulocyte (FEG) and coarse eosinophilic granulocyte (CEG) will be indicated as cross-reference
when practical.

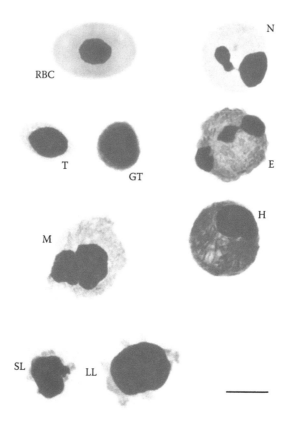

Figure 5.2   Romanowsky stained peripheral blood cells typical for the sandtiger shark, *Carcharius taurus*. Cell preparation by two-slide method. RBC, red blood cell; T, thrombocyte; GT, granulated thrombocyte/ G4 (Mainwaring); M, monocyte; SL, small lymphocyte; LL, large lymphocyte; N, neutrophil/G2 (Mainwaring); E, eosinophil/coarse eosinophilic granulocyte (Hine *et al.*, 1987)/G1 (Mainwaring); and H, heterophil/fine eosinophilic granulocyte (Hine *et al.*, 1987)/G3 (Mainwaring). Mark = 10 μm.

DESCRIPTION OF ELASMOBRANCH BLOOD CELL MORPHOLOGY

Erythrocyte

Elasmobranch erythrocyte morphology resembles those of other nonmammalian vertebrates. Mature cells are oval with abundant smooth eosinophilic cytoplasm and a central, oval-shaped condensed nucleus (Figures 5.1 through 5.5, RBC). Ultrastructurally, erythrocytes demonstrate a finely textured cytoplasm with very few cell organelles (Figure 5.6, RBC). Cell size varies with fish species; this was extensively demonstrated by D. C. Saunders' work in the 1960s where she measured mature erythrocytes of over 600 specimens of marine fish from Puerto Rico and over 200 specimens of fish from the Red Sea; both articles report the significantly larger size found in elasmobranchs (~20 × 15 μm) compared to those of teleosts (~10 × 5 μm) (Saunders, 1966, 1968). A small percentage of polychromatophilic and immature erythrocytes, as well as mitotic cells, are commonly observed in elasmobranchs (Glomski, 1992; Saunders, 1966, 1968) (Figures 5.3 and 5.4; iRBC). The ultrastructure of erythrocyte development had been described, where early stage cells were typically round rather than oval and contained a large round nucleus showing less condensed chromatin and a patent nucleolus (Hyder *et al.*, 1983; Zapata, 1980). The role of elasmobranch erythrocytes in gas exchange and acid–base balance has been reviewed widely (Butler and Metcalfe, 1988; Claiborne, 1998; Claiborne *et al.*, 2002; Henry and Heming, 1998; Jensen *et al.*, 1998; Nikinmaa and Salama, 1998).

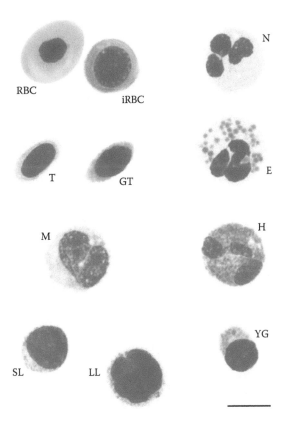

Figure 5.3   Romanowsky stained peripheral blood cells typical for the sandbar shark, *Carcharhinus plumbeus.*
Cell preparation by two-slide method. RBC, red blood cell; iRBC, immature RBC; T, thrombocyte; GT, granulated thrombocyte/G4 (Mainwaring); M, monocyte; SL, small lymphocyte; LL, large
lymphocyte; N, neutrophil/G2 (Mainwaring); E, eosinophil/coarse eosinophilic granulocyte (Hine
et al., 1987)/G1 (Mainwaring); H, heterophil/fine eosinophilic granulocyte (Hine *et al.,* 1987)/G3
(Mainwaring); and YG, young granulocyte. Mark = 10 μm.

Thrombocyte

Typical elasmobranch thrombocytes also resemble closely those found in nonmammalian
species; size varies with species but typically they are small (10–15 × 5–8 μm) oval cells with
clear, colorless cytoplasm and a central oval, condensed nucleus (Figures 5.1 through 5.5; T). By
transmission electron microscopy, they have a coarse cytoplasm of moderate electron density that
distinguishes them from erythrocytes, despite a somewhat similar shape (Figure 5.6; T). On peripheral blood smears, as in the hemacytometer, thrombocytes can appear in the typical oval shape,
or as spindle shaped or round. The variations in shape may be due to cell maturation or degree of
activation. By light microscopy, thrombocytes are often confused with the similarly sized lymphocyte; this similarity can contribute to error in the total white cell count and in the differential
count. On Romanowsky stained blood smears, the main features that distinguish thrombocytes
from lymphocytes are the cytoplasm color (colorless vs. light blue, respectively) and the N (nucleus):
C (cytoplasm) ratio (higher in the lymphocyte).

Multiple authors have described a second population of thrombocytes in some elasmobranch
species that is identical to the typical thrombocyte, with the exception that the cytoplasm is
filled with slender rod-shaped eosinophilic granules (Arnold, 2005). This cell is Mainwaring and

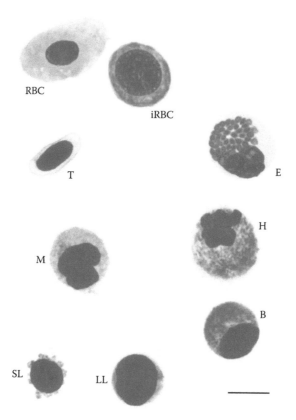

Figure 5.4   Romanowsky stained peripheral blood cells typical for the clearnose skate, *Raja eglanteria*. Cell preparation by two-slide method. RBC, red blood cell; iRBC, immature RBC; T, thrombocyte; M, monocyte; SL, small lymphocyte; LL, large lymphocyte; E, eosinophil/coarse eosinophilic granulocyte (Hine *et al.*, 1987)/G1 (Mainwaring); H, heterophil/fine eosinophilic granulocyte (Hine *et al.*, 1987)/G3 (Mainwaring); and B, basophil. Mark = 10 μm.

Rowley's (1985) G4, and the function and any clinical significance of this cell is not yet understood (Arnold, 2005; Campbell, 1988). Table 5.1 shows that these G4 (GT) cells were not detected in the skates and rays but were present in some, but not all, shark species. This cell type has been reported to represent a significant component of the leukocyte differential if included in the white blood cell count: 17% in *Scyliorhinus canicula* and 21.8% average in 42 wild caught sandbar sharks, *C. plumbeus*, representing the second-most prevalent cell type for this group (Arnold, 1997; Parish *et al.*, 1985). Hine (1992) stated that the ultrastructure of these *type 4 granulocytes* resembles basophil/mast cells and reported them as basophils. Campbell (2007) includes an image of elasmobranch thrombocytes containing granules *indicating reactivity*. Any immune function of the GT (G4) remains to be described.

Elasmobranch thrombocytes have been described as the equivalent of mammalian platelets (Pica *et al.*, 1990). They aggregate *in vitro* and adhere to glass, hence their presumed function in blood clotting. Functional differences from mammalian platelets have been reported, including temperature reversibility and aggregation independent of thrombin and adenosine diphosphate (ADP) (Lewis, 1972; Stokes and Firkin, 1971). Importantly, platelets may also have an immune function because they have been shown to have some phagocytic capability (Stokes and Firkin, 1971; Walsh and Luer, 1998).

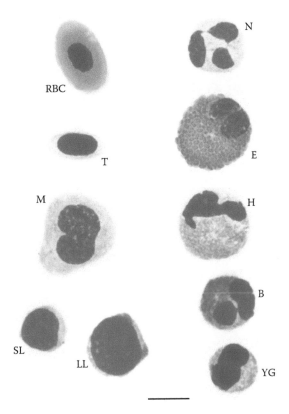

Figure 5.5 Romanowsky stained peripheral blood cells typical for the southern stingray, *Dasyatis americana*. Cell preparation by two-slide method. RBC, red blood cell; T, thrombocyte; M, monocyte; SL, small lymphocyte; LL, large lymphocyte; N, neutrophil/G2 (Mainwaring); E, eosinophil/coarse eosinophilic granulocyte (Hine *et al.*, 1987)/G1 (Mainwaring); H, heterophil/fine eosinophilic granulocyte (Hine *et al.*, 1987)/G3 (Mainwaring); B, basophil; and YG, young granulocyte. Mark = 10 μm.

Monocyte

The morphology of fish monocytes is comparable to those found in mammals, birds, and reptiles. They are large (10–18 μm), usually round cells but may have irregular margins. The blue cytoplasm is abundant, coarse in texture compared to the lymphocyte cytoplasm, and often contains vacuoles. By contrast, the cytoplasm of the similarly sized neutrophil has little or no color. The shape of the monocyte nucleus can be round, oval, or multilobed, with loosely compact chromatin (Figures 5.1 through 5.5; M). Ultrastructurally, the texture of the cytoplasm is grainier (due to the presence of rough endoplasmic reticulum and vacuoles) than thrombocytes or erythrocytes, and the nucleus is more euchromatic (Figure 5.6; M). Hyder *et al.* (1983) report abundant mitochondria as a predominant feature. Monocytes have a clearly demonstrated phagocytic capability and hence a presumed immune function. Hyder *et al.* (1983) describe the monocyte of the nurse shark as monocyte-macrophage, because they observed *fully differentiated macrophage-like* cells in the *peripheral blood*. In our hands, the differentiation to a macrophage can be observed as a result of preparation on a glass or plastic surface.

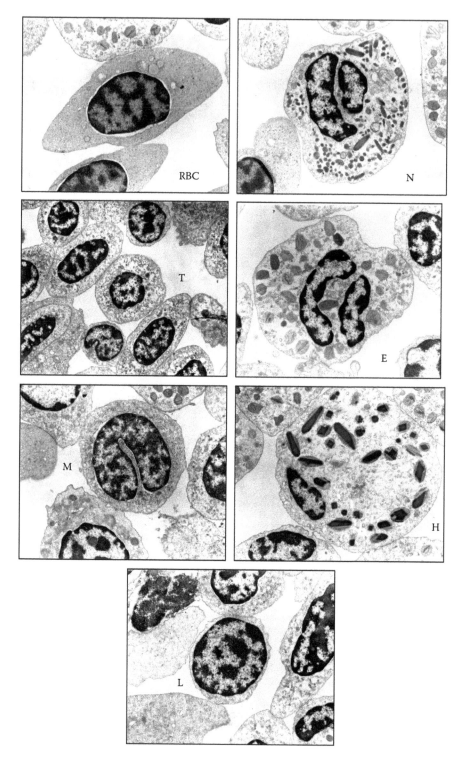

Figure 5.6   Transmission electron microscopy of the nurse shark peripheral blood cells. RBC, red blood
cell; T, thrombocyte; M, monocyte; L, lymphocyte; N, neutrophil; E, eosinophil; H, heterophil.
Magnification: 8000X.

Table 5.1 Elasmobranch Blood Cell List

Common Name (n)	Monocyte	Lymphocyte	Neutrophil	Heterophil	Eosinophil	GT	Basophil
Elasmobranch Cell List: Sharks							
Order: Heterodontiformes—Bullhead Sharks							
Heterodontus francisci — Horn shark[a] (>5)	+	+	+	+	a	+	ND
Heterodontus portusjacksoni — Port Jackson shark[b] (1)	ND	+	+	+	a	+	ND
Order: Orectolobiformes—Carpet Sharks							
Eucrossorhinus dasypogon — Tassled wobbegong[a] (1)	+	+	+	+	ND	ND	ND
Orectolobus maculatus — Spotted wobbegong[b] (1)	+	+	+	+	ND	ND	ND
Orectolobus ornatus — Ornate wobbegong[a] (2)	+	+	+	+	a	ND	ND
Chiloscyllium plagiosum — Whitespotted bamboo shark[a] (5)	+	+	+	+	a	ND	+
Chiloscyllium punctatum — Brownbanded bamboo shark[b] (1)	+	+	+	+	a	ND	ND
Ginglymostoma cirratum — Nurse shark[a] (>5)	+	+	+	+	c	ND	ND
Stegostoma fasciatum — Zebra shark[a] (>5)	+	+	+	+	a	ND	ND
Rhincodon typus — Whale shark[c] (5)	+	+	+	+	a	ND	+
Order: Lamniformes—Mackerel Sharks							
Carcharius Taurus — Sandtiger shark[a] (>5)	+	+	+	+	a	+	ND
Order: Carcharhiniformes—Ground Sharks							
Cephaloscyllium ventriosum — Swell shark[a] (5)	+	+	+	+	a	+	ND
Mustelus canis — Dusky smooth hound[a] (>5)	+	+	+	+	c	+	ND
Triakis semifasciata — Leopard shark[a] (>5)	+	+	+	+	b	+	ND
Carcharhinus acronatus — Blacknose shark[a] (5)	+	+	+	+	b	+	ND
Carcharhinus brevipinna — Spinner shark[a] (1)	+	+	+	+	b	+	ND
Carcharhinus leucas — Bull shark[a] (1)	+	+	+	+	b	+	ND
Carcharhinus limbatus — Blacktip shark[a] (1)	+	+	+	+	b	+	ND
Carharhinus melanopterus — Blacktip reef shark[a] (5)	+	+	+	+	b	+	ND
Carcharhinus obscurus — Dusky shark[a] (1)	+	+	+	+	b	+	ND
Carcharhinus plumbeus — Sandbar shark[a] (>5)	+	+	+	+	b	+	ND

(Continued)

Table 5.1 (Continued) Elasmobranch Blood Cell List

	Common Name (n)	Monocyte	Lymphocyte	Neutrophil	Heterophil	Eosinophil	GT	Basophil
Galeocerdo cuvier	Tiger shark[a] (1)	+	+	+	+	b	+	+
Negaprion brevirostris	Shortnosed lemon shark[a] (>5)	+	+	+	+	c	+	ND
Rhizoprionodon terranovae	Atlantic sharpnose shark[a] (1)	+	+	+	+	a	+	ND
Sphyrna tiburo	Bonnethead shark[a] (5)	+	+	+	+	c	+	ND
Triaenodon obesus	Whitetip reef shark[a] (>5)	+	+	+	+	b	+	ND
Order: Chimaeriformes								
Hydrolagus collei	Spotted ratfish[d] (1)	+	+	+	+	a	NF	ND
Elasmobranch Cell List: Skates and Rays								
Order: Pristiformes—Sawfishes								
Pristis microdon	Greattooth or freshwater sawfish[a] (>5)	+	+	rare	+	a	ND	ND
Pristis pectinata	Smalltooth sawfish[a] (3)	+	+	rare	+	a	ND	+
Rhynchobatus laevis	Smoothnose wedgefish[a] (>5)	+	+	rare	+	a	ND	+
Order: Rhiniformes—Wedgefishes								
Rhina ancylostoma	Bowmouth guitarfish[d] (4)	+	+	ND	+	a	ND	+
Order: Rhinobatiformes—Guitarfishes								
Trygonorrhina fasciata	Southern fiddler ray[d] (5)	+	+	+	+	a	ND	+
Platyrhinoidis triseriata	Thornback guitarfish[a] (4)	+	+	+	+	a	ND	ND
Order: Rajiformes—Skates								
Raja eglanteria	Clearnose skate[a] (3)	+	+	ND	+	a	ND	+

Order: Myliobatiformes—Stingrays

Species	Common name						
Urobatis jamaicensis	Yellow stingray[d] (5)	+	ND	+	a	ND	ND
Paratrygon aiereba	Discus/Ceja freshwater stingray[d] (4)	+	ND	+	a	ND	+
Potamotrygon castexi	Vermiculate river stingray[d] (3)	+	rare	+	a	ND	+
Potamotrygon henli	Bigtooth river stingray[d] (5)	+	ND	+	a	ND	+
Potamotrygon leopoldi	White-blotched river stingray[d] (5)	+	ND	+	a	ND	+
Potamotrygon tigrina	Tiger stingray[d] (3)	+	ND	+	a	ND	ND
Dasyatis americana	Southern ray[e] (>5)	+	rare	+	a	ND	+
Dasyatis centroura	Roughtail ray[a] (>5)	+	rare	+	a	ND	+
Dasyatis say	Bluntnose stingray[a] (1)	+	ND	+	a	ND	+
Dasyatis violacea	Pelagic stingray[a] (5)	+	rare	+	a	ND	+
Himantura chaophraya	Freshwater whipray[a] (2)	+	ND	+	a	ND	ND
Himantura dalyensis	Freshwater whipray[a] (3)	+	ND	+	a	ND	ND
Himantura granulata	Mangrove whipray[d] (2)	+	ND	+	a	ND	+
Himantura uarnak	Honeycomb stingray[a] (5)	+	ND	+	a	ND	+
Taeniura lymma	Blue-spotted ribbontail stingray[d] (5)	+	ND	+	a	ND	ND
Taeniura meyeni	Round ribbontail stingray[a] (2)	+	ND	+	a	ND	+
Gymnura altavela	Spiny butterfly ray[a] (>5)	+	rare	+	a	ND	+
Aetobatus narinari	Spotted eagle ray[a] (>5)	+	rare	+	a	ND	+
Myliobatis freminvillei	Bullnose eagle ray[a] (>5)	+	ND	+	a	ND	+
Rhinoptera bonasus	Cownose ray[a] (>5)	+	rare	+	a	ND	+

Notes: +, Indicates presence; a, Eosinophil granules brighter than heterophil; b, Eosinophil granules paler than heterophil; c, Eosinophil granules similar in tinctorial properties to heterophil but differ in shape; ND, None detected on observed slides.

Slides provided by:
a National Aquarium;
b Underwater Adventures;
c Georgia Aquarium;
d John G Shedd Aquarium.

Lymphocyte

Similar to those found in birds and reptiles, elasmobranch lymphocytes are small (7–9 μm) round cells with a narrow rim of smooth light blue cytoplasm around a large oval-round condensed nucleus (Figures 5.1 through 5.5; SL, LL). On blood films prepared by the slide-to-slide method, the lymphocytes often appear pleomorphic with irregular margins, rather than round, and this can be helpful in distinguishing them from thrombocytes, which retain their oval-round shape with smooth margins (Figure 5.2; SL, LL). Smaller, presumably, inactivated cells are more numerous, whereas larger lymphocytes (>10 μm) often display cytoplasmic projections or *blebs* and more euchromatin (Figures 5.1 through 5.3; SL, LL). Ultrastructurally, few organelles are seen aside from numerous ribosomes and a few mitochondria in larger cells (Figure 5.6; L).

The role of lymphocytes in the adaptive immunity of elasmobranchs has been widely studied and is discussed elsewhere in this volume (Chapters 2 and 3). Both B and T lymphocytes exist. B cells produce immunoglobulin of three classes: IgM, IgW, and IgNAR (Flajnik and Rumfelt, 2000; Zapata and Amemiya, 2000). Gene rearrangement of both immunoglobulin and T-cell receptor genes has been described in Criscitiello *et al.* (2010), Hsu *et al.* (2006), and Rumfelt *et al.* (2001).

Neutrophil (G2)

Elasmobranch neutrophils are large (13–18 μm) round cells with smooth margins. Various developmental stages are observed commonly on the same blood smear, as is true for the other granulocytes. The nucleus may appear as eccentric, condensed, and round in shape (Figure 5.1; N) or multilobed (Figures 5.2, 5.3, 5.5, and 5.6; N). The cytoplasm is often light gray to colorless and, in some cells, contains very fine azurophilic granules. Ultrastructurally, the cytoplasm contains round to oblong granules of similar electron density to those in eosinophils, but distinguishable by their smaller size and lack of fibrillar components (Figure 5.6; N).

This cell type is sometimes reported as a heterophil in the literature, but to those familiar with avian and reptilian blood, the term *heterophil* indicates a different cell type, one with abundant, distinct rod-shaped red cytoplasmic granules (Campbell, 2007).

Pacheco *et al.* (2002) describe the structure and ultrastructure of the neutrophil in the Brazilian sharpnose shark, *Rhizoprionodon lalandii*. Immature neutrophils observed in the hemopoietic tissues had a central, or eccentric, nucleus with fine chromatin and frequently with a nucleolus, whereas mature cells contained more elongate nucleus and the cytoplasm of some contained slightly stained granules. Walsh and Luer (2004) describe a granulocyte that matches this description, where the nucleus is spherical or may be multilobed, but the granules do not take up stain and indicate that this may be another cell type or may be a staining artifact. Table 5.1 shows that cells matching this description were observed in each of the shark species and in the *Rhinobatiformes*, suggesting that staining artifact may not fully explain the presence of this cell. In comparison, neutrophils were found only rarely or not detected in the other orders of skates and rays (Table 5.1).

Neutrophil function has been demonstrated to include phagocytosis; additional data indicate that they engage in antibody-mediated phagocytosis (Hyder *et al.*, 1983; McKinney and Flajnik, 1997; Stokes and Firkin, 1971). The cytochemistry of neutrophils, an implied indicator of enzyme-based killing mechanisms, has been reviewed (Hine, 1992). However, other functions of neutrophils are less well known.

Heterophil (FEG/G3)

The elasmobranch heterophil closely resembles the avian heterophil. It typically presents as a large round cell (13–18 μm) with clear to pale gray smooth cytoplasm filled with slender, rod-shaped, reddish granules. The shape of the condensed nucleus may be round, indented (Figures 5.2 and

5.4; H), or multilobed and typically polar (Figures 5.1, 5.3, and 5.5; H). As shown in Table 5.1, this cell type was observed in all species listed. Comparable to birds and reptiles, the heterophil is often the most abundant of the granulocytes on peripheral blood smears. Ultrastructurally, heterophils have prominent electron dense granules containing crystalline inclusions that are often rectangular or rhomboid in shape with a less electron dense matrix surrounding them that is smooth or fibrillar in nature (Figure 5.6; H).

Eosinophil (CEG/G1)

The elasmobranch eosinophil is distinguished from the similarly sized heterophil in the same blood smear by the difference in color of the cytoplasm and the granule morphology and tinctorial properties. Eosinophils are large round cells (13–18 μm) with pale blue cytoplasm containing typically round to oval granules. As with the other granulocytes, the shape of the condensed nucleus may appear as round or multilobed. By comparison, the number of granules is sometimes less than in the heterophil (Figures 5.3 and 5.4; E, H). The morphology of these granules varies with species, as indicated in Table 5.1 and Figures 5.1 through 5.5, and is described here in three categories: (1) brighter than the heterophil granules, (2) paler than the heterophil granules, or (3) similar tinctorial properties to the heterophil but different in shape. For example, the granules of *R. eglanteria* (clearnosed skate) are very bright orange (e.g., type a; Figure 5.4; E) compared with those found in the *Carcharhinus* species, which are pale pink (e.g., type b; Figure 5.3; E), whereas the color difference is more subtle in *G. cirratum* (e.g., type c; Figure 5.1; E). Ultrastructurally, eosinophils have a heterochromatic cytoplasm containing round to ellipsoidal granules often with a central fibrillar component (Figure 5.6; E). The granules are moderately electron dense and do not contain the crystalline inclusions seen in the heterophils of some species.

Young Granulocyte

Small numbers of immature granulocytic (YG) cells are found in some species, as shown in Figures 5.1, 5.3, and 5.5. Typically, these oval to round cells are smaller (~10 μm) than the mature eosinophils and heterophils, with cytoplasm containing azurophilic granules of indistinct shape. It is unknown which cell line they represent. Hine and Wain (1987) described the *type RB* cell observed in the epigonal organ of stingrays as having sparse granules, an unlobed, irregular nucleus with small condensed nucleoli, suggesting immaturity; this cell represented ~2% of the eosinophils.

Basophil

The morphology of the elasmobranch basophil is similar to those of other vertebrates. On Wright's stained smears, the granules are blue–black, whereas the granules may appear pale or as empty vacuoles on three-step quick Romanowsky-type stains. Basophils are similar in size to the other granulocytes (13–18 μm) and typically round. The cytoplasm is light gray and smooth. The nucleus can be round to multilobed and may be difficult to observe due to the abundance of granules (Figures 5.4 and 5.5; B). Rarely reported in teleosts, basophils are commonly observed in low numbers on peripheral blood smears in stingray species (Table 5.1). This observation was reported in seven species of rays by Hine and Wain (1987). The function of elasmobranch basophils remains unclear. Early authors presumed a similar function to mammalian basophils, which would include a role in inflammation due to histamine release from the granules. However, this has not been demonstrated experimentally (Ainsworth, 1992).

CONCLUSION

Although the blood cell morphology in sharks, skates, and rays is quite variable, the information in this chapter should provide sufficient detail to serve as a good reference for interpreting cell types in many species. The micrographs and table of cell types are, to our knowledge, the first example of multiple species being presented together in color, enabling researchers to readily identify the cell types they wish to study. Finally, and most importantly, the primary goal of this chapter is to offer a standardized nomenclature, so that researchers are speaking a common language as we continue to learn about the form and function of elasmobranch peripheral blood cells.

REFERENCES

Ainsworth, A.J. Fish granulocytes: Morphology, distribution, and function. *Annual Review of Fish Diseases* 1992; 2:123–148.

Arnold, J.E. Preliminary hematology study of the sandbar shark, *Carcharhinus plumbeus*. *Proceedings: International Association of Aquatic Animal Medicine 28th Annual Conference*, Harderwijk, the Netherlands, 1997, pp. 53–54.

Arnold, J.E. Hematology of the sandbar shark, *Carcharhinus plumbeus*: Standardization of complete blood count techniques for elasmobranchs. *Veterinary Clinical Pathology* 2005; 34(2):115–123.

Butler, P.J., and Metcalfe, J.D. Cardiovascular and respiratory system. In: Shuttleworth, T.J., editor. *Physiology of Elasmobranch Fishes*. Berlin, Germany: Springer-Verlag, 1988, pp. 1–48.

Campbell, T.W. Tropical fish medicine. Fish cytology and hematology. *Veterinary Clinics of North America (Small Animal Practice)* 1988; 18(2):347–364.

Campbell, T.W. Hematology of fish. In: Campbell, T.W., and Ellis, C.K., editors. *Avian and Exotic Animal Hematology and Cytology*, 3rd ed. Ames, IA: Blackwell Publishing, 2007, pp. 100–105.

Claiborne, J.B. Acid-base regulation. In: Evans, D.H., and Claiborne, J.B., editors. *The Physiology of Fishes*. Boca Raton, FL: CRC Press, 1998, pp. 177–198.

Claiborne, J.B., Edwards, S.L., and Morrison-Shetlar, A.I. Acid–base regulation in fishes: Cellular and molecular mechanisms. *Journal of Experimental Zoology* 2002; 293:302–319.

Clauss, T.M., Dove, A.D.M., and Arnold, J.E. Hematologic disorders of fish. *Veterinary Clinics of North America (Exotic Animal Practice)* 2008; 11:445–462.

Compango, L.J. Systematics and body form. In: Hamlett, H.C., editor. *Sharks, Skates and Rays, the Biology of Elasmobranch Fish*. Baltimore, MD: The Johns Hopkins University Press, 1999, p. 2.

Criscitiello, M.F., Ohta, Y., Saltis, M.E., McKinney, C., and Flajnik, M.F. Evolutionarily conserved TCR binding sites, identification of T cells in primary lymphoid tissues, and surprising trans-rearrangements in nurse shark. *Journal of Immunology* 2010; 184:6950–6960.

Dove, A.D.M., Arnold, J., and Clauss, T.M. Blood cells and serum chemistry in the world's largest fish: The whale shark *Rhincodon typus*. *Aquatic Biology* 2010; 10:177–184.

Ellis, A.E. The leukocytes of fish: A review. *Journal of Fish Biology* 1977; 11:453–491.

Fänge, R. The formation of eosinophilic granulocytes in the oesophageal lymphomyeloid tissue of the elasmobranch. *Acta Zoologica* 1968; 49:155–161.

Fänge, R. Size relations of lymphomyeloid organs in some cartilaginous fish. *Acta Zoologica* 1977; 58:125–128.

Fänge, R. Fish blood cells. In: Hoar, W.S., Randall, D.J., and Farrell A.P., editors. *Fish Physiology*. San Diego, CA: Academic Press Inc, 1992, Vol. 12B, pp. 1–54.

Fänge, R., and Mattisson, A. The lymphomyeloid (hemapoietic) system of the Atlantic nurse shark (*Ginglymostoma cirratum*). *Biological Bulletin* 1981; 160:240–249.

Flajnik, M.F., and Rumfelt, L.L. Early and natural antibodies in non-mammalian vertebrates. *Current Topics in Microbiology and Immunology* 2000; 252:233–240.

Glomski, C.A., Tamburlin, J., and Chainani, M. The phylogenetic odyssey of the erythrocytes. III. Fish, the lower vertebrate experience. *Histology and Histopathology* 1992; 7:501–528.

Henry, R.P., and Heming, T.A. Carbonic anhydrase and respiratory gas exchange. In: Perry, S.F., and Tufts, B., editors. *Fish Respiration*. San Diego, CA: Academic Press, 1998, Vol. 17, pp. 75–111.

Hine, P.M. The granulocytes of fish. *Fish and Shellfish Immunology* 1992; 2:79–82.

Hine, P.M., and Wain, J.M. Composition and ultrastructure of elasmobranch granulocytes. II. Rays (Rajiformes). *Journal of Fish Biology* 1987; 30:557–565.

Hine, P.M., Wain, J.M., and Boustead, N.C. The leucocyte enzyme cytochemistry of fish. *New Zealand Research Bulletin* 1987; 28:5–75.

Hsu, E., Pulham, N., Rumfelt, L.L., and Flajnik, M.F. The plasticity of immunoglobulin gene systems in evolution. *Immunological Reviews* 2006; 210(1):8–26.

Hyder, S.L., Cayer, M.L., and Pettey, C.L. Cell types in the peripheral blood of the nurse shark: An approach to structure and function. *Tissue and Cell* 1983; 15:437–455.

Jensen, F.B., Fago, A., and Weber, R.E. Hemoglobin structure and function. In: Perry, S.F., and Tufts, B., editors. *Fish Respiration*. San Diego, CA: Academic Press, 1998, Vol. 17. pp. 1–40.

Lewis, J.H. Comparative hemostasis: Studies on elasmobranchs. *Comparative Biochemistry and Physiology Part A: Physiology* 1972; 42:233–236.

Lucas, A.M., and Jamroz, C. Granular leukocytes. In: Lucas, A.M., and Jamroz, C., editors. *Atlas of Avian Hematology*. Lansing, MI: Regional Poultry Research Laboratory, USDA, 1961, pp. 73–87.

Mainwaring, G., and Rowley, A.F. Studies on granulocyte heterogeneity in elasmobranchs. In: Manning, M.J., and Tatner, M.F., editors. *Fish Immunology*. New York: Academic Press, 1985, pp. 57–69.

Mattisson, A., and Fänge, R. The cellular structure of the Leydig Organ in the shark, *Etmopterus spinax*. *Biological Bulletin* 1982; 162:182–194.

Mattisson, A., and Fänge, R. The cellular structure of lymphomyeloid tissues in *Chimera monstrosa* (Pisces, Holocephali). *Biological Bulletin* 1986; 171:660–671.

McKinney, E.C., and Flajnik, M.F. IgM-mediated opsonization and cytotoxicity in the shark. *Journal of Leukocyte Biology* 1997; 61:141–146.

Nikinmaa, M., and Salama, A. Oxygen transport in fish. In: Perry, S.F., and Tufts, B., editors. *Fish Respiration*. San Diego, CA: Academic Press, 1998, Vol. 17, pp. 141–184.

Pacheco, F.J., Pacheco, S.O.S., Segreto, H.R.C., Segreto, R.A., Silva, M.R.R., and Egami, M.I. Development of granulocytes in haematopoietic tissues of *Rhizoprionodon lalandii*. *Journal of Fish Biology* 2002; 61:888–898.

Parish, N., Wrathmell, A., and Harris, J.E. Phagocytic cells in the dogfish (*Scyliorhinus canicula L.*). In: Manning, H.J., and Tatner, M.F., editors. *Fish Immunology*. London: Academic Press, 1985, pp. 71–83.

Pica, A., Lodato, A., Grimaldi, M.C., and Corte, F.D. Morphology, origin and functions of the thrombocytes of elasmobranchs. *Archivio Italiano di Anatomia e Embriologia (Italian Journal of Anatomy and Embryology)* 1990; 95(3–4):187–207.

Rumfelt, L.L., Avila, D., Diaz, M., Bartl, S., McKinney, E.C., and Flajnik, M.F. A shark antibody heavy chain encoded by a nonsomatically rearranged VDJ is preferentially expressed in early development and is convergent with mammalian IgG. *Proceedings of the National Academy of Sciences USA* 2001; 98:1775–1780.

Saunders, D.C. Differential blood cell counts of 121 species of marine fishes of Puerto Rico. *Transactions of the American Microscopical Society* 1966; 85(3):427–449.

Saunders, D.C. Differential blood cell counts of 50 species of fishes from the Red Sea. *Copeia* 1968; 3:491–498.

Stokes, E.E., and Firkin, B.G. Studies of the peripheral blood of the Port Jackson shark (*Heterodontus portusjacksoni*) with particular reference to the thrombocyte. *British Journal of Haematology* 1971; 20:427–435.

Stoskopf, M.K. (editor). Clinical pathology of sharks, skates and rays. In: *Fish Medicine*. Philadelphia, PA: WB Saunders Co., 1993, pp. 754–757.

Van Rijn, A.J., and Reina, R.D. Distribution of leukocytes as indicators of stress in Australian swellshark, *Cephaloscyllium laticeps*. *Fish and Shellfish Immunology* 2010; 29:534–538.

Walsh, C.J., and Luer, C.A. Comparative phagocytic and pinocytic activities of leucocytes from peripheral blood and lymphomyeloid tissues of the nurse shark (*Ginglymostoma cirratum Bonaterre*) and the clearnose skate (*Raja eglanteria Bosc*). *Fish and Shellfish Immunology* 1998; 8:197–215.

Walsh, C.J., and Luer, C.A. Elasmobranch hematology: Identification of cell types and practical applications. In: Smith, M., Warmolts, D., Thoney, D., and Hueter, R. editors. *The Elasmobranch Husbandry Manual: Captive Care of Sharks, Rays and Their Relatives*. Columbus, OH: Ohio Biological Survey, 2004, pp. 307–323.

Zapata, A. Splenic erythropoiesis and thrombopoiesis in elasmobranchs: An ultrastructural study. *Acta Zoologica* 1980; 61:59–64.

Zapata, A., and Amemiya, C.T. Phylogeny of lower vertebrates and their immunological structures. *Current Topics in Microbiology and Immunology* 2000; 248:67–107.

CHAPTER 6

Leukocyte Function in Elasmobranch Species
Phagocytosis, Chemotaxis, and Cytotoxicity

Catherine J. Walsh and Carl A. Luer

CONTENTS

SUMMARY

Phagocytosis, pinocytosis, chemotaxis, and cytotoxicity are important immune-defense functions of leukocytes. Many studies confirm the role of immune cells from elasmobranch fishes participating in these important processes. Taken together, existing studies point to the multiple cell types participating in phagocytosis and pinocytosis, but with granulocytic cells primarily responsible for phagocytic and pinocytic activity in the elasmobranch immune system, although thrombocytes and lymphocytes appear to participate in phagocytosis to some degree. A macrophage-like effector cell appears to be responsible for spontaneous cytotoxicity, as well as cytotoxicity that is elicited with mitogen stimulation or that requires antibody-dependent mechanisms. A lymphoid-like cell appears to be involved in regulating spontaneous cytotoxicity through a mechanism that is temperature dependent. Chemotactic migration and accumulation of leukocytes have been observed in the elasmobranch immune system and appear to occur in response to complement or other chemoattractants, leading to induction of inflammatory processes.

INTRODUCTION

Phagocytosis and chemotaxis are two closely related processes that are important functions of leukocytes. Phagocytosis is the internalization of cells or particles and functions as an essential component of innate immune defense. A primary role of phagocytosis is to participate in physical clearance of foreign material. A similar process, ingestion of nutrients and fluid, is also important in this role and is known as pinocytosis (Silverstein *et al.*, 1977). Phagocytic cells are principally dedicated to recognition and elimination of invading organisms and damaged tissue. Phagocytic processes trigger the cascade of events involved in immune responses to foreign antigen through antigen uptake, presentation, and cytokine release (Hiemstra, 1993). Phagocytosis of foreign material serves both as a defense mechanism in itself and as an initial step in the onset of the specific immune response. In mammals, cells of the immune system respond to microbial invasion and/or tissue injury through inflammation. Chemotaxis is an important mechanism by which immune effector cells localize to inflammation sites and involves directional migration of leukocytes to a site of inflammation along chemical gradients. In higher vertebrates, phagocytic granulocytes initially migrate toward the site of tissue injury in response to locally generated chemostimulants or chemotactic factors. The movement of phagocytic cells to sites of microbial invasion occurs early during the inflammatory response. Chemotaxis and the subsequent phagocytosis of foreign particles are two integral components of the inflammatory response.

Natural killer (NK) cells and cytotoxic cells also function in immune surveillance. NK cells were discovered more than 30 years ago and cytolytic effector cells resembling NK cells were recognized as part of innate immune defense well before the arrival of T and B cells of the adaptive immune system approximately 500 million years ago (MYA) (Cooper and Alder, 2006).

By way of clarification, cell types mentioned in this chapter are referred to in general terms, such as *phagocytic* or *granulocytic* cells. The term *granulocytic cell* can imply either neutrophilic, heterophilic, or eosinophilic cells, as described by Haines and Arnold (2014). In this chapter, except for Figures 6.4 and 6.5, the phagocytic cells are not stained, thus preventing identification of specific granulocytic cell subtypes. The terms *monocyte-macrophage* and *mononuclear phagocyte* are used to identify a mononuclear leukocyte type.

PHAGOCYTOSIS

Cell Types Involved in Phagocytosis

Phagocytosis can be mediated by a number of cell types. In mammals, mononuclear phagocytes, as well as cells belonging to the granulocyte series, possess phagocytic capacity. In most vertebrates, neutrophils or heterophils are highly phagocytic, whereas eosinophils are reported to have little phagocytic activity (Hiemstra, 1993). This also seems to be true with granulocytic cells of elasmobranch species—although eosinophils are capable of phagocytizing and killing ingested microorganisms, this is not considered a primary role of this cell type (Hiemstra, 1993). In fish, cells described as phagocytic include neutrophilic granulocytes, or neutrophils, which are often referred to as heterophils in lower vertebrates, and mononuclear phagocytes such as tissue macrophages and circulating monocytes (Secombes and Fletcher, 1992). Phagocytic cells in elasmobranch species have been described in several studies, with most phagocytic activity attributed to granulocytic cells. Studies reporting *in vitro* and *in vivo* phagocytosis in elasmobranch species are summarized in Tables 6.1 and 6.2.

Several studies report observations of phagocytosis occurring in elasmobranch immune cells *in vitro* in response to various types of challenges. These *in vitro* experiments are summarized in Table 6.1. In the Port Jackson shark, *Heterodontus portusjacksoni*, granulocytes and thrombocytes were observed to ingest yeast particles, with heterophils ingesting greater numbers of yeast cells than eosinophils (Stokes and Firkin, 1971). In the smooth dogfish, *Mustelus canis*, phagocytic cells were described as resembling polymorphonuclear leukocytes (Weissman *et al.*, 1978), which are likely the equivalent of neutrophils, as described by Haines and Arnold (2014). In contrast to most other reports in elasmobranchs, eosinophils were determined to be the primary phagocytic cell type in the small spotted catshark, *Scyliorhinus canicula* (Fänge and Pulsford, 1983). In studies conducted using cells from the nurse shark, *Ginglymostoma cirratum*, heterophils, but not eosinophils,

Table 6.1 *In Vitro* Studies Reporting Phagocytosis/Pinocytosis in Elasmobranch Species

Challenge	Species	Cell Types Active in Process	Reference
Yeast	*Heterodontus portusjacksoni*	Heterophils, thrombocytes	Stokes and Firkin (1971)
Sea urchin eggs	*Mustelus canis*	Polymorphonuclear leukocytes	Weissman *et al.* (1978)
Erythrocytes	*Scyliorhinus canicula*	Eosinophils	Fänge and Pulsford (1983)
Yeast	*Ginglymostoma cirratum*	Heterophils, monocyte/macrophage	Hyder *et al.* (1983)
Colloidal carbon, yeast	*Scyliorhinus canicula*	Monocytes, granulocytes	Parish *et al.* (1985)
Yeast, mammalian erythrocytes, latex, carbon, bacteria, keyhole limpet hemocyanin (KLH), bovine gamma globulin (BGG)	*Scyliorhinus canicula*	Monocytes, heterophil-like granulocytes	Parish *et al.* (1986a)
Congo red-stained yeast; neutral red dye	*Ginglymostoma cirratum*	Granulocytes, thrombocytes, monocytes/macrophages	Walsh and Luer (1998)
Congo red-stained yeast; neutral red dye	*Raja eglanteria*	Granulocytes, thrombocytes, monocytes/macrophages	Walsh and Luer (1998)

Table 6.2 *In Vivo* Studies Reporting Phagocytosis in Elasmobranch Species

Challenge	Species	Measurement	Reference
T_1 bacteriophage (*E. coli*)	*Scyliorhinus canicula*	Clearance rate	Nelstrop *et al.* (1968)
T_2 bacteriophage	*Negaprion brevirostris*	Clearance rate	Sigel *et al.* (1968a)
T_2 bacteriophage	*Ginglymostoma cirratum*	Clearance rate	Russell *et al.* (1976)
Colloidal carbon	*Scyliorhinus canicula*	Particle uptake (monocytes, thrombocytes, granulocytes)	Morrow and Pulsford (1980)
Bacteria, yeast, keyhole limpet hemocyanin (KLH)	*Scyliorhinus canicula*	Particle uptake (monocytes)	Parish *et al.* (1986b)

were observed to be phagocytic (Hyder *et al.*, 1983), with phagocytic activity of neutrophilic granulocytes not described.

Walsh and Luer (1998) described heterophils as the most active phagocytic and pinocytic cells
in the nurse shark as well as in the clearnose skate, *R. eglanteria*. Examples of phagocytosis and
pinocytosis by nurse shark immune cells are shown in Figures 6.1 through 6.3. Phagocytosis was
demonstrated by uptake of yeast particles stained with Congo red. Pinocytosis, the ingestion of
fluid, was demonstrated by uptake of neutral red dye. Cells of the epigonal organ, a lymphomyeloid organ comprised primarily of granulocytic cells (Luer *et al.*, 2004), are shown demonstrating
phagocytosis and pinocytosis in Figure 6.1. Granulocyte cell types are not distinguished from each
other in these assays and are only referred to as *epigonal cells*. Figure 6.1a shows empty epigonal
cells before being exposed to either Congo red-stained yeast or neutral red dye. In Figure 6.1b, nurse
shark epigonal cells can be seen engorged with yeast cells (arrows), with the number of yeast cells
per individual epigonal cell generally greater than five. Pinocytosis of neutral red by a single epigonal cell can be seen in Figure 6.1c. Microscope views using a lower magnification of phagocytic
and pinocytic processes in nurse shark epigonal cells are shown in Figure 6.2. Figure 6.2a and b
shows uptake of Congo red-stained yeast, and Figure 6.2c and d allows visualization of neutral red
dye uptake. The number of cells active in these processes can be seen in these photomicrographs.
Many of the cells in Figure 6.2c and d demonstrate elongation and formation of pseudopodia. In
Figure 6.3, pinocytosis of neutral red dye can be observed at a high magnification (1000X) in nurse
shark epigonal cells (Figure 6.3a) and in nurse shark peripheral blood leukocytes (Figure 6.3b).
Distinct pockets of neutral red dye within the cells are visible. Most cells appear to be granulocytic,
but specific cell types were not identified in these experiments.

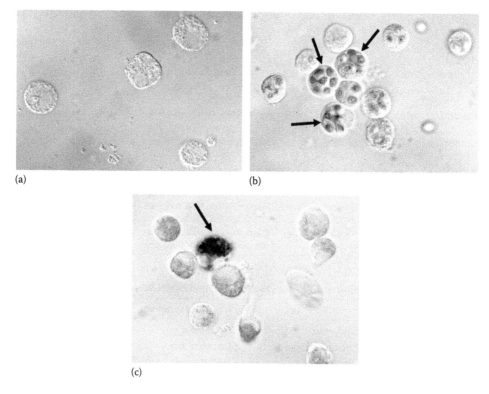

(a)   (b)

(c)

Figure 6.1   Phagocytosis and pinocytosis by nurse shark, *G. cirratum*, epigonal cells viewed with differen-
            tial interference contrast (DIC) optics. (a) Epigonal cells without phagocytosis or pinocytosis;
            (b) phagocytosis of Congo red-stained yeast (indicated by arrows) by epigonal cells; (c) pinocytosis
            of neutral red dye (arrow) by epigonal cells. Magnification: 1000X.

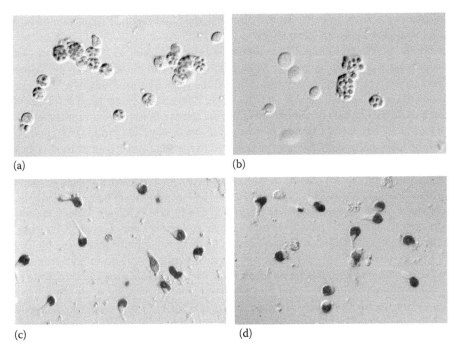

Figure 6.2   Phagocytosis and pinocytosis by nurse shark, *G. cirratum*, epigonal cells viewed with DIC optics. (a) and (b) Phagocytic cells containing engulfed Congo red-stained yeast. (c) and (d) Pinocytic cells containing neutral red dye. Magnification: 400X.

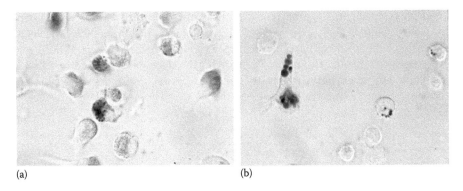

Figure 6.3   Pinocytosis by nurse shark, *G. cirratum*, epigonal cells (a) and peripheral blood leukocytes (b) viewed with DIC optics. Pinocytic cells have ingested neutral red dye. Magnification: 1000X

In addition to heterophils or neutrophils, monocyte-macrophages have been shown to be involved in phagocytic processes in elasmobranchs. Hyder *et al.* (1983), studying phagocytosis in the peripheral blood of the nurse shark, demonstrated *in vitro* phagocytosis of yeast particles by monocyte-macrophage cells. Parish *et al.* (1985) demonstrated monocytes as phagocytic through uptake of particulate material (i.e., colloidal carbon and yeast). In this study, monocytes were less discriminatory than granulocytes, but granulocytes were more avidly phagocytic. Walsh and Luer (1998) observed phagocytic ability in macrophages of the clearnose skate and nurse shark. Monocytes from peripheral blood of *S. canicula* were shown to be phagocytic by both *in vitro* (Parish *et al.*, 1986a) and *in vivo* (Parish *et al.*, 1986b) studies. Walsh and Luer (1998) described greater phagocytic activity in cells isolated from epigonal and Leydig organs compared with the adherent monocyte-macrophage population.

Isolated reports (Parish *et al.*, 1986a, 1986b; Stokes and Firkin, 1971; Walsh and Luer, 1998) indicate a possible role for thrombocytes in both phagocytic and pinocytic activity, although their involvement in these processes may be limited, and the capacity of thrombocytes for intracellular digestion and degradation is not clear (Secombes *et al.*, 1996). Some examples of thrombocyte ingestion of foreign material include accumulation of neutral red dye as well as ingestion of latex beads by thrombocytes in the peripheral blood of *H. portusjacksoni* (Stokes and Firkin, 1971) and occasional engulfment of yeast and accumulation of neutral red dye in shark and skate thrombocytes (Walsh and Luer, 1998). Thrombocyte phagocytosis has also been reported in other fish species (Ellis, 1977; Parish *et al.*, 1986a, 1986b). Phagocytosis may also play a role in removing apoptotic immune cells in the peripheral circulation and lymphomyeloid tissues of elasmobranch fishes (Walsh *et al.*, 2002).

Phagocytosis of microorganisms, including *E. coli* and *Vibrio harveyi*, was demonstrated by cells of the peripheral blood of the nurse shark (Figure 6.4; Haines and Arnold, 2014) using *in vitro* experimentation. Figure 6.4a shows *E. coli* ingested by both heterophilic and eosinophilic granulocytes after 1-h incubation. Figure 6.4b shows *V. harveyi* ingested by an eosinophil from the peripheral blood of a nurse shark after 3.5-h incubation. Arrows in Figure 6.4a and b indicate uptake of microbes into phagosomes.

In vitro experiments have also demonstrated uptake of both fluorescent and nonfluorescent bovine serum albumin-coated latex beads (Figure 6.5a and b) by nurse shark peripheral blood leukocytes (Haines and Arnold, 2014). Figure 6.5a shows uptake of nonfluorescently labeled beads by eosinophilic granulocytes and lymphocytes from nurse shark peripheral blood. Uptake of fluorescently labeled latex beads by eosinophilic granulocytes and lymphocytes is shown in Figure 6.5b.

Recently, lymphocyte phagocytosis was observed in some teleost species and is primarily attributed to B lymphocytes (Li *et al.*, 2006; Sunyer, 2012). Moreover, recent findings have also suggested that lymphocytes from amphibians, reptiles, and mammals are also capable of phagocytosis, and provide support for a current hypothesis that B cells and macrophages may have originated from a common phagocytic ancestor (Kawamoto and Katsura, 2009). An evolutionarily close relationship between B cells and macrophages was indicated by Katsura (2002), implying a common predecessor for both cells in species predating the emergence of jawed vertebrates. These studies represent a paradigm shift from restriction of *professional* phagocytes to cells of myeloid

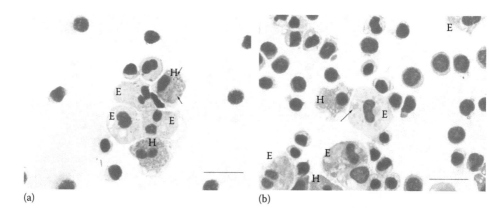

(a)   (b)

Figure 6.4 Phagocytosis of microorganisms by nurse shark, *G. cirratum*, peripheral blood leukocytes. (a) Uptake of formalin-fixed *E. coli* by eosinophilic (E) and heterophilic (H) granulocytes. Incubation time = 1 h; Bar = 20 μm. (b) Uptake of formalin-fixed *V. harveyi* by eosinophilic granulocytes (E) and lymphocytes (L). Incubation time = 3.5 h; Bar = 20 μm. Images were taken using DIC optics. Arrows show bacteria or remnants of bacteria inside phagosomes. E, eosinophil; H, heterophil; and N, neutrophil. (Courtesy of A. Haines and C. Gargan.)

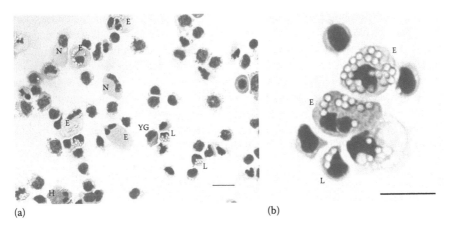

(a)   (b)

Figure 6.5 Phagocytosis of BSA-coated latex beads (1 μm) by nurse shark (*G. cirratum*) peripheral blood leukocytes. (a) Uptake of unstained latex beads by eosinophilic (E) granulocytes and lymphocytes (L). Incubation time = 1 h; Bar = 20 μm. (b) Uptake of yellow–green fluorescent latex beads by eosinophilic (E) granulocytes and lymphocytes (L). Incubation time = 1 h; Bar = 20 μm. Images were taken using DIC optics. E, eosinophil; H, heterophil; N, neutrophil; L, lymphocyte; and YG, young granulocyte. (Courtesy of A. Haines and C. Gargan.)

origin expanded to include cells of lymphocyte origin. It is exciting to see the same trend in elasmobranch species as well.

In vivo antigen clearance experiments with elasmobranchs. Although phagocytosis was readily demonstrated *in vitro, in vivo* clearance studies in elasmobranchs have not demonstrated significant depletion of bacterial numbers (Parish *et al.*, 1985). Ellis *et al.* (1976) also concluded that phagocytic properties of elasmobranch blood were relatively inactive, based on observations of only a few carbon particles per cell. Morrow and Pulsford (1980) determined that circulating phagocytes generally showed little activity in clearing particulate matter *in vivo*. Parish *et al.* (1986b) examined phagocytic properties of the spleen and observed poor clearance from the circulation, leading to the assumption that elasmobranchs had a poorly developed phagocytic system. Russell *et al.* (1976) reported temperature-dependent T_2 bacteriophage clearance in nurse shark that paralleled that of a bony fish, and Sigel *et al.* (1968a) reported T_2 bacteriophage clearance from lemon shark, *Negaprion brevirostris*, with phage particles completely eliminated from circulation by days 4–5.

Complement

Because nurse shark monocyte-macrophages are phagocytic when presented with yeast particles (Hyder *et al.*, 1983), they may also rely on a complement receptor to facilitate phagocytic activity. In addition, shark monocyte-macrophages may possess surface lectins, such as the mannose receptor, which may participate in binding and subsequent ingestion of particles (Sharon, 1984; Smith, 2000). Proteins considered homologues of C1q, C3, and C4 (C2n) of the mammalian complement system have been isolated from nurse shark serum (Smith, 1998). Both classical and alternative pathways of complement activation have been established for the nurse shark, *G. cirratum*. Isolation of a cDNA clone encoding a mannan-binding protein-associated serine protease (MASP)-1-like protein from Japanese dogfish, *Triakis scyllia* suggests the presence of a lectin pathway (For further details on shark complement refer to Smith and Nonaka Chapter 8 in this book).

Opsonization

Opsonins function to mediate recognition between phagocyte and particle. Antibody-mediated target cell death through phagocytosis is an important receptor-mediated immune process. In elasmobranchs, the heterophil is believed to be responsible for antibody-mediated target cell death through phagocytosis (McKinney and Flajnik, 1997). Phagocytic activity, however, can occur without opsonization in elasmobranch species (Walsh and Luer, 1998), indicating involvement of direct cell to particle interaction. As a defense against fungi, vertebrates have developed various recognition mechanisms for β-glucan, the major structural component of yeast cell walls, including the activation of nonspecific defense mechanisms such as the alternative complement pathway (Czop and Kay, 1991). Consequently, phagocytosis of yeast by elasmobranch immune cells may occur through binding to β-glucan receptors and activating the alternative complement pathway as it does in other vertebrate species (Culbreath et al., 1991).

Fcμ Receptor

Receptors for the Fc portion of immunoglobulin molecules have been described on a number of lymphomyeloid cells in mammals, including T and B lymphocytes, monocyte-macrophages, null cells, and granulocytes. On macrophages and granulocytes, these receptors enhance phagocytosis of antibody-coated foreign antigens, whereas on killer cells, they are essential for antibody-dependent cellular cytotoxicity (Roitt et al., 1985). Fc receptors also serve to regulate immune responses. In general, Fc receptors are believed to generate signals for phagocytosis, to aid in binding of immune complexes, to direct cytotoxic potential of receptor-bearing populations, and to possibly function in lymphocyte regulation. Among mammals, Fcγ receptors dominate the functions ascribed to phagocytic cells, and the presence and relative importance of Fcμ receptors on macrophages and neutrophils is controversial in many mammalian species. Throughout evolution, IgM reacts with particulate antigen and binds to Fcμ receptors and thus enhances phagocytosis (McKinney and Flajnik, 1997), although the ability to stimulate phagocytosis after binding of the Fcμ receptor varies widely. The neutrophil appears most frequently associated with IgM-mediated phagocytosis. Fc receptors for shark IgM have been demonstrated on shark leukocytes (Haynes et al., 1988). In elasmobranchs, the shark heterophil is believed to be responsible for antibody-mediated target cell death, with the mechanism of killing phagocytosis (McKinney and Flajnik, 1997). Receptors for the Fcμ region of shark IgM are expressed on nurse shark heterophils (Haynes et al., 1988; McKinney and Flajnik, 1997) and are likely comparable in function to the role of these receptors on the surface of immune cells in higher vertebrates. Shark monocyte-macrophages, however, do not express Fcμ receptors and do not appear to participate in antibody-mediated reactions (McKinney and Flajnik, 1997).

CHEMOTAXIS

In mammals and other higher vertebrates, an integral component of nonspecific innate immunity is the ability to initiate an inflammatory reaction in response to the presence of foreign objects such as microorganisms. Such an inflammatory response involves migration of leukocytes followed by accumulation of leukocytes at sites of tissue injury or infection. Chemotaxis and subsequent phagocytosis are two integral components of the inflammatory response. Chemotactic migration of leukocytes to the site of inflammation is active and directed with the leukocytes moving from the blood to inflamed tissue along chemical gradients originating at the site of inflammation. Chemotaxis is essential for selectively attracting cells of the immune system to the site of inflammation. Migration can occur in response to microbial factors of host origin. Although a wide variety of substances can stimulate chemotactic migration, complement appears to be the most important biological source of chemotactic activity for leukocytes in higher vertebrates. Chemotactic migration was little studied in lower vertebrates. Several studies, summarized in Table 6.3, have demonstrated that elasmobranch leukocytes are capable of chemotactic migration toward a variety of chemoattractants, such as activated serum, complement components, and leukotriene.

Table 6.3 *In Vitro* Studies Measuring Chemotaxis by Elasmobranch Leukocytes

Chemoattractant	Species	Assay	References
LPS-activated rat serum, activated shark serum, shark complement components	*Ginglymostoma cirratum*	Blind well chemotaxis chamber	Obenauf and Smith (1985)
Leukotriene B$_4$	*Scyliorhinus canicula*	Migration under agarose, bipolar shape formation	Hunt and Rowley (1986b)
LPS-activated rat serum	*Ginglymostoma cirratum*	Blind well chemotaxis chamber	Smith *et al.* (1989)
Porcine C5a, LPS-activated guinea pig serum	*Ginglymostoma cirratum*	Blind well chemotaxis chamber	Obenauf and Smith (1992)

In mammals, chemotaxis is generally a property associated with neutrophils and macrophages, which are both phagocytic cell types and important participants in the inflammatory process. In mammals, phagocytic granulocytes initially migrate toward the site of tissue injury in response to locally generated chemostimulants, including complement-derived chemotactic factors such as C5a, the primary serum-derived agent that attracts leukocytes in mammalian systems. In mammalian systems, C5a promotes neutrophil chemotaxis through direct interaction with specific receptors on the cell (Chenoweth and Hugli, 1978). Specific cell receptors exist for other chemoattractants as well (Snyderman and Goetzl, 1981). Binding of various ligands to such surface receptors results in the morphological and cytoskeletal changes in the cell that are associated with directed cell movement.

Role of Complement

Activated complement affects the inflammatory response through release of inflammatory mediators, such as anaphylatoxins C3a, C4a, and C5a, which interact with cellular components through specific receptor–ligand interactions. In mammals, the chemotactic migration seen with activated serum is induced by C5a, a chemotactic peptide fragment released from the fifth complement component (C5) in activated serum (Shin *et al.*, 1968; Ward and Newman, 1969). C5a is a small, cationic peptide that is a potent anaphylatoxin and chemoattractant that attracts a broad range of cells and enhances the ability of neutrophils and monocytes to adhere to vessel walls, migrate, and phagocytize particles (Gerard and Hugli, 1981). In mammals, phagocytic granulocytes initially migrate toward the site of tissue injury in response to a local generation of chemostimulants that include complement-derived chemotactic factors such as C5a. Lipopolysaccharide (LPS)-activated shark serum generates a ligand that chemotactically attracts human leukocytes *in vitro* (Smith *et al.*, 1997).

Shark Leukocyte Migration

Chemotaxis of leukocytes was reported in elasmobranch species. Functional evidence suggests that complement-related anaphylatoxin activity is present in shark serum. The first published study reporting chemotaxis in an elasmobranch species was conducted by Obenauf and Hyder Smith (1985). Using a micropore filter chemotaxis assay, nurse shark leukocytes demonstrated *in vitro* migration in response to different chemoattractants. Active migration of shark leukocytes was observed in response to homologous attractants such as activated shark serum and purified shark complement. A bipolar shape formation (BSF) assay was used to determine chemotactic migration potential of dogfish, *S. canicula*, leukocytes (Hunt and Rowley, 1986a). Migration was also observed in response to heterologous attractants such as LPS-activated rat and guinea pig sera and to porcine C5a desArg (Obenauf and Hyder Smith, 1985, 1992; Hyder Smith *et al.*, 1989). Using checkerboard analysis, which can be used to distinguish between chemotaxis and chemokinesis, shark leukocytes were shown to exhibit both chemokinetic (i.e., random migration) as well as chemotactic (i.e., directional migration) movement in response to activated rat serum (Obenauf and Hyder Smith, 1985). Chemotactic migration was not observed in response to unactivated shark serum, indicating that migration likely occurred in response to complement components. Shark leukocytes migrated in response to purified porcine C5a in levels comparable to activated rat serum (Obenauf and Hyder Smith, 1992). Because shark leukocytes did respond to activated shark serum, complement proteins analogous to C5a likely exist in shark serum (Obenauf and Hyder Smith, 1992). Attraction of heterophils and mononuclear leukocytes to sites of inflammation by C5a, or in the nurse shark, an equivalent to C5a, appears to be a function of the immune response that was conserved over a long period of evolution (Obenauf and Hyder Smith, 1992). Mammalian cells also respond to factors generated in activated shark serum, suggesting that chemotactic and anaphylactic activity in shark serum might be due to biologically active peptide(s) derived from the cleavage of complement component(s) that may be analogous to mammalian C3a and/or C5a (Smith *et al.*, 1997).

From these observations, it is likely that the response of shark cells probably involves some form of receptor present on shark leukocytes that interacts with mammalian complement-derived chemotactic ligand. Chemotaxis of shark leukocytes in response to porcine C5a and LPS-activated mammalian serum, which is a source of C5a, suggests shark leukocytes may possess surface receptors which recognize C5a and that ligand–receptor binding induces subsequent cellular changes associated with migrating cells (Obenauf and Hyder Smith, 1985, 1992; Hyder Smith *et al.*, 1989). For mammalian cells, the chemotactic response to C5a was shown to be mediated by a specific cell surface receptor for C5a (Chenoweth and Hugli, 1978). Taken together, existing observations indicate that activated shark serum produces biologically active fragments with anaphylatoxin-like activity that may be analogous to C5a and that shark cells likely possess surface reactive groups or receptor-like surface molecules that respond to mammalian complement-derived factors. In fact, a full-length cDNA encoding putative shark C5 protein (GcC5) was isolated from shark liver (Graham *et al.*, 2009).

In an *in vitro* system, cells which were previously shown to be phagocytic (Hyder *et al.*, 1983), granulocytes and monocyte-macrophages were found capable of initiating a chemotactic migratory response to serum-derived chemotactic factors in the nurse shark (Smith *et al.*, 1989). Electron micrographs of shark leukocytes migrating through filters, in response to purified porcine C5a desArg and complement-derived chemotactic factors produced in activated mammalian serum, show that in elasmobranchs, macrophages and granulocytes are chemotactic (Smith *et al.*, 1989). The same two cell types were observed to migrate in response to chemotactic factors generated in activated shark serum. Similar to chemotactically responding mammalian leukocytes, shark white blood cells alter their shape and become polarized and oriented with respect to the stimulus gradient when exposed to a chemotactic gradient. Hyder Smith *et al.* (1989) demonstrated cell migration that involves the production of lamellipodial processes from the leading edge of the cell.

Migration in Response to Leukotriene

Leukotrienes function as inflammatory mediators and immunomediators. These molecules are involved in the mammalian inflammatory response and are responsible for the accumulation of leukocytes at the site of inflammation (Simmons *et al.*, 1983). Leukotriene B_4 (LTB_4) is one of the most widely studied leukotrienes and is an inflammatory mediator produced by a number of mammalian cell types with a number of known properties, including chemotactic/chemokinetic action (Ford-Hutchinson *et al.*, 1980; Goetzl and Pikett, 1980; Palmer *et al.*, 1980). Cell migration of eosinophilic granulocytes in response to leukotriene B_4 was described for the lesser spotted dogfish (*S. canicula*) (Hunt and Rowley, 1986b), although higher levels of LTB_4 were required to generate this effect than required by mammalian neutrophils under similar conditions. It is postulated that this may be due to dogfish granulocytes possessing fewer receptors for LTB_4 than their mammalian counterparts. The effect of LTB_4 on dogfish granulocytes was also monitored with the BSF assay, in which cells adopt a bipolar configuration if they are capable of responding to chemotactic/chemokinetic factors. In this assay, LTB_4 induced BSF in both granulocyte types.

Inflammation

Chemotaxis is a fundamental component of inflammation. Based on currently available data, the elasmobranch immune system likely has the potential to mount an inflammatory response. Studies to support this include anaphylatoxin activity in activated shark serum (Hyder Smith, 1998) and evidence for the acute-phase protein, hemopexin (Dooley *et al.*, 2010). Also, Karsten and Rice (2004) have reported C-reactive protein, a biomarker of inflammation, in the Atlantic sharpnose shark, *Rhizoprionodon terraenovae*.

CYTOTOXICITY

Natural Killer Cells and Cytotoxic T Lymphocytes

NK cells are a type of cytotoxic lymphocyte critical to the innate immune system. Abnormal cells, such as those that are virus-infected or tumorigenic, are targeted and killed by NK cells. Cytotoxic activity of NK cells is mediated by lytic molecules stored in secretory lysosomes. NK cells are unique in that they have the ability to recognize stressed cells without antibodies or major histocompatibility complex (MHC) and thus facilitate a more rapid immune response. Initially, both histological and functional definitions of NK cells were limited to a large granular lymphocyte that could kill a target cell *naturally*, that is, spontaneously without priming and unrestricted by target cell MHC expression (Herberman *et al.*, 1975; Kiessling *et al.*, 1975; Lanier and Phillips, 1986), and the name *natural killer* arose from the observation that these cells did not require activation in order to kill cells that lack MHC class I. Although originally associated with the innate immune system, NK cells are now known to function in the adaptive immune response with recent research demonstrating multiple inhibitory and activating receptors on the surface of NK cells called NK receptors (NKR). These activating and inhibitory receptors recognize MHC class I molecules, MHC class I-like molecules, and non-MHC molecules and regulate cytotoxicity of NK cells. Receptor diversity is crucial in allowing NK cells to respond effectively to a variety of different pathogens, and NK receptors evolve extremely rapidly in order to adapt to rapidly changing pathogens.

Another type of cytotoxic cell, the cytotoxic T lymphocyte (CTL), is recognized as part of the adaptive immune response. A major difference between CTL and NK cells is that CTLs are antigen-specific and recognize peptides derived from virus and tumor antigens presented by MHC class I. CTLs kill virally infected cells, preventing them from being the source of more viral pathogen, and are thought to provide some protection against spontaneous malignant tumors. NK cells and cytolytic T cells have complementary roles in target recognition and host defense and utilize similar mechanisms of cytolysis. Specifically, NK cells and CTLs perform complementary roles in immune responses directed against viruses and tumors (Topham and Hewitt, 2009). These observations suggest that each of the two cell types may have evolved from a common ancestral cytolytic effector cell.

B7 and NKp30

B7 ligands are expressed on the cell surface of many cell types, including antigen-presenting cells (APCs), with the B7 family of genes recognized as being essential in regulating the adaptive immune system. The interaction of these ligands with receptor molecules on T cells provides activating and/or inhibitory signals involved in regulating T-cell activation and tolerance (Collins *et al.*, 2005). In humans, one of the B7 family members, B7H6, binds to the activating NK receptor,

NKp30. Upregulation of B7H6 occurs during tumor transformation or cell stress—situations in which these cells are eliminated by NK cells either directly through cytotoxicity or indirectly through secretion of cytokines (Baratin and Vivier, 2010).

Shark Cytotoxic Cells

Primitive vertebrates have well-developed innate host defense systems that appear to rely heavily on nonspecific cytotoxic mechanisms. The shark immune system also has the capacity for antigen nonspecific cellular cytotoxicity similar to that observed with NK and CTLs of higher vertebrates (Clem *et al.*, 1967; Jensen *et al.*, 1981; Pettey and McKinney, 1983). Shark peripheral blood leukocytes have been shown to mediate three types of cytotoxic reactions: immunoglobulin-induced, spontaneous (McKinney *et al.*, 1986; Pettey and McKinney, 1983), and mitogen-induced cytotoxicity (Pettey and McKinney, 1981).

Nurse shark peripheral blood contains two interacting leukocyte populations that can be separated by adherence to glass. The three cytotoxic reactions found in shark blood, antibody mediated, spontaneous, and mitogen-induced, are all mediated by glass-adherent leukocytes. Different effector cells within the adherent population appear to be responsible for different types of cytotoxic activity. A spontaneously cytotoxic cell resembling an activated mammalian macrophage is present in the adherent cell population (Pettey and McKinney, 1983). The cytotoxic activity of macrophages from the nurse shark was reported to resemble the activity of mammalian tumoricidal macrophages (McKinney, 1990) in their ability to display selective recognition of target cells.

Antibody-Dependent Cell-Mediated Cytotoxicity

Plasma from unimmunized nurse sharks can mediate a reaction similar to antibody-dependent cell-mediated cytotoxicity (ADCC; Pettey and McKinney, 1988) observed in higher vertebrates. Normal shark plasma contains large quantities of IgM—*natural* antibodies that react with a variety of antigens to which the animal presumably has not been exposed (Sigel *et al.*, 1968b, 1970). In the presence of shark complement or appropriate effector cells, these IgM antibodies have been shown to function in lytic, opsonic, and ADCC-like reactions (Jensen *et al.*, 1981; Pettey and McKinney, 1988).

In addition to natural antibodies in the plasma, the shark immune system also possesses leukocytes with the capability of mediating ADCC-like reactions. In contrast to other systems where NK and ADCC reactions are carried out by similar cell populations (Fleischer, 1980; Pape *et al.*, 1979), separate leukocyte populations are responsible for NK and ADCC reactions in sharks. In the shark immune system, antibody-induced cytotoxicity appears to be mediated by nonphagocytic adherent cells, whereas spontaneous effector cells are present in a small subset of phagocytic adherent macrophages (McKinney *et al.*, 1986). These leukocytes, in concert with those mediating spontaneous cytotoxicity, likely provide the shark with an effective immunosurveillance system. Although little is known about evolution of cells with NK or ADCC function, these observations support the hypothesis that ADCC-like mechanisms, with IgM as the primary effector molecule, appeared early in evolution (Pettey and McKinney, 1988) and are carried out by separate, distinct cell populations.

Shark leukocytes are able to mediate an ADCC-like reaction in the presence of either shark plasma or purified 19S IgM. In sharks, ADCC-like reactions are mediated by 19S IgM antibodies (Pettey and McKinney, 1988), in contrast to IgG that participates in these reactions in mammalian systems. Also in contrast to NK and ADCC effectors in homeothermic vertebrates, shark spontaneous cytotoxicity and plasma-induced cytotoxicity are mediated by separate populations of effector cells. Adherent effector populations contain Fcμ receptor-bearing leukocytes, which could serve as effectors of an ADCC-like reaction (Pettey and McKinney, 1988). These observations suggest that ADCC mechanisms, with IgM as the primary effector molecule, and spontaneous cytotoxicity carried out by separate, distinct cell populations appeared early in vertebrate evolution (Pettey and McKinney, 1988).

Spontaneous Cytotoxicity

In mammalian systems, spontaneous extracellular cytotoxicity is primarily a function of NK cells, with other types of cytotoxicity requiring stimulation either by antigen for immunologically specific responses or by lymphokines for nonspecific macrophage activation. The effector cell responsible for spontaneous cytotoxicity in shark peripheral blood was shown to be a macrophage-like cell (Hyder *et al.*, 1983; McKinney *et al.*, 1986), with cellular properties including adherence to glass and phagocytic activity. Macrophage-mediated cytotoxic activity in the shark resembles that of mammalian tumoricidal macrophages in reaction time, requirement for cell contact, dose-dependent inhibition of target lysis by homologous targets, and no requirement for antibody (McKinney *et al.*, 1986; Meltzer *et al.*, 1982). More importantly, cytotoxic macrophages from the nurse shark differentially recognize targets similar to mammalian tumoricidal macrophages (McKinney *et al.*, 1986). Similar to mammalian tumoricidal macrophages, shark macrophages display selective recognition of target cells. Although nurse sharks have natural antibody to trinitrophenol (TNP), spontaneously cytotoxic cells are incapable of killing TNP-modified targets, which indicates that natural antibody is not required for reactivity and also that natural antibody and spontaneous effector cells do not have the same repertoire (McKinney *et al.*, 1986). In studies using hapten-modification of target membranes, it was demonstrated that shark macrophages interact with amino-containing groups on target cell membranes, indicating that, similar to studies in mammals, shark target cell killing is selective and relies on interaction with specific membrane ligands to initiate lytic events (McKinney, 1990). In contrast to mammalian systems, however, spontaneous cytotoxic cells from the shark do not appear to require *in vitro* activation (McKinney, 1990; McKinney *et al.*, 1986; Pettey and McKinney, 1983). These findings suggest that the shark macrophage-like cell may be an evolutionary precursor giving rise to NK cells and perhaps to other cytotoxic effector cells as well.

A regulatory population that controls activity of cytotoxic effector cells also appears to be present in the shark immune system. Although the adherent cell population contains a spontaneously cytotoxic cell that resembles an activated mammalian macrophage, the nonadherent population contains a temperature sensitive regulatory cell that has a suppressive effect on cytotoxic activity (Pettey and McKinney, 1983). An unusual feature of spontaneous activity of shark peripheral blood leukocytes is that it is observed only during periods when environmental temperature falls below 23°C (McKinney *et al.*, 1986; Pettey and McKinney, 1983); when temperature rises above 26°C, the regulatory, suppressor cells downregulate cytotoxic activity (Pettey and McKinney, 1983). This temperature-dependent regulation inhibits only spontaneous killing and does not affect either mitogen-induced or ADCC-like response (Pettey and McKinney, 1988). The shark suppressor cell population may be unrestricted with regard to MHC antigens, similar to observations with regard to regulation of NK cells in mammalian species (Haynes and McKinney, 1991). Because approximately one-third of the circulating leukocytes are in mitosis (Hyder *et al.*, 1983), a natural suppressor cell may exist in the peripheral blood of these primitive vertebrates to regulate blood cell proliferation and differentiation, rather than as a primary regulator of cytotoxic macrophages. Morphology of the regulatory population in nurse shark indicates that either lymphocyte-like cells or neutrophils are responsible for downregulation of spontaneous cytotoxicity, with most evidence indicating lymphoid regulation.

Taken together, these data provide evidence that by the time of emergence of the nurse shark, a temperature-dependent mechanism had evolved for cellular regulation of at least one immune function, spontaneous cytotoxicity (Pettey and McKinney, 1983). In the shark system, monocyte-macrophages may serve as a major host defense function during periods of decreased environmental temperature when immunologically specific immune responses are impaired (Avtalion *et al.*, 1973).

Host defense mechanisms of sharks appear to rely on antigen nonspecific cellular effector systems, and it was hypothesized that macrophage-mediated cytotoxicity plays a dominant role in host protection during periods of decreased environmental temperatures when lymphocyte responses of poikilothermic vertebrates are compromised.

Mitogen-Induced Cytotoxic Activity

Nurse shark peripheral blood leukocytes also demonstrate cytotoxic activity that can be induced by mitogens such as PHA, ConA, and LPS (Pettey and McKinney, 1981). Mitogen-induced shark leukocytes were demonstrated to have cytotoxic activity toward xenogeneic marine erythrocyte targets (gray snapper) (Pettey and McKinney, 1981). When separated by adherence to glass, differential cytotoxic effects were observed, with only the adherent cell population showing significant cytotoxicity toward target cells in the presence of mitogen. Mitogen-stimulated shark leukocytes were also able to distinguish autologous or allogeneic targets from xenogeneic targets. The mitogen-induced cytotoxicity observed in the nurse shark resembles that of mammalian species, suggesting the shark leukocyte population may contain phylogenetic precursors to the cytotoxic effector cells in immune systems of higher vertebrates (Pettey and McKinney, 1981).

NK Receptors in Elasmobranchs

Comparative analysis of NK receptors among vertebrates has demonstrated remarkable plasticity in these receptors. Although some receptor families have experienced significant expansion, other receptor families have undergone species-specific losses. NK receptors evolve rapidly, which makes it challenging to find orthologues among divergent taxa (Yoder and Litman, 2011), and rapid expansion and contraction of NK receptor families in a species-specific manner is common, particularly with activating receptors (Yoder and Litman, 2011). One of the NK receptors, NKp30, is considered the most conserved and oldest NK receptor in jawed vertebrates, but was initially only reported in mammals (Kaifu et al., 2011). Flajnik et al. (2012), however, reported the presence of NKp30, as well as B7H6, in both *Xenopus* and nurse shark, but not in teleost species. Identification of both B7H6 and NKp30 genes in a cartilaginous fish species indicates that this system is evolutionarily very old, and in fact, the most ancient among the NK receptor families (Flajnik et al., 2012). The selection pressure on NKp30 may be exceptional and quite different from most NK receptors, possibly as a result of coevolution with the self-ligand, B7H6 (Flajnik et al., 2012). In fact, in all species examined so far, from sharks to mammals, the correlation between the number of NKp30 and AB7H6 genes is consistent with receptor–ligand coevolution (Flajnik et al., 2012).

Results strongly suggest that immune regulation associated with most B7 family members arose with the emergence of the Ig/TCR/MHC-based adaptive immune system, because B7 ligands are expressed on the cell surface of many different cell types including APCs. Their interaction with receptor molecules on T cells provides activating and/or inhibitory signals that regulate T-cell activation and tolerance (Collins et al., 2005). In addition, B7H6 is upregulated on cells under tumor transformation or stress conditions, and NK cells eliminate such cells either directly through cytotoxicity or indirectly by cytokine secretion (Baratin and Vivier, 2010).

REFERENCES

Avtalion, R.R., Wojdani, A., Malik, Z., Shahrabani, R., and Duczyminer, M. Influence of environmental temperature on the immune response in fish. *Current Topics in Microbiology and Immunology* 1973; 61:1–35.
Baratin, M., and Vivier, E. B7-H6: A novel alert signal for NK cells. *Médecine Sciences (Paris)* 2010; 26(2):119–120.

Chenoweth, D.E., and Hugli, T.E. Demonstration of specific C5a receptor on intact human polymorphonuclear leukocytes. *Proceedings of the National Academy of Sciences USA* 1978; 75(8):3943–3947.

Clem, L.W., DeBoutand, F., and Sigel, M.M. Phylogeny of immunoglobulin structure and function. II. Immunoglobulins in the nurse shark. *Journal of Immunology* 1967; 99:1226–1235.

Collins, M., Ling, V., and Carreno, B.M. The B7 family of immune regulatory ligands. *Genome Biology* 2005; 6:223.

Cooper, M.D., and Alder, M.N. The evolution of the adaptive immune systems. *Cell* 2006; 124(4):815–822.

Culbreath, L., Smith, S.L., and Obenauf, S.D. Alternative complement pathway activity in nurse shark serum. *American Zoologist* 1991; 31:131A.

Czop, J.K., and Kay, J. Isolation and characterization of β-glucan receptors on human mononuclear phagocytes. *Journal of Experimental Medicine* 1991; 173:1511–1520.

Dooley, H., Buckinham, E.B., Criscitiello, M.F., and Flajnik, M.F. Emergence of the acute-phase protein hemopexin in jawed vertebrates. *Molecular Immunology* 2010; 48(1–3):147–152.

Ellis, A.E. The leukocytes of fish: A review. *Journal of Fish Biology* 1977; 11:453–491.

Ellis, A.E., Monroe, L.S., and Roberts, R.J. Defence mechanisms in fish. A study of the phagocytic system and the fate of intraperitoneally injected particulate material in plaice (*Pleuronectes platessa* L.). *Journal of Fish Biology* 1976; 8:67–78.

Fänge, R., and Pulsford, A. Structural studies on lymphomyeloid tissues of the dogfish, *Scyliorhinus canicula* L. *Cell and Tissue Research* 1983; 230(2):337–351.

Flajnik, M.F., Tlapakova, T., Criscitiello, M.F., Krylov, V., and Ohta, Y. Evolution of the B7 family: Co-evolution of B7H6 and NKp30, identification of a new B7 family member, B7H7, and of B7's historical relationship with the MHC. *Immunogenetics* 2012; 64(8):571–590.

Fleischer, B. Effector cells in avian spontaneous and antibody-dependent cell-mediated cytotoxicity. *Journal of Immunology* 1980; 125(3):1161–1168.

Ford-Hutchinson, A.W., Bray, M.A., Doig, M.V., Shipley, M.E., and Smith, M.J.H. Leukotriene B, a potent chemokinetic and aggregating substance released from polymorphonuclear leukocytes. *Nature* 1980; 286:264–265.

Gerard, C., and Hugli, T.E. C5a: A mediator of chemotaxis and cellular release reactions. *Kroc Foundation Series* 1981; 14:147–160.

Goetzl, E.J., and Pickett, W.C. The human PMN leukocyte chemotactic activity of complex hydroxyl-eicosatetraenoic acids (HETEs). *Journal of Immunology* 1980; 125(4):1789–1791.

Graham, M., Shin, D.H., and Smith, S.L. Molecular expression analysis of complement component C5 in the nurse shark (*Ginglymostoma cirratum*) and its predicted functional role. *Fish and Shellfish Immunology* 2009; 27(1):40–49.

Haines, A.N., and Arnold, J. Elasmobranch blood cells. Ch. 5. In: *Immunobiology of the Shark*, Smith, S.L., Sim, R.B., and Flajnik, M.F. (eds.), CRC Press, Boca Raton, FL, 2014.

Haynes, L., Fuller, L., and McKinney, E.C. Fc receptor for shark immunoglobulin. *Developmental and Comparative Immunology* 1988; 12:561–571.

Haynes, L., and McKinney, E.C. Shark spontaneous cytotoxicity: Characterization of the regulatory cell. *Developmental and Comparative Immunology* 1991; 15(3):123–134.

Herberman, R.B., Nunn, M.E., and Lavrin, D.H. Natural cytotoxic reactivity of mouse lymphoid cells against syngeneic and allogeneic tumors. I. Distribution of reactivity and specificity. *International Journal of Cancer* 1975; 16(2):216–229.

Hiemstra, P.S. Role of neutrophils and mononuclear phagocytes in host defense and inflammation. *Journal of the International Federation of Clinical Chemistry* 1993; 5:94–99.

Hunt, T.C., and Rowley, A.F. Preliminary studies on the chemotactic potential of dogfish (*Scyliorhinus canicula*) leucocytes using the bipolar shape formation assay. *Veterinary Immunology and Immunopathology* 1986a; 12(1–4):75–82.

Hunt, T.C., and Rowley, A.F. Leukotriene B$_4$ induces enhanced migration of fish leucocytes *in vitro*. *Immunology* 1986b; 59:563–568.

Hyder, S.L., Cayer, M.L., and Pettey, C.L. Cell types in peripheral blood of the nurse shark: An approach to structure and function. *Tissue and Cell* 1983; 15:437–455.

Hyder Smith, S., Obenauf, S.D., and Smith, D.S. Fine structure of shark leucocytes during chemotactic migration. *Tissue and Cell* 1989; 21(1):47–58.

Jensen, J.A., Festa, E., Smith, D.L., and Cayer, M. The complement system of the nurse shark: Hemolytic and comparative characteristics. *Science* 1981; 214:5566–5569.

Kaifu, T., Escaliere, B., Gastinel, N.L., Vivier, E., and Baratin, M. B7-H6/NKp30 interaction: A mechanism of alerting NK cells against tumors. *Cellular and Molecular Life Sciences* 2011; 68:3531–3539.

Karsten, A.H., and Rice, C.D. C-Reactive protein levels as a biomarker of inflammation and stress in the Atlantic sharpnose shark (*Rhizoprionodon terraenovae*) from three southeastern USA estuaries. *Marine Environmental Research* 2004; 58:747–751.

Katsura, Y. Redefinition of lymphoid progenitors. *Nature Reviews Immunology* 2002; 2(2):127–132.

Kawamoto, H., and Katsura, Y. A new paradigm for hematopoietic cell lineages: Revision of the classical concept of the myeloid-lymphoid dichotomy. *Trends in Immunology* 2009; 30(5):193–200.

Kiessling, R., Klein, E., Pross, H., and Wigzell, H. "Natural" killer cells in the mouse. II. Cytotoxic cells with specificity for mouse Moloney leukemia cells. Characteristics of the killer cell. *European Journal of Immunology* 1975; 5(2):117–121.

Lanier, L.L., and Phillips, J.H. Evidence for three types of human cytotoxic lymphocyte. *Immunology Today* 1986; 7:132–134.

Li, J., Barreda, D.R., Zhang, Y.A., Boshra, H., Gelman, A.E., Lapatra, O., Tort, L., and Sunyer, J.O. B lymphocytes from early vertebrates have potent phagocytic and microbicidal abilities. *Nature Immunology* 2006; 7(10):1116–1124.

Luer, C.A., Walsh, C.J., and Bodine, A.B. The immune system of Sharks, Skates, and Rays. In: *Biology of Sharks and Their Relatives*, Carrier, J.C., Musick, J.A., and Heithaus, M.R. (eds.), CRC Press, Boca Raton, FL, 2004, pp. 369–395.

McKinney, E.C. Shark cytotoxic macrophages interact with target membrane amino groups. *Cellular Immunology* 1990; 127(2):506–513.

McKinney, E.C., and Flajnik, M.F. IgM-mediated opsonization and cytotoxicity in the shark. *Journal of Leukocyte Biology* 1997; 61:141–146.

McKinney, E.C., Haynes, L., and Droese, A.L. Macrophage-like effector of spontaneous cytotoxicity from the shark. *Developmental and Comparative Immunology* 1986; 10(4):497–508.

Meltzer, M.S., Occhionero, M., and Ruco, L.P. Macrophage activation for tumor cytotoxicity: Regulatory mechanisms for induction and control of cytotoxic activity. *Federation Proceedings* 1982; 41:2198.

Morrow, W.J.W., and Pulsford, A. Identification of peripheral blood leucocytes of the dogfish (*Scyliorhinus canicula* L.) by electron microscopy. *Journal of Fish Physiology* 1980; 17:461–475.

Nelstrop, A.E., Taylor, G., and Collard, P. Antigen clearance studies in invertebrates and poikilothermic vertebrates. *Immunology* 1968; 14:347–356.

Obenauf, S.D., and Hyder Smith, S. Chemotaxis of nurse shark leukocytes. *Developmental and Comparative Immunology* 1985; 9(2):221–230.

Obenauf, S.D., and Hyder Smith, S. Migratory response of nurse shark leucocytes to activated mammalian sera and porcine C5a. *Fish and Shellfish Immunology* 1992; 2:173–181.

Palmer, R.M.J., Stepney, R.J., Higgs, G.A., and Eakins, K.E. Chemokinetic activity of arachidonic acid lipoxygenase products on leucocytes of different species. *Prostaglandins* 1980; 29(2):411–418.

Pape, G.R., Troye, M., and Perlmann, P. In: *Natural and Induced Cell-Mediated Cytotoxicity*, Riethmuller, G., Wernet, P., and Cudkowicz, G. (eds.), Academic Press, New York, 1979, p. 37.

Parish, N., Wrathmell, A., and Harris, J.E. Phagocytic cells in the dogfish (*Scyliorhinus canicula* L.). In: *Fish Immunology*, Manning, M.J., and Tatner, M.F. (eds.), Academic Press, New York, 1985, pp. 71–83.

Parish, N., Wrathmell, A., Hart, S., and Harris, J.E. Phagocytic cells in the dogfish, *Scyliorhinus canicula* L. I. *In vitro* studies. *Acta Zoologica (Stockholm)* 1986a; 67:215–224.

Parish, N., Wrathmell, A., Hart, S., and Harris, J.E. Phagocytic cells in the peripheral blood of the dogfish, *Scyliorhinus canicula* L. II. *in vivo* studies. *Acta Zoologica (Stockholm)* 1986b; 67:225–234.

Pettey, C.L., and McKinney, E.C. Mitogen induced cytotoxicity in the nurse shark. *Developmental and Comparative Immunology* 1981; 5(1):53–64.

Pettey, C.L., and McKinney, E.C. Temperature and cellular regulation of spontaneous cytotoxicity in the shark. *European Journal of Immunology* 1983; 13(2):133–138.

Pettey, C.L., and McKinney, E.C. Induction of cell-mediated cytotoxicity by shark 19S IgM. *Cellular Immunology* 1988; 111(1):28–38.

Roitt, I.M., Brostoff, J., and Male, D.K. *Immunology*. Gower Medical Publishers, New York, 1985.

Russell, W.J., Taylor, S.A., and Sigel, M.M. Clearance of bacteriophage in poikilothermic vertebrates and the effect on temperature. *Journal of the Reticuloendothelial Society* 1976; 19(2):91–96.

Secombes, C.J., and Fletcher, T.C. The role of phagocytes in the protective mechanisms of fish. *Annual Review of Fish Diseases* 1992; 2:53–71.

Secombes, C.J., Hardie, L.J., and Daniels, G. Cytokines in fish: An update. *Fish and Shellfish Immunology* 1996; 6(4):291–304.

Sharon, N. Carbohydrates as recognition determinants in phagocytosis and in lectin-mediated killing of target cells. *Biology of the Cell* 1984; 51(2):239–245.

Shin, H.S., Snyderman, R., Friedman, E., Mellorss, A., and Mayer, M.M. Chemotactic and anaphylatoxic fragment cleaved from the fifth component of guinea pig complement. *Science* 1968; 162(3851):361–363.

Sigel, M.M., Acton, R.T., Evans, E.E., Russell, W.J., Wells, T.G., Painter, B., and Lucas, A.H. T2 bacteriophage clearance in the lemon shark. *Proceedings of the Society for Experimental Biology and Medicine* 1968a; 128:977.

Sigel, M.M., Russell, W.J., Jensen, J.A., and Beasley, A.R. Natural immunity in marine fishes. *Bulletin de l'Office International des Epizooties* 1968b; 69:1349–1351.

Sigel, M.M., Voss, E.W., Rudikoff, S., Lichter, W., and Jensen, J.A. Natural antibodies in primitive vertebrates. The sharks. In: *Miami Winter Symposium 2*, Whelan, W., and Schultz, J. (eds.), North Holland Publishing Company, Amsterdam, the Netherlands, 1970, p. 409.

Silverstein, S.C., Steinman, R.M., and Cohn, Z.A. Endocytosis. *Annual Review of Biochemistry* 1977; 46:669–722.

Simmons, P.M., Salmon, J.A., and Moncada, S. The release of leukotriene B_4 during experimental inflammation. *Biochemical Pharmacology* 1983; 32(8):1353–1359.

Smith, S.L. Shark complement: An assessment. *Immunological Reviews* 1998; 166:67–78.

Smith, S.L. Cellular and humoral aspects of innate immunity in the shark. *Fish and Shellfish Immunology* 2000; 10(3):287.

Smith, S.L., Riesgo, M., Obenauf, S.D., and Woody, C.J. Anaphylactic and chemotactic response of mammalian cells to zymosan activated shark serum. *Fish and Shellfish Immunology* 1997; 7:503–514.

Snyderman, R., and Goetzl, E.J. Molecular and cellular mechanisms of leukocyte chemotaxis. *Science* 1981; 213(4510):830–837.

Stokes, E.E., and Firkin, B.G. Studies of the peripheral blood of the Port Jackson shark (*Heterodontus portusjacksoni*) with particular reference to the thrombocytes. *British Journal of Haematology* 1971; 20:427–435.

Sunyer, J.O. Evolutionary and functional relationships of B cells from fish and mammals: Insights into their novel roles in phagocytosis and presentation of particulate antigen. *Infectious Disorders Drug Targets* 2012; 12(3):200–212.

Topham, N.J., and Hewitt, E.W. Natural killer cell cytotoxicity: How do they pull the trigger? *Immunology* 2009; 128(1):7–15.

Walsh, C.J., and Luer, C.A. Comparative phagocytic and pinocytic activities of leucocytes from peripheral blood and lymphomyeloid tissues of the nurse shark (*Ginglymostoma cirratum* Bonaterre) and the clearnose skate (*Raja eglanteria* Bosc). *Fish and Shellfish Immunology* 1998; 8:197–215.

Walsh, C.J., Wyffels, J.T., Bodine, A.C., and Luer, C.A. Dexamethasone-induced apoptosis in immune cells from peripheral circulation and lymphomyeloid tissues of juvenile clearnose skates, *Raja eglanteria*. *Developmental and Comparative Immunology* 2002; 26:623–633.

Ward, P.A., and Newman, L.J. A neutrophil chemotactic factor from human C'5. *Journal of Immunology* 1969; 102(1):93–99.

Weissman, G., Finkelstein, M.C., Csernanasky, J., Quigley, J.P., Quinn, R.S., Techner, L., Troll, W., and Dunham, P.B. Attack of sea urchin eggs by dogfish phagocytes: Model of phagocyte-mediated cellular toxicity. *Proceedings of the National Academy of Sciences USA* 1978; 75:1825–1829.

Yoder, J.A., and Litman, G.W. The phylogenetic origins of natural killer receptors and recognition: Relationships, possibilities, and realities. *Immunogenetics* 2011; 63:123–141.

CHAPTER 7

Cytokines of Cartilaginous Fish

C.J. Secombes, J. Zou, and S. Bird

CONTENTS

SUMMARY

The initiation and regulation of the Chondrichthyan immune system will require intercellular communication, with secretion of signaling molecules that will bind to specific receptors on target cells, to elicit a downstream intracellular cascade and response. In mammals, an important group of such signaling molecules are the cytokines, which are generally small proteins (or glycoproteins), that are secreted by different leucocyte types, and belong to a number of cytokine families, including the interleukins, interferons, tumor necrosis factors, transforming growth factors, chemokines, and colony-stimulating factors. Although it is still early days for the determination of cytokine function in Chondrichthyans, significant progress was made in cytokine gene discovery, and this chapter summarizes the work that was carried out to date on characterizing these molecules, mainly discovered by exploiting expressed sequence tag (EST) sequences from selected cartilaginous fish species and by searching the elephant shark, *Callorhinchus milii*, and little skate, *Leucoraja erinacea*, genome databases.

INTRODUCTION

Over the last few decades, there was an increasing amount of research undertaken on cartilaginous fish (Chondrichthyes), to elucidate their immune system components and function. This has shown that they possess a complex immune system typical of all jawed vertebrates, consisting innate and adaptive compartments, with the latter effected by lymphocytes (T cells and B cells) with antigen-specific receptors (see Chapters 10 through 12). Jawless vertebrates (lampreys and hagfish) also possess a form of adaptive immunity, but it is fundamentally different in terms of the antigen receptors used and secreted (Herrin and Cooper, 2010; McCurley *et al.*, 2012). Hence, modern day Chondrichthyan fish are derived from an ancestral Gnathostome that had already developed classical adaptive immunity prior to their divergence. Although they are likely to have developed many remarkable and unique features since their first appearance hundreds of millions of years ago, they may still shed light on the quintessential features necessary for optimal functioning of such complex immune systems, which are clearly highly effective and have allowed cartilaginous fish to stay ahead of the biological arms race.

Cartilaginous fish appear to be very resistant to infections, cancers, and circulatory diseases, and it was reported that they can heal rapidly and recover from severe injuries (Ballantyne, 1997; Senior, 2000). This fact suggests that they have well-coordinated and orchestrated immune responses, the ability to rapidly detect infection and mobilize antimicrobial defenses, and the capability of undertaking effective immune surveillance of self for tumor-associated molecules. Although some of the effector molecules involved in such responses are beginning to be elucidated, we still know very little about the molecules involved in the immune system coordination in this group of vertebrates. The initiation and regulation of the Chondrichthyan immune system will require intercellular communication, with secretion of signaling molecules that will bind to specific receptors on target cells, to elicit a downstream intracellular cascade and response. In mammals, an important group of such molecules are the cytokines, which are generally small proteins (or glycoproteins), that are secreted by different leucocyte types, and belong to a number of cytokine families, including the interleukins, interferons, tumor necrosis factors, transforming growth factors, chemokines, and colony-stimulating factors. They play a very important role in nearly all aspects of inflammation and immunity and activate the appropriate innate and adaptive responses to protect an infected or injured organism (Thomas and Lotze, 2003).

Early work on cartilaginous fish suggested that cytokines are present, but these conclusions were based on observations of unpurified molecules in cross-species experiments (Grogan and Lund, 1991) or cross-reactivity of antibodies against human cytokines (Cateni *et al.*, 2003). However, in the absence of cloned genes or sequenced proteins, it was not clear if the molecules being detected were homologous to known cytokines, or were unique, or were not cytokines at all. Recently, a large number of cytokine homologues were discovered within bony fish (Secombes *et al.*, 2011) and that suggest they are necessary for the coordination of complex immune systems in vertebrates and hence are indeed likely to exist in cartilaginous fish. The bottleneck to cytokine discovery in cartilaginous fish was the difficulty in isolating cytokine gene homologues, due to the low homology they typically share with molecules in other vertebrates groups (O'Connell and McInerney, 2005). However, with recent advances in molecular techniques, the sequencing of genomes and transcriptomes is now faster and more affordable, and this problem is beginning to be overcome. Although it is still early days for cytokine gene discovery in Chondrichthyans, significant progress was already made, and this chapter summarizes the work that was carried out to date on characterizing these molecules, mainly discovered by exploiting expressed sequence tag (EST) sequences that were deposited in GenBank from selected cartilaginous fish species (Tan *et al.*, 2012) and by searching the elephant shark, *C. milii*, and little skate, *L. erinacea*, whole genome shotgun (WGS) nucleotide databases (Venkatesh *et al.*, 2007, 2014).

INTERLEUKINS

The term *interleukin* (IL) is used to describe a group of cytokines produced not only by cells of the immune system, but by a wide variety of cell types. They exhibit complex immunomodulatory functions that affect cell proliferation, maturation, migration, and adhesion, pro- and anti-inflammatory effects, and immune cell differentiation and activation (Brocker *et al.*, 2010). This group of cytokines is one of the largest, with currently 37 characterized in humans (Banchereau *et al.*, 2012), although this number may continue to rise in the future with new discoveries. The control of the immune system is highly dependent on the interleukins, and where rare deficiencies in them were found, it leads to autoimmune diseases or immune deficiencies. The molecules are grouped into a number of families, including the IL-1 family, the IL-2 family, the IL-6 family, the IL-10 family, and the IL-17 family. In addition, IL-8 is a member of the chemokine family of cytokines, which is discussed in a later section, and IL-14 and IL-16 are somewhat anomalous and not typical cytokines.

IL-1 Family

In mammals, the IL-1 family (IL-1F) of proteins contains 11 family members (Boraschi *et al.*, 2011) and includes IL-1α (IL-1F1), IL-1β (IL-1F2), IL-1Ra (IL-1F3), IL-18 (IL-1F4), IL-36Ra (IL-1F5), IL-36α (IL-1F6), IL-37 (IL-1F7), IL-36β (IL-1F8), IL-36γ (IL-1F9), IL-1Hy2 (IL-1F10), and IL-33 (IL-1F11). This family of proteins plays a key role in mediating the activation of innate immunity, which is the first line of defense against pathogenic microorganisms and physical damage/stress. Studies in bony fish have discovered three IL-1F to date, that is, IL-1β, IL-18, and a novel member nIL-1F, which acts as an IL-1β antagonist (Husain *et al.*, 2012; Wang *et al.*, 2009; Zou *et al.*, 2004). IL-1β was the first cytokine to be discovered in cartilaginous fish where it was reported in the lesser spotted catshark, *Scyliorhinus canicula* (Bird *et al.*, 2002a), and the banded dogfish, *Triakis scyllium* (Inoue *et al.*, 2003b). The discovery of this gene in *S. canicula* was by *homology cloning*, an approach that in general has had limited success. Nevertheless, in this case, relatively high transcript levels combined with relatively higher homology of some cytokines involved in innate immunity allowed this method to be successful.

Although the overall amino acid (aa) identity of this gene was low (29% to human IL-1β and ~32% to trout IL-1β), a number of important features of the *S. canicula* IL-1β sequence were found to be conserved when compared with other known vertebrate IL-1β homologues. Looking at features such as secondary structure and the presence of the IL-1 family signature provided good evidence as to the identity of this molecule. Members of the IL-1 family have a β-trefoil structure (Nicola, 1994) where the secondary structure consists of 12 β-sheets and dictates how the molecule will fold, giving the molecule its tertiary structure. This is extremely important if it is to interact correctly with its receptor and induce a signal in a target cell. Interestingly, on alignment with selected teleost fish, bird, and mammalian sequences, *S. canicula* aa homology was found to be the highest in regions important for the secondary structure of the human molecule, suggesting the *S. canicula* protein will also form a β-trefoil (Bird *et al.*, 2002a). Another critical conserved feature present in the *S. canicula* IL-1β sequence was the IL-1 family signature pattern or motif that is located in a selected conserved region in the C-terminal section of IL-1β sequences, as found in PROSITE database (Hofmann *et al.*, 1999). No apparent signal peptide could be determined in *S. canicula* IL-1β, which was also in common with other IL-1β sequences (Bird *et al.*, 2002b), indicating that the molecule is secreted through a nonclassical pathway, not involving the golgi/endoplasmic reticulum route (Rubartelli *et al.*, 1990). In all mammalian IL-1β homologues, interleukin-converting enzyme (ICE, or caspase-1) processes the inactive precursor, which is cut at an aspartic acid (Asp^{116} in human IL-1β) to release the active mature peptide (Fantuzzi and Dinarello, 1999; Howard *et al.*, 1991). However, as with most nonmammalian IL-1β sequences characterized to date, *S. canicula* IL-1β has no identifiable ICE cut site in the region where mammalian IL-1β is cleaved (Secombes *et al.*, 2011), and so it remains to be determined experimentally if the precursor molecule is active or must be cleaved. It is possible that cartilaginous fish IL-1β could be cut by an enzyme distinct from an ICE homologue, as other enzymes in mammals can produce a biologically active IL-1β protein (Black *et al.*, 1988; Irmler *et al.*, 1995; Mizutani *et al.*, 1991; Schonbeck *et al.*, 1998), and is an area that warrants future investigation.

Evidence to show that the *S. canicula* IL-1β is involved in this organism's immune response was gained initially by gene expression analysis. In mammals, IL-1β is mainly expressed by monocytes and macrophages, but also by endothelial cells, fibroblasts, and epidermal cells. It is produced in response to many stimuli, including bacterial lipopolysaccharides (LPS) and other microbial products, cytokines (TNF, IFN-γ, GM-CSF, and IL-2), T-cell/antigen-presenting cell interactions, and immune complexes (Stylianou and Saklatvala, 1998). The expression studies in *S. canicula* found that the IL-1β molecule is biologically relevant to cartilaginous fish immune responses to Gram-negative bacteria (Bird *et al.*, 2002a). The *S. canicula* IL-1β transcript was induced *in vivo* following injection with LPS, 24 h earlier. Expression was clearly detectable in the spleen and testes, tissues known to contain a variety of leucocyte types (Zapata *et al.*, 1996). Stimulation of splenocytes *in vitro* with LPS also resulted in a significant increase in expression with a maximal response seen after 5 h, with northern blot analysis confirming there was a seven fold increase in transcript level. Inflammatory responses can also harm host tissues, it is important that they resolve quickly, and to this end, most key proinflammatory cytokines are short lived and have post-transcriptional regulation effected by adenylate- and uridylate-rich (AU-rich) elements in the 3′ untranslated regions (UTR) of the transcript. Eight such elements were identified in the *S. canicula* IL-1β 3′UTR, and when compared with the 3′ UTRs of IL-1β genes in a variety of bony fish and mammals, the *S. canicula* 3′ UTR was found to have the greatest negative impact on the expression of a reporter gene (Roca *et al.*, 2007). Thus, it is likely that cartilaginous fish cytokines will be controlled at different levels to ensure appropriate kinetics of expression.

To verify that *S. canicula* IL-1β was able to induce a response in immune cells, the recombinant protein was produced in *E. coli* (unpublished results). Primers were designed to amplify the predicted mature IL-1β coding sequence from Ala^{145}, and the product was cloned into the PQE30 expression vector. The mature protein was next expressed in *E. coli* M15 cells and purified under native conditions. When added to *S. canicula* splenocytes for 5 h, a clear increase in the expression level of several proinflammatory genes was apparent (Figure 7.1), with 10 ng/ml or higher required for significant upregulation of these genes.

To date, IL-1β is one of the most characterized IL-1 family members in vertebrates, but the evolution of this family still remains to be determined. This is something that was investigated in bony fish, where along with IL-1β, a homologue of IL-18 was found (Zou *et al.*, 2004). However, it is now known that there are additional IL-1 family members present in bony fish (Husain *et al.*, 2012; Wang *et al.*, 2009) one with no direct homologue in mammals, leaving the possibility of additional family members being found in cartilaginous fish, some of which may be unique to this group of organisms. This possibility was hinted due to the presence of a second smaller transcript of approximately 900 bp, which followed a similar pattern of expression as the expected IL-1β 1.35 kb transcript (Bird *et al.*, 2002a), mentioned above. The identity of this transcript remains to be determined, but could indicate the presence of another member of the IL-1 family within cartilaginous fish, perhaps a homologue of a mammalian IL-1 family member or a novel family member, found only in cartilaginous fish.

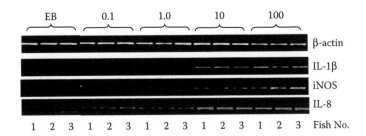

Figure 7.1   Expression of β-actin, IL-1β, iNOS, and IL-8 in *S. canicula* splenocytes 5 h after stimulation with recombinant *S. canicula* IL-1β at 0.1, 1, 10 and 100 ng/ml. Splenocytes were also exposed to elution buffer (EB) alone, used for recombinant protein purification, as a control. Cells from three fish (1–3) were used.

IL-2/IL-6 Family

To date, no members of the IL-2 or IL-6 cytokine family were found in cartilaginous fish. These molecules are referred to as short and long type I cytokines, respectively (Boulay *et al.*, 2003), and have a typical secondary structure consisting four alpha helices. However, two pieces of evidence suggest that such cytokines may exist. First, the heterodimeric cytokines IL-12, IL-23, IL-27, and IL-35 are formed between an alpha chain (p19, p28/IL-30, or p35) and a beta chain consisting either p40 or Epstein–Barr virus-induced gene 3 (EBI3). The alpha chains are the equivalent of long type I cytokines, whereas the beta chains are more like long type I cytokine receptors. Both p40 and EBI3 can be found in cartilaginous fish (in *C. milii*: scaffold_198, Accession no. KI636052.1; scaffold_19, Accession no. KI635873.1 for p40; and scaffold_85, Accession no. KI635939.1 for EBI3), with two p40 chains appearing to exist in this species. As for the alpha chain, p35 can be found (scaffold_1; Accession no. KI635855.1), allowing the formation of an active IL-12 molecule. In addition, a potential p28/IL30 appears to exist (scaffold_47; Accession no. KI635901.1), which would allow the possibility of a functional IL-27 molecule, important for regulating the activity of B and T lymphocytes (Larousserie *et al.*, 2006); however, this will require further investigation. Second, the type I cytokines are also very similar to several hormones such as growth hormone, prolactin, and erythropoietin (EPO), sometimes called prototypic class I cytokines (Boulay *et al.*, 2003), and again at least the latter two can be found in the *C. milii* genome (scaffold_16, Accession no. KI635870.1 for prolactin; scaffold_67, Accession no. KI635921.1 for EPO).

IL-10 Family

Although no papers were published on IL-10 cytokine family members in cartilaginous fish, sequences with homology to IL-10 can be found in the spiny dogfish, *Squalus acanthias* (Accession no. CX789095) and *C. milii* (scaffold_70; Accession no. KI635924.1). In addition, two sequences with homology to IL-19/20/24 can be found in this scaffold next to IL-10. These molecules are members of the type II cytokine family, along with interferons, which also exist in cartilaginous fish (see the following text), and are again helical proteins. In mammals, IL-10, IL-19, IL-20, and IL-24 are clustered together in the genome (on Chr 1q32 in humans). Although a clear homologue of IL-10 exists in bony fish (Secombes *et al.*, 2011), a second gene colocated at the same locus appears to have homology to IL-19/20/24 and may be related to an ancestral gene that subsequently diverged later in vertebrate evolution (Wang *et al.*, 2010a). Phylogenetic analysis (data not shown) groups the IL-10- and IL-20-like genes with their corresponding bony fish homologue, indicating the same situation may also be true in cartilaginous fish, as presently no other genes exist for IL-19/20/24.

IL-17 Family

The IL-17 cytokine family is a group of six molecules in mammals (IL-17A-F), which have a cysteine knot structure similar to that of transforming growth factor-βs and nerve growth factors, with the difference being that only four cysteines form the knot in the IL-17 molecules (vs. six in TGF-βs and NGFs) (Weaver *et al.*, 2007). Several IL-17 family members can be found in the *C. milii* genome, including sequences with apparent homology to IL-17A/F (scaffold_2515; Accession no. KI637887), IL-17B (scaffold_161; Accession no. KI636015.1), IL-17C (scaffold_197; Accession no. KI636051.1), and IL-17D (scaffold_223; Accession nos. KI636077 and JX052591). That IL-17 family members would be found in cartilaginous fish is no surprise in that IL-17D homologues were discovered in Agnathans (Tsutsui *et al.*, 2007) and even in invertebrates homologues of IL-17 exist. Although the exact relationship of the invertebrate molecules to vertebrate IL-17 family members is not clear, the Pacific oyster, *Crasso streagigas*, IL-17 molecule does have highest homology to IL-17D (Roberts *et al.*, 2008). IL-17B has only recently been found in bony fish (Wang, Scottish Fish Immunology Research Centre (SFIRC), pers. comm.), and so its presence in cartilaginous fish is again not unexpected. Lastly, IL-17C has also been discovered in bony fish, but as is the case for IL-20 above, it appears to be related to several known mammalian genes, in this case to both IL-17C and IL-17E (IL-25) (Wang *et al.*, 2010b). In mammals, IL-17C is produced by epithelial cells in response to bacterial challenge and inflammatory stimuli and induces expression of proinflammatory cytokines, chemokines, and antimicrobial peptides (Ramirez-Carrozzi *et al.*, 2011), whereas IL-17E is produced by a broader range of cell types (e.g., mucosal epithelial cells, Th2 cells, mast cells, macrophages, eosinophils, and NKT cells) and is involved in Th2 cell differentiation (Kawaguchi *et al.*, 2004). Whether an ancestral molecule with both of these functions exists in fish will be interesting to determine.

OTHER INTERLEUKINS

IL-14

IL-14 is described as a B-cell growth factor able to induce B-cell proliferation but inhibits antibody secretion. Several IL-14 homologous sequences are present in the *C. milii* genome (scaffold_121; Accession no. KI635975.1), and it seems likely this molecule will have some role to play in the cartilaginous fish immune system.

IL-16

IL-16 was initially described as lymphocyte chemoattractant factor (LCF) but is now known to be a proinflammatory cytokine that attracts a variety of cell types, including T cells, monocytes, eosinophils, and dendritic cells that express CD4. The molecule is produced as a precursor that is cleaved by caspase 3 to release the active C-terminal cytokine. Curiously, IL-16 was the first known extracellular protein to contain a PDZ-like fold (Mühlhahn *et al.*, 1998), named after three proteins with this domain, namely, mammalian postsynaptic density protein 95 kD (PSD-95), Drosophila disc large tumor suppressor (DlgA), and mammalian tight junction protein ZO1. As with IL-14, two IL-16 homologous sequences are present in the *C. milii* genome (scaffold_5; Accession no. KI635859.1) and perhaps this molecule will have a role in T-cell activation in cartilaginous fish.

IL-34

IL-34 is a recently described cytokine that binds to the same receptor as colony-stimulating factor 1 (CSF-1 or macrophage-CSF) and promotes the formation and proliferation of macrophages in a similar way to CSF-1 (Lin *et al.*, 2008; Wei *et al.*, 2010). This is quite a unique situation, as these two cytokines have very little sequence similarity. However, conformational adaptations were recently discovered that can explain the cross-reactivity of CSF-1R for these two distantly related ligands (Ma *et al.*, 2012). In addition, although CSF-1 and IL-34 bind to the same receptor, they have different abilities to induce chemokine production in primary macrophages, indicating differences in signaling through CSF-1R (Liu *et al.*, 2012).

Discovery of IL-34 in vertebrate species other than mammals has only occurred recently, in birds and bony fish (Garceau *et al.*, 2010; Wang *et al.*, 2013), highlighting the importance of this gene through evolution and indicating its possible existence in cartilaginous fish. Searching the available cartilaginous fish sequence databases and using the available genome allowed an IL-34 homologue to be identified in *C. milii* (scaffold_148; Accession no. KI636002.1) and *S. acanthias* (Accession no. ES606032). Alignment of these sequences with known vertebrate IL-34 sequences highlighted a number of conserved features. Similar to mammalian IL-34, the cartilaginous fish sequences possess a signal peptide, a conserved glycosylation site (N^{79}-R^{82} in humans), and appear to have a four-helix bundle structure (Liu *et al.*, 2012; Ma *et al.*, 2012), evidenced by the high similarity found between the sequences in the regions where these helices are predicted. In addition, three cysteines are conserved in all known IL-34 sequences. In human IL-34, there are six cysteines in total, with C^{35} binding to C^{180} and C^{177} binding to C^{191}, to stabilize the tertiary structure (Liu *et al.*, 2012). Only C^{35} and C^{180} are conserved in the cartilaginous fish as in birds (Garceau *et al.*, 2010) and may mean only one disulphide bond is needed in these molecules. Lastly, using the sequence predicted from the genome, it is possible to predict the gene organization, which was found to be very similar to the known mammalian, fish, and bird IL-34 genes (Wang *et al.*, 2013) in having seven exons and six introns (Figure 7.2). The role of this cytokine within cartilaginous fish remains to be determined.

TUMOR NECROSIS FACTORS

Tumor necrosis factors are members of a superfamily (TNFSF) of proteins that normally exist as type II transmembrane proteins (except TNF-β and APRIL), where the extracellular domains are cleaved by metalloproteinases to release the bioactive soluble cytokines. TNF-α and TNF-β were the founding members of the group that consists of 19 ligands in humans. These two molecules signal through the same receptors (TNFRI and TNFRII) and are released mainly by macrophages and T cells, respectively. The molecules form a *jellyroll* structure composed of 10 β-strands and are typically homotrimers or heterotrimers (Wiens and Glenney, 2011).

Within cartilaginous fish, different members of the TNFSF have been discovered to date and potentially include TNF-a (TNFSF2), TNF-b (TNFSF1), LT-b (TNFSF3) as well as BAFF (TNFSF13B) and CD40L (TNFSF5/CD154). TNF-a has been identified in the nurse shark (Accession no. AGQ17907.1) and little skate genome (Accession no. AESE010953997). In addition, there are two other sequences in the little skate genome, showing relatedness to TNF-β (Accession nos. AESE011886057) and LT-b (AESE010094715), however, this is based on partial sequences and no functional studies have been performed. The two little skate sequences are two distinct sequences, different to each other and to the TNF-a. However, sequencing of the full transcripts is required to allow characterization of these two genes, as to date no homologue for TNF-β has been identified in bony fish. CD40L was cloned in *S. acanthias* and *S. canicula* (Li, SFIRC, pers. comm.) and is known to be expressed on activated T cells, where it serves as a co-stimulatory molecule to activate antigen-presenting cells, including B cells. Studies in bony fish have shown a good conservation of function throughout the vertebrates (Lagos *et al.*, 2012). Also recently, BAFF was cloned in *S. acanthias* (Li *et al.*, 2012) and the white-spotted catshark, *Chiloscyllium plagiosum* (Ren *et al.*, 2011). In *S. acanthias*, the BAFF gene was found to consist of seven exons/six introns in contrast to other vertebrates where six exons are present. This results in an insertion into the transcript that increases the

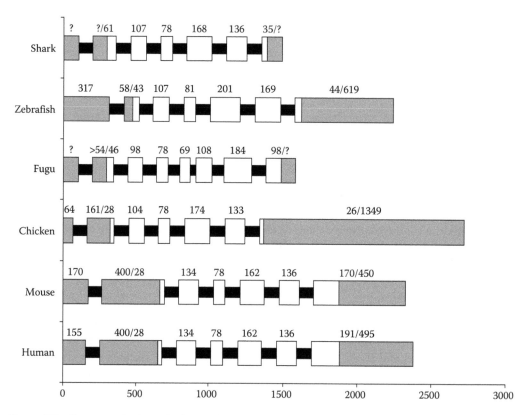

Figure 7.2 Gene organizations of selected vertebrate IL-34 molecules. The gene organization of the elephant shark was predicted using Genscan (Data from Burge, C.B., and Karlin, S., *Curr. Opin. Struct. Biol.,* 8, 346–354, 1998) with possible coding regions within the contig (GenBank AAVX01015196.1) analyzed using BLAST (Data from Altschul, S.F. *et al., J. Mol. Biol.,* 215, 403–410, 1990) and FASTA (Data from Pearson, W.R., and Lipman, D.I., *Proc. Natl. Acad. Sci. USA,* 1988; 85:2444–2448). The following GenBank Accession numbers were used to obtain the IL-34 gene organizations for: human, AC020763; mouse, AC139245; and chicken, AADN03006282. (Fugu and zebrafish sequences from Wang, T. *et al., Mol. Immunol.,* 2013; 53:398–409.)

translated protein by 28 aa between the a and a′ β-sheets, which significantly increases the size of this loop. This loop is near the DE-loop that borders the receptor-binding groove and so could impact on the binding specificity. The insertion was not present in the reported *C. plagiosum* transcript, but was present in both *C. milii* and *S. canicula* BAFF (unpublished). Interestingly, a minor transcript, a splice variant lacking the fifth exon, was also detectable in *S. acanthis* in epigonal, gut, and liver tissue. BAFF is highly expressed in the spleen and also in the pancreas in *S. acanthias*, potential sites of B-cell development in cartilaginous fish. BAFF expression in *S. acanthis* blood leucocytes was significantly increased by stimulation with pokeweed mitogen, in a time- and dose-dependent manner.

CHEMOKINES

Chemokines are a group of small (8–10 kDa) *chemo*tactic cyto*kines* that serve to ensure different leucocyte types appear in the correct location in the right circumstances. This may be as a consequence of infection where, for example, phagocytes are needed to be attracted to a site of inflammation to help combat and kill invading microbes. It may also be during maturation and differentiation of different leucocyte types that require to be in a particular cellular environment to develop. Chemokines are subdivided on the basis of the cysteine residues they possess, as to whether there are amino acid residues between the first pair (i.e., CC, CXC, and CX₃C) or whether one of the residues is missing (i.e., XC and CX). The cysteines (three or four) allow folding of the protein into a characteristic Greek key structure. The majority of chemokines found to date belong to the CXC (α-chemokines) or CC (β-chemokines) subgroups. They bind to G-protein-linked receptors to bring about a biological effect in the target cells.

IL-8 and CXC Chemokines

Although named an *interleukin*, IL-8 is actually a CXC chemokine (also known as CXCL8), which is produced by both immune and nonimmune cells in response to inflammatory stimulants (Baggiolini and Clark-Lewis, 1992). Its main role is to attract neutrophils to sites of infection or injury and induce them to release lysozomal enzymes, undergo a change in shape, trigger the respiratory burst, and increase the expression of adhesion molecules on cell surfaces (Peveri *et al.*, 1988). The presence of an IL-8 homologue in cartilaginous fish was not unexpected, as it was already found in an Agnathan, the lamprey, *Lampetra fluviatilis* (Najakshin *et al.*, 1999). This gene was cloned in three cartilaginous fish to date, *T. scyllium* (Inoue *et al.*, 2003b), the silver chimaera, *Chimaera phantasma* (Inoue *et al.*, 2003a), and *S. canicula* (unpublished). The aa sequences from these IL-8 homologues showed good identity with the other known vertebrate IL-8 genes, along with the presence of a signal peptide and conservation of the four cysteine residues that are essential for the tertiary structure (Figure 7.3). In addition, CXCL8-like sequences are identifiable in the genomes and/or ESTs of *C. milii*, *L. erinacea*, *S. acanthias*, and the Pacific electric ray, *Torpedo californica*. As found with many of the bony fish IL-8 sequences (Corripio-Miyar *et al.*, 2007), the cartilaginous

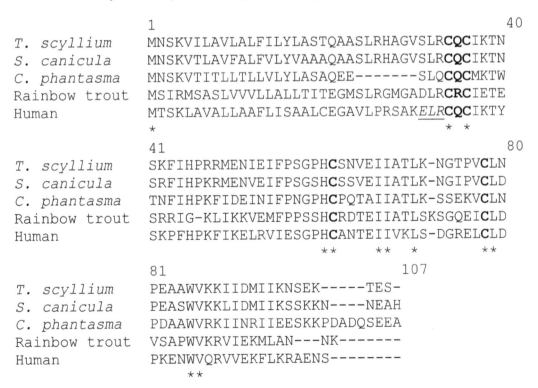

Figure 7.3 Multiple sequence alignment of the predicted cartilaginous fish IL-8 proteins with a representative bony fish and mammalian sequence. Asterisks indicate conserved residues, and include the four cysteines typical of CXC chemokines. The *ELR* motif in human IL-8 is underlined and in italics.

fish IL-8 sequences lack an important ELR motif found in mammalian and bird IL-8 genes immediately prior to the CXC motif. Generally, this motif is needed to recruit granulocytes during inflammation and promote angiogenesis, whereas those CXC chemokines lacking this motif specifically attract lymphocytes and monocytes and inhibit angiogenesis (Strieter *et al.*, 1995). However, even in the absence of this motif, bony fish recombinant proteins demonstrate chemotactic activity for neutrophils and macrophages (van der Aa *et al.*, 2010), suggesting it is not an absolute prerequisite for chemotaxis of at least some fish leucocytes. In *S. canicula*, the IL-8 transcript is induced following stimulation of splenocytes with LPS (unpublished) or rIL-1β (Figure 7.1), indicating its involvement within inflammatory events in cartilaginous fish.

In addition to the IL-8-like molecules, cartilaginous fish possess homologues of CXCL12, CXCL13, and CXCL14, as seen in bony fish, with the former clearly divided into two clades (Figure 7.4). Homologues of CXCL9-11 are also present, and again multiple genes are present within some species. Curiously, genes with apparent relatedness to fish-specific clades, recently termed *CXCL-F* (Chen *et al.*, 2013), are present with homologues of the CXCL-F3-F5 genes identified. Clearly, many CXC chemokine types are present in cartilaginous fish.

A number of CXC receptor (CXCR) genes were also discovered in cartilaginous fish. Again it had been reported before that a CXCR was present in an Agnathan, the sea lamprey, *Petromyzon marinus* (Kuroda *et al.*, 2003), of the CXCR4 type (the receptor for CXCL12), and hence, the identification of this molecule in the salmon shark, *Lamna ditropis*, could be anticipated (Goostrey *et al.*, 2005). Of the remaining six known CXCR genes in mammals, two genes with similar homology to CXCR1 and CXCR2 were found in the great white shark, *Carcharodon carcharias*, basking shark, *Cetorhinus maximus*, and cuckoo ray, *Raja naevus*, and one gene in *S. caniculus*. The molecules were termed *CXCR1/2a* and *CXCR1/2b* where two existed, had 79%–89% nucleotide identity, and in the case of the *C. carcharias* and *C. maximus* sequences, branched together (i.e., a with a and b with b) in phylogenetic tree analysis, suggesting that lineage-specific duplication has occurred independently in sharks and rays post their divergence. CXCR1 and CXCR2 are receptors for IL-8 (CXCL8) in mammals, although they can also bind CXCL6 or CXCL2/3/5/6, respectively.

CC Chemokines

A number of CC chemokine genes were also discovered in cartilaginous fish. A first sequence was reported in *S. canicula* by Kuroda *et al.* (2003) and termed *SCYA107* (Note. SCY stands for Small CYtokines, a term initially used for chemokine genes, with SCYA being CC chemokines and SCYB CXC chemokines). Subsequently, SCYA107 and two further CC chemokine genes, macrophage inflammatory protein (MIP) 3α1 and MIP3α2, were found in *T. scyllium* (Inoue *et al.*, 2005). MIP3α is nowadays referred to as CCL20 and is a potent chemoattractant for lymphocytes. A number of other chemokine-related sequences are in the database, which include a *C. milii* sequence with homology to CCL20 (Accession no. AFK10939) and two *S. acanthias* sequences with homology to CCL2 and CCL11 (Accession nos. EB688023 and CX663036, respectively). Lastly, a CCL14-like sequence exists for the cloudy catshark, *Scyliorhinus torazame* (Accession no. FY418051).

The three *T. scyllium* genes all consist of four exons/three introns, a typical organization of many CC chemokines, including human CCL20. The two SCYA107 genes had 60.3% nucleotide identity and were ~55% identical with the rainbow trout CC chemokine CK4A. The two MIP3α genes were 69% identical and had 53% identity to human CCL20. Phylogenetic tree analysis confirmed the homology of the MIP3α genes to members of the CCL20 clade, but as with many bony fish CC genes, the homology of SCYA107 was not clear (Nomiyama *et al.*, 2008), and indeed, it may be a member of a subgroup present only in fish. After injection of *T. scyllium* with LPS for 24 h, strong upregulation of SCYA107 was seen in the gill, heart, Leydig organ, spleen, and testes, but MIP3α expression levels were unaltered. However, MIP3α1 was strongly induced by the stimulation of blood leucocytes in culture with phorbol 12-myristate 13-acetate (PMA) for 3–12 h.

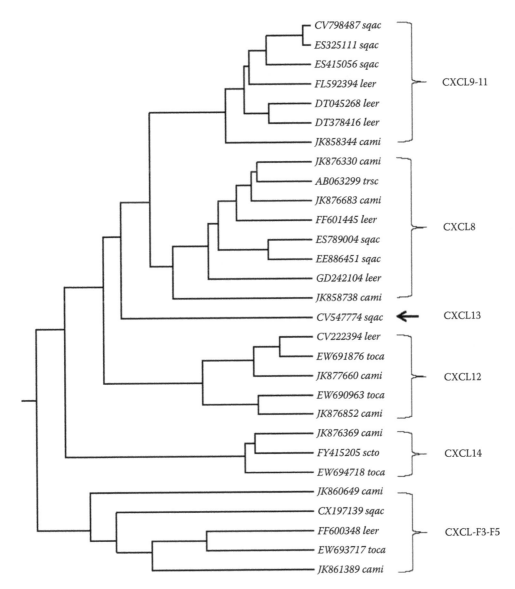

Figure 7.4 A rooted UPMGA phylogenetic tree of cartilaginous fish CXC chemokines. The nucleotide sequences of putative CXC chemokine homologues were obtained from the cartilaginous EST data by BLASTN search using protein sequences of known CXC chemokines (http://blast.ncbi.nlm.nih.gov/) and the translated proteins analyzed to confirm their phylogenetic relationships with known vertebrate CXC chemokines. The cartilaginous fish molecules were aligned by the MAFFT program (version 6.864) and the corresponding rooted phylogenetic tree was constructed using the UPGMA method using the online tools on the website (http://www.genome.jp/tools/mafft). cami, *C. milii*; leer, *L. erinacea*; sqac, *S. acanthias*; toca, *T. californica*.

TRANSFORMING GROWTH FACTORS

Transforming growth factor-β is a cytokine that controls cell proliferation and differentiation. It exists in three isoforms in bony fish, birds, and mammals (TGF-β1-3), although a fourth form was identified in bony fish (called TGF-β6) (Funkenstein *et al.*, 2010). TGF-β1 is the founding member of the TGF-β superfamily of proteins that include, among others, bone morphogenetic proteins, growth differentiation factors, activins, and inhibins, with invertebrate as well as vertebrate members. TGF-βs are produced as secreted precursors that are cleaved by cellular proteases to release the C-terminal mature peptide (or 112–114 aa). The mature peptide adopts a cysteine knot secondary structure, typical of the superfamily, and is active as a homodimer. TGF-β1 is normally associated with immune function, and although its release from regulatory T cells has received much recent attention, in fact, it can be released from most leucocytes. However, recent work indicates that TGF-β2 can also be important for immunity (Rautava et al., 2012; Tohidi et al., 2012). In cartilaginous fish, all of the TGF-β genes appear to be present, with TGF-β1 in C. milii (Accession no. KI638058.1), TGF-β2 in C. milii (Accession no. KI635979.1), TGF-β3 in C. milii (Accession no. AAVX02063563.1) and S. acanthias (Accession no. DV496228), and a homologue of the fourth form found in fish (TGF-β6) in C. milii (Accession no. KI638086.1).

INTERFERONS

The interferon (IFN) system is the hallmark of vertebrate antiviral defenses. It consists of type I, II, and III IFN molecules, with type I and III IFNs involved in the activation of an antiviral state in the host after viral infection, whereas type II IFNs mainly regulate specific cell-meditated immunity. To date, type I and II IFNs were found in bony fish, amphibians, reptiles, birds, and mammals, whereas type III IFNs appear restricted to the tetrapods.

In our search for cartilaginous fish IFN, we analyzed the WGS nucleotide databases of C. milii and L. erinacea. Previously, we reported a partial sequence of a type I IFN gene in C. milii discovered by in silico analysis (Zou et al., 2007). Here we report the discovery of additional contigs containing putative IFN sequences. In total, four different IFN molecules were identified, translating into one full length protein of 185 aa (IFN1) and three partial sequences (IFN2-4) of 107–142 aa. Surprisingly, the four IFN proteins are highly divergent and share only low sequence identity (Table 7.1). A similar bioinformatics approach was used to identify the IFN homologues from L. erinacea using the obtained C. milii sequences as bait, and 46 DNA contigs and one transcript (encoding a protein of 186 aa) were found. In general, the DNA contigs are short and all encode partial IFN genes. Alignment of the amino acid sequences revealed that at least 10 IFN genes are present.

Overall the cartilaginous fish IFN molecules have a comparative protein length of ~185–186 aa and all contain a signal peptide, suggesting they are secreted cytokines. Based on the formation of disulphide bonds, type I IFNs can be divided into two and four cysteine-containing proteins, with the former retaining Cys 1 and Cys 3 in bony fish, or Cys 2 and Cys 4 in mammalian forms (i.e., IFN-β and IFN-ε). All the cartilaginous fish IFN found to date have four cysteines in the predicted mature peptide, demonstrating that this type of IFN is present in all groups of jawed vertebrates, although some ray-finned fish appear to have lost this form. Hence, in teleost fish, the four cysteine-containing type I IFNs (teleost group II IFN) was found in species of ostariophysii (i.e., cyprinids), protacanthopterygii (i.e., salmonids), and paracanthopterygii (i.e., gadoids) but not in

Table 7.1 **Amino Acid Identity (Upper Right Quadrant) and Amino Acid Similarity (Lower Left Quadrant) between Four Type I Interferon Genes (IFN1 to IFN4) in C. milii**

	IFN1	IFN2	IFN3	IFN4
IFN1	100	56.2	32.6	21.0
IFN2	58.9	100	32.2	32.7
IFN3	49.7	50.0	100	26.4
IFN4	33.5	55.4	40.8	100

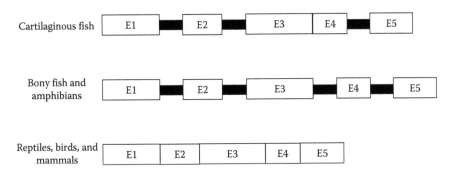

Figure 7.5 The genomic organization of interferon genes in cartilaginous fish compared to interferon genes in other vertebrate groups.

the acanthopterygian species, which only possess the two cysteine group (teleost group I IFN). The distinctive CA(s)WE(a) motif, previously identified in the C terminal region of the four cysteine-containing IFNs, is also well conserved in cartilaginous fish IFNs. It is worth noting that the third cysteine residue is absent in one of the four IFNs in *C. milii*, suggesting two cysteine-containing forms may exist in this species. However, further sequencing is needed to confirm this finding.

Type I IFNs are encoded by intron-lacking genes in reptiles, birds, and mammals, whereas teleost and amphibian IFN genes contain five exons and four introns, a genomic organization shared by type III IFNs and IL-10 family members. It was hypothesized that these genes originated from a common ancestor with a genomic organization of five exons and four introns (Lutfalla *et al.*, 2003). Surprisingly, all the IFN genes found in cartilaginous fish have a genomic structure of four exons and three introns (Figure 7.5). Analysis of the intron positions in sequence alignments of the IFN proteins from (cartilaginous and bony) fish and amphibians found that the intron positions are conserved and are all phase 0. It appears that the third exon in the cartilaginous fish IFN genes is equivalent to exons 3 and 4 in teleost and amphibian IFN genes, indicating this exon either split into two exons in teleosts and amphibians or has fused in cartilaginous fish.

Expansion of IFNs by gene duplication within lineages or species during evolution seems apparent. Multiple copies can be found in all vertebrates and copy number varies between different species. In most vertebrates with sequenced genomes, the IFN genes are clustered in a single locus in the same chromosome. One exception is zebrafish where the four IFN genes are located in two separate chromosomes (Chr 3 and Chr 12) (genome version ZV9). It remains to be determined whether the IFN genes found in *C. milii* and *L. erinacea* are colocated in their genomes.

HEMATOPOIETIC CYTOKINES

Hematopoiesis (formation of cellular blood components derived from stem cells) is governed by a number of cytokines that promote survival, proliferation, and differentiation of hematopoietic stem cells and progenitor cells. Colony-stimulating factors (CSF) are key players in this process, with four identified in mammals and named according to the most numerous white blood cells they stimulate; macrophage CSF (M-CSF or CSF1), granulocyte/macrophage CSF (GM-CSF or CSF2), granulocyte CSF (G-CSF or CSF3), and multi-CSF, more commonly termed *interleukin 3* (IL-3), which stimulate a broad range of hematopoietic cell colony types (Metcalf, 2010). To date, none of these were identified in cartilaginous fish, but CSF1 and CSF3 do have homologues in bony fish (Santos *et al.*, 2006; Wang *et al.*, 2013). However, a cytokine that shares a number of features with CSF1, stem cell factor (SCF) can be found in a *C. milii* EST (Accession no. JK963148). In mammals, it has a wide range of activities with direct effects on myeloid and lymphoid cell development and synergistic effects with other growth factors such as CSF2, IL-7, and EPO (Broudy, 1997). EPO, also known as hematopoietin, is an essential cytokine for red blood cell production, having its primary effect on red blood cell progenitors and precursors, which are found in the bone marrow in humans and has more recently been found to play important roles within infection and inflammation (Nairz *et al.*, 2012). Due to its important role in red blood cell formation, it would be expected to be present in all vertebrates and a sequence can be found within the *C. milii* genome (Accession no. AAVX01038459) with homology to this cytokine. The roles of SCF and EPO in blood cell formation in cartilaginous fish and whether other CSF's exist in this group need further investigation.

RECENT CYTOKINE DISCOVERY

The availability and quality of genomes within different vertebrate organisms is changing rapidly, due to the continued development of new technologies within molecular biology. Although this chapter was being submitted for publication, a paper was published on the elephant shark genome that incorporated some interesting findings related to the presence of members of the IL-2 and IL-6 family (e.g., IL-6, IL-7, and IL-15), along with the Type II IFN, IFN-gamma (Venkatesh *et al.*, 2014). This publication led to a further investigation of the updated draft genome, with our interest focused on unidentified cytokines, some of which are produced by T helper (Th) cells. These cells play a key role in adaptive immunity by secreting cytokines that initiate and activate downstream effector mechanisms (Jiang and Dong, 2013; Oh and Gosh, 2013). In mammals, Th cells can be divided into a number of effector subpopulations, which include Th1, Th2, Th9, Th17, and Th22, that elicit appropriate immune responses to different pathogen/antigen types by releasing different repertoires of cytokines. Many of the cytokines released by these cells have only been characterized within bony fish relatively recently (Secombes *et al.*, 2011), opening up the possibility that similar subsets of Th cells exist within this group, regulating adaptive immunity in a similar complex way.

An analysis on the new draft of the genome has identified some additional cytokines that includes IL-2-, IL-21-, and IL-4/13-like molecules. IL-2 was initially described as a T-cell growth factor and is secreted primarily by Th cells that were activated by stimulation with certain mitogens or by interaction of the T-cell receptor complex with antigen/MHC complexes on the surface of antigen-presenting cells (Hatakeyama and Taniguchi, 1990). IL-21 is also a T-cell-derived cytokine, acting as a lineage stabilizing factor for Th17 cells, and is essential for the generation of B-cell responses induced directly by activated T cells (Deenick and Tangye, 2007; Kuchen *et al.*, 2007). IL-2 and IL-21 are closely related cytokines that are located next to each other, not only within the mammalian and fish genomes (Bird *et al.*, 2005), but this also appears to be the case within the elephant shark genome (Scaffold 208; Accession no. KI636062).

IL-4 has many biological roles, including the stimulation of activated B-cell and T-cell proliferation, the differentiation of B cells into plasma cells, and the differentiation of naive helper T cells (Th0 cells) into Th2 cells (Sokol *et al.*, 2008). This gene is found close to two other cytokines (IL-13 and IL-5) in the mammalian genome and this region was termed the *Th2 cytokine locus*, as these cytokines also play a role in Th2 responses (Lee *et al.*, 2003). Two IL-4-like genes were discovered within selected bony fish sequences, called IL-4/13A and IL-4/13B, as it was unclear which gene they are truly related to (Secombes *et al.*, 2011). Their position within the fish genomes also does not provide help to their exact identity, as these genes are found on separate chromosomes. However, one gene is found next to RAD50 and the other is next to KIF3A, both of which are also part of the Th2 cytokine locus. It was possible to find a similar Th2 cytokine locus within the elephant shark genome (scaffold_92; Accession no. KI635946.1), which has both RAD50 and KIF3A in the same locus with two IL-4-like genes between them that share a high identity to each other (Figure 7.6). This is a significant finding as it indicates that the separation of RAD50 and KIF3A into two different loci is a bony fish-specific event.

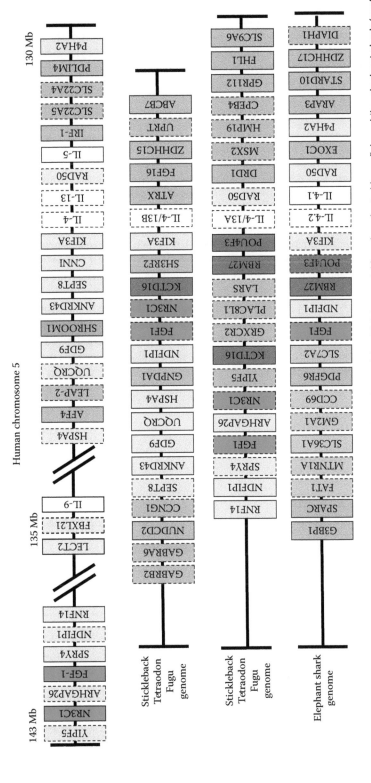

Figure 7.6 Synteny analysis of the Th2 cytokine locus in humans with the loci containing Th2-type cytokines in selected bony fish and the elephant shark (scaffold 92; Accession no. KI635946.1). The solid or dashed outline of the boxes indicates the transcriptional orientation.

This quick look at the newly updated genome provides further evidence that the repertoire of cytokines in cartilaginous fish is just as complex as that of other jawed vertebrates. However, caution needs to be taken as the genome still requires a lot more work before it is fully complete, and it is still too early to make any final conclusions as to the exact components of the immune system this group of vertebrates truly has. It is anticipated that a number of other cytokine homologues are yet to be found, potentially along with novel molecules found only within cartilaginous fish.

PHYLOGENETIC PERSPECTIVE

It would seem that complex immune systems require complex regulatory mechanisms to ensure appropriate immune responses are initiated to effect disease resistance and then subsequently downregulated to avoid host damage. Initial analysis of cartilaginous fish genomes and ESTs suggests that a significant number of regulatory cytokines are present. For example, proinflammatory mediators such as IL-1β, TNF, IL-17, IL-20L, and chemokines are present to effect antimicrobial defenses. Antiviral defenses are also present and include a diverse range of interferon molecules. Molecules to promote adaptive immunity include BAFF and IL-16 to promote B-cell responses, and activators of T cells such as CD40L and cytokines to drive Th-cell responses. Negative regulators such as TGF-βs and IL-10 are also present to complete the necessary negative feedback loop. In comparison with what can be hypothesized as the cytokine network present in the Chondrichthyan ancestor (by comparison of cytokines present in mammals/birds and bony fish) (Figure 7.7), it is clear that the full repertoire found in Chondrichthyan species to date can be included as potentially present in the Gnathostome ancestor from which cartilaginous fish are derived. As we learn more about the cytokine genes present in cartilaginous fish, we will also determine the minimal cytokine repertoire that was needed to coordinate adaptive and innate immunity in early Gnathostomes.

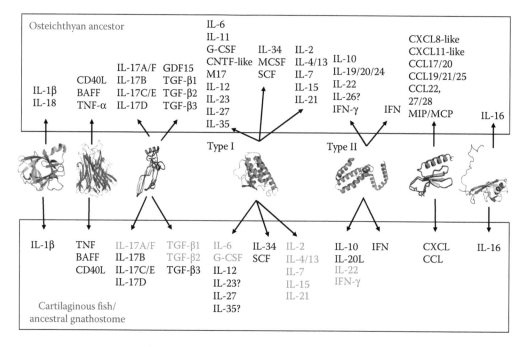

Figure 7.7   Comparison of known cytokines in cartilaginous fish with those predicted to were present in ancestral Osteichthyan fish. Cytokines shown in red are those discovered from the recent analysis of the updated elephant shark genome.

ACKNOWLEDGMENTS

The authors acknowledge the help of Dr. A. Goostrey (deceased) and Mr. W. Quinn in the production and testing of *S. canicula* rIL-1β and the cloning of *S. canicula* IL-8. The previously unpublished figures in this chapter were from the work supported by the BBSRC (grant 1/S15536).

REFERENCES

Altschul, S.F., Gish, W., Miller, W., Myers, E., and Lipman, D.J. Basic local alignment search tool. *Journal of Molecular Biology* 1990; 215:403–410.

Baggiolini, M., and Clark-Lewis, I. Interleukin-8, a chemotactic and inflammatory cytokine. *FEBS Letters* 1992; 307:97–101.

Ballantyne, J. Jaws: The inside story. The metabolism of elasmobranch fishes. *Comparative Biochemistry & Physiology* 1997; 118B:703–742.

Banchereau, J., Pascual, V., and O'Garra, A. From IL-2 to IL-37: The expanding spectrum of anti-inflammatory cytokines. *Nature Immunology* 2012; 13:925–931.

Bird, S., Wang, T., Zou, J., Cunningham, C., and Secombes, C.J. The first cytokine sequence within cartilaginous fish: IL-1 beta in the small spotted catshark (*Scyliorhinus canicula*). *Journal of Immunology* 2002a; 168:3329–3340.

Bird, S., Zou, J., Kono, T., Sakai, M., Dijkstra, J.M., and Secombes, C. Characterisation and expression analysis of interleukin 2 (IL-2) and IL-21 homologues in the Japanese pufferfish, *Fugu rubripes*, following their discovery by synteny. *Immunogenetics* 2005; 56:909–923.

Bird, S., Zou, J., Wang, T., Munday, B., Cunningham, C., and Secombes, C.J. Evolution of interleukin-1beta. *Cytokine & Growth Factor Reviews* 2002b; 13:483–502.

Black, R., Kronheim, S., Cantrell, M., Deeley, M., March, C., Prickett, K., Wignall. J. *et al.* Generation of biologically active interleukin-1β by proteolytic cleavage of the inactive precursor. *Journal of Biological Chemistry* 1988; 263:9437–9442.

Boraschi, D., Lucchesi, D., Hainzl, S., Leitner, M., Maier, E., Mangelberger, D., Oostingh, G.J. *et al.* IL-37: A new anti-inflammatory cytokine of the IL-1 family. *European Cytokine Network* 2011; 22:127–147.

Boulay, J.L., O'Shea, J.J., and Paul, W.E. Molecular phylogeny within type I cytokines and their cognate receptors. *Immunity* 2003; 19:159–163.

Brocker, C., Thompson, D., Matsumoto, A., Nebert, D.W., and Vasiliou, V. Evolutionary divergence and functions of the human interleukin (IL) gene family. *Human Genomics* 2010; 5:30–55.

Broudy, V.C. Stem cell factor and hematopoiesis. *Blood* 1997; 90:1345–1364.

Burge, C.B., and Karlin, S. Finding the genes in genomic DNA. *Current Opinion in Structural Biology* 1998; 8:346–354.

Cateni, C., Paulesu, L., Bigliardi, E., and Hamlett, W.C. The interleukin 1 (IL-1) system in the uteroplacental complex of a cartilaginous fish, the smoothhound shark, *Mustelus canis*. *Reproductive Biology and Endocrinology* 2003; 1:25.

Chen, J., Xu, Q., Wang, T., Collet, B., Corripio-Miyar, Y., Bird, S., Xie, P., Nie, P., Secombes, C.J. *et al.* Phylogenetic analysis of vertebrate CXC chemokines reveals novel lineage specific groups in teleost fish. *Developmental and Comparative Immunology* 2013; 41:137–152.

Corripio-Miyar, Y., Bird, S., Tsamopoulos, K., and Secombes, C.J. Cloning and expression analysis of two pro-inflammatory cytokines, IL-1 beta and IL-8, in haddock (*Melanogrammus aeglefinus*). *Molecular Immunology* 2007; 44:1361–1373.

Deenick, E.K., and Tangye, S.G. Autoimmunity: IL-21: A new player in Th17-cell differentiation. *Immunology and Cell Biology* 2007; 85:503–505.

Fantuzzi, G., and Dinarello, C. Interleukin-18 and interleukin-1β: Two cytokine substrates for ICE (Caspase-1). *Journal of Clinical Immunology* 1999; 19:1–11.

Funkenstein, B., Olekh, E., and Jakowlew, S.B. Identification of a novel transforming growth factor-beta (TGF-β 6) gene in fish: Regulation in skeletal muscle by nutritional state. *BMC Molecular Biology* 2010; 11:37.

Garceau, V., Smith, J., Paton, I.R., Davey, M., Fares, M.A., Sester, D.P., Burt, D.W., and Hume, D.A. Pivotal advance: Avian colony-stimulating factor 1 (CSF-1), interleukin-34 (IL-34), and CSF-1 receptor genes and gene products. *Journal of Leukocyte Biology* 2010; 87:753–764.

Goostrey, A., Jones, G., and Secombes, C.J. Isolation and characterization of CXC receptor genes in a range of elasmobranchs. *Developmental and Comparative Immunology* 2005; 29:229–242.

Grogan, E., and Lund, R. Reactivity of human white blood cells to factors of elasmobranch origin. *Copeia* 1991; 2:402–408.

Hatakeyama, M., and Taniguchi, T. Interleukin-2. In: Sporn, M.B., and Roberts A.B. (eds.) *Peptide Growth Factors and Their Receptors I*. Springer, Berlin, Heidelberg, Germany; New York, 1990, p. 523.

Herrin, B.R., and Cooper, M.D. Alternative adaptive immunity in jawless vertebrates. *Journal of Immunology* 2010; 185:1367–1374.

Hofmann, K., Bucher, P., Falquet, L., and Bairoch, A. The PROSITE database, its status in 1999. *Nucleic Acids Research* 1999; 27:215–219.

Howard, A., Kostura, M., Thornberry, N., Ding, G., Limjuco, G., Weidner, J., Salley, J. *et al.* IL-1 converting enzyme requires aspartic acid residues for processing of the IL-1β precursor at two distinct sites and does not cleave 31-kDa IL-1β. *Journal of Immunology* 1991; 9:2964–2969.

Husain, M., Bird, S., Zwieten, R., Secombes, C.J., and Wang, T. Cloning of the IL-1β3 gene and IL-1β4 pseudo-gene in salmonids uncovers a second type of IL-1β gene in teleost fish. *Developmental and Comparative Immunology* 2012; 38:431–446.

Inoue, Y., Endo, M., Haruta, C., Taniuchi, T., Moritomo, T., and Nakanishi, T. Molecular cloning and sequencing of the silver chimaera (*Chimaera phantasma*) interleukin-8 cDNA. *Fish & Shellfish Immunology* 2003a; 15:269–274.

Inoue, Y., Haruta, C., Usui, K., Moritomo, T., and Nakanishi, T. Molecular cloning and sequencing of the banded dogfish (*Triakis scyllia*) interleukin-8 cDNA. *Fish & Shellfish Immunology* 2003b; 14:275–281.

Inoue, Y., Saito, T., Endo, M., Haruta, C., Nakai, T., Moritomo, T., and Nakanishi, T. Molecular cloning and preliminary expression analysis of banded dogfish (*Triakis scyllia*) CC chemokine cDNAs by use of suppression subtractive hybridization. *Immunogenetics* 2005; 56:722–734.

Irmler, M., Hertig, S., MacDonald, H., Sadoul, R., Bercherer, J., Proudfoot, A., Solari, R., and Tschopp, J. Granzyme A is an interleukin-1β converting enzyme. *Journal of Experimental Medicine* 1995; 181:1917–1922.

Jiang, S., and Dong, C. A complex issue on CD4(+) T-cell subsets. *Immunological Reviews* 2013; 252:5–11.

Kawaguchi, M., Adachi, M., Oda, N., Kokubu, F., and Huang, S.K. IL-17 cytokine family. *Journal of Allergy and Clinical Immunology* 2004; 114:1265–1273.

Kuchen, S., Robbins, R., Sims, G.P., Sheng, C., Phillips, T.M., Lipsky, P.E., and Ettinger, R. Essential role of IL-21 in B cell activation, expansion, and plasma cell generation during CD4 +T Cell-B cell collaboration. *Journal of Immunology* 2007; 179:5886–5896.

Kuroda, N., Uinuk-ool, T.S., Sato, A., Samonte, I.E., Figueroa, F., Mayer, W.E., and Klein, J. Identification of chemokines and a chemokine receptor in cichlid fish, shark, and lamprey. *Immunogenetics* 2003; 54:884–895.

Lagos, L.X., Iliev, D.B., Helland, R., Rosemblatt, M., and Jorgensen, J.B. CD40L—A costimulatory molecule involved in the maturation of antigen presenting cells in Atlantic salmon (*Salmo salar*). *Developmental and Comparative Immunology* 2012; 38:416–420.

Larousserie, F., Charlot, P., Bardel, E., Froger, J., Kastelein, R.A., and Devergne, O. Differential effects of IL-27 on human B cell subsets. *Journal of Immunology* 2006; 176:5890–5897.

Lee, G.R., Fields, P.E., Griffin, T.J., and Flavell, R.A. Regulation of the Th2 cytokine locus by a locus control region. *Immunity* 2003; 19:145–153.

Li, R., Dooley, H., Wang, T., Secombes, C.J., and Bird, S. Characterisation and expression analysis of B-cell activating factor (BAFF) in spiny dogfish (*Squalus acanthias*): Cartilaginous fish BAFF has a unique extra exon that may impact receptor binding. *Developmental and Comparative Immunology* 2012; 36:707–717.

Lin, H., Lee, E., Hestir, K., Leo, C., Huang, M., Bosch, E., Halenbeck, R. *et al.* Discovery of a cytokine and its receptor by functional screening of the extracellular proteome. *Science* 2008; 320:807–811.

Liu, H., Leo, C., Chen, X., Wong, B.R., Williams, L.T., Lin, H., and He, X. The mechanism of shared but distinct CSF-1R signaling by the non-homologous cytokines IL-34 and CSF-1. *Biochimica et Biophysica Acta* 2012; 1824:938–945.

Lutfalla, G., Crollius, H.R., Stange-thomann, N., Jaillon, O., Mogensen, K., and Monneron, D. Comparative genomic analysis reveals independent expansion of lineage-specific gene family in vertebrates: The class II cytokine receptors and their ligands in mammals and fish. *BMC Genomics* 2003; 4:29.

Ma, X.L., Lin, W.Y., Chen, Y.M., Stawicki, S., Mukhyala, K., Wu, Y., Martin, F., Bazan, J.F., and Starovasnik, M.A. Structural basis for the dual recognition of helical cytokines IL-34 and CSF-1 by CSF-1R. *Structure* 2012; 20:676–687.

McCurley, N., Hirano, M., Das, S., and Cooper, M.D. Immune related genes underpin the evolution of adaptive immunity in jawless vertebrates. *Current Genomics* 2012; 13:86–94.

Metcalf, D. The colony-stimulating factors and cancer. *Nature Reviews Cancer* 2010; 10:425–434.

Mizutani, H., Schechter, N., Lazarus, G., Black, R., and Kupper, T. Rapid and specific conversion of precursor interleukin-1 beta (IL-1 beta) to an active IL-1 species by human mast cell chymase. *Journal of Experimental Medicine* 1991; 174:821–825.

Mühlhahn, P., Zweckstetter, M., Georgescu, J., Ciosto, C., Renner, C., Lanzendörfer, M., Lang, K. *et al.* Structure of interleukin 16 resembles a PDZ domain with an occluded peptide binding site. *Nature Structural Biology* 1998; 5:682–686.

Nairz, M., Sonnweber, T., Schroll, A., Theurl, I., and Weiss, G. The pleiotropic effects of erythropoietin in infection and inflammation. *Microbes and Infection* 2012; 14:238–246.

Najakshin, A.M., Mechetina, L.V., Alabyev, B.Y., and Taranin, A.V. Identification of an IL-8 homolog in lamprey (*Lampetra fluviatilis*): Early evolutionary divergence of chemokines. *European Journal of Immunology* 1999; 29:375–382.

Nicola, N.A. (ed.) *Guidebook to Cytokines and Their Receptors*. Oxford University Press/Sambrook and Tooze, Oxford, 1994.

Nomiyama, H., Hieshima, K., Osada, N., Kato-Unoki, Y., Otsuka-Ono, K., Takegawa, S., Azawa, T. *et al.* Extensive expansion and diversification of the chemokine gene family in zebrafish: Identification of novel chemokine subfamily CX. *BMC Genomics* 2008; 9:222.

O'Connell, M.J., and McInerney. J.O. Gamma chain receptor interleukins: Evidence for positive selection driving the evolution of cell-to-cell communicators in the mammalian immune system. *Journal of Molecular Evolution* 2005; 61:608–619.

Oh, H., and Ghosh, S. NF-kappaB: Roles and regulation in different CD4(+) T-cell subsets. *Immunological Reviews* 2013; 252:41–51.

Pearson, W.R., and Lipman, D.I. Improved tools for biological sequence comparison. *Proceedings of the National Academy of Sciences USA* 1988; 85:2444–2448.

Peveri, P., Walz, A., Dewald, B., and Baggiolini, M. A novel neutrophil-activating factor produced by human mononuclear phagocytes. *Journal of Experimental Medicine* 1988; 167:1547–1559.

Ramirez-Carrozzi, V., Sambandam, A., Luis, E., Lin, Z., Jeet, S., Lesch, J., Hackney, J. *et al.* IL-17C regulates the innate immune function of epithelial cells in an autocrine manner. *Nature Immunology* 2011; 12:1159–1166.

Rautava, S., Lu, L., Nanthakumar, N.N., Dubert-Ferrandon, A., and Walker, W.A. TGF-beta2 induces maturation of immature human intestinal epithelial cells and inhibits inflammatory cytokine responses induced via the NK-kappa B pathway. *Journal of Pediatric Gastroenterology and Nutrition* 2012; 54:630–638.

Ren, W., Pang, S., You, F., Zhou, L., and Zhang, S. The first BAFF gene cloned from the cartilaginous fish. *Fish and Shellfish Immunology* 2011; 31:1088–1096.

Roberts, S., Gueguen, Y., de Lorgeril, J., and Goetz, F. Rapid accumulation of an interleukin 17 homolog transcript in *Crasso streagigas* hemocytes following bacterial exposure. *Developmental and Comparative Immunology* 2008; 32:1099–1104.

Roca, F.J., Cayuela, M.L., Secombes, C.J., Meseguer, J., and Mulero, V. Post-transcriptional regulation of cytokine genes in fish: A role for conserved AU-rich elements located in the 3'-untranslated region of their mRNAs. *Molecular Immunology* 2007; 44:472–478.

Rubartelli, A., Cozzolino, F., Talio, M., and Sitia, R. A novel secretory pathway for interleukin-1 beta, a protein lacking a signal sequence. *EMBO Journal* 1990; 9:1503–1510.

Santos, M.D., Yasuike, M., Hirono, I., and Aoki, T. The granulocyte colony-stimulating factors (CSF3s) of fish and chicken. *Immunogenetics* 2006; 58:422–432.

Schonbeck, U., Mach, F., and Libby, P. Generation of biologically active IL-1β by matrix metalloproteinases: A novel caspase-1 independent pathway of IL-1β processing. *Journal of Immunology* 1998; 161:3340–3346.

Secombes, C.J., Wang, T., and Bird, S. The interleukins of fish. *Developmental and Comparative Immunology* 2011; 35:1336–1345.

Senior, K. New aminosterols from the dogfish shark. *Drug Discovery Today* 2000; 5:267–268.

Sokol, C.L., Barton, G.M., Farr, A.G., and Medzhitov, R. A mechanism for the initiation of allergen-induced T helper type 2 responses. *Nature Immunology* 2008; 9:310–318.

Strieter, R.M., Polverini, P.J., Kunkel, S.L., Arenberg, D.A., Burdick, M.D., Kasper, J., Dzuiba, J. *et al.* The functional role of the ELR motif in CXC chemokine-mediated angiogenesis. *Journal of Biological Chemistry* 1995; 270:27348–27357.

Stylianou, E., and Saklatvala, J. Interleukin-1. *International Journal of Biochemistry and Cell Biology* 1998; 30:1075–1079.

Tan, Y.Y., Kodzius, R., Tay, B., Tay, A., Brenner, S., and Venkatesh, B. Sequencing and analysis of full-length cDNAs, 5-ESTs and 3-ESTs from a cartilaginous fish, the elephant shark (*Callorhinchus milii*). *PLoS One* 2012; 7:e47174.

Thomas, A.W., and Lotze, M.T. *The Cytokine Handbook*, 4th Edition. Elsevier Science Publishing Company: London, 2003.

Tohidi, R., Idris, I.B., Panandam, J.M., and Bejo, M.H. The effects of polymorphisms in IL-2, IFN-gamma, TGF-beta2, IgL, TLR-4, MD-2, and iNOS genes on resistance to *Salmonella enteritidis* in indigenous chickens. *Avian Pathology* 2012; 41:605–612.

Tsutsui, S., Nakamura, O., and Watanabe, T. Lamprey (*Lethenteron japonicum*) IL-17 upregulated by LPS-stimulation in the skin cells. *Immunogenetics* 2007; 59:873–882.

van der Aa, L.M., Chadzinska, M., Tijhaar, E., Boudinot, P., and Verburg-van Kemenade, B.M. CXCL8 chemokines in teleost fish: Two lineages with distinct expression profiles during early phases of inflammation. *PLoS One* 2010; 5:e12384.

Venkatesh, B., Kirkness, E.F., Loh, Y.H., Halpern, A.L., Lee, A.P., Johnson, J., Dandona, N. *et al.* Survey sequencing and comparative analysis of the elephant shark (*Callorhinchus milii*) genome. *PLoS Biology* 2007; 5:e101.

Venkatesh, B., Lee, A.P., Ravi, V., Maurya, A.K., Lian, M.M., Swann, J.B., Ohta, Y. *et al.* Elephant shark genome provides unique insights into gnathostome evolution. *Nature* 2014; 505(7482):174–179.

Wang, T., Bird, S., Koussounadis, A., Holland, J.W., Carrington, A., Zou, J., and Secombes, C.J. Identification of a novel IL-1 cytokine family member in teleost fish. *Journal of Immunology* 2009; 183:962–974.

Wang, T., Diaz-Rosales, P., Martin, S.A.M., and Secombes, C.J. Cloning of a novel interleukin (IL)-20 like gene in rainbow trout *Oncorhynchus mykiss* gives an insight into the evolution of the IL-10 family. *Developmental and Comparative Immunology* 2010a; 34:158–167.

Wang, T., Kono, T., Monte, M.M., Kuse, H., Costa, M.M., Korenaga, H., Maehr, T., Husain, M., Sakai, M., and Secombes, C.J. Identification of IL-34 in teleost fish: Differential expression of rainbow trout IL-34, MCSF1 and MCSF2, ligands of the MCSF receptor. *Molecular Immunology* 2013; 53:398–409.

Wang, T., Martin, S.A.M., and Secombes, C.J. Two interleukin-17C-like genes exist in rainbow trout *Oncorhynchus mykiss* that are differentially expressed and modulated. *Developmental and Comparative Immunology* 2010b; 34:491–500.

Weaver, C.T., Hatton, R.D., Mangan, P.R., and Harrington, L.E. IL-17 family cytokines and the expanding diversity of effector T cell lineages. *Annual Review of Immunology* 2007; 25:821–852.

Wei, S., Nandi, S., Chitu, S., Yeung, Y.G., Yu, W., Huang, M., Williams, M.T., Lin, H., and Stanley, E.R. Functional overlap but differential expression of CSF-1 and IL-34 in their CSF-1 receptor-mediated regulation of myeloid cells. *Journal of Leukocyte Biology* 2010; 88:495–505.

Wiens, G.D., and Glenney, G.W. Origin and evolution of TNF and TNF receptor superfamilies. *Developmental and Comparative Immunology* 2011; 35:1324–1335.

Zapata, A., Torroba, M., Sacedon, R., Varas, A., and Vicente, A. Structure of the lymphoid organs of elasmobranchs. *Journal of Experimental Zoology* 1996; 275:125–143.

Zou, J., Bird, S., Truckle, J., Bols, N., Horne, M., and Secombes, C.J. Identification and expression analysis of an IL-18 homologue and its alternatively spliced form in rainbow trout (*Oncorhynchus mykiss*). *European Journal of Biochemistry* 2004; 271:1913–1923.

Zou, J., Tafalla, C., Truckle, J., and Secombes, C.J. Identification of a second group of type I IFNs in fish sheds light on IFN evolution in vertebrates. *Journal of Immunology* 2007; 179:3859–3871.

CHAPTER 8

Shark Complement
Genes, Proteins and Function

Sylvia L. Smith and Masaru Nonaka

CONTENTS

SUMMARY

Complement, an integral part of the innate immune system, is a complex cascade of interacting proteins, some of which are soluble serum proteins, whereas others are membrane-anchored receptor proteins (Fearon, 1997, 1998; Hoffmann et al., 1999; Mayilyan et al., 2008). Complement can be activated by a variety of complex molecules, such as antigen–antibody complexes, aggregated antibody, endotoxin (lipopolysaccharide), complex sugars, and nucleic acids. The activation of complement can occur through one or more of three activation pathways, the classical pathway (CP), alternative pathway (AP), and lectin pathway (LP), and leads to an inflammatory response in the host. Activation involves enzymatic cleavage of several complement proteins with the generation of bioactive peptides that initiate a variety of immune reactions, including opsonization of targets and enhancement of phagocytosis, removal of antigen–antibody complexes, lysis of target cells, and stimulation of antibody production by B lymphocytes. The last is the mechanism by which complement is the bridge between the innate and adaptive immune system. The activation of complement is regulated and limited by a complex system of control proteins. Many complement components and their regulators are structurally and functionally similar and have arisen by gene duplication and in many instances contain structural domains/motifs that are also found in noncomplement proteins. A complement system and/or one or more complement components were described for vertebrate and invertebrate species across the phylogenetic spectrum. A lytic complement system composed of functionally distinct components and with activation properties of the mammalian CP was first demonstrated in serum of the nurse shark, *Ginglymostoma cirratum* (Jensen et al., 1981). Later, functional studies showed the involvement of complement-derived peptides in opsonization, chemotaxis, and anaphylaxis. Over the last decade, several shark complement proteins have been isolated and characterized, and these include homologues of C1, C3, and C4. In addition, genes for C3, C4, C5, C8α, C8β, C9, factor B/C2, and factor I have been cloned from shark species. Unfortunately, there is very little information on shark complement receptors and most of the control proteins. Complement of rays and skates has not been studied. With the recent publication of the genome of a chimaera, the elephant shark, *Callorhinchus milii*, it is hoped that the available sequence information will serve as a blue print for investigators in their efforts to identify and characterize additional complement components of cartilaginous fish and to define their functional role in innate immunity of these lower vertebrates.

HISTORICAL AND EVOLUTIONARY PERSPECTIVE OF COMPLEMENT

It was more than a century ago that the lytic properties of mammalian plasma were reported and the term *complement* was first coined to denote the factor in serum that complemented antibodies in the lysis of antibody-coated bacterial targets (Meyer, 1984). Our understanding of the functional role of complement, which is a complex system of soluble and cell-associated proteins, has expanded considerably from the early concept of a four-component system with a primary role of lysis of foreign targets. Most of what we currently know of complement genes and proteins has been acquired from the study of mammalian systems, particularly that of human (Hugh-Jones, 1986; Law and Reid, 1995; Müller-Eberhard, 1988; Pangburn and Müller-Eberhard, 1984; Volanakis and Frank, 1998; Walport, 2001a, 2001b), mouse (Andrews and Theofilopoulos, 1978; Odink et al., 1981), rabbit (Giclas et al., 1981), guinea pig (Linscott and Nishioka, 1963; Nelson et al., 1966; Nishioka and Linscott, 1963), and goat (Moreno-Indias et al., 2012). However, in the last four decades, immunologists have examined nonmammalian vertebrates for complement components and activity and significant information has accumulated on the structure and function of complement genes and proteins of birds (Koch, 1986; Koch et al., 1983; Koppenheffer et al., 1999), reptiles (Kuo et al., 2000), amphibians (Kunnath-Muglia et al., 1993; Weinheimer et al., 1971), fish

(Boshra *et al.*, 2006; Holland and Lambris, 2002; Nonaka and Smith, 1999; Sunyer and Lambris, 1998; Sunyer and Tort, 1994), and primitive chordates such as the lamprey (Nonaka *et al.,* 1994; Nonaka *et al.*, 1984), amphioxus (Suzuki *et al.*, 2000), and *Ciona* (Azumi *et al.*, 2003; Marino *et al.*, 2002; Nonaka and Kimura, 2006; Yoshizaki *et al.*, 2005). In addition, investigators interested in the evolutionary origins of this complex system have examined several invertebrate species (Smith *et al.*, 1999) for evidence of complement-related genes and/or proteins, and C3 orthologues have been cloned from nonchordate deuterostome sea urchin (Al-Sharif *et al.*, 1998) and sea cucumber (Zhou *et al.*, 2011), protostome horseshoe crab (Ariki *et al.*, 2008; Zhu *et al.*, 2005), spider (Sekiguchi *et al.*, 2011), clam (Prado-Alvarez *et al.*, 2009), squid (Castillo *et al.*, 2009), cnidaria coral (Dishaw *et al.*, 2005), and sea anemone (Fujito *et al.*, 2009; Kimura *et al.*, 2009). No C3 gene is present in the published genome information for a sponge, *Amphimedon queenslandica* (Srivastava *et al.*, 2010) and a choanoflagellate *Monosiga brevicollis* (King *et al.*, 2008), suggesting that the C3 gene arose in the eumetazoan lineage. Although the C3 gene was conserved in all deuterostome species analyzed thus far, it is not present in some deciphered genome sequences of protostome species such as *Drosophila melanogaster* (Adams *et al.*, 2000) and *Caenorhabditis elegans* (Consortium, 1998), indicating that loss of the C3 gene occurred in the protostome lineage. It has become apparent that although there is considerable evidence of significant structural conservation of complement genes/ proteins (Reid and Campbell, 1993), there is also diversity and variability in complement across species not only in the structural organization of components but also in the number of isoforms present. For example, in humans, C3 and C5 are each encoded by a single gene, whereas in rainbow trout, several forms of C3 (a pivotal protein in complement activation) are present each encoded by more than one gene (Sunyer *et al.*, 1996, 1997), and several isoforms of C5 have been described in the carp (Kato *et al.*, 2003). Similarly, three properdin factor P isoforms were reported in trout (Chondrou *et al.*, 2008) and four factor I isoforms in the nurse shark (Shin *et al.*, 2009).

Convincing evidence from studies on invertebrate and lower vertebrate species supports the hypothesis that contemporary mammalian complement originated from a much simpler system with a limited function and narrower functional role (Al-Sharif *et al.*, 1998; Dodds and Day, 1993; Farries and Atkinson, 1991; Nonaka *et al.*, 1984; Kasahara *et al.*, 1997). Furthermore, it is believed that the system grew in complexity by increasing its components by gene duplications and thus expanded its range of function and component interaction (Grossherger *et al.*, 1989). If one accepts this premise, then it would be reasonable, given what we know of the system to date, to consider the shark complement system to represent a simpler stage in the evolution of this complex system. The phylogenetic position of the shark makes it a key subject in the search for clues to the evolution of the complement system as it is the most primitive vertebrate to contain elements that characterize both innate and adaptive immunity (Bartl and Weissman, 1994; Clem *et al.*, 1967; Marchalonis and Edelman, 1965, 1966; Marchalonis *et al.*, 1988; Schluter and Marchalonis, 2003; for further details refer to Dooley, Chapter 2 in the book). It is clear that the immune system of the shark is distinctly different in several respects not only from mammals but also from the teleosts (bony fish) (Nakao *et al.*, 1998; Nakao *et al.*, 2003; Sunyer and Lambris, 1998; Sunyer and Tort, 1994; Sunyer *et al.*, 1997). Its adaptive immune system was the focus of study of several investigators who have shown, at the protein and gene level, distinct differences in its immunoglobulins, TCR, and MHC Class I and II from that of the well-characterized mammalian system (Bartl *et al.*, 1997; Criscitiello *et al.*, 2006; Dooley and Flajnik, 2006; Hsu, 1998; Hsu and Criscitiello, 2006; Kasahara *et al.*, 1993, 1997; Lee *et al.*, 2008; Rumfelt *et al.*, 2004). The innate (nonadaptive) immune mechanisms operating in the shark, however, were less studied. Shark innate humoral immunity is best represented by its very effective complement system, which is the focus here. Moreover, sharks (elasmobranchs) represent the first appearance of a lytic complement system. It is generally believed based on phylogenetic studies that of the three complement activation pathways identified in mammals, the CP arose by gene duplications from a simpler, more ancient LP and predates the AP. However, up until the LP was discovered, the AP was originally assumed to be the more primitive pathway (Dodds, 2002;

Dodds and Matsushita, 2007). Many important gene systems have evolved through the mechanism of gene duplication, and the establishment of the complement system is a prime example of gene duplications and exon-shuffling. For example, the terminal complement components C6, C7, C8 alpha and beta, and C9 are considered to be manifestation of sequential addition or deletion of different types of domains (Volanakis, 1998). In several teleost species, some complement genes are present in several isoforms (Chondrou et al., 2008; Kato et al., 2003; Sunyer et al., 1996, 1997; Nakao et al., 2003). Similarly, in elasmobranchs, certain complement genes are present in multiple forms, such as C3, Bf/C2, and factor I genes (Shin et al., 2007, 2009; Terado et al., 2002). It was suggested that in bony fish, the multiple forms of C3 and factor I serve a functional advantage by providing diversity because in ancient forms of adaptive immunity the antibody response is not well developed, although shark immunoglobulin accounts for almost 45% of total serum protein, and there is a considerable repertoire of natural antibody to provide diversity. To appreciate the complexities and evolution of complement, one must understand gene and protein structural relationships, and thus, shark complement will be described taking a comparative approach.

Studies of complement in cartilaginous fish have shown functional parallels to mammals sharing analogous components (Legler and Evans, 1967). Early functional studies identified a six-component complement system that had activation and functional properties similar to those of the CP of mammals. The six functionally distinct serum fractions when added sequentially to sensitized sheep erythrocytes resulted in hemolysis (Jensen et al., 1981). Later, it was reported that unsensitized rabbit erythrocytes could be lysed by shark complement through an alternative activation mechanism distinct from that of the CP (Culbreath et al., 1991; Webb, 1994). Electron microscopy revealed holes in the membrane of lysed erythrocytes very similar to those caused by mammalian complement suggesting that complement lysis by either one or both activation pathways most likely involved assembly of a MAC-like complex (Esser, 1994; Humphrey and Dourmashkin, 1969; Podack and Tschopp, 1984). Our understanding of the genes and proteins of shark complement has significantly increased since the topic was last reviewed (Smith, 1998). Although several complement genes were cloned from three species of shark, there is a significant lack of corresponding functional information, which makes it difficult to ascribe a reliable accurate functional role based on deduced primary structure alone. Here we provide details of our current understanding of shark complement.

MAMMALIAN COMPLEMENT: AN OVERVIEW

Because many of us initially learn about complement by the study of the system in humans and other mammals, we present here an overview of the human system (Figure 8.1) that will be used as the foundation upon which we will make comparisons with the system in sharks. Mammalian complement is a complex system of interacting activation proteins, control mechanisms, and cofactors, all of which behave harmoniously to form a unique system that is a major component of the innate immune response and also modulates certain aspects of the adaptive immune response (Carroll, 2004; Fearon and Carroll, 2000; Liszewski et al., 1996). It plays a significant role in the host's immediate response to foreign substances and is composed of soluble and membrane-bound proteins. A variety of substances can activate complement by one or more of the three activation pathways, CP, AP, and LP (Fujita, 2002). Each pathway involves the sequential proteolytic cleavage of specific components by serine proteases to generate active peptide fragments (Figure 8.1). Each pathway leads to the activation of two pivotal complement components, C3 and C5, through the specific action of multicomponent serine protease complexes, the C3 convertases (C4b2a; C3bBb) and C5 convertases (C4b2a3b; C3bBbC3b) (Reid and Porter, 1981; Sim and Laich, 2000). Unlike the activation pathways, the terminal lytic pathway does not involve proteolytic cleavage of components C6 through C9, but rather activation involves conformational changes following the sequential binding of components C6 and C7 to C5b, while the latter is still attached to the C3b moiety of

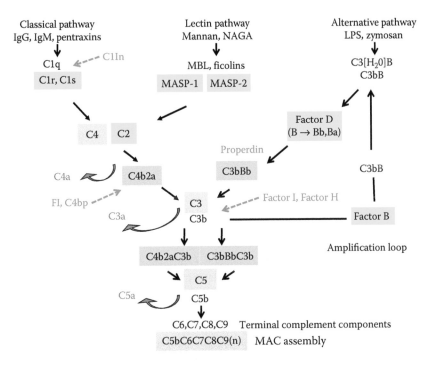

Figure 8.1 Components of complement activation pathways and terminal lytic components in humans. Shown for each pathway are activators and recognition molecules (black), serine proteases (yellow blocks), convertases (tan blocks), thiolester proteins (blue blocks), anaphylatoxins (blue lettering) regulator/control proteins (red lettering), MAC (light green block).

C5 convertase and the binding is thought to occur through sites on the α/β-chain of the C5b fragment (Podack, 1986; Thai and Ogata, 2003; Vogt *et al.*, 1978).

Lectin Pathway

The existence of the LP was recognized by the isolation and identification of specific components: the mannan-binding lectin (MBL) and MBL-associated serine proteases (MASPs) (Matsushita and Fujita, 1992, 1995; Ohta *et al.*, 1990; Sato *et al.*, 1994). The LP in humans has several recognition proteins. It can be initiated by the binding of MBL or collectin-11 to neutral sugar structures, such as mannan, mannose, or GlcNAc (*N*-acetylglucosamine)-terminating glycans, in the presence of Ca^{++}. MBL and collectin-11 are C-type lectins (collectins) with a collagen-like domain and a carbohydrate recognition domain (CRD) (Hansen *et al.*, 2010). Following the binding of these to a target molecule, the MBL-associated serine proteases (MASP-1 and MASP-2) subsequently activate C4 and C2 (Figure 8.1). The LP can also be activated by ficolins that contain a collagen and fibrinogen-like domain, with a binding specificity for acetylated structures, such as N-acetylated sugars (Krarup *et al.*, 2008). Ficolins also associate with MASPs and activate C4 and C2. Both MBL and ficolins function as pattern recognition molecules of the innate immune system.

Classical Pathway

The CP is activated in mammals when C1, the first complement component in the sequence, binds, through its C1q subcomponent, to charge clusters on surfaces (e.g., anionic phospholipids, lipid A) or to antibody- or pentraxin-coated targets and initiates the sequential activation of C4, C2, C3, and C5 (Figure 8.1). The activation of C3 followed by C5 is common to all three activation pathways. C1 is a macromolecular complex of three subunits, a single C1q molecule, and two molecules each of C1r and C1s associated in a Ca^{++}-dependent manner (Gigli et al., 1976; Ziccardi and Cooper, 1977). Structurally, C1q resembles a bunch of tulips with six globular heads connected to collagen-like stalks forming a stem. It is considered a pattern recognition molecule and binds by its globular heads to charge clusters, such as those on IgG and IgM immune complexes, and by doing so undergoes a conformational change that activates C1r which in turn activates C1s (Dodds et al., 1978; Naff and Ratnoff, 1968). C1r and C1s are serine proteases that are very similar in structure and by a series of gene duplication events likely arose from a common ancestral protein, which was also an ancestor of the MASPs. C1s cleaves and activates C4 into two fragments: a small C4a peptide and a larger C4b fragment, which through the thiolester bond in its α-chain is able to bind to amino and hydroxyl groups on the surface of complement activators and thus behaves like an opsonin and facilitates phagocytosis. When native C2 (a serine protease proenzyme) associates with C4b, it is cleaved by C1s into two fragments: a small peptide C2b that is released and a larger fragment C2a that remains associated with C4b to form the enzymatic complex C4b2a, the C3 convertase of CP. The cleavage of C3 by the convertase releases the small anaphylatoxin fragment, C3a, and generates C3b, a major opsonic molecule, which like C4b can bind to the surface of complement activators through its thiolester. C3b when bound to the C3 convertase complex changes its substrate specificity from C3 to C5 and forms the C5 convertase (C4b2aC3b) of the CP. The C5 convertase activates C5 by cleaving its α-chain and releasing a highly potent anaphyatoxin (C5a) fragment and a larger C5b fragment that initiates the assembly of the terminal components (C6 through C9) into the MAC.

Alternative Pathway

Unlike the CP and LP, the activation of the AP of complement activation in mammals is probably not triggered by the binding of pattern recognition molecules such as MBL and C1q to PAMPs (pathogen-associated molecular patterns) on targets, but involves a "tick-over" mechanism involving the slow hydrolysis of native C3 and exposure of an internal thiolester bond (Figure 8.1) (Pangburn, 1998). The formation of $C3(H_2O)$ in the fluid phase resembles C3b and permits binding of factor B that is subsequently cleaved into Ba and Bb fragments by factor D to form the first C3 convertase $[C3(H_2O)Bb]$ of the AP. Cleavage of native C3 by this initial convertase generates C3b with an activated thiolester by which it binds to hydroxyl or amino groups on the surface of targets and allows for the assembly of an anchored C3 convertase (C3bBb) that in turn cleaves more molecules of C3 and forms an amplification loop. The addition of a C3b molecule to the C3 convertase converts it to a C5 convertase (C3bBbC3b) with specificity for C5 cleavage (Figure 8.1). We should also consider the recent controversy over whether properdin could be an initiator of the AP. Some recent studies have suggested that properdin may be an initiating molecule (pattern recognition molecule) for the AP, in that properdin may bind directly to the target surface and then recruit binding of C3b, then FB to form the C3 convertase C3bBbP. The C3b could be derived from the activity of $C3(H_2O)Bb$ or of C4b2a. Direct binding of properdin to zymosan and rabbit erythrocytes (both AP activators) and to bacteria and apoptotic and necrotic cells was reported (Spitzer et al., 2007; Xu et al., 2008). That properdin may be a recognition molecule for initiating the AP was implied by Wedgewood and Pillemer (1958). Although properdin is not similar in structure to C1q, it is multimeric and has functional similarity in that it has multiple binding sites that recognize charge clusters and perhaps other motifs.

Terminal Pathway

Lysis of microorganisms and target cells is an effector function of mammalian complement and is accomplished through the terminal lytic pathway by the assembly and insertion of the membrane attack complex (MAC) into target membranes. MAC is a supramolecular organization of molecules that contains C5b, C6, C7, C8 (α, β, and γ), together with several molecules of C9, and is responsible for creating the transmembrane channels that disrupt the integrity of target membranes making them *leaky* with subsequent outpouring of cellular metabolites and cell lysis (Podack *et al.*, 1991). The terminal complement components (C6 through C9) in humans arose by a series of gene duplications and are structurally similar modular proteins (Figure 8.2), with each containing a perforin-like domain making them members of the perforin gene family (Hobart, 1998; Hobart *et al.*, 1995; Mondragon-Palomino *et al.*, 1999). What sets the components of the lytic pathway apart from components of the activation pathways (LP, CP, and AP) is that the activation of C6 through C9 does not involve proteolytic cleavage of the molecules but rather a sequential nonenzymatic conformational change involving hydrophilic-amphiphilic molecular transition with exposure of hydrophobic domains. This makes the molecular complex able to interact with the lipid bilayer of target membranes. The assembly of MAC is initiated with the formation of C5b which associates first with C6 and C7 followed by C8 and the polymerization of several molecules of C9 (Bhakdi and Tranum-Jensen, 1991).

Functionally, complement when activated triggers an inflammatory response that involves the attraction and chemotactic migration of leukocytes to the site of the triggering target. It promotes phagocytosis by opsonization, initiates lysis of target cells, clears antigen–antibody complexes, and stimulates antibody production by B lymphocytes (Carroll, 1998, 2004; Fearon and Locksley, 1996). Enhancement of phagocytosis of microorganisms by neutrophils and macrophages is accomplished by coating the surface of targets with C3b and C4b (i.e., opsonins), which are ligands for C3 receptors (CR1, CR3, CR4) on phagocytic cells. C3b is also the source of iC3b and C3d or C3dg, fragments formed by factor I cleavage of bound C3b while bound to the target. When iC3b is cleaved to C3dg, the remainder of the molecule, C3c, is released (see Figure 8.8a and b). The bound iC3b is a major opsonin, stimulating phagocytic uptake through the receptors CR3 and CR4, whereas C3d/C3dg modulates B- and T-cell response and promotes Ab production through the receptor CR2 (Carroll, 2004). By the binding of C3b to antigen–antibody immune complexes, they are solubilized and removed from circulation. The release of anaphylatoxins (C3a, C4a, C5a) leads to smooth muscle contraction and degranulation of mast cells with histamine release. C5a is a powerful chemotactic factor, attracting leukocytes. Lysis of target cells/bacteria is through the assembly and insertion of MAC. Complement has also been shown to be involved in wound healing and tissue/limb regeneration (Cazander *et al.*, 2012; Mastellos *et al.*, 2013).

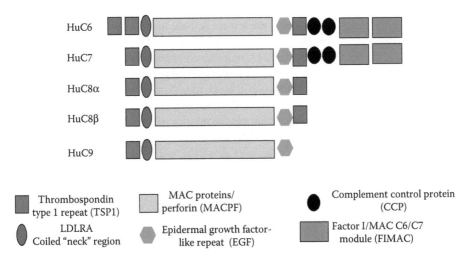

Figure 8.2 Modular structure of the terminal components of human complement. A linear comparison of the modular components of C6, C7, C8, and C9 is shown.

NONMAMMALIAN VERTEBRATE COMPLEMENT

Both functional and molecular evidence shows that AP, CP, and/or LP pathway(s) exist in non-mammalian higher vertebrates. Functional studies at the protein level were described only for a few species (Kato *et al.*, 2004; Sunyer *et al.*, 2005). In the absence of functional data at the protein level, it is unwise to endow putative proteins a functional role based on amino acid (aa) sequence alone. Structures (i.e., proteins and pathways) must have preceded function, and although genes reveal descent and ancestry, it is the functional capability resulting from these structures that came under selection pressure in the past. In evolutionary terms, structure comes before function, for example, Ig domains existed before immunoglobulins, and in terms of complement, thiolester proteins and serine proteases come before complement systems. Therefore, a protein resembling a complement-like protein could initially have evolved to have a function distinct from that of its role in a complement system. This might explain the presence of C6-like protein in amphioxus (a cephalochordate), whereas a MAC (C-related lytic activity) appears to be absent in lamprey (agnathan) (Nonaka and Kimura, 2006).

SHARK COMPLEMENT ACTIVATION AND LYTIC PATHWAYS

Complement-mediated hemolytic activity in elasmobranchs was first reported by Legler and Evans (1967) although no activation pathway was defined. Functionally, complement activity through the LP was not reported in the shark, nor has an MBL-like molecule been isolated from shark serum. Furthermore, the MBL gene was not cloned. However, one MASP cDNA was cloned and characterized from *Triakis scyllium*. Evolutionary analysis of the MASP, C1r, and C1s genes indicated that they are classified into two groups: one comprises only MASP1 and the other comprises MASP2, C1r, and C1s. The serine protease domain of the latter group is unique in that (1) the cysteine residues involved in formation of the disulfide bond called the histidine loop are missing, (2) the active site serine is encoded not by the TCN codon but by the AGY codon, and (3) the entire serine protease domain is encoded by a single exon. The serine protease domain of the shark MASP possessed all these three characteristics, suggesting that it is a member of the MASP2, C1r, and C1s group. However, the entire aa sequence of shark MASP showed a higher similarity to human MASP1 than to human MASP2. Thus, the evolutionary history of MASPs does not seem to be straightforward (Endo *et al.*, 1998).

Jensen *et al.* (1981) described a CP of complement activation in the shark. It involved at least six functionally distinct components reacting sequentially, namely, C1n, C2n, C3n, C4n, C8n, and C9n (n = nurse shark). The activity was initiated by immune complexes, was Mg^{++} and Ca^{++} dependent, and led to the hemolysis of sensitized target cells (sheep erythrocytes sensitized with shark natural antibody). Components C1n, C8n, and C9n were shown to be functionally analogous to mammalian C1, C8, and C9, respectively (Jensen *et al.*, 1973; Ross and Jensen, 1973a, 1973b). A later study described the physicochemical and functional characteristics of C2n showing it to be a functional analogue of mammalian C4 (Hyder Smith and Jensen, 1986). Dodds *et al.* (1998) isolated and purified from shark serum two thiolester-containing proteins, which, based upon the N-terminal aa sequence of their α and β chains, were shown to be homologues of mammalian C3 and C4. It should be noted that one or more of the "single" middle components (C3n and C4n) could yield additional activities (i.e., components) that remained undetected under earlier experimental conditions (a story similar to that of mammalian complement pathway elucidation). The isolation of factor B/C2, C3, C4, C5, and factor I cDNA clones from the shark suggests that additional components are present, although their position in the reaction sequence and exact role in complement activation remain to be defined functionally. Functional evidence for an AP of complement activation in the shark was first provided by the demonstration that shark serum could be activated by a mechanism distinct from that of the CP (Culbreath *et al.*, 1991; Webb, 1994). This alternative complement activity does not require immune complex to trigger it. It requires Mg^{++} but not Ca^{++} and is depleted when serum is treated with zymosan, inulin, or lipopolysaccharide (endotoxin). The pathway can be activated by unsensitized rabbit erythrocytes, and activity is not lost when serum natural antibody is absorbed with rabbit erythrocyte stroma. The pathway is more sensitive to heat inactivation (inactivated in 15 minutes at 48°C) than the CP (Smith unpublished data).

SHARK COMPLEMENT COMPONENTS

C1n

A shark C1 homologue (C1n) was the first shark complement protein to be isolated and shown to be the first in the activation sequence of CP by interacting with shark antibody bound to antigen. Ross and Jensen (1973a, 1973b) isolated, purified, and characterized the hemolytic and physicochemical properties of C1n obtained from low ionic strength euglobulin precipitation of shark serum. When purified C1n was subjected to immunoelectrophoresis in the presence of urea and using anti-C1n antiserum, three precipitated proteins were revealed suggesting that C1n is a complex of three subunits, likely analogues of mammalian C1q, C1r, and C1s. C1n was found to be incompatible with rabbit IgG or IgM but could bring about lysis of sheep erythrocytes (E) in C1-depleted guinea pig (gp) serum if erythrocytes were sensitized with shark natural antibody (EAn), indicating the compatibility of the cellular intermediate EAnC1n with gp C4 and C2. Interestingly, EAn is incompatible with gp C1. C1n is irreversibly inactivated by EDTA and when examined by immunoelectrophoresis, following chelation, separated into three subunits suggesting that Ca^{++} and/or Mg^{++} was required to hold the functional complex together. Later studies revealed the protein structure of shark C1(n)q purified from shark plasma (Dodds *et al.*, 1992; Smith, 1998). SDS-PAGE analysis of the protein under reducing conditions showed its disulfide-linked chain structure, and N-terminal aa sequence of two of the chains was similar to that of human C1q. Shark C1q-1 and C1q-2 showed 50% identity with human and mouse C1q-A and C1q-B chain, respectively, over the first 20 aa with a collagen-like sequence. N-terminal sequence for the globular domain was obtained after collagenase digestion and aligned with mammalian C1q.

Serine Protease: Bf/C2

The serine proteases of the complement cascade occur in serum as zymogens with the exceptions of factors D and I, which are present in active form. C2 and factor B, once activated by cleavage to C2a and Bb, form the catalytic moiety of the C3 and C5 convertases of the CP and AP, respectively. The two molecules have identical modular structures: in mammals, starting from the N-terminus there are three complement control protein (CCP) modules, one linker segment (LS), one von Willebrand factor type A (vWF-A) module, a second linker segment (LS), and one serine protease (SP) domain at the C-terminal. The CCP and VWFA modules contain binding sites for C4b or C3b (Horiuchi *et al.*, 1993; Hourcade *et al.*, 1995; Oglesby *et al.*, 1988; Tuckwell *et al.*, 1997; Xu and Volanakis, 1997). Although structurally similar, the two proteins are functionally distinct and each is encoded by a single gene. In humans, the two genes differ in size of introns with C2 gene being larger. C2 and factor B genes are believed to have evolved from a common ancestor by gene duplication (Reid and Campbell, 1993).

Separate genes for C2 and factor B were not cloned from sharks. However, cloning for factor B and C2 homologues yielded two distinct Bf/C2 cDNA clones, GcBf/C2-1 and GcBf/C2-2, from the nurse shark and TrscBf-1 and TrscBf-2 from the banded houndshark, *T. scyllium* (Shin *et al.*, 2007; Terado *et al.*, 2001). More recently, two sequences, Bf/C2 a-type and Bf/C2 b-type cDNA sequences,

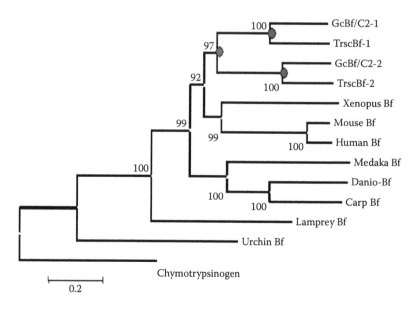

Figure 8.3 Phylogenetic tree of factor B/C2 of selected species. The numbers on branches are bootstrap percentages. This minimum evolution distance tree was constructed with 10,000 bootstrap replicates, using NNI branch swapping algorithm, and rooted on chymotrypsinogen. The bar indicates a genetic distance of 0.2. Yellow branch point = orthologous copies (duplication separated by speciation). Blue branch point = paralogous copies (duplication of genes).

were identified in the genome of the elephant shark, *Callorhinchus millii* (Venkatesh *et al.*, 2014). GcBf/C2-1 and GcBf/C2-2 genes of *G. cirratum* are orthologous to TrscBf-1 and TrscBf-2 genes of *T. scyllium*, respectively (Figure 8.3). The cDNA sequences of both shark genes show the same level of sequence similarity to both C2 and factor B of mammals, and consequently, cDNA sequence alone fails to distinguish between Bf and C2. In the absence of correlating protein functional data, it is not possible to definitively identify the clones as Bf or C2, hence the designation Bf/C2-1 and Bf/C2-2. The deduced primary structure of nurse shark Bf/C2 is similar to that of mammals, and sequence analysis has revealed a modular structure consisting three typical modules: CCP, vWF-A, and the serine protease (SP) domain. In the serine protease domain, the critical residues His[57], Asp[102], and Ser[195] (chymotrypsin numbering) (Kraut, 1977) that form the catalytic triad are present in both molecules but at slightly different positions. The presence and position of these residues is compatible with C3/C5 cleaving activity when associated with bimolecular (C3 convertase) or trimolecular (C5 convertase) enzyme complexes. Both shark proteins have putative cleavage sites for C1s and/or factor D, which suggests that the activation peptides released in case of both molecules are of comparable size. A detailed structural analysis of shark Bf/C2 homologues and their putative function was reported (Shin *et al.*, 2007). Analysis of N-linked glycosylation sites within the SP domain suggests that the shark molecules are organized more like C2 than Bf. Whether or not these glycosylation sites are occupied in the native shark proteins is not known. More recent genome analysis has shown that the intron/exon organization of GcBf/C2-1 and GcBf/C2-2 is more like that of C2 than factor B (Shin and Smith unpublished data). Interestingly, previously functional studies have shown the formation of a labile intermediate complex, C1nC2nC3n, in activated shark serum (through the CP) with properties similar to the labile mammalian complex, C142, suggesting that the yet-to-be characterized shark component, C3n, is a functional analogue of mammalian C2 (Jensen *et al.*, 1981). Functional studies have also revealed that AP activity requires a heat labile component, akin to mammalian factor B (Smith unpublished data). Taken together, these observations suggest that the enzymatic activity of AP and CP convertases likely resides in two distinct molecules, that is, factor B and C2 homologues. However, phylogenetic analysis indicated that the gene duplication event between shark Bf/C2-1 and Bf/C2-2 is independent from that between mammalian factor B and C2. It is of interest to clarify whether shark Bf/C2-1 and Bf/C2-2 show a similar functional diversification as human factor B and C2.

Thiolester Proteins and Genes: C3, C4, and C5

Complement proteins C3, C4, and C5 are thought to have arisen by gene duplication and belong to a family of proteins known as TEPs (thiolester proteins), which includes noncomplement protein α2M (alpha 2 macroglobulin, a protease inhibitor) and GPI-anchored CD109. All except C5 have the characteristic identifying thiolester bond (Dodds and Law, 1998; Fernandez and Hugli, 1977; Law and Dodds, 1997; Sottrup-Jensen et al., 1985) (Figure 8.4). In spite of the lack of a thiolester bond, C5 is included in the family because of the considerable structural similarity that has been conserved (Lambris et al., 1998). A major distinction and defining characteristic of complement thiolester proteins is in the C-terminal region of C3, a region that is absent from α2M and CD109. C3, C4, and C5 are each synthesized as a single chain precursor molecule that undergoes posttranslational modification involving glycosylation and specific cleavage into two chains (α and β) in case of C3 and C5, and three chains (α, β, and γ) in case of C4, linked together by disulfide bonds (Figure 8.4). All three molecules are present in serum in inactive form and are activated by specific cleavage of their α-chains into a small peptide (C3a, C4a, and C5a) released from the amino terminal end of their α-chain and a large fragment (C3b, C4b, and C5b) (Hugli, 1984, 1986).

Component C3

In contrast to the factor B/C2 gene duplication, which does not seem to have preceded the divergence of sharks from bony vertebrates, the C3/C4/C5 gene duplication certainly preceded this

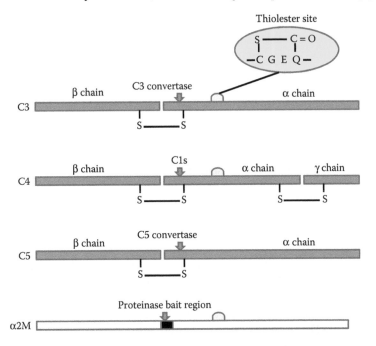

Figure 8.4 Structural organization of thiolester proteins (TEPs). Complement proteins C3, C4, and C5 showing α, β, and γ chains with thiolester and enzyme cleavage sites, and the noncomplement serum protein alpha-2-macroglobulin (α2M) showing the proteinase bait region and thiolester site.

divergence, and sharks possess the orthologues of mammalian C3, C4, and C5. In the shark, both activation pathways (CP and AP) result in the activation of C3. When complement in serum is activated with zymosan in the presence of EGTA, supplemented with Mg^{++}, a C3b fragment binds to zymosan. Subsequent elution and protein sequencing of zymosan-eluted proteins (ZEP) revealed peptide fragments with N-terminal sequence identical to the β-chain of C3 molecule previously isolated and purified from shark serum (Dodds *et al.*, 1998). In addition, analysis of ZEP fragments by Western blotting using specific antibody raised in mice identified shark C3-derived peptide fragments confirming the opsonic property of this protein (Smith unpublished data). Since these earlier studies, two C3-like cDNA clones (GenBank Accession nos: GcC3-1 KF670856; GcC3-2 KF670857) have been isolated from the nurse shark. Comparison of the deduced aa sequence corresponding to the N-terminal sequence obtained for the α and β chains of shark C3 protein isolated from serum has confirmed that clone GcC3-1 encodes for the serum protein. Whether this protein corresponds to the C4n component originally described by Jensen *et al.* (1981) was not determined. Comparison of the deduced aa sequence of GcC3-1 and GcC3-2 reveals that aa substitutions have occurred in specific regions throughout the two proteins, yet the physiochemical properties are retained in almost all replacements. This can be interpreted as structural conservation with "fine-tuning" of specific regions. GcC3-1 contains the thiolester-specific catalytic histidine (characteristic of hemolytic C3 proteins). Functional studies and protein isolation have confirmed in the nurse shark that the hemolytically active form is GcC3-1. GcC3-2 does not possess the catalytic histidine (substituted with tyrosine) and appears to have additional N-linked glycosylation sites including one near the putative properdin-binding site. The two C3 genes, GcC3-1 and GcC3-2, are 69% identical and 83% similar to each other and are 48% and 46% identical, and 68% and 65% similar to human C3, respectively (Smith unpublished data).

Component C4

Evidence for the presence of C4 in the shark was obtained from functional studies, which showed that nurse shark serum contained a component C2n that restored hemolytic activity to C4-deficient guinea pig serum (Hyder Smith and Jensen, 1986). The protein's molecular size was estimated to be 200 kDa. These studies were followed by the isolation and purification of nurse shark C4 from plasma (Dodds *et al.*, 1998). SDS-PAGE analysis of the reduced protein revealed it to be composed of three peptide chains, and N-terminal aa sequence data identified them as the α, β, and γ chains of shark C4. A C4 cDNA was cloned from the banded houndshark *T. scyllium* (Terado *et al.*, 2003). Sequence analysis confirmed the basic structure of shark C4 predicted by the protein-level analysis. However, shark C4 lacked the histidine residue catalytic for the thioester bond like human C4A. Linkage analysis indicated that the shark C4 gene is linked with two shark factor B genes as well as the shark MHC class I and II genes, suggesting that the basic structure of MHC was established before the emergence of cartilaginous fish more than 460 million years ago.

Component C5

Mammalian C5 is a glycoprotein of 190 kDa, and the native protein is composed of two polypeptide chains, an α-chain (115 kDa) and a β-chain (75 kDa) linked by a disulfide bond (Tack *et al.*, 1979; Wetsel *et al.*, 1988). Activation occurs by cleavage of the α-chain at a specific site by C5 convertase of either the AP or CP to generate C5b (180 kDa) and an anaphylotoxic and chemotactic peptide, C5a (9 kDa). Although the presence of shark C3 and C4 was confirmed by isolation of the proteins from shark serum, direct evidence for a C5 protein in serum was lacking (Dodds *et al.*, 1998; Hyder Smith and Jensen, 1986). Its existence, however, was indicated by the generation of potent complement-derived chemoattractant and spasmogenic activity in zymosan-activated shark serum (Smith *et al.*, 1997). Moreover, shark serum effectively lysed erythrocytes, and this ability most likely involved the assembly of a MAC-like macromolecular complex consistent with the involvement of a C5-like molecule (Jensen *et al.*, 1981; Smith, 1998). The existence of shark C5 was confirmed with the isolation and characterization of the complete cDNA sequence of C5 orthologue (GcC5) (Graham *et al.*, 2009). Analysis of the derived protein sequence shows that the molecule has structural elements necessary to function as a lead-in molecule for the assembly of a MAC-like complex. Furthermore, the molecule contains cleavage sites suggesting that upon activation a peptide with structural features of C5a anaphylatoxin is likely released. Structural analysis also revealed a domain structure similar to human and other vertebrate C5 molecules (Franchini *et al.*, 2001; Kato *et al.*, 2003; Kumar *et al.*, 2004; Lambris *et al.*, 1998; Ooi and Colten, 1979; Tack *et al.*, 1979; Wetsel *et al.*, 1987; Woods *et al.*, 1985). Because GcC5 lacks the sequence CGEQ that denotes thiolester bond in its α-chain, the GcC5b fragment, therefore, like mammalian C5, lacks the ability to covalently link to target surfaces, a feature seen with related opsonic molecules C3b and C4b. Three potential N-linked glycosylation sites suggest the native molecule is most likely a glycoprotein. Complement TEPs, C3, C4, and C5 have a unique feature in common that distinguishes them from other members of the TEP superfamily of proteins. Common to all three molecules is a ~150-residue-long C-terminal segment that contains the characteristic C345C module (Low *et al.*, 1999; Thai and Ogata, 2003). In C5, the module is considered important for its activation by C5 convertase (Sandoval *et al.*, 2000), and it was shown to play an important role in binding of C5 with C6 and C7 (Discipio, 1992b). C5b once formed is labile and decays (refolds) unless it quickly associates with C6 (Cooper and Müller-Ebearhard, 1970) through a metastable binding site for C6 followed by C7. Whether GcC5-C345C plays a similar role is uncertain because the existence of C6 and C7 orthologues in shark remains speculative. The constitutive expression of GcC5 was seen in most tissues examined including blood leukocytes with the highest level being in the liver. The GcC5 gene is present as a single copy, unlike the several C5 isotypes reported for carp (Kato *et al.*, 2003). C5 cDNA clones with a close sequence similarity to GcC5 have been isolated from two shark species, *Mustelus manazo* and *Squalus acanthias* (Nagumo and Nonaka unpublished data). However, only aberrant C5 cDNA clones lacking a long open reading frame were isolated from *T. scyllium*. When nucleotide sequences of the *T. scyllium* C5 cDNA clones were compared, none of them were identical possessing many insertions/deletions at various positions. These insertions/deletions were explained by unspliced intron and skip of exons, suggesting that the *T. scyllium* C5 gene is a novel type pseudogene. In accord with these observations, the AP hemolytic activity in serum was detected from *M. manazo* and *S. acanthias*, but not from *T. scyllium* (Nagumo and Nonaka unpublished data).

Anaphylatoxins: C3a, C4a, and C5a

Mammalian anaphylatoxins (C3a, C4a, and C5a) are generated by the cleavage of C3, C4, and C5 by specific serine proteases (MASP2, C1s, C3, and C5 convertases) (Figure 8.1). The peptides are highly cationic and characterized by cysteine residues that form three intra-chain disulfide bonds imparting significant chemical and physical stability to the molecules. Human anaphylatoxins are similar in structure and size, approximately 74–77 aa in length (Hugli, 1986; Hugli and Müller-Eberhard, 1978). Despite the variability in primary structure of C3a, C4a, and C5a of human and nonhuman species, complement anaphylatoxins appear very similar in overall tertiary structure. Physiologically, they mediate smooth muscle contraction, release histamine from mast cells, and increase vascular permeability (Gerard and Hugli, 1981). C5a is both a potent anaphylatoxin and a chemoattractant that enhances the ability of neutrophils and monocytes to adhere to vessel walls, migrate, and phagocytize particles (Gerard and Gerard, 1994; Gerard and Hugli, 1981; Hugli, 1986). A similar role for complement-derived anaphylatoxin was reported in several teleost fish (Kato *et al.*, 2004; Sunyer and Tort, 1994; Sunyer *et al.*, 2005). The activity of anaphylatoxins is controlled by a plasma enzyme, carboxypeptidase N, that cleaves the C-terminal arginine of the molecule and in case of C3a and C4a renders them inactive, whereas C5adesArg retains approximately 1% of its anaphylotoxic activity even though overall bioactivity is significantly reduced (Gerard and Hugli, 1981).

Functional evidence for the presence of complement-derived anaphylatoxins in activated shark serum was obtained when a low molecular weight fraction of zymosan-activated serum induced a spasmogenic (smooth muscle contraction of rat ileum) and chemotactic response (migration of human leukocytes) in mammalian cells in a manner similar to that of activated guinea pig and rat serum (Smith et al., 1997). Although the bioactive molecule(s) was not characterized, the absence of spasmogenic and chemotaxin activity in nonactivated or heat-inactivated serum strongly suggested that the activity was complement derived. The lack of response of shark leucocytes to activated human serum noted could be due to human C5a being a heavily glycosylated molecule causing steric hindrance, unlike porcine C5a, which is unglycosylated. Glycosylation of proteins, including a variety of immune molecules, affects their functional properties (Harvey et al., 2009; Lis and Sharon, 1993; Ritchie et al., 2002). Cloning of GcC5 cDNA shows that the deduced peptide sequence of putative GcC5a has the necessary structural elements including the last five residues at the C-terminus MQLGR considered essential for biological activity. The putative shark GcC5a peptide is 77 aa residues in length and lacks a glycosylation site. The conserved cysteine residues suggest preservation of a preferred folding pattern that will provide molecular stability required for maximal functional activity. The C-terminal putative effector region of GcC5a contains the sequence LTLGR that includes the critical terminal arginine residue that is considered essential for anaphylatoxic activity. The C-terminal sequence of GcC5a is distinct from the C-terminal sequences, SQMTLAR and TQMTLAR, of GcC3a-1 and GcC3a-2 (anaphylatoxin fragments of shark GcC3-1 and GcC3-2), respectively, and KVDSIAR of TrscC4a of the banded houndshark *T. scyllium*. Taken together, this suggests that complement-derived anaphylatoxins can be generated in activated shark serum.

Terminal Complement Components

Hemolytic activity of shark serum was known for decades (Legler and Evans, 1967). Jensen et al. (1981) reported that the lytic activity of shark serum involved two complement components, namely, t1 and t2, later referred to as C8n and C9n based on their terminal position in the activation sequence that resulted in the formation of holes in the membrane with subsequent lysis of the target sheep erythrocyte in a manner similar to that seen with human complement on sheep erythrocytes (Humphrey and Dourmashkin, 1969; Jensen et al., 1981). C8n and C9n are believed to be analogous to mammalian C8 and C9. It is highly likely, based on recent evidence and cloning of C5 C6, C8α, and C8β orthologues from shark species (Aybar et al., 2009; Kimura and Nonaka, 2009; Wang et al., 2013a, 2013b), that the proteins involved in the assembly of shark MAC are similar to those of mammals but this has not been conclusively shown. Although C6-like molecules have been described for *Amphioxus* (Suzuki et al., 2000) and *Ciona* (Wakoh et al., 2004), their functional role as complement proteins in these species was not demonstrated. Because shark complement very effectively lyses targets, the current hypothesis is that the assembly of a MAC most likely involves C6 and C7 functional analogues, although their presence in shark serum is unresolved. The presence of functional analogues of C8 and C9 was indicated when it was shown that EAC1-7hu cells (E = erythrocyte; A = antibody) could be lysed by two functionally distinct proteins in shark serum referred to as the terminal components C8n and C9n (Jensen et al., 1973). Western blotting has since revealed the presence in shark serum of proteins that cross react with antiserum to human C8 and C9. In reduced serum sample, reactive bands of ~70 and ~60 kDa, respectively, were present (Smith unpublished data).

Components C8 and C9

Human C8 is a trimeric complex composed of three nonidentical subunits α, β, and γ chains, each encoded by a separate gene (Ng *et al.*, 1987; Stekel *et al.*, 1980; Sodetz, 1989). C8α, C8β, and C8γ are 65, 65, and 22 kDa, respectively (Morley and Walport, 2000). Unlike human C8 in which the three C8 subunits collectively total 152 kDa, C8n appears to be considerably larger, approximately 190 kDa. This suggests structural differences in the C8n subunits. The difference in size could be explained by increased glycosylation of the putative subunits (Tsiftsoglou and Sim, 2004).

A C8α homologue (GcC8α) was the first MAC gene to be cloned from a shark (Aybar *et al.*, 2009). Based on the size of the coding region and not taking into account potential glycosylation of the molecule, the predicted molecular weight (589 aa residues) is likely to be higher than that of human C8α (554 aa residues), which also has fewer N-linked glycosylation sites (Figure 8.5). Recently, a C8α homologue (CpC8α) was cloned from a second shark species, the whitespotted bamboo shark, *Chiloscyllium plagiosum*, that shows similar conservation of modular structure (Wang *et al.*, 2013b). Four putative N-linked glycosylation sites and three mannosylation sites have been identified at similar positions in both species. The glycosylation sites of human C8α are at different positions (Figure 8.5). In the shark, the functional significance of a potential glycosylation site in the leader peptide (UTR) is unclear. The potential C-mannosylation sites are highly conserved in all orthologues examined. The distribution and position of hydrophobic residues through the entire coding region indicates that GcC8α and CpC8α most likely participate in hydrophilic-amphiphilic transition and contribute to the assembly and anchoring of a MAC-like macromolecule into target membranes. The region corresponding to human indel site that permits C8α to form a disulfide linked α–γ dimer with C8γ is present in GcC8α and contains the conserved cysteine residue suggesting the molecule may occur as a dimer. There is a single gene copy of GcC8α and the protein is synthesized in several tissues including erythrocytes and leukocytes, with the highest expression in the liver. In addition, C6 and C8β cDNAs have been isolated and characterized from a shark *M. manazo* and a chimaera *Chimaera phantasma*, respectively (Kimura and Nonaka, 2009). Shark C6 and chimaera C8β possessed exactly the same

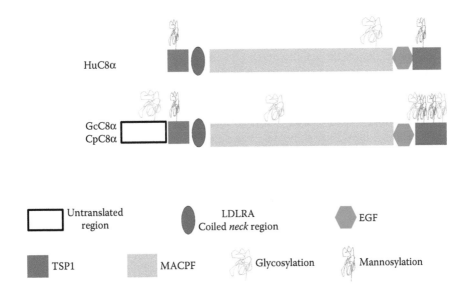

Figure 8.5   Comparison of human and shark C8α. Modular structure shows putative glycosylation and mannosylation sites in C8α from two species of shark: GcC8α of nurse shark, *G. cirratum*, and CpC8α of whitespotted bamboo shark, *C. plagiosum*.

domain structure as mammalian counterparts, TSP/TSP/LDLa/MACPF/EGF/TSP/CCP/CCP/FIM/FIM and TSP/LDLa/MACPF/EGF/TSP, respectively.

The polymerization of C9 completes the assembly of the MAC that forms the transmembrane channels, which appear in electron micrographs as membrane pore lesions (Humphrey and Dourmashkin, 1969). Sheep erythrocytes, when treated with shark complement components, exhibit similar lesions, however, of smaller diameter size (Jensen et al., 1981). A full length cDNA of C9 (CpC9) was cloned from the whitespotted bamboo shark, which encodes for a protein of 603 aa with three N-linked glycosylation sites positioned in the MACPF domain (Wang et al., 2013a). The protein was not purified from shark serum.

MISCELLANEOUS COMPLEMENT CONTROL/REGULATOR PROTEINS

Mammalian complement activation through any one of the pathways is tightly controlled by several regulator proteins that limit the extent of activation (Pangburn, 1998; Sim et al., 1993). In doing so, regulator proteins localize the activity of complement to specific foreign targets and protect self-cells from complement's lytic effect (Discipio, 1992a). Moreover, by restricting the extent of complement activation, they limit the systemic effect of circulating complement anaphylatoxins that, if unchecked, can lead to anaphylactic shock. Although the control of complement activity occurs at different stages of activation, a major target of complement control is the inactivation or limitation of C3 and C5 convertase activity (Figure 8.1). C1-inactivator inhibits C1 activity and regulates CP activation. It is a serine protease inhibitor (serpin) that targets C1r and C1s and also MASPs 1 and 2. Decay accelerating factor (DAF) is a membrane intrinsic protein that accelerates the decay of C3 and C5 convertases of the CP and AP and protects host cells from autologous complement lysis. Membrane cofactor protein (MCP, CD46), similarly, is present in the cell membrane and is a cofactor for factor I cleavage of C3b and C4b that deposit on self-cells. Factor I is an essential regulatory serine protease that functions to limit complement activity by cleaving sequentially several peptide bonds in the α'-chain of C3b and C4b (cleavage fragments of C3 and C4) (Davis and Harrison, 1982; Pangburn, 1977; Sim et al., 1993). C4b-binding protein (C4BP) is a cofactor for factor I cleavage and inactivation of C4b (Scharfstein et al., 1978). It binds to C4b and accelerates the decay of the C3 and C5 convertase of the CP and acts as a cofactor for the factor I-mediated cleavage of C4b to C4c$^+$ C4d. Factor H controls the activity of the C3 and C5 convertase of the AP by competing with factor B for C3b binding. It displaces the Bb subunit from the convertase and also serves as a cofactor for factor I cleavage of C3b to iC3b (the inactive form) (Discipio, 1992a; Pangburn et al., 2000; Ripoche et al., 1988; Sim and Discipio, 1982). Properdin of the AP is the only positive regulator; by binding to C3b, it protects it from factor I cleavage and enhances the binding of factor B to C3b and consequently stabilizes the C3bBb convertase complex. By doing so, it promotes the positive amplification of C3b deposition on activating surfaces. CD59 inhibits the formation of MAC by binding to C8 and blocking the subsequent polymerization of C9 molecules necessary for the formation of the MAC pore.

Factor I

The mature human factor I protein consists of an N-terminal heavy chain and a C-terminal light chain covalently linked through a disulfide bond (Catterall et al., 1987; Goldberg et al., 1987). The site for posttranslational cleavage of FI into heavy and light chains consists of four aa RKKR; in two shark species, G. cirratum and T. scyllium, it is RSKR suggesting a similar two-chain structure. There is a strong correlation between the exonic organization of the gene and the modular structure of the protein (Vyse et al., 1994). The arrangement of the protein modules starting from N-terminal is a leader peptide (LP), the FIMAC (factor I-membrane attack complex) domain, CD5 domain

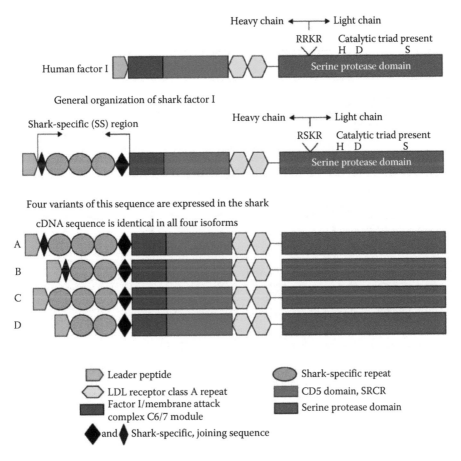

Figure 8.6 Modular organization of human and shark factor I. Comparison of structural organization of the four isoforms of shark factor I with that of human factor I showing the shark-specific region (SSR) that is absent from human factor I and that consists of repeat sequences.

(also known as SRCR, scavenger receptor cysteine-rich domain), two LDLRA (low-density lipo-protein receptor class A) domains, and the SP (serine protease) domain (Morley and Walport, 2000) (Figure 8.6). A factor I homologue in shark was first reported by Terado *et al.* (2002), who deduced the primary structure of shark factor I from cDNA cloned from the banded houndshark *Triakis*. They identified an insertion of repeat cDNA sequences that extended between the LP and FIMAC sequence and are not found in other vertebrates (Nakao *et al.*, 2003; Terado *et al.*, 2002). These insertions were further characterized when four factor I cDNAs (GcIf-1, GcIf-2, GcIf-3, and GcIf-4) were isolated from the nurse shark (Shin *et al.*, 2009). Sequence analysis of the four cDNAs revealed them to be identical except for an additional novel shark-specific region (SSR) that is absent from mammalian factor I and is likely formed by insertion, deletion, and/or duplication of small defined sequences (Figures 8.6 and 8.7). The SSR has been shown to be derived from differential organization in the shark genome of additional shark-specific (SS) sequence consisting up to three identical, tandem, domains. The aa sequence identities between the repeat sequence (RP) region originally identified in Trsc FI and the RS1~RS3 of the GcIf isotypes are not significantly high; however, the number of aa residues making up the repeat sequence is 16 in both species, suggesting that both GcIf and Trsc FI have evolved from an ancestor having a similar pattern of gene organiza-tion. If we interpret the boundaries between SS1 and RS1, and RS2 and RS3 (shown in Figure 8.7) as splicing acceptor sites, the different domain composition of the SSR of the four GcIf isotypes

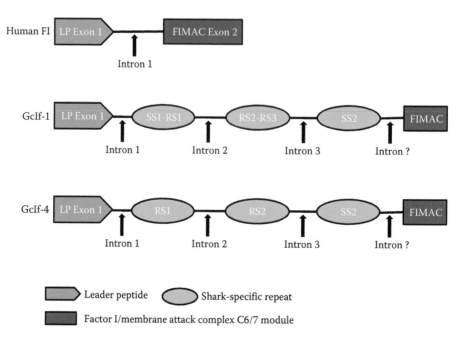

Figure 8.7   Genomic organization of the shark-specific region (SSR) of shark factor I isoforms (GcIf). GcIf-1
and GcIf-4 represent two of four distinct isoforms formed from differential organization of shark-
specific repeat sequences. Shown are repeat sequences (RS) and shark-specific (SS) sequences
in the SSR region. The differences of the different isoforms are primarily based on the presence or
absence of a domain. The SS joining sequences (shown in Figure 8.6) are not shown.

could be generated from a single gene by alternative splicing. However, we cannot emphatically
rule out the possibility that future genomic studies will show that the GcIf-1–4 isolates reflect four
genes, although in case of *T. scyllium* Southern blots indicate a single gene. The presence of additional gene inserts in the four highly similar factor I isotypes suggests different domain structure
that might reflect unique functional roles when expressed in different tissues. Taken together, the
findings of exons and introns in the region close to the N-terminus in the heavy chain, particularly
between the LP and FIMAC, suggest that it plays a key role in generating structural and functional
diversity of complement factor I in lower vertebrates. The absence of any corresponding insertion
between the LP and FIMAC domains in human and mouse indicates that this region of diversity in
cartilaginous fish has been lost at a specific point prior to the emergence of mammals.

     In the banded houndshark, the liver is the site of factor I synthesis. In the nurse shark, all four
isotypes (GcIf-1–4) are expressed in the liver and kidney; however, isotypes GcIf-2 and GcIf-3 were
also expressed in the brain and muscle. This differential pattern of expression was reported for two
factor I isotypes in the carp, a bony fish (Nakao *et al.*, 2003). The only structural difference between
GcIf-1/-2 and GcIf-3/-4 is the presence of SS1, which suggests that the SS1 may have a role in factor I expression. Currently, it is not known which GcIf isotype is functionally active in nurse shark
complement regulation. Why the nurse shark has four highly similar factor I molecules is a question
that remains to be answered. Furthermore, in the absence of functional studies, it is premature to
assume that shark C3b is the substrate for shark factor I. C3 and C4 proteins were isolated from
shark serum and two distinct C3 cDNAs and one C4 cDNA were cloned from sharks. In mammals,
a single factor I protein cleaves C3b and C4b (Figure 8.8a and b); it is plausible that the two shark
proteins are cleaved by different factor I isotypes in the shark. Furthermore, the two site-specific
cleavages of C3b by mammalian factor I to generate C3d and C3dg fragments may be the function
of separate factor I molecules in the shark. The inactivation of C3b and C4b by factor I requires

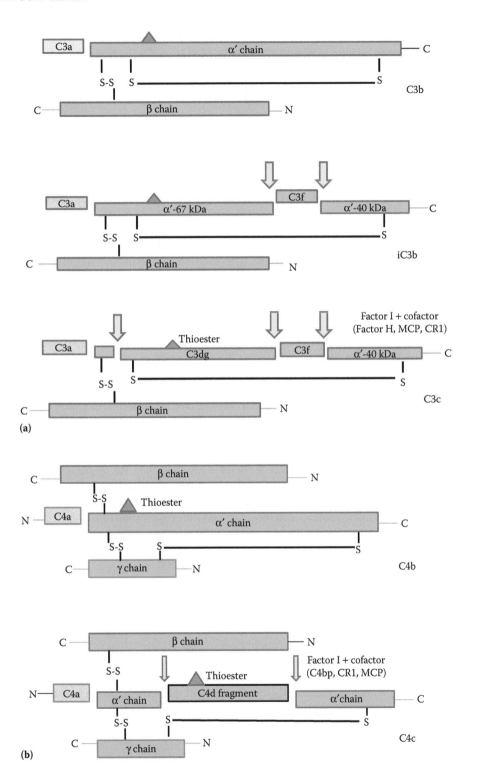

Figure 8.8 Cleavage of human C3 and C4 by factor I. (a) Sequential factor I cleavage of C3b releasing fragment C3dg from α'-chain. (b) Factor I cleavage of C4b releasing fragment C4d from α'-chain. Factor I cleavage sites shown by yellow arrows.

certain cofactors, such as factor H, C4b-binding protein (C4BP), complement receptor type 1 (CR1/CD35), and membrane cofactor protein (MCP/CD46) (Blon *et al.*, 2003; Nagasawa and Stroud, 1977; Pangburn *et al.*, 1977). It has, however, been shown that human factor I does not require cofactors to cleave synthetic substrates (Tsiftsogolou and Sim, 2004). Such cofactors were not identified in the shark.

COMPLEMENT RECEPTORS

Several mammalian complement receptors, which are found on a variety of cells (Ross, 1980), have been described. They are complement receptor type 1(CR1, CD35), type 2 (CR2, CD21), type 3 (CR3, CD11b, 18), and type 4 (CR4, CD11c, 18) that primarily bind peptides released from the cleavage of C3 and C4. CR1g is another complement receptor present on macrophages and involved in the phagocytosis of circulating pathogens (Helmy *et al.*, 2006). Cells also have receptors for C3a (C3aR), C5a (C5aR), and C1q (C1qR). Upon interaction with their complement ligands, cells can engage in chemotactic migration, immune adherence, phagocytosis, immune complex removal, and release of histamine and other metabolites such as cytokines and chemokines. It is through the specific interaction of C3d and CR2 on B cells that complement modulates adaptive immune response by forming a link between the innate and adaptive system. Unfortunately, information on shark complement receptors is scarce. Much of our understanding of complement receptors in the shark comes from functional studies that suggest the existence of receptors for certain complement peptides such as C5a. Shark leukocytes migrate chemotactically in response to LPS-activated rat, guinea pig, and shark serum (Obenauf and Hyder Smith, 1985; Hyder Smith *et al.*, 1989), suggesting the presence of a receptor for a complement-generated peptide in activated serum because the response was absent in untreated or heat-inactivated serum. Because no measures were taken to inhibit carboxypeptidase in activated rat or guinea pig serum, it was assumed that the anaphylatoxin was most likely C5a desArg. The chemotactic response was also induced by purified porcine C5a desArg further pointing to the presence of a receptor for the mammalian ligand on shark leukocytes (Obenauf and Hyder Smith, 1992). Another study illustrated the migrating cells to be the shark granulocyte and macrophage (Hyder Smith *et al.*, 1989). Whether the receptor is analogous to mammalian C5aR (CD88) remains undetermined. Such studies illustrate the interspecies functional compatibility of complement proteins and receptors and the conservation of key structural features.

CONCLUDING REMARKS

Contemporary mammalian complement is believed to have originated from a much simpler system with limited components and narrower functional role. The system grew in complexity by increasing its components by gene duplications and thus expanded its range of function and component interaction. The shark, a primitive vertebrate, is an extant animal model representing a stage in complement evolution where for the first time the complement system is functioning in an environment that also has a functional adaptive immune system. Given what we know to date of shark complement, it is likely that further investigation will reveal that some, if not all, essential structural elements necessary for complement-associated activities seen in higher vertebrates had evolved to some level at the time sharks diverged (about 450 million years ago) and that the fundamental and crucial interaction and communication between innate and adaptive immune system through complement was already established, albeit perhaps not necessarily in the form we see in contemporary mammals. It is likely that early molecules may have had greater molecular flexibility to permit a broader functional role, with narrower specific activity (such as that exhibited by complement serine proteases) evolving later. Defining what evolutionary changes evolved in the system when compared to lower vertebrates, such as hagfish, lamprey and amphioxus, is important (Nonaka,

1994; Suzuki *et al.*, 2002). It should be noted that sharks are not *living fossils* and have likely faced evolutionary pressures from pathogens different from those faced by mammals, and which may have caused increases in complexity different from those seen in mammals. The questions that can be asked are "what stage in the evolution of the complement system does the shark system represent and what evolutionary changes occur in the shark system that make it different from that found in lower vertebrates, such as hagfish, lamprey and Amphioxus (which lack components of an adaptive immune system), yet are similar to those found in mammals that possess a highly developed adaptive system? Also, can conservation of ancestral structural elements, that are highly similar in shark and human complement proteins and their cellular receptors, permit interspecies functional compatibility, particularly at the interface of innate and adaptive immune system?" The recent publication of the genome of the chaemera *C. milii* has revealed the presence of several complement genes, suggesting that this is likely true for other elasmobranchs (Venkatesh *et al.*, 2014). However, while most appear to be present, genes for only two C1q chains have been identified, similar to that previously seen in the nurse shark, in which only two C1(n)q chains were found; see previous section on C1n and Dodds *et al.* (1992). No factor D, MBL Ficolin, or MASP was found (based on their annotation), and many of the control proteins and receptors also seem to be missing. To obtain a more comprehensive picture of mammalian and other vertebrate and invertebrate complement systems related to the early pioneering work on complement cited in this chapter, we refer the reader to informative reviews for further details (Carroll and Sim, 2011; Lambris *et al.*, 1999; Litman, 1996; Nonaka and Kimura, 2006; Song *et al.*, 2000; Zhu *et al.*, 2005).

ACKNOWLEDGMENTS

Our thanks to Dr. Robert Sim for his constructive suggestions and to Dr. Alister Dodds for reviewing the manuscript. Thanks also to Dr. Dong-Ho Shin for his assistance with some of the figures.

REFERENCES

Adams, M.D.S.E., Celniker, R.A., Holt, C.A., Evans, J.D., Gocayne, P.G., Amanatides, S.E., Scherer, P.W. *et al.* The genome sequence of *Drosophila melanogaster*. *Science* 2000; 287:2185–2195.

Al-Sharif, W.Z., Sunyer, G.O., Lambris, J.D., and Smith, L.C. Sea urchin coelomocytes specifically express a homologue of the complement component C3 [published erratum appears in *J. Immunol.* 1999 Mar 1;162(5):3105]. *J. Immunol.* 1998; 160:2983–2997.

Andrews, B.S., and Theofilopoulos, A.N. A microassay for the determination of hemolytic complement activity in mouse serum. *J. Immunol. Methods* 1978:22(3–4):273–278.

Ariki, S., Takahara, S., Shibata, T., Fukuoka, T., Ozaki, A., Endo, Y., Fujita, T., Koshiba, T., and Kawabata, S. Factor C acts as a lipopolysaccharide-responsive C3 convertase in horseshoe crab complement activation. *J. Immunol.* 2008; 181:7994–8001.

Aybar, L., Shin, D.-H., and Smith, S. L. Molecular characterization of the alpha subunit of complement component C8 (GcC8α) in the nurse shark (*Ginglymostoma cirratum*). *Fish Shellfish Immunol.* 2009; 27(3):397–406.

Azumi, K., Santis, R.D., Tomaso, A.D., Rigoutsos, I., Yoshizaki, F., Pinto, M.R., Marino, R. *et al.* Genomic analysis of immunity in a Urochordate and the emergence of the vertebrate immune system: "Waiting for Godot." *Immunogenetics* 2003; 55:570–581.

Bartl, S., Baish, M.A., Flajnik, M.F., and Ohta, Y. Identification of class I genes in cartilaginous fish, the most ancient group of vertebrates displaying an adaptive immune response. *J. Immunol.* 1997; 159:6097–6104.

Bartl, S., and Weissman, I.L. Isolation and characterization of major histocompatibility complex class IIB genes from the nurse shark. *Proc. Natl. Acad. Sci. USA* 1994; 91:262–266.

Bhakdi, S., and Tranum-Jensen, J. Complement lysis: A hole is a hole. *Immunol. Today* 1991; 12(9):318–320.

Blon, A.M., Kask, L., and Dahlback, B. CCP1-4 of the C4b-binding protein alpha-chain are required for factor I mediated cleavage of complement factor C3b. *Mol. Immunol.* 2003; 39:547–556.

Boshra, H., Li, J., and Sunyer, J.O. Recent advances in the complement system of teleost fish. *Fish Shellfish Immunol.* 2006; 20:239–262.

Carroll, M.C. The role of complement and complement receptors in induction and regulation of immunity. *Annu. Rev. Immunol.* 1998; 16:545.

Carroll, M.C. The complement system in regulation of adaptive immunity. *Nat. Immunol.* 2004; 4:981–986.

Carroll, M.V., and Sim, R.B. Complement in health and disease. *Adv. Drug Deliv. Rev.* 2011; 63(12):965–975.

Castillo, M.G., Goodson, M.S., and McFall-Ngai, M. Identification and molecular characterization of a complement C3 molecule in a lophotrochozoan, the Hawaiian bobtail squid *Euprymna scolopes. Dev. Comp. Immunol.* 2009; 33:69–76.

Catterall, C.F., Lyons, A., Sim, R.B., Day, A.J., and Harris, T.J. Characterization of the primary amino acid sequence of human complement factor I from an analysis of cDNA clones. *Biochem. J.* 1987; 242:849–856.

Cazander, G., Jukema, G.N., and Nibbering, P.H. Complement activation and inhibition in wound healing. *Clin. Dev. Immunol.* 2012; 2012:534291, doi:10.1155/2012/534291.

Chondrou, M., Papanastasiou, A.D., Spyroulias, G.A., and Zakardis, I.K. Three isoforms of complement properdin factor P in trout? Cloning, expression, gene organization and constrained modeling. *Dev. Comp. Immunol.* 2008; 32:1454–1466.

Clem, I.W., De Boutaud, F., and Sigel, M.M. Phylogeny of immunoglobulin structure and function. II. Immunoglobulins of the nurse shark. *J. Immunol.* 1967; 99:1226–1235.

Consortium, T.C.E.S. Genome sequence of the nematode C. elegans: A platform for investigating biology. *Science* 1998; 282:2012–2018.

Criscitiello, M.F., Saltis, M., and Flajnik, M.F. The evolutionary mobile antigen receptor variable region gene: Doubly rearranging NAR-TcR genes in the shark. *Proc. Natl. Acad. Sci. USA* 2006; 28:36–41.

Culbreath, L., Smith, S.L., and Obenauf, S.D. Alternative complement pathway activity in nurse shark serum. *Am. Zool.* 1991; 31:131A.

Davis, A.E., and Harrison, R.A. Structural characterization of factor I mediated cleavage of the third component of complement. *Biochemistry* 1982; 21:5745–5749.

Discipio, R.G. Ultrastructures and interactions of complement factors H and I. *J. Immunol.* 1992a; 149:2592–2599.

Discipio, R.G. Formation and structure of the C5b-7 complex of the lytic pathway of complement. *J. Biol. Chem.* 1992b; 267:17087.

Dishaw, L.J., Smith, S.L., and Bigger, C.H. Characterization of a C3-like cDNA in a coral: Phylogenetic implications. *Immunogenetics* 2005; 57:535–548.

Dodds, A., and Law, S. The phylogeny and evolution of the thiolester bond-containing proteins C3, C4 and α2-macroglobulin. *Immunol. Rev.* 1998; 166:15–26.

Dodds, A.W. Which came first, the lectin/classical pathway or the alternative pathway of complement? *Immunobiology* 2002; 205:340–354.

Dodds, A.W., and Day, A.J. The phylogeny and evolution of the complement system. In: Whaley, K., Loos, M., and Weiler, J.M., editors. *Complement in Health and Disease.* The Netherlands: Kluwer Academic Publishers; 1993. pp. 39–88.

Dodds, A.W., and Matsushita, M. The phylogeny of the complement system and the origins of the classical pathway. *Immunobiology* 2007; 212:233–243.

Dodds, A.W., Sim, R.B., Porter, R.R., and Kerr, M.A. Activation of the first component of human complement (C1) by antibody-antigen aggregates. *Biochem. J.* 1978; 175(2):383–390.

Dodds, A.W., Smith, S.L., Levine, R.P., and Willis, A.C. Purification and initial characterization of complement component C1q from the channel catfish and the nurse shark. *III International Meeting on CI;* 1992 November 6–9. Mainz, Germany.

Dodds, A.W., Smith, S.L., Levine, R.P., and Willis, A.C. Isolation and initial characterization of complement components C3 and C4 of the nurse shark and the channel catfish. *Dev. Comp. Immunol.* 1998; 22:207–216.

Dooley, H., and Flajnik, M.F. Antibody repertoire development in cartilaginous fish. *Dev. Comp. Immunol.* 2006; 30:43–56.

Endo, Y., Takahashi, M., Nakao, M., Saiga, H., Sekine, H., Matsushita, M., Nonaka, M., and Fujita, T. Two lineages of mannose-binding lectin-associated serine protease (MASP) in vertebrates [published erratum appears in *J. Immunol.* 2000 May 15;164(10):5330]. *J. Immunol.* 1998; 161:4924–4930.

Esser, A.F. The membrane attack complex of complement: Assembly, structure and cytotoxic activity. *Toxicology* 1994; 87:229–247.

Farries, T.C., and Atkinson, J.R. Evolution of the complement system. *Immunol. Today* 1991; 12:295–300.

Fearon, D.T. Seeking wisdom in innate immunity. *Nature* 1997; 388:323.

Fearon, D.T. The complement system and adaptive immunity. *Seminars in Immunology.* Elsevier; 1998; 10(5): 355–361.

Fearon, D.T., and Carroll, M.C. Regulation of B lymphocyte responses to foreign and self-antigens by the CD19/CD21 complex. *Annu. Rev. Immunol.* 2000; 18:393–422.

Fearon, D.T., and Locksley, R.M. The instructive role of innate immunity in the acquired immune response. *Science* 1996; 272:50.

Fernandez, H.N., and Hugli, T.E. Chemical evidence for genetic ancestry of complement components C3 and C4. *J. Biol. Chem.* 1977; 252:1826–1828.

Franchini, S., Zarkadis, I.K., Sfyroera, G., Sahu, A., Moore,W.T., Mastellos, D., LaPatra, S.E., and Lambris, J.D. Cloning and purification of the rainbow trout fifth component of complement (C5). *Dev. Comp. Immunol.* 2001; 25:419–430.

Fujita, T. Evolution of the lectin-complement pathway and its role in innate immunity. *Nat. Rev. Immunol.* 2002; 2:346–352.

Fujito, N.T., Sugimoto, S., and Nonaka, M. Evolution of thioester-containing proteins revealed by cloning and characterization of their genes from a cnidarian sea anemone, *Haliplanella lineate. Dev. Comp. Immunol.* 2009; 34:775–784.

Gerard, C., and Gerard, N. C5a anaphylatoxin and its seven transmembrane segment receptor. *Annu. Rev. Immunol.* 1994; 12:775–808.

Gerard, C., and Hugli, T.E. C5a: A mediator of chemotaxis and cellular release reactions. *Kroc Found. Ser.* 1981; 14:147–160.

Giclas, P.C., Keeling, P.J., and Henson, P.M. Isolation and characterization of the third and fifth components of rabbit complement. *Mol. Immunol.* 1981; 18:113–123.

Gigli, I., Porter, R.R., and Sim, R.B. The unactivated form of the first component of human complement, C1. *Biochem. J.* 1976; 157(3):541–548.

Goldberger, G., Bruns, G.A.P., Rits, M., Edge, M.D., and Kwiatkowski, D.J. Human complement factor I: Analysis of cDNA-derived primary structure and assignment of its gene to chromosome 4. *J. Biol. Chem.* 1987; 262:10065–10071.

Graham, M., Shin, D.-H., and Smith, S.L. Molecular and expression analysis of complement component C5 in the nurse shark (*Ginglymostoma cirratum*) and its predicted functional role. *Fish Shellfish Immunol.* 2009; 27:40–49.

Grossherger, D., Marcuz, A., Dupasquier, L., and Lambris, J.D. Conservation of structural and functional domains in complement component C3 of *Xenopus* and mammals. *Proc. Natl. Acad. Sci. USA* 1989; 86:1323–1327.

Hansen, S., Selman, L., Palaniyar, N., Ziegler, K., Brandt, J., Kliem, A., Jonasson, M. *et al.* Collect in 11 (CL-11, CL-K1) is a MASP-1/3-associated plasma collect in with microbial-binding activity. *J. Immunol.* 2010; 185(10):6096–6104, doi:10.4049/jimmunol.1002185.

Harvey, D.J., Crispin, M., Moffat, B.E., Smith, S.L., Sim, R.B., Rudd, P.M., and Dwek, R.A. Identification of high mannose and multiantennary complex-type N-linked glycans containing alpha-galatose epitopes from the nurse shark IgM heavy chain. *Glycoconj. J.* 2009; 26(8):1055–1064.

Helmy, K.Y., Katschke, K.J., Jr., Gorgani, N.N., Kljavin, N.M., Elliott, J.M., Diehl, L., Scales, S.J., Ghilardi, N., and van Lookeren Campagne, M. CRIg: A macrophage complement receptor required for phagocytosis of circulating pathogens. *Cell* 2006; 124(5):915–927.

Hobart, M. Evolution of the terminal complement genes: ancient and modern. *Exp. Clin. Immunogenet.* 1998; 15:235e43.

Hobart, M.J., Fernie, B.A., and DiScipio, R.G. Structure of the human C7 gene and comparison with the C6, C8a, C8b, and C9 genes. *J. Immunol.* 1995; 154:5188–5194.

Hoffmann, J.A., Kafatos, F.C., Janeway, C.A., Jr., and Ezekowitz, R.A.B. Phylogenetic perspectives in innate immunity. *Science* 1999; 284:1313.

Holland, M.H., and Lambris, J.D. The complement system in teleosts. *Fish Shellfish Immunol.* 2002:12:399–420.

Horiuchi, T., Kim, S., Matsumoto, M., Watanabe, I., Fujita, S., and Volanakis, J.E. Human complement factor B: cDNA cloning, nucleotide sequencing, phenotype conversion by site-directed mutagenesis and expression. *Mol. Immunol.* 1993; 30:1587–1592.

Hourcade, D.E., Wagner, L.M., and Oglesby, T.J. Analysis of the short consensus repeats of human complement factor B by directed mutagenesis. *J. Biol. Chem.* 1995; 270:19716–19722.

Hsu, E. Maturation, selection and memory in B lymphocytes of ectothermic vertebrates. *Immunol. Rev.* 1998; 162:25–36.

Hsu, E., and Criscitiello, M.F. Diverse immunoglobulin light chain organization in fish retain potential to revise B cell receptor specificities. *J. Immunol.* 2006; 177:2452–2462.

Hugh-Jones, N. The classical pathway. In: Ross, G.D., editor. *Immunobiology of the Complement System.* New York: Academic Press; 1986. pp. 21–44.

Hugli, T. Biochemistry and biology of anaphylatoxin. *Complement* 1986; 3:111–127.

Hugli, T.E. Structure and function of the anaphylatoxins. *Semin. Immunopathol.* 1984; 7:193–219.

Hugli, T.E., and Müller-Eberhard, H.J. Anaphylatoxins: C3a and C5a. *Adv. Immunol.* 1978; 25:1–55.

Humphrey, J.H., and Dourmashkin, R.R. The lesions in cell membranes caused by complement. *Adv. Immunol.* 1969; 11:75–115.

Hyder Smith, S., and Jensen, J.A. The second component (C2n) of the nurse shark complement system: Purification, physico-chemical characterization and functional comparison with guinea pig C4. *Dev. Comp. Immunol.* 1986; 10:191–206.

Hyder Smith, S., and Obenauf, S.D. Fine structure of shark leukocytes during chemotactic migration. *Tissue Cell* 1989; 21:47–58.

Jensen, J.A., Festa, E., Cayer, M., and Smith, D.S. The complement system of the nurse shark: Hemolytic and comparative characteristics. *Science* 1981; 214:566–569.

Jensen, J.A., Fuller, L., and Iglasias, E. The terminal components of the nurse shark C system. *J. Immunol.* 1973; 111:306–307.

Kasahara, M., McKinney, E.C., Flajnik, M.F., and Ishihashi, T. The evolutionary origin of the major histocompatibility complex; polymorphism of class II alpha chain genes in cartilaginous fish. *Eur. J. Immunol.* 1993; 23:2160–2165.

Kasahara, M., Nakaya, J., Satta, Y., and Takahata, N. Chromosomal duplication and the emergence of the adaptive immune system. *Trends Genet.* 1997; 13:90–92.

Kato, Y., Nakao, M., Mutsuro, J., Zarkadis, I.K., and Yano, T. The complement component C5 of the common carp (*Cyprinus carpio*): cDNA cloning of two distinct isotypes that differ in a functional site. *Immunogenetics* 2003; 54:807–815.

Kato, Y., Nakao, M., Shimizu, M., Wariishi, H., and Yano, T. Purification and functional assessment of C3a, C4a and C5a of the common carp (*C. caprio*) complement. *Dev. Comp. Immunol.* 2004; 28:901–910.

Kimura, A., and Nonaka, M. Molecular cloning of the terminal complement components C6 and C8beta of cartilaginous fish. *Fish Shellfish Immunol.* 2009; 27:768–772.

Kimura, A., Sakaguchi, E., and Nonaka, M. Multi-component complement system of Cnidaria: C3, Bf, and MASP genes expressed in the endodermal tissues of a sea anemone, *Nematostella vectensis*. *Immunobiology* 2009; 214:165–178.

King, N., Westbrook, M.J., Young, S.L., Kuo, A., Abedin, M., Chapman, J., Fairclough, S. *et al*. The genome of the choanoflagellate *Monosiga brevicollis* and the origin of metazoans. *Nature* 2008; 451:783–788.

Koch, C. Complement system in avian species. In: Toivanen, A., and Toivanen, P., editors. *Avian Immunology: Basis and Practice.* Boca Raton, FL: CRC Press; 1986. pp. 43–55.

Koch, C., Kongerslev, L., and Jensen, L.B. The alternative complement pathway in chickens. *Dev. Comp. Immunol.* 1983; 7:785–786.

Koppenheffer, T.L., Chan, S.W.S., and Higgins, D.A. The complement system of the duck. *Avian Pathol.* 1999; 28:17–25.

Krarup, A., Mitchell, D.A., and Sim, R.B. Recognition of acetylated oligosaccharides by human L-ficolin. *Immunol. Lett.* 2008; 118(2):152–156, doi:10.1016/j.imlet.2008.03.014.

Kraut, J. Serine proteases: Structure and mechanism of catalysis. *Annu. Rev. Biochem.* 1977; 46:331–358.

Kumar, K.G., Ponsuksili, S., Schellander, K., and Wimmers, K. Molecular cloning and sequencing of porcine C5 gene and its association with immunological traits. *Immunogenetics* 2004; 55:811–817.

Kunnath-Muglia, L.M., Chang, G.H., Sim, R.B., Day, A.J., and Ezekowitz, R.A. Characterization of *Xenopus laevis* complement factor I structure—Conservation of modular structure except for an unusual insert not present in human factor I. *Mol. Immunol.* 1993; 30(14):1249–1256.

Kuo, M.M., Lane, R.S., and Giclas, P.C. A comparative study of mammalian and reptilian alternative pathway of complement-mediated killing of the lyme disease spirochete (*Borrelia burgdorferi*). *J. Parasitol.* 2000; 86:1223–1228.

Lambris, J.D., Reid, K., and Volanakis, J. The evolution, structure, biology and pathophysiology of complement. *Immunol. Today* 1999; 20:207–211.

Lambris, J.D., Sahu, A., and Wetsel, R. The chemistry and biology of C3, C4 and C5. In: Volanakis, J.E., and Frank, M., editors. *The Human Complement System in Health and Disease*. New York: Marcel Dekker; 1998. pp. 83–118.

Law, S.K.A., and Dodds, A.W. The internal thioester and the covalent binding properties of the complement proteins C3 and C4. *Protein Sci.* 1997; 6:263.

Law, S.K.A., and Reid, K.B.M. Complement. In: Male, D., editor. *Complement*, 2nd edn. Oxford: IRL Press, 1995.

Lee, V., Huang, J.L., Lui, M.F., Malecek, K., Ohta, Y., Mooers, A., and Hsu, E. The evolution of multiple isotypic IgM heavy chain genes in the shark. *J. Immunol.* 2008; 180:7461–7470.

Legler, D.W., and Evans, E.E. Comparative immunology: Hemolytic complement in elasmobranchs. *Proc. Soc. Exp. Biol. Med.* 1967; 124:30–34.

Linscott, W.D., and Nishioka, K. Components of guinea pig complement. II. Separation of serum fractions essential for immune hemolysis and immune adherence. *J. Exp. Med.* 1963; 118:795–815.

Lis, H., and Sharon, N. Protein glycosylation: Structural and functional aspects. *Eur. J. Biochem.* 1993; 218:1–27.

Liszewski, M.K., Farries, T.C., Lublin, D.M., Rooney, I.A., and Atkinson, J.P. Control of the complement system. *Adv. Immunol.* 1996; 61:201.

Litman, G.W. Sharks and the origins of vertebrate immunity. *Sci. Am.* 1996; 275:67–71.

Low, P.J., Ai, R., and Ogata, L.R.T. Active sites in complement components C5 and C3 identified by proximity to indels in the C3/4/5 protein family. *J. Immunol.* 1999; 162:6580.

Lu, J., Thiel, S., Miederman, H., Timpl, R., and Reid, K.B.M. Binding of the pentamer/hexamer forms of mannan binding protein to zymogen activates the pro enzyme C1r2C1s2 complex of the classical pathway of complement without involvement of Clq. *J. Immunol.* 1990; 144:2287–2294.

Marchalonis, J., and Edelman, G.M. Phylogenetic origins of antibody structure. I. Multichain structure of immunoglobulins in the smooth dogfish (*Mustelus canis*). *J. Exp. Med.* 1965; 122:601–618.

Marchalonis, J., and Edelman, G.M. Polypeptide chains of immunoglobulins from the smooth dogfish (*Mustelus canis*). *Science* 1966; 154:1567–1568.

Marchalonis, J.J., Schluter, S.F., Rosenshein, I.L., and Wang, A.C. Partial characterization of immunoglobulin light chains of carcharhine sharks: Evidence for phylogenetic conservation of variable region and divergence of constant region structure. *Dev. Comp. Immunol.* 1988; 12:65–74.

Marino, R., Kimura, Y., De Santis, R., Lambris, J.D., and Pinto, M.R. Complement in urochordates: Cloning and characterization of two C3-like genes in the ascidian *Ciona intestinalis*. *Immunogenetics* 2002; 53:1055–1064.

Mastellos, D.C., Deangelis, R.A., and Lambris, J.D. Complement-triggered pathways orchestrate regenerative responses throughout phylogenesis. *Semin. Immunol.* 2013; 25(1):29–38.

Matsushita, M., and Fujita, T. Activation of the classical complement pathway by mannose binding protein with a novel C1s-like serine protease. *J. Exp. Med.* 1992; 176:1497–1502.

Matsushita, M., and Fujita, T. Cleavage of the third component of complement (C3) by mannose-binding protein-associated serine protease (MASP) with subsequent complement activation. *Immunobiology* 1995; 194:443–448.

Mayilyan, K.R., Kang, Y.H., Dodds, A.W., and Sim, R.B. The complement system in innate immunity. In: Heine, H., editor. *Innate Immunity of Plants, Animals and Humans*. Berlin, Germany: Springer-Verlag; 2008. pp. 219–236. 21.

Meyer, M.M. Complement: Historical perspectives and some current issues. *Complement* 1984; l:2–26.

Mondragon-Palomino, M., Pinero, D., Nicholson-Weller, A., and Lacette, J.P. Phylogenetic analysis of the homologous proteins of the terminal complement complex supports the emergence of C6 and C7 followed by C8 and C9. *J. Mol. Evol.* 1999; 49:282–289.

Moreno-Indias, I., Dodds, A.W., Argüello, A., Castro, N., and Sim, R.B. The complement system of the goat: Haemolytic assays and isolation of major proteins. *BMC Vet. Res.* 2012; 8:91 http://www.biomedcentral .com/1746-6148/8/91.

Morley, B.J., and Walport, M.J. editors. *The Complement Facts Book*, 1st edn. London: Academic Press; 2000. pp. 105–120.

Müller-Eberhard, H.J. Molecular organization and function of the complement system. *Annu. Rev. Biochem.* 1988; 57:321.

Naff, G.B., and Ratnoff, O.S. The enzymatic nature of C'1r. Conversion of C'1s to C'1 esterase and digestion of amino acid esters by C'1r. *J. Exp. Med.* 1968; 128(4):571–593.

Nagasawa, S., and Stroud, R.M. Mechanism of action of the C3b inactivator: Requirement for a high molecular weight cofactor (C3b-C4bINA cofactor) and production of a new C3b derivative (C3b'). *Immunochemistry* 1977; 14:749–756.

Nakao, M., Fushitani, Y., Fujiki, K., Nonaka, M., and Yano, T. Two diverged complement factor B/C2-like cDNA sequences from a teleost, the common carp (*Cyprinus carpio*). *J. Immunol.* 1998; 161:4811–4818.

Nakao, M., Hisamatsu, S., Nakahara, M., Kato, Y., Smith, S.L., and Yano, T. Molecular cloning of the complement regulatory factor I isotypes from the common carp (*Cyprinus carpio*). *Immunogenetics* 2003; 54:801–806.

Nakao, M., Matsumoto, M., Nakazawa, M., Fujiki, K., and Yano, T. Diversity of complement factor B/C2 in the common carp (*Cyprinus carpio*): Three isotopes of B/C2-A expressed in different tissues. *Dev. Comp. Immunol.* 2002; 26:533–541.

Nelson, R.A., Jensen, J., Gigli, I., and Tainura, N. Methods for the separation, purification and measurement of nine components of haemolytic complement in guinea pig complement. *Immunochemisiry* 1966; 3:111–135.

Ng, S.C., Rao, A.G., Howard, O.M., and Sodetz, J.M. The eighth component of human complement (C8): Evidence that it is an oligomeric serum protein assembled from products of three different genes. *Biochemistry* 1987; 26: 5229–5233.

Nishioka, K., and Linscott, W.D. Components of guinea pig complement. I. Separation of a serum fraction essential for immune hemolysis and immune adherence, *J. Exp. Med.* 1963; 118:767.

Nonaka, M., Fujii, T., Kaidoh, T., Natsuume-Sakai, S., Yamaguchi, N., and Takahashi, M. Purification of a lamprey complement protein homologous to the third component of the mammalian complement system. *J. Immunol.* 1984; 133:3242–3249.

Nonaka, M., and Kimura, A. Genomic view of the evolution of the complement system. *Immunogenetics* 2006; 58:701–713.

Nonaka, M., and Smith, S.L. Complement system of bony and cartilaginous fish. *Fish Shellfish Immunol.* 1999; 10:215–228.

Nonaka, M., Takahashi, M., and Sasaki, M. Molecular cloning of a lamprey homologue of the mammalian MHC class III gene, complement factor B. *J. Immunol.* 1994; 152:2263–2269.

Obenauf, S.D., and Hyder Smith, S. Chemotaxis of nurse shark leukocytes. *Dev. Comp. Immunol.* 1985; 9:221–230.

Obenauf, S.D., and Hyder Smith, S. Migratory response of nurse shark leucocytes to activated mammalian sera and porcine C5a. *Fish Shellfish Immunol.* 1992; 2:173–181.

Odink, K.G., Fey, G., Wiebauer, K., and Diggelmann, H. Mouse complement components C3 and C4. Characterization of their messenger RNA and molecular cloning of complementary DNA for C3. *J. Biol. Chem.* 1981; 256(3):1453–1458.

Oglesby, T.J., Accavitti, M.A., and Volanakis, J.E. Evidence for a C4b binding site on the C2b domain of C2. *J. Immunol.* 1988; 141:926–931.

Ohta, M., Okada, M., Yamashina, I., and Kawasaki, T. The mechanisms of carbohydrate mediated complement activation by the serum mannose-binding protein. *J. Biol. Chem.* 1990; 265:1980–1984.

Ooi, Y.M., and Colten, H.R. Bliosynthesis and post-synthetic modification of a precursor (Pro-C5) of the fifth component of mouse complement (C5). *J. lmmunol.* 1979; 123:2494.

Pangburn, M.K. The alternative pathway: Activation and regulation. In: Rother, K., and Till, G.O., editors. *The Complement System*, 2nd edn. New York: Springer-Verlag; 1998. p. 93.

Pangburn, M.K., and Müller-Eberhard, H.J. The alternative pathway of complement. *Semin. Immunopathol.* 1984; 7:63–192.

Pangburn, M.K., Pangburn, K.L.W., Koistinen, V., Meri, S., and Sharma, A.K. Interactions among Factor H, C3b, and target in the alternative pathway of human complement. *J. Immunol.* 2000; 164:4742–4751.

Pangburn, M.K., Schreiber, R.D., and Müller-Eberhard, H.J. Human complement C3b inactivator: Isolation, characterization, and demonstration of an absolute requirement for the serum protein beta 1H for cleavage of C3b and C4b in solution. *J. Exp. Med.* 1977; 146:257–270.

Podack, E.R. Assembly and functions of the terminal components. In: Ross, G.D., editor. *Immunobiology of the Complement System.* New York: Academic Press; 1986. pp. 115–137.

Podack, E.R., Hengartner, H., and Lichtenheld, M.G. A central role of perforin in cytolysis? *Annu. Rev. Immunol.* 1991; 9:129–147.

Podack, E.R., and Tschopp, J. Membrane attack by complement. *Mol. Immunol.* 1984; 7:589–603.

Prado-Alvarez, M., Rotllant, J., Gestal, C., Novoa, B., and Figueras, A. Characterization of a C3 and a factor B-like in the carpet-shell clam, Ruditapes decussatus. *Fish Shellfish Immunol.* 2009; 26:305–315.

Reid, K.B.M., and Campbell, R.D. Structure and organization of complement genes. In: Whaley, K., Loos, M., and Weiler, J.M., editors. *Complement in Health and Disease.* Lancaster, UK: Kluwer Academic Publishers; 1993. pp. 89–125.

Reid, K.B.M., and Porter, R.R. The proteolytic activation systems of complement. *Annu. Rev. Biochem.* 1981; 50:433–464.

Ripoche, J., Day, A.J., Harris, T.J.R., and Sim, R.B. The complete amino acid sequence of human complement factor H. *Biochem. J.* 1988; 249:593.

Ritchie, G.E., Moffatt, B.E., Sim, R.B., Morgan, B.P., Dwek, R.A., and Rudd, P.M. Glycosylation and the complement system. *Chem. Rev.* 2002; 102:305–319.

Ross, G.D. Analysis of the different types of leukocyte membrane complement receptors and their interaction with the complement system. *J. Immunol. Methods* 1980; 37:197–211.

Ross, G.D., and Jensen, J.A. The first component (C1n) of the complement system of the nurse shark (*Ginglymostoma cirratum*) I. Hemolytic characteristics of partially purified C1n. *J. Immunol.* 1973a; 110:175–182.

Ross, G.D., and Jensen, J.A. The first component (C1n) of the complement system of the nurse shark (*Ginglymostoma cirratum*) II. Purification of the first component by ultracentrifugation and studies of its physicochemical properties. *J. Immunol.* 1973b; 110:911–918.

Rumfelt, L.L., Lohn, R.L., Dooley, H., and Flajnik, M.F. Diversity and repertoire of IgW and IgM VH families in the newborn nurse shark. *BMC Immunology* 2004; 5:8.

Sandoval, A., Ai, R., Ostresh, J.M., and Ogata, R.T. Distal recognition site for classical pathway convertase located in the C345C/netrin module of complement component C5. *J. Immunol.* 2000; 165:1066.

Sato, T., Endo, Y., Matsushita, M., and Fujita, T. Molecular characterization of a novel serine protease involved in activation of the complement system by mannose-binding protein. *Int. Immunol.* 1994; 6:665–669.

Scharfstein, J., Ferreira, A., Gigli, I., and Nussenzweig, V. Human C4-binding protein. I. Isolation and characterization. *J. Exp. Med.* 1978; 148(1):207–222.

Schluter, S.F., and Marchalonis, J.J. Cloning of shark RAG2 and characterization of the RAG1/RAG2 gene locus. *FASEB J.* 2003; 17:470–472.

Sekiguchi, R., Fujito, N.T., and Nonaka, M. Evolution of the thioester-containing proteins (TEPs) of the arthropoda, revealed by molecular cloning of TEP genes from a spider, *Hasarius adansoni. Dev. Comp. Immunol.* 2011; 36:483–489.

Shin, D.-H., Webb, B., Nakao, M., and Smith, S.L. Molecular cloning, structural analysis and expression of complement component Bf/C2 genes in the nurse shark, *Ginglymostoma cirratum. Dev. Comp. Immunol.* 2007; 31:1168–1182.

Shin, D.-H., Webb, M., Nakao, M., and Smith, S.L. Characterization of shark factor I gene(s): Genomic analysis of novel shark-specific sequence. *Mol. Immunol.* 2009; 46(11–12):2299–2308.

Sim, R.B., Day, A.J., Moffatt, B.E., and Fontaine, M. Complement factor I and cofactors in control of complement system convertase enzymes. *Meth. Enzymol.* 1993; 223:13–35.

Sim, R.B., and Discipio, R.G. Purification and structural studies on the complement-system control protein β1H (Factor H). *Biochem. J.* 1982; 205:285–293.

Sim, R.B., and Laich, A. Innate immunity: Serine proteases of the complement system. *Biochem. Soc. Trans.* 2000; 28:545–550.

Smith, L.C., Azumi, K., and Nonaka, M. Complement system in invertebrates. The ancient alternative and lectin pathways. *Immunopharmacology* 1999; 42:107–120.

Smith, S.L. Shark complement: An assessment. *Immunol. Rev.* 1998; 166:67–78.

Smith, S.L., Riesgo, M., Obenauf, S., and Woody, C. Anaphylactic and chemotactic response of mammalian cells to zymosan activated shark serum. *Fish Shellfish Immunol.* 1997; 7:503–514.

Sodetz, J.M. Structure and function of C8 in the membrane attack sequence of complement. *Curr. Top. Microbiol. Immunol.* 1989; 140:19–31.

Song, W.-C., Sarrias, M.R., and Lambris, J.D. Complement and innate immunity. *Immunopharmcology* 2000; 49:187–198.

Sottrup-Jensen, L., Stepanik, T.M., Kristensen, T., Lonblad, P.B., Jones, C.M., Wierzbicki, D.M., Magnusson, S. *et al.* Common evolutionary origin of alpha 2-macroglobulin and complement components C3 and C4. *Proc. Natl. Acad. Sci. USA* 1985; 82:9–13.

Spitzer, D., Mitchell, L.M., Atkinson, J.P., and Hourcade, D.E. Properdin can initiate complement activation by binding specific target surfaces and providing a platform for *de novo* convertase assembly. *J. Immunol.* 2007; 179:2600–2608.

Srivastava, M., Simakov, O., Chapman, J., Fahey, B., Gauthier, M.E., Mitros, T., Richards, G.S. *et al.* The Amphimedon queenslandica genome and the evolution of animal complexity. *Nature* 2010; 466:720–726.

Steckel, E.W., York, R.G., Monahan, J.B., and Sodetz, J. The eighth component of human complement. Purification and physicochemical characterization of its unusual subunit structure. *J Biol Chem* 1980; 255:11997e2005.

Sunyer, J.O., Boshra, H., and Li, J. Evolution of anaphylatoxins their diversity and novel roles in innate immunity: Insights from the study of fish complement. *Vet. Immunol. Immunopathol.* 2005; 108:77–89.

Sunyer, J.O., and Lambris, J.D. Evolution and diversity of the complement system of poikilothermic vertebrates. *Immunol. Rev.* 1998; 166:39.

Sunyer, J.O., and Tort, L. The complement system of the teleost fish *Sparus aurata. Ann. NY Acad. Sci.* 1994; 712:371.

Sunyer, J.O., Tort, L., and Lambris, J.D. Structural C3 diversity in fish: Characterization of five forms of C3 in the diploid fish *Sparus aurata. J. Immunol.* 1997; 158:2813–2821.

Sunyer, J.O., Zarkadis, I.K., Sahu, A., and Lambris, J.D. Multiple forms of complement C3 in trout that differ in binding to complement activators. *Proc. Natl. Acad. Sci. USA* 1996; 93:8546–8551.

Suzuki, N.M., Satoh, N., and Nonaka, M. C6-like and C3-like molecules from the cephalochordate, amphioxus, suggest a cytolytic complement system in invertebrates. *J. Mol. Evol.* 2000; 54:671–679.

Tack, B.F., Harrison, R.A., Janatova, J., Thomas, M.L., and Prahl, J.W. Evidence for presence of an internal thiolester bond in third component of human complement. *Proc. Natl. Acad. Sci. USA* 1980; 77:5764–5768.

Tack, B.F., Morris, S.C., and Prahl, J.W. Fifth component of human complement: Purification from plasma and polypeptide chain structure. *Biochemistry* 1979; 18:1490–1497.

Terado, T., Nonaka, M.I., Nonaka, M., and Kimura, H. Conservation of the modular structure of complement factor I through vertebrate evolution. *Dev. Comp. Immunol.* 2002; 26:403–413.

Terado, T., Okamura, K., Ohta, Y., Shin, D.H., Smith, S.L., Hashimoto, K., Takemoto, T. *et al.* Molecular cloning of C4 gene and identification of the class III complement region in the shark MHC. *J. Immunol.* 2003; 171:2461–2466.

Terado, T., Smith, S.L., Nakanhishi, T., Nomaka, M.I., Kimura, H., and Nonaka, M. Occurrence of structural specialization of the serine protease domain of complement factor B at the emergence of jawed vertebrates and adaptive immunity. *Immunogenetics* 2001; 53:250–254.

Thai, C.T., and Ogata, R.T. Expression and characterization of the C345C/NTR domains of complement components C3 and C5. *J. Immunol.* 2003; 171:6565–6657.

Tomlinson, S. Complement defense mechanisms. *Curr. Opin. Immunol.* 1993; 5:83–89.

Tsiftsoglou, S.A., Arnold, J.N., Roversi, P., Crispin, M.D., Radcliffe, C., Lea, S.M., Dwek, R.A., Rudd, P.M., and Sim, R.B. Human complement factor I glycosylation: Structural and functional characterization of the N-linked oligosaccharides. *Biochim. Biophys. Acta* 2006; 1764:1757–1766.

Tsiftsoglou, S.A., and Sim, R.B. Human complement factor I does not require cofactors for cleavage of synthetic substrates. *J. Immunol.* 2004; 173:367–375.

Tuckwell, D.S., Xu, Y., Newham, P., Humphries, M.J., and Volanakis, J.E. Surface loops adjacent to the cation-binding site of the complement factor B-von Willebrand factor type A module determine C3b binding specificity. *Biochemistry* 1997; 36:6605–6613.

Venkatesh, B., Lee, A.P., Ravi, V., Maurya, A.K., Lian, M.M., Swann, J.B., Ohta, Y. *et al.* Elephant shark genome provides unique insights into gnathostome evolution. *Nature* 2014; 505(7482):174–179.

Vogt, W., Schmidt, E., Von Buttlar, B., and Dieminger, L.A. A new function of the activated third component of complement: Binding to C5, an essential step for C5 activation. *Immunology* 1978; 34:29–30.

Volanakis, J.E., editor. Overview of the complement system. In: *The Human Complement System in Health and Disease Cop.* New York: Marcel Dekker, Inc.; 1998. pp. 9–31.

Volanakis, J.E., and Frank, M.M. *The Human Complement System in Health and Disease.* New York: Marcel Dekker, Inc; 1998.

Vyse, T.J., Bates, G.P., Walport, M.J., and Marley, B.J. The organization of the human complement factor I gene (IF): A member of the serine proteases gene family. *Genomics* 1994; 24:90–98.

Wakoh, T., Ikeda, M., Uchino, R., Azumi, K., Nonaka, M., Kohara, Y., Metoki, H., Satou, Y., Satou, M., and Sataka, M. Identification of transcripts expressed preferentially in hemocytes of *Ciona intestinalis* that can be used as molecular markers. *DNA Research.* 2004; 11:345–352.

Walport, M.J. Advances in immunology: Complement part 1. *New Eng. J. Med.* 2001a; 344:1058–1066.

Walport, M.J. Advances in Immunology: Complement part 2. *New Eng. J. Med.* 2001b; 344:1140–1144.

Wang, Y., Xu, S., Su, Y., Ye, B., and Hua, Z. Molecular characterization and expression analysis of complement component C9 gene in the whitespotted bambooshark, *Chiloscyllium plagiosum. Fish Shellfish Immunol.* 2013a; 35:599–606.

Wang, Y., Zhang, M., Wang, C., Ye, B., and Hua, Z. Molecular cloning of the alpha subunit of complement component C8 (CpC8α) of whitespotted bamboo shark (*Chiloscyllium plagiosum*). *Fish Shellfish Immunol.* 2013b; 35:1993–2000.

Webb, B.M. Characterization of the complement alternative pathway in the shark and identification of a factor B-like component. Medical laboratory Sciences, Master's thesis. 1994. Florida International University, Miami, FL.

Wedgewood, R.J., and Pillemer, L. The nature and interactions of the properdin system. *Acta Haematol.* 1958; 20:253–259.

Weinheimer, P.F., Evans, E.E., and Acton, R.T. Comparative immunology: The hemolytic complement system of the anuran amphibian, *Bufo marinus. Comp. Biochem. Physiol. A Comp. Physiol.* 1971; 38(3):483–488.

Wetsel, R.A., Lemons, R.S., Le Beau, M.M., Barnum, S.R., Noack, D., and Tack, B.F. Molecular analysis of human complement component C5: Localization of the structural gene to chromosome 9. *Biochemistry* 1988; 27:1474–1482.

Wetsel, R.A., Ogata, R.T., and Tack B.F. Primary structure of the fifth component of murine complement. *Biochemistry* 1987; 26:737–743.

Woods, E., Ogden, R.C., Colten, H.R., and Tack, B.F. Isolation and sequence of a cDNA clone encoding the fifth complement component. *J. Biolog. Chem.* 1985; 260:2108.

Xu, Y., and Volanakis, J.E. Contribution of the complement control protein modules of C2 in C4b binding assessed by analysis of C2/Bf chimeras. *J. Immunol.* 1997; 158:5958–5965.

Xu, W., Berger, S.P., Trouw, L.A., de Boer, H.C., Schlagwein, N., Mutsaers, C., Daha, M.R., and van Kooten, C. Properdin binds to late apoptotic and necrotic cells independently of C3b and regulates alternative pathway complement activation. *J. Immunol.* 2008; 180:7613–7621.

Yoshizaki, F.Y., Ikawa, S., Satake, M., Satou, N., and Nonaka, M. Structure and the evolutionary implication of the triplicated complement factor B genes of a urochordate ascidian, *Ciona intestinalis. Immunogenetics* 2005; 56:930–942.

Zhou, Z., Sun, D., Yang, A., Dong, Y., Chen, Z., Wang, X., Guan, X., Jiang, B., and Wang, B. Molecular characterization and expression analysis of a complement component 3 in the sea cucumber (*Apostichopus japonicus*). *Fish Shellfish Immunol.* 2011; 31:540–547.

Zhu, Y., Thangamani, S., Ho, B., and Ding, J.L. The ancient origin of the complement system. *EMBO J.* 2005; 24:382–394.

Ziccardi, R.J., and Cooper, N.R. The subunit composition and sedimentation properties of human C1. *J. Immunol.* 1977; 118(6):2047–2052.

CHAPTER 9

MHC Molecules of Cartilaginous Fishes

Simona Bartl and Masaru Nonaka

CONTENTS

SUMMARY

The major histocompatibility complex (MHC) is a deoxyribonucleic acid (DNA) region that contains genes involved in antigen processing and presentation, though not exclusively. Evidence for MHC genes has been found in all jawed vertebrates studied, but not in jawless fish or invertebrates. Orthologous genes for the antigen presentation proteins, namely, class Iα/β-2-microglobulin (*mhc1a/b2m*) and class IIα/β (*mhc2a/b*), have been isolated in many cartilaginous fishes. The genes code for proteins with similar structures, levels of polymorphism, expression patterns, and linkage, though the number of loci appears reduced. Nonclassical class I genes (*mhc1b*), which have distinguishing features and functions, appear to exist in sharks as well. Two separate antigen-processing pathways provide the peptides that are ultimately loaded onto class I and class II molecules for presentation to

T cells. Some of the components are MHC linked, namely, proteosome (*psmb*) and transporter (*tap*) subunits from the class I pathway and a chaperone (DM) from the class II pathway, and some are not, such as class II invariant chain (Ii). The shark MHC appears to contain at least one copy of *mhc1a*, *mhc2a*, *mhc2b* as well as two *tap* and two *psmb* genes, two genes for components of the complement system (*c4* and *cfb*), and a *brd2* gene, all genes that are MHC linked in most other vertebrates. Unique to the shark MHC is the presence of *b2m*, an ancient *psmb8l* gene, and two copies of *cfb*. Sharks have two other genes, *cd74* and *ciita* (class II transactivator), that are important for class II protein function but are not MHC linked. Obviously missing from all fishes is DM, a nonclassical class II protein that serves as a chaperone in the class II antigen-processing pathway.

INTRODUCTION

For much of the twentieth century, evidence accumulated for genetic components that not only controlled the rejection of tissue allografts (transplants between members of the same species) but mapped to a single DNA region in mice and humans that became generically known as the major histocompatibility complex (MHC). Independently, studies of T-cell responses found that the recognition of foreign agents (antigens) required an association with proteins encoded in the MHC, a process termed *MHC restriction*. Studies of experimental models suggested that MHC-restricted antigens were proteins that were processed into peptides and then in some way presented to T cells, but the nature of this three-way interaction was elusive until 1987. That year, when the first MHC protein was crystallized, immunologists first viewed a groove filled with a diffuse mass that was the result of a pool of diverse peptides (Figure 9.1) (Bjorkman *et al.*, 1987). It became clear that *antigen presentation* was the placement of MHC proteins, each loaded with a peptide, on the cell surface to await binding by T cells (for review see Penn and Ilmonen, 2005).

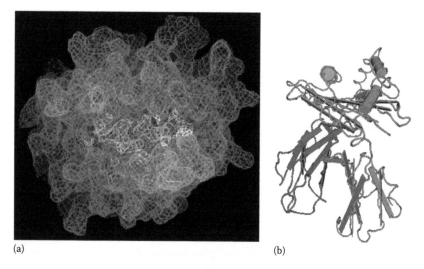

(a)   (b)

Figure 9.1   The three-dimensional structure of the first crystallized MHC protein. (a) View of the *antigen-binding cleft* (ABC) of an MHC protein (blue) with bound peptide (red). (From Bjorkman, P.J. *et al.*, *Nature*, 329, 506–512, 1987.) This view looks down on the top of the molecule if it were membrane bound. (b) Side view of the same protein shows α-helical sides and a β-plated sheet floor that create the ABC in the membrane distal portion of the protein. The membrane proximal region forms two immunoglobulin-fold domains. This protein is composed of a class Iα subunit (pink) and β-2-microglobulin (blue). (MMDB ID: 1202 viewed using Cn3D.)

More recent work to map genes in the MHC revealed an extremely gene dense region that varies in size and gene content across taxa. The human MHC (historically called HLA, for human leukocyte antigen) appears to be the largest spanning >4 Mb with 224 identified genes of which 128 are predicted to be expressed and of those about 40% have immune functions (The MHC Sequencing Consortium 1999). In sharp contrast, the chicken MHC may be the smallest with only 19 genes (Kaufman et al., 1999b). With time and greater scrutiny, additional MHC-relevant genes were found elsewhere. First, conspicuous examples of clusters of MHC-paralogous genes were documented on other chromosomes in humans and other vertebrates (Flajnik and Kasahara, 2010; Kasahara et al., 1996a; Katsanis et al., 1996). These MHC paralogous regions are thought to have originated in early vertebrate evolution from two or three rounds of full genome duplication. Second, MHC-relevant genes were mapped outside the boundaries that defined the original MHC leading to a defined extended MHC (xMHC) in humans, that is, 7.6 Mb (Horton et al., 2004). Third, the core genes found in most MHC regions have been lost or dispersed in some taxonomic groups, such as birds and bony fish, during their evolutionary history (e.g., see Bannai and Nonaka, 2013; Hansen et al., 1999; Kaufman et al., 1999a; Kulski et al., 2002; Nonaka et al., 2011; Sato et al., 2000; Shiina et al., 1999).

The MHC contains the *classical* MHC genes that present antigens to T cells and fall into two types, class I and class II. Additional structurally similar but functionally different, or so-called nonclassical, class I and class II genes are found within some MHC regions as well as elsewhere in the genome. A region within the mammalian MHC that contains additional genes that appeared different from class I and II was originally called class III. Examples include genes that encode proteins with clear immune functions, such as components of the complement system and cytokines, and also proteins involved in inflammation and stress responses (Gruen and Weissman, 1997). Now that some MHC regions have been fully mapped, categories of genes based on function, rather than order of discovery, are more meaningful: antigen processing and presentation (class I, class II, and others); other immune functions (many in the original class III region); and functions possibly unrelated to immunity (reviewed in Kulski et al., 2002).

MHC Class I and Class II Molecules

MHC molecules have been well studied in several model vertebrate systems. The structures and functions of classical class I and class II proteins are well defined (Neefjes et al., 2011; Vyas et al., 2008). Though the subunit structure differs, the three-dimensional structure of classical class I and class II molecules is remarkably similar (Figure 9.2). Hundreds of these proteins have been crystallized although the vast majority is from mammals (Adams and Luoma, 2012). Domains that are conserved and can be modeled on existing structures are found across the jawed vertebrates. Each protein has two membrane proximal domains and two membrane distal domains. Immunoglobulin fold regions near the cell membrane support the outermost two halves of the antigen-binding cleft (ABC). Class I genes consist of a heavy chain and a light chain. The membrane-bound heavy chain (class Iα) contains the entire ABC made up of two domains and the membrane proximal immunoglobulin-fold region. The light chain is β-2-microglobulin, a single immunoglobulin-fold region that maps outside the mammalian MHC. Class II proteins are a heterodimer of two membrane-bound chains (α and β) each contributing half of the ABC and one immunoglobulin-fold region (for review see Murphy, 2011).

Each cell-surface MHC protein has an ABC that holds a single short peptide and is bound by receptors on T cells or NK cells. This antigen presentation is the result of intracellular antigen processing and loading of peptides onto MHC molecules via two different pathways, one for each class (Neefjes et al., 2011). Class I molecules usually receive antigenic peptides from proteins that originate from within the cell and are expressed on almost all cells of the body. These complexes

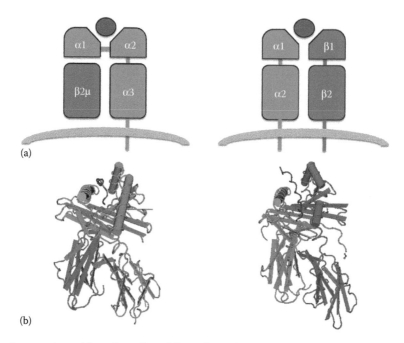

Figure 9.2   A comparison of the schematic and three-dimensional diagrams for class I (left) and class II (right) proteins. (a) The overall structure of membrane-bound (gray) MHC proteins with peptides (brown) in the ABC. Class I (left) is composed of a class I α-chain (pink) with three domains (α1, α2, and α3), a connecting peptide, transmembrane region, and a cytoplasmic tail. β-2-microglobulin (blue) noncovalently binds to domains α1, α2, and α3 and has no transmembrane piece. Class II (right) is a heterodimer of an α-chain (pink) and a β-chain (blue) each with two extracellular domains, a connecting peptide, transmembrane region, and a cytoplasmic tail. (b) Corresponding three-dimensional structures for human class I (HLA-A0201 + insulin peptide; PDB ID: 3UTQ) and human class II (HLA-DR + clip peptide; PDB ID: 3PDO) are shown at the bottom. Colors are the same as the top figure. The left–right orientation corresponds to domains of the greatest structural similarity. The left domains of class I (α1 and β2μ) are more like the left domains of class II (α1 and α2) and right side domains (class I α2 and α3 and class II β1 and β2) are also more similar.

are most often ligands for cytotoxic T cells but are also bound by some NK cells. Class II proteins usually bind peptides from proteins sampled from extracellular spaces and are expressed on the antigen-presenting subset of cells. They become bound by helper T cells. A *cross priming* process allows for some peptide loading across classes (Bevan, 1976, 2006; Huang *et al.*, 1996; Rock, 1996). Nonclassical molecules differ from their classical counterparts in a variety of ways (Adams and Luoma, 2012). Each type of MHC protein has a defined structure, expression pattern, source of peptides, and cellular partner for binding. Common features of each group of MHC protein are summarized in Table 9.1.

Antigen Processing and Presentation

Classical class I and class II proteins have unique features and separate antigen-processing pathways; however, some commonalities exist. First, antigens are proteins that are cleaved within the cell. Second, peptides are inserted into the cleft of MHC proteins before the complex is placed on the cell surface. Third, each MHC protein binds just one peptide but has the potential to display a repertoire of many peptides. The classes of MHC proteins differ in their sampling of separate populations of antigens that move through separate processing pathways. In addition, a set of unique

Table 9.1 **A Comparison of the Most Common Features of MHC Class I and Class II Proteins**

Protein	Structure	Expressed	Bound Antigen	Presented To
Classical class I (class Ia)	α chain (α1, α2, α3, TM, CY) and β2 microglobulin	Almost all cell surfaces	Short peptides of cytosolic proteins (synthesized by cell)	CD8+ cytotoxic T cells and some NK cells
Classical class II	α chain (α1, α2, TM, CY) and β chain (β1, β2, TM, CY)	On antigen-presenting cells (APCs)	Short peptides of proteins in endosomes or lysosomes (from extracellular spaces)	CD4+ helper T cells
Nonclassical class I (class Ib)	α chain (α1, α2, α3, TM, CY) and β2 microglobulin	Limited cell surface expression	Varies—peptide, non-peptide, or unknown	CD8+ T cells, NKT cells, or NK cells
Nonclassical class II	α chain (α1, α2, TM, CY) and β chain (β1, β2, TM, CY)	Only intracellular in APCs	Peptides that are then loaded onto classical class II molecules or unknown[a]	

[a] Fish have many nonclassical class II genes with unknown functions.
CY, cytoplasmic tail; TM, transmembrane region.

proteins is found in each pathway and some of the encoding genes are found in the MHC. The differences in the antigen processing and presentation pathways are shown in Figure 9.3 (reviewed in Neefjes *et al.*, 2011).

In the mammalian MHC class I antigen presentation pathway, specialized proteasomes degrade endogenous proteins. The resultant peptides, of a particular size and with certain amino acids, are carried into the lumen of the endoplasmic reticulum (ER) by TAP (Transporter associated with Antigen Processing) to bind to the MHC class I molecules (Rock and Goldberg, 1999). Although constitutively expressed proteasomes can process peptides, it is the *immunoproteasomes* that appear to produce peptides that are optimal for binding class I proteins. Of the seven α and seven β subunits of the constitutively expressed 20S proteasome, only three β subunits have proteolytic activity. They are PSMB5 (X), PSMB6 (Y), and PSMB7 (Z), with chymotrypsin-like, caspase-like, and trypsin-like peptidase activities, respectively. In the immunoproteasome, these β subunits are replaced by the interferon γ (IFN-γ)-inducible PSMB8 (LMP7), PSMB9 (LMP2), and PSMB10 (MECL-1), respectively (Tanaka and Kasahara, 1998). These changes in subunit composition are believed to increase the chymotrypsin-like activity and to generate peptides with hydrophobic C-terminal residues suitable for binding to the clefts of MHC class I molecules (Rock *et al.*, 2004). Once folded correctly and loaded with a peptide, the MHC protein moves through the Golgi to the cell surface.

The so-called professional antigen-presenting cells (APCs), such as macrophages and dendritic cells, express MHC class II molecules that sample antigens from extracellular spaces. They take up various particles by phagocytosis or receptor-mediated endocytosis. These exogenous proteins are degraded by acid-dependent proteases in endosomes and phagosomes (Vyas *et al.*, 2008). As class II proteins are synthesized in the ER, they bind invariant chain (Ii), which allows for proper folding while filling the ABC and regulates effective trafficking (Zhong *et al.*, 1996). The Ii-class II complex is transported through the Golgi to the late endosomal compartment, which contains degraded exogenous proteins and is termed the *MHC class II containing compartment* (MIIC). Here, Ii is digested leaving a small Ii peptide (called Class II-associated invariant chain peptide or CLIP) in the ABC. The nonclassical class II protein DM then binds the MHC class II protein, releasing CLIP and allowing other peptides in the MIIC to bind. The peptide-class II complex then moves to the cell membrane to be displayed and potentially bound by CD4 T cells (Neefjes *et al.*, 2011).

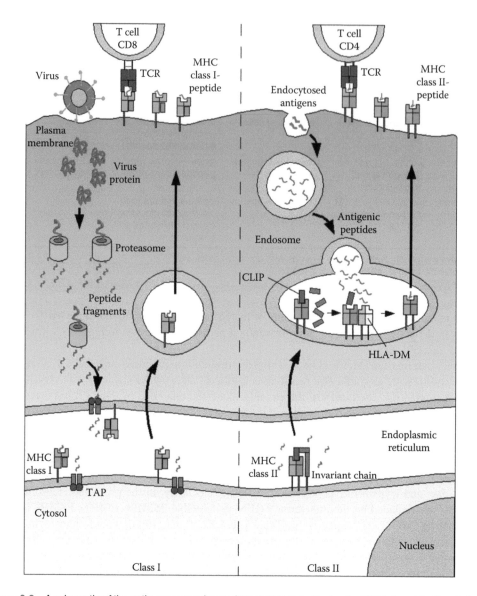

Figure 9.3 A schematic of the antigen processing and presentation pathways for MHC class I (left) and class II (right) proteins. (From http://www.lesc.ic.ac.uk/projects/appp.html.)

In addition to the MHC class Iα and IIα/β molecules that present antigenic peptides to T cells and thus play the pivotal role in adaptive immunity, the molecules directly involved in MHC class I and class II antigen processing/presentation, namely, immunoproteasome beta subunits (*PSMB8* and *PSMB9*), TAP transporters (*TAP1* and *TAP2*), TAP-binding protein (*TAPBP*), and DM (*HLA-DM*), are also encoded in the human MHC (Figure 9.3). Gene names are written in italics and are according to the HUGO Gene Nomenclature Committee and other species-specific resources. TAPBP allows optimal peptide loading of the MHC protein during class I antigen processing (not shown in Figure 9.3) (Raghuraman *et al.*, 2002). Although antigen processing and presentation genes show an intimate functional linkage, there is almost no structural similarity among the *mhc1a*, *PSMB*, *TAP*, and *TAPBP* genes. This is not the case for DM, which is a nonclassical class II protein with an α-chain and a β-chain that are both MHC encoded. Invariant chain (Ii or *CD74*) is not MHC linked in humans or other species.

Polymorphism

Early work on graft rejection found that vertebrates rejected tissues from all but the most closely related animals (Borysenko and Hildemann, 1970; McCumber *et al.*, 1982). Once the genes controlling rejection were isolated, the mechanism for this became clear. MHC genes, and the proteins they encode, are highly polymorphic for both the number of alleles and the genetic distances between alleles, more so than almost all other loci studied (Penn and Ilmonen, 2005). Thus, only closely related individuals share alleles and will not recognize grafted tissues as foreign. Up to three class I loci and hundreds of alleles have been found for most vertebrates examined (Kulski *et al.*, 2002; Potts and Wakeland, 1990). The mechanism that created this high level of polymorphism was thought to be rapid intraspecific accumulation of mutations. However, nucleotide substitution rates of primate MHC genes are not higher than those of non-MHC genes (Satta *et al.*, 1993). The concept of trans-species polymorphism (TSP), the passage of allelic lineages from ancestral to descendant species, has been proposed as an entirely different explanation for the MHC polymorphism (Klein *et al.*, 2007). Alleles, that are older than the species in which they are found, would have sufficient time to accumulate multiple mutations that would be seen as inter-allelic differences. The first molecular evidence for TSP was reported for rodent MHC class II genes and the primate MHC class I genes (Figueroa *et al.*, 1988; Lawlor *et al.*, 1988; McConnell *et al.*, 1988). Later, TSP was reported for various non-MHC genes such as immunoglobulins, virus resistance genes, and plant self-incompatibility genes that prevent self-fertilization and thus promote out-crossing (Esteves *et al.*, 2005; Ferguson *et al.*, 2008; Hiscock *et al.*, 1996; Su and Nei, 1999). These trans-species polymorphisms that persist for up to tens of millions of years are believed to be maintained by long-term balancing selection such as overdominant selection or negative frequency-dependent selection (Sutton *et al.*, 2011).

The mechanisms that maintain the high level of polymorphism at MHC loci were discussed and studied for many years. Due to their important central role in immunity, MHC diversity has long been considered to be pathogen driven (Bernatchez and Landry, 2003; Doherty and Zinkernagel, 1975; Jeffery and Bangham, 2000; Spurgin and Richardson, 2010). MHC-dependent disease resistance and susceptibility to infection were reported in most vertebrate taxa from bony fish to mammals and for various infectious agents (i.e., viruses, bacteria, and parasites) (Bernatchez and Landry, 2003; Hill, 1991; Piertney and Oliver, 2006; Sommer, 2005). However, MHC genes were also associated with autoimmunity in humans and mice, and, adding more complexity, an MHC haplotype that provides resistance to HIV (preventing the progression to AIDS) is also associated with an increased risk for autoimmunity (Fujisawa *et al.*, 2006; Kosmrlj *et al.*, 2010; Zenewicz *et al.*, 2010). Our current understanding considers overdominant selection, negative frequency-dependent selection, and fluctuating selection as being most important. The overdominance model predicts that heterozygous individuals are favored because they can mount effective immune responses to a broader array of pathogens resulting in a maximization of MHC diversity (Doherty and Zinkernagel, 1975; Hughes and Nei, 1988). In HIV-infected humans, extended survival time was associated with heterozygosity at all class I loci (Carrington *et al.*, 1999). The negative frequency-dependent selection hypothesis considers rapid coevolution between hosts and pathogens that causes the relative fitness of common MHC genotypes to decline, thus making rare alleles more advantageous and creating a dynamic turnover of beneficial alleles (Clarke and Kirby, 1966; Slade and McCallum, 1992). Lastly, fluctuating selection is driven by a continual change in time and place of the type and abundance of pathogens (Hill, 1991). Most studies on nonmodel vertebrates have confirmed that MHC variation is strongly influenced by selection but were inconsistent with traditional models that find overdominant balancing selection as a driver of high MHC diversity (Piertney and Oliver, 2006).

Expression

Classical class I and class II proteins and their nonclassical counterparts all differ in their function and their patterns of expression (Table 9.1 and reviewed in Murphy, 2011). In mammals, classical class I genes are found on almost all cells except for those with *immune privilege* (i.e., brain, eyes, placenta, and fetus). Classical class II genes are only expressed on antigen-presenting cells including macrophages, dendritic cells, and cells in the thymus but can be induced on other cells by interferon γ. Nonclassical protein expression patterns are quite variable as are the functions, which include peptide loading onto classical molecules, presentation of formylated peptide or nonpeptide antigens, interaction with natural killer cell receptors, and transport of IgG (Beckman and Brenner, 1995; Parham, 1994).

Other Genes with Immune Functions

Besides antigen processing and presentation, other proteins encoded within the MHC have a variety of functions. Many have known or inferred immune functions but for some the connection to or exclusion from the immune system remains unknown (Trowsdale, 2011). Those with known immune functions include proteins associated with the inflammatory response, complement components, and receptors that bind class I and II MHC molecules (Deakin *et al.*, 2006; Kaufman, 2010; Terado *et al.*, 2003). The complement system assists antibodies and phagocytic cells in clearing pathogens from the body. In humans, complement genes *C2*, *CBF*, *C4A*, and *C4B* are within the MHC (The MHC Sequencing Consortium, 1999).

MHC GENES OF CARTILAGINOUS FISHES

MHC class I and II genes were found in all jawed vertebrates examined from humans to shark but are absent from the recently sequenced lamprey, sea squirt, amphioxus, and sea urchin genomes (Azumi *et al.*, 2003; Hibino *et al.*, 2006; Holland *et al.*, 2008; Smith *et al.*, 2013), indicating that these genes arose in the common ancestor of jawed vertebrates. Before, 1990, MHC genes and their functions were mostly identified in higher vertebrates. Acute graft recognition, a hallmark of MHC function, was demonstrated in bony fishes as the organisms most distantly related to mammals (Hildemann, 1958). Early efforts to document in cartilaginous fishes hallmarks of MHC function, such as acute graft rejection, mixed leukocyte reactions, or antibody production resulting in T/B-cell collaboration, yielded unconvincing results (Borysenko and Hildemann, 1970; McCumber *et al.*, 1982; Tam *et al.*, 1976). Therefore, it was considered that MHC probably evolved in a more recent ancestor shared only by bony fish and tetrapods (Smith and Davidson, 1992). However, in the 1990s, MHC genes were isolated from sharks (Bartl and Weissman, 1994; Bartl *et al.*, 1997; Hashimoto *et al.*, 1992; Kasahara *et al.*, 1992). Though a bit of a surprise, this discovery led to an in-depth characterization of the genes, the proteins, and the locus in which they resided. In this chapter, we will discuss our current understanding of MHC genes in cartilaginous fish including their structure, expression, function, polymorphism, and chromosomal linkage.

MHC Class I and Class II Molecules

All four chains that make up class I and class II molecules have been isolated from a variety of chondrichthyan species (Table 9.2). Sequence comparisons find many conserved features such as inferred three-dimensional structure, disulfide bonds, salt bridges, and primary sequence similarity that are consistent across taxa from sharks to mammals (see references in Table 9.2). Based on the deduced amino acid sequences, it is predicted that the overall structure as seen in Figure 9.2 is conserved.

Table 9.2   **List of Species for Which Class I or Class II Sequences (Partial or Complete)**
Were Deposited in Genbank as of January 2013

Sequence Identity	Chondrichthyan Species Reported in Genbank	Genes Reported
mhc1a	Callorhinchus milii	1 gene (Tan et al., 2012)
	Chiloscyllium plagiosum	2 genes (unpublished Shen et al., 2010)
	Ginglymostoma cirratum	1 classical locus-8 alleles, 3 nonclassical genes (Bartl, 2001; Bartl et al., 1997; Ohta et al., 2000, 2002)
	Heterodontus francisci	2 genes (Bartl et al., 1997)
	Hydrolagus colliei	1 gene (reported here and unpublished Mochon-Collura et al., 2013)
	Leucoraja erinacea	28 ESTs (unpublished Towle and Smith, 2004)
	Squalus acanthias	2 genes (Wang et al., 2003)
	Torpedo californiensis[a]	2 genes (Venkatesh et al., 1999)
	Triakis scyllium	UAA-29 alleles, UBA-6 alleles (Hashimoto et al., 1992; Okamura et al., 1997), 1 gene (unpublished Hashimoto and Okamura, 2003)
	Raja eglanteria	2 genes (reported here and unpublished Bartl, 2012)
	Rhinobatos productus	1 gene (reported here and unpublished Mochon-Collura et al., 2013)
b2m	Carcharhinus plumbeus	2 alleles (Chen et al., 2010)
	Triakis scyllium	2 genes (unpublished Okamura and Hashimoto, 2010)
	Ginglymostoma cirratum	1 gene (Chen et al., 2010; Ohta et al., 2011)
	Raja eglanteria	1 gene (Cannon et al., 2002)
mhc2a	Callorhinchus milii	13 genes (Tan et al., 2012)
	Ginglymostoma cirratum	UAA-11 alleles, UBA-8 alleles (may be pseudoalleles rather than 2 loci) (Gersch, 2012; Kasahara et al., 1992, 1993; Ohta et al., 2000)
	Leucoraja erinacea	4 ESTs (unpublished Towle and Smith, 2004)
	Scyliorhinus canicula	7 gene fragments (Yang et al., 1997)
mhc2b	Chiloscyllium plagiosum	8 alleles (Ma et al., 2013)
	Ginglymostoma cirratum	9 genes, possibly 2 loci (Bartl, 2001; Bartl and Weissman, 1994; Ohta et al., 2000)
	Leucoraja erinacea	15 ESTs (unpublished Towle and Smith, 2004)

Note: When reported, proposed alleles or loci are listed. Gene names are according to the criteria of HGNC (genenames.org) and Xenbase (xenbase.org), access date July 2013 (Bowes et al., 2008; Gray et al., 2013).

[a] These gene fragments were reported as MHC class II but nucleotide and protein BLAST searches suggest they are class I.

ESTs, expressed sequence tags.

Polymorphism

Many polymorphic features of MHC genes found in other vertebrates are present in cartilaginous fishes. All the reports listed in Table 9.2 find commonalities including a predicted overall molecular structure with an ABC that contains most of the allelic polymorphism. The best characterized shark MHC locus is the *Triakis scyllium* class Iα with allele numbers that are comparable to humans (Okamura et al., 1997). When larger pools of alleles were analyzed, they could be grouped into two clades based on regions outside the ABC, suggesting the possibility of two loci for *mhc1a*, *mhc2a*, and *mhc2b* (Bartl, 2001; Ohta et al., 2000; Okamura et al., 1997). Only TSP remains as a missing piece. More alleles in more cartilaginous fish species may need to be analyzed before evidence for TSP is documented. Thus far, TSP in several bony fish taxa was reported including MHC alleles that have persisted for 184 million years, the oldest to date (Graser et al., 1996; Miller and Withler, 1997; Ottova et al., 2005; Wang et al., 2010a).

Polymorphic features of MHC molecules can be exemplified by comparing genes of one MHC chain in one species of shark. The polymorphic sites are located in the floor and along the alpha helical sides of the ABC where they influence peptide and TCR binding (Garcia *et al.*, 1999; Robey and Fowles, 1994). A comparison of the deduced proteins of four *mhc2b* genes for nurse shark, *Ginglymostoma cirratum*, finds that most of the allelic variation is located in the β1 domain that creates one side of the ABC, as shown in Figure 9.4 part 1 (a and b) (Bartl, 2001). However, one of the four genes also exhibits significant variation in the other domains, including the usually well-conserved β2 domain. This gene, named *mhc2b.b* *01 here, also has a 3′ untranslated region that is 57% different from genes *mhc2b.a* *01, *mhc2b.a* *02, and *mhc2b.a* *03, which may indicate that it represents a separate locus (Bartl, 2001). A GenBank search of the Protein Data Bank (with three-dimensional structure files) of human proteins found a best match of *mhc2b.a*09* with HLA-DQB and a best match of *mhc2b.b*01* with HLA-DRB. These cross-species alignments display greater identity (red) in the membrane proximal (β2) domain than the membrane distal (β1) with the shark allelic differences (yellow) found mostly within the ABC (Figure 9.4). Comparisons to

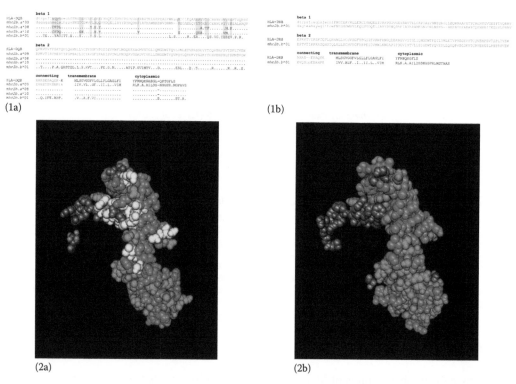

Figure 9.4   The deduced amino acid sequence of four *class IIB* genes from nurse shark aligned with representative human proteins. Clones are named using a convention consistent with others. (From Ohta, Y. *et al.*, *Proc. Natl. Acad. Sci. USA*, 97, 4712–4717, 2000.) Domains are displayed as defined in Figure 9.2 for the class II β subunit. (1) Best matches in the Protein Data Bank of human proteins for *mhc2b.a*09*(a) with HLA-DQB (3PDO_B) and *mhc2b.b*01*(b) with HLA-DRB (1YMM_B) are shown. Aligned amino acids are colored either red for identical or blue for not identical or gray for unaligned. Additional shark sequences that were only aligned with the top shark sequence are shown in black with identity indicated by dots. Potential allelic differences within the *mhc2b.a* locus are highlighted in yellow. (2) The three-dimensional space filling models for the human proteins with colored amino acids corresponding to the aligned sequences are shown. The reader's view corresponds to the orientation of the class IIB (blue) structure in Figure 9.2b. Genbank accession numbers are L20274 (*mhc2b.a*09*), L20275 (*mhc2b.a* *08), AF1894146 (*mhc2b.a* *10), and AF194147 (*mhc2b.b* *01).

other sequences in the nucleotide database find that *mhc2b.b* *01 is more similar to all the reported *mhc2b* genes from whitespotted bamboo shark, *Chiloscyllium plagiosum*, than to the other nurse shark sequences (Bartl unpublished data). However, genes with observed sequence differences need to be mapped to confirm the presence of two loci. In horn shark, *mhc2a* genes that formed two groups based on sequence did not map to separate loci leading the researchers to conclude that they were pseudoalleles (Ohta *et al.*, 2000). With genome sequencing progressing on two elasmobranch species, these issues should be resolved in the next few years. Taken together, the data on MHC polymorphism strongly support shark and skate MHC molecules encoded by 1–2 loci and having a function in peptide presentation to T cells as documented in other vertebrate systems.

In addition to its role in immune recognition, the MHC plays a role in social and reproductive interactions by mediating mate choice, pregnancy block, and behavior among relatives. Originally discovered to regulate mating preference in laboratory mice, data for MHC-mediated social signaling are now found across vertebrate taxa from mammals to teleost fish (Ruff *et al.*, 2011). Both fish and amphibian tadpoles form schools, a strategy that is associated with higher survival rates, with relatives that share MHC haplotypes (Olsen *et al.*, 2002; Rajakaruna *et al.*, 2006; Villinger and Waldman, 2008). The actual signal appears to be an odor associated with the peptides presented on MHC molecules as evidenced in studies of both mice and stickleback fish (Leinders-Zufall *et al.*, 2004; Milinski *et al.*, 2005; Spehr *et al.*, 2006). Lab experiments that test orientation by sharks in response to olfactory cues might be manipulated to assess responses to MHC stimuli (Gardiner and Atema, 2010). Studies across vertebrate taxa support both the effects of parasites and mate choice in maintaining MHC diversity, making it likely that these processes are operating in cartilaginous fish and will be revealed by formal studies.

Classical versus Nonclassical

The diversity displayed by MHC proteins allows for diversity in antigen presentation, but the ability to bind peptides also depends on anchoring amino acids that are particularly well conserved in classical class I proteins (Madden, 1995). This conservation has been used to infer peptide-binding function in the shark molecules (Bartl, 1998; Ohta *et al.*, 2000; Okamura *et al.*, 1997). Because nonclassical proteins have functions other than peptide binding and presentation to T cells, the anchoring site amino acids differ. Table 9.3 shows an analysis of the sites on class I molecules that anchor the N-terminus and C-terminus of bound peptides. By these criteria, both classical and nonclassical genes are found in teleost and cartilaginous fishes. In addition, nonclassical genes display little polymorphism, limited or different patterns of expression, and, in most cases, code for proteins that have specialized functions different from classical MHC genes (Beckman and Brenner, 1995; Parham, 1994). Some are also located outside the MHC, which is the case for (nurse shark) *mhc1b.a* that nonetheless has conserved peptide anchors (Ohta *et al.*, 2002).

Both nonclassical class I and class II genes have been found in vertebrates. In contrast to nonclassical class I molecules with very diverse functions, the nonclassical class II molecules are intracellular and function in antigen processing (Hughes, 2008). Orthologues of nonclassical class II genes have yet to be found in any fish. Phylogenetic analyses have determined that some nonclassical genes have ancient origins (i.e., *CD1*), whereas others have evolved recently (i.e., *HLA-G* and *H-2M3*) (Beckman and Brenner, 1995; Kasahara *et al.*, 1996b). Thus far, well-conserved *CD1* orthologues have been found in birds and well-conserved *DM* orthologues have been isolated from frogs (Miller *et al.*, 2005; Ohta *et al.*, 2006). An analysis of published bony and cartilaginous fish genomes finds no evidence of DM (Flajnik, MF, personal communication). In the primates and rodents, there is evidence for rapid turnover of classical and nonclassical genes (Cadavid *et al.*, 1996). This may be a more recent phenomenon, as evidence from amphibian, bony fish, and shark genes finds that these nonclassical genes have deep well-conserved lineages (Bartl, 1998; Goyos *et al.*, 2011; Nonaka *et al.*, 2011; Ohta *et al.*, 2006).

Table 9.3 The Deduced Peptide-Anchoring Amino Acids of Class I Genes Listed by Inferred Groups

Organism	Gene Symbol	Clone Name	Accession #	N-terminus				C-terminus			
				7	59	159	171	84	143	146	147
Mammals				Y	Y	Y	Y	Y	T	K	W
Other tetrapods				Y	Y	Y	Y	R	T	K	W
Salmon	*mhc1a.b*101*	*UBA-0101*	AF504019	Y	Y	Y	Y	R	T	K	W
Coelacanth	*mhc1a.a*02*	*Lach-UA*02*	U08033	–	–	Y	Y	–	T	K	W
Houndshark	*mhc1a.a*101*	*Trsc-UAA*101*	AF034316	Y	Y	Y	Y	R	T	K	W
Houndshark	*mhc1a.a*201*	*Trsc-UAA*201*	AF034352	Y	Y	Y	Y	R	T	K	W
Houndshark	*mhc1a.b*202*	*Trsc-UBA*202*	AF034346	Y	Y	Y	Y	R	T	K	W
Dogfish	*mhc1a.a*01*	*Sqac-UAA*01*	AY150811	Y	Y	Y	Y	R	T	K	W
Skate	*mhc1a.a*01*	*Rael-10*	KC335152	Y	Y	Y	Y	R	T	K	W
Skate	*mhc1a.a*02*	*Rael-3*	KC335153	Y	Y	Y	Y	R	T	K	W
Guitarfish	*mhc1a.a*01*		KC469286	Y	Y	Y	Y	R	T	K	W
Ratfish	*mhc1a.a*01*		KC469287	Y	Y	–	–	R	–	–	–
Nurse shark	*mhc1b.a*	*UAA-NC1*[a]	AF220360	Y	Y	Y	Y	R	T	K	W
Nurse shark	*mhc1a.a*01*	*UAA01*	AF220063	Y	–	Y	Y	R	T	K	W
Nurse shark	*mhc1a.a*05*	*UAA05*	AF357925	–	Y	Y	Y	R	T	K	W

Species	Gene symbol	Allele	Accession								
Carp	mhc1b.a*101	ZE-0101	AJ420958	Y	Y	Y	F	R	T	K	W
Zebrafish	mhc1b.a	Dare-ZE	AJ420953	Y	Y	Y	F	R	T	K	W
Zebrafish	mhc1b.b	Dare-UFA	AF137534	V	F	Y	Y	R	T	K	W
Bamboo shark	mhc1a.a*01	chpIUAA	HQ023243	Y	F	Y	Y	R	T	K	W
Bamboo shark	mhc1a.a*02	chpIUAA	ADM21329	Y	F	Y	Y	R	T	K	W
Horn shark	mhc1a.a	Hefr-20	AF028559	Y	F	Y	Y	R	T	K	W
Nurse shark	mhc1b.b	UAA-NC2	AF357926	–	Y	Y	Y	R	S	K	W
Carp	mhc1b.b	Cyca-Zr4	AJ007851	–	D	Y	F	C	T	R	W
Coelacanth	mhc1b.a	Lach-UA*01	LCU08034	Y	S	D	Y	E	R	V	C
Coelacanth	mhc1b.b	Lach-UB*01	U08034	Y	S	D	Y	K	R	V	C
Horn shark	mhc1b.a	Hefr-19	AF028558	Y	A	Y	Y	R	I	N	W
Nurse shark	mhc1b.a	Gici-11(UCA01)	AF028557	Y	F	F	Y	R	I	R	W

Source: Numbered according to the HLA-A2 structure (Bjorkman et al., 1967).

Note: By the developed convention, species names are given in parentheses before the gene symbol (here in column one) or after the full gene name. Gene symbols assigned using HGNC and Xenbase guidelines (Bowes *et al.*, 2008; Gray *et al.*, 2013): *mhc1a* = classical class I gene; *mhc1b* = nonclassical (as previously reported or in this analysis); period + letter = homologs within species; asterisk + number = reported allele designations. Red letters = not conserved.

[a] Reported as nonclassical due to its location outside the MHC locus. (From Ohta, Y. *et al.*, *J. Immunol.*, 2002:168, 771–781.)

Expression

Studies of MHC expression in sharks have used northern blots, *in situ* hybridization, immuno-histochemical staining, and quantitative PCR. Expression of classical and nonclassical class I genes was assayed by northern blotting in nurse shark (Ohta *et al.*, 2002). Classical class I appeared highest in gill, intestine, peripheral blood leukocytes (PBLs), and spleen as would be expected because these tissues have an important role in immunity (Luer *et al.*, 2004). Levels were intermediate in kidney, liver, testis, and thymus and low in brain, epigonal gland, heart, and pancreas. The positive signal in the testis is not surprising because both spermatozoa and interstitial tissues express MHC (Hedger, 2007). A blood–brain barrier that maintains immune privilege in the brain may not exist in sharks (Hedger, 2007). A nonclassical class I probe, (nurse shark) *mhc1b.a*, showed a similar tissue distribution as the classical class I, but at lower levels of expression overall. A separate study of β-2-microglobulin found an expression pattern nearly identical to class I in nurse shark (Ohta *et al.*, 2011).

MHC class II gene expression in adult bamboo shark, *C. plagiosum*, was assayed using quantitative real-time PCR (Ma *et al.*, 2013). High levels of *mhc2b* mRNA were detected in the spleen and gill with moderate levels in the intestine, stomach, liver and heart and low levels in blood, brain, skin, and muscle. Liver, spleen, and gill were examined after bacterial challenge, and significant increases in the expression were found at 4 h in all three organs.

A more extensive examination of the nurse shark thymus was done using *in situ* hybridization with MHC class I and II probes (Criscitiello *et al.*, 2010). In the adult thymus, both classes were highly expressed in the medulla with a more punctuate pattern seen in the cortex, presumably due to the staining of the epithelium and accessory cells. This expression pattern is similar to that of the mammalian thymus.

MHC class II antiserum was used to assess surface expression in nurse shark spleen and epigonal gland at different life stages (Rumfelt *et al.*, 2001). In adult splenic white pulp, positive staining was seen on B cells and putative dendritic cells. In newborns, splenic B cells in the white pulp had very low class II expression similar to neonatal mice but high levels on some red pulp cells were similar to reticuloendothelium cells of frog tadpoles. Class II high cells were dispersed in newborn epigonal gland. By 2.5 months, expression is more like the adult with clusters of MHC class II high cells.

The especially high levels of MHC class I and II reported in spleen and gill indicate the importance of these tissues in immune surveillance. High class II expression in thymus and epigonal gland, especially in young sharks is consistent with their lymphoid role (Luer *et al.*, 2004). The upregulation of class II expression after challenge is uniformly seen in other vertebrates. Thus, reported expression patterns for shark MHC class I and class II genes and proteins support a role that is consistent across all jawed vertebrates.

Antigen Presentation and Processing Genes

Comparative analyses of the MHC genomic structure were performed in various jawed vertebrates, first by classical linkage analysis, and then by genome sequence analysis. The genomic organization of the *Xenopus* MHC, determined by linkage analysis and draft sequence analysis, indicates a high degree of conserved synteny with the human MHC (Ohta *et al.*, 2006). In contrast, the teleost MHC genome shows a uniquely derived configuration, with teleost orthologues of the mammalian MHC-encoded genes dispersed on several chromosomes (Bingulac-Popovic *et al.*, 1997; Naruse *et al.*, 2000). However, an extremely tight linkage is observed among the MHC class Iα genes (*mhc1a*) and the genes involved in class I antigen processing/presentation: *tap2*, *psmb8*, *psmb9*, *psmb9l*, *psmb10*, and *tapbp* (Clark *et al.*, 2001; Matsuo *et al.*, 2002; Michalova *et al.*, 2000). The tight linkage among these genes was conserved not only in nonmammalian jawed vertebrates but also in the marsupial opossum and the monotreme platypus (Belov *et al.*, 2006; Dohm *et al.*, 2007). Only in the placental mammals analyzed thus far and in the marsupial wallaby is this tight linkage not observed (Siddle *et al.*, 2009). These results suggest that this tight linkage was present in the common ancestor of mammals and was lost at least twice independently in the wallaby and the placental mammal lineages.

Psmb8 Genes

Analyses of genes related to *psmb8* reveal an interesting history. Highly divergent dichotomous types of *psmb8* have been reported in amphibians (Nonaka *et al.*, 2000). They are also found in Actinopterygian species belonging to Cypriniformes (zebrafish and loach), Salmoniformes (salmon and trout), Beloniformes (medaka), Tetraodontiformes (pufferfish), and Polypteriformes (polypterus) and in two species of cartilaginous fishes (Fujito and Nonaka, 2012; Kandil *et al.*, 1996; Miura *et al.*, 2010; Ohta *et al.*, 2002; Tsukamoto *et al.*, 2009, 2012). These highly divergent types, designated A and F type, share a deduced amino acid sequence with an Ala/Val^{31}Phe/Tyr substitution that probably affects cleaving specificity. Moreover, phylogenetic analysis identified two lineages that were called *PSMB8A* and *PSMB8F* (Tsukamoto *et al.*, 2012). The A type *psmb8* genes of Salmoniformes, Cypriniformes, polypterus, and the shark *psmb8* genes belong to the *PSMB8A* lineage, whereas the F type *psmb8* genes of these fishes and the shark *psmb8l* genes belong to the *PSMB8F* lineage, indicating that these two lineages were established before the divergence of sharks from the bony vertebrates (Figure 9.5). The *psmb8* gene itself is believed to have arisen by a gene duplication of *psmb5* in the common ancestor of jawed vertebrates, simultaneously with the appearance of MHC genes and other features of adaptive immunity (Kasahara, 1997). The two *PSMB8* lineages appear to have persisted for 500 million years, splitting soon after the *psmb8* gene originated and being retained by basal Actinopterygii and sharks until today. In *Xenopus* and higher actinopterygian teleosts (i.e., *Oryzias* and pufferfish), the *PSMB8F* lineage genes were lost, and new F type genes, functional equivalents of the *PSMB8F* gene, were independently revived as alleles within the *PSMB8A* lineage. Once established, however, these dimorphic alleles were passed from ancestral to descendent species for more than 30–60 million years in *Oryzias* (Miura *et al.*, 2010) and more than 80 million years in *Xenopus* (Nonaka *et al.*, 2000). These dimorphisms are examples of TSPs maintained by long-term balancing selection. Unlike all bony vertebrates analyzed thus far, sharks have paralogous genes in the *PSMB8A* and *PSMB8F* lineages, and they are reportedly pseudoalleles in nurse sharks (Kandil *et al.*, 1996; Ohta *et al.*, 2002). Thus, these two lineages have experienced the conversion between alleles and paralogs at least once during their evolution. From an evolutionary viewpoint, their status as real or pseudoalleles makes little difference for balancing selection if the paralogous loci of the pseudoalleles are close and thus having effectively zero possibility of recombination. In such a case, the pseudoalleles can also be regarded as an example of a long-maintained polymorphism. Therefore, the TSP of the *psmb8* genes of sharks and basal Actinopterygii has been maintained for more than 500 million years by an extremely long-term balancing selection.

Other MHC-Linked Genes with Immune Functions

Genes with known immune functions other than antigen processing and presentation are found within the MHC of a variety of taxa. First, a closely linked set of structurally distinct genes within MHC of humans and other species, including the tumor necrosis factor (*TNF*) family, are associated with the inflammatory response and are now referred to as the Class IV or inflammatory region (Deakin *et al.*, 2006). Second, complement genes, *C2* and *C4*, are closely linked in the mammalian MHC. Although structurally distinct, the C4 protein is functionally linked with C2 protein in the complement classical pathway (Terado *et al.*, 2003). Third, receptors that bind class I and II MHC molecules are located in the MHC of chicken (NK cell receptor and *CD1* genes) and

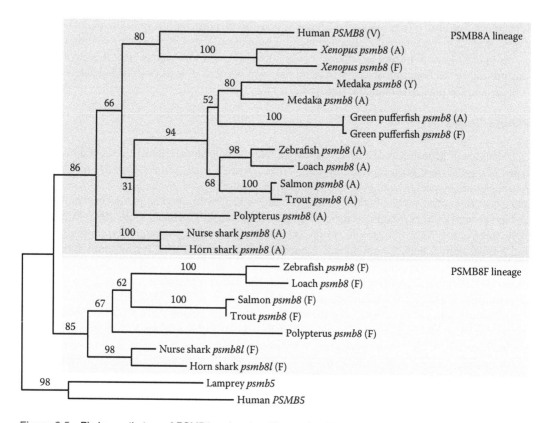

Figure 9.5 Phylogenetic tree of PSMB8 molecules. The nucleotide sequences of A and F types of PSMB8 mature proteins were aligned with ClustalX, and the phylogenetic trees were constructed based on this alignment using the maximum likelihood method based on Kimura's two-parameter model. PSMB5 sequences of human and lamprey (*P. marinus*) were used as the out-group. Evolutionary analyses were conducted in MEGA5. (From Tamura, K. *et al.*, *Mol. Biol. Evol.*, 28, 2731–2739, 2011.) Capital letters in parentheses indicate the single letter code for the amino acid at the 31st position of the mature protein. Colored boxes delineate two ancient lineages of psmb8 genes. Accession numbers for the genes are human *PSMB8* (NP_004150), *Xenopus psmb8* (A) (BAA07954), *Xenopus psmb8* (F) (BAA07945), medaka *psmb8* (Y) (BA000027), medaka *psmb8* (A) (AB183488), green pufferfish *psmb8* (A) (CR697191), green pufferfish *psmb8* (F) (CR691449), zebrafish *psmb8* (A) (BC066288), loach *psmb8* (A) (BJ830309), salmon *psmb8* (A) (DY733578), trout *psmb8* (A) (CA360946), polypterus *psmb8* (A) (AB686528), nurse shark *psmb8* (D64057), horn shark *psmb8* (AF363583), zebrafish *psmb8* (F) (BC092889), loach *psmb8* (F) (BJ838416), salmon *psmb8* (F) (DW555236), trout *psmb8* (F) (CA361317), polypterus *psmb8* (F) (AB686530), nurse shark *psmb8l* (D64056), horn shark *psmb8l* (AF363582), lamprey *psmb5* (D64055), and human *PSMB5* (NM_002797).

frog (TCR-like gene), suggesting an ancestral MHC containing genes that coevolved into receptor–ligand pairs (Kaufman, 2010). Comparative analyses support the idea that present day MHC regions are the vestiges of a single primordial *immune supercomplex* of genes that contained MHC antigen presentation genes (Class I and II), antigen processing genes *(TAP* and *PSMB), CD1,* and C-type and Ig-type NK receptor genes (Belov *et al.*, 2006).

The evolutionary origin of the complement system is very ancient, predating the divergence of Cnidaria and Bilateria, because the central component C3 was identified in an anthozoan Cnidarian coral and the sea anemone (Dishaw *et al.*, 2005; Kimura *et al.*, 2009). Additional components, factor B (*cfb*) and mannan-binding lectin-associated serine protease (*masp*), were also identified in sea anemone, indicating that a multicomponent complement system was established at the early stage of the metazoa evolution (Kimura *et al.*, 2009). The genome analysis of the ascidian *Ciona intestinalis* indicates that some complement gene families that are characterized by unique domain architectures, namely, the *c3*, *cfb*, *masp*, and terminal complement complex (TCC) families, are present in ascidians (Azumi *et al.*, 2003). Gene duplications within each of these families, namely, *c3/c4/c5*, *cfb/c2*, *masp1/masp2/c1r/c1s*, and *c6/c7/c8a/c8b/c9*, established the pathways of the mammalian complement system. Evidence for these gene duplications is obvious in the teleost genome but lacking in the ascidian genome, indicating that these duplications occurred in the early stage of vertebrate evolution (Nonaka and Kimura, 2006). As described in Chapter 8, the gene duplication that gave rise to *c2* and *cfb* occurred before the appearance of amphibians, because *Xenopus* has clear orthologues. An apparent expansion and diversification of *cfb/c2-like* genes in bony fishes makes it difficult to assign these genes to either the *cfb* or the *c2* lineage (Nakao *et al.*, 2003). Two *cfb/c2l* genes are present in nurse shark (Shin *et al.*, 2007); however, they are not as divergent as *cfb* and *c2* of tetrapods.

In contrast to the lack of clarity for *cfb* and *c2* gene evolution, the *c3/c4/c5* gene duplication must have predated the divergence of cartilaginous and bony vertebrates, because obvious *c3*, *c4*, and *c5* orthologues exist in sharks and bony fishes (GcC3-1, GcC3-2; Smith unpublished; Graham *et al.*, 2009; Nonaka and Kimura, 2006). Lamprey and hagfish have only *c3l* genes, suggesting that the *c3* gene may be ancestral (Escriva *et al.*, 2002; Kuraku *et al.*, 1999). Although the *c4* and *cfb/c2* genes are not linked to each other nor to the *mhc1* or *mhc2* genes in bony fish, they are linked in the shark MHC, which contains two *cfb/c2* genes (Terado *et al.*, 2003). Thus, the linkage between the *c4* and *cfb/c2* genes was most likely established in the common ancestor of the jawed vertebrates and was secondarily lost in the teleost lineage.

Genomic Organization of the MHC

The curious linkage found in the MHC region of many species is among many functionally linked but structurally unrelated genes (Figure 9.6). Two totally dissimilar ideas have been proposed to explain this linkage. One suggests that the close linkage facilitates simultaneous regulation of gene expression. The other proposes that the close linkage allows for the coevolution of functionally linked genes as stable haplotypes (Kelley *et al.*, 2005). There is no conclusive evidence for either of these theories thus far, and the physiological and/or evolutionary meaning of this peculiar linkage awaits further study.

Several close linkages observed in all tetrapod MHC regions examined to date are broken in all bony fish species analyzed thus far, with relevant genes being found on different chromosomes (Kelley *et al.*, 2005). Studies of cartilaginous fish MHC regions can clarify linkages and elucidate an ancestral configuration of the MHC. The first linkage analysis with shark MHC genes was performed with the classical class Iα and class IIα and β (*mhc1a*, *mhc2a*, and *mhc2b*) genes. Segregation analyses of two shark species, nurse shark and banded houndshark, *T. scyllium*, found these genes to be closely linked (Ohta *et al.*, 2000), indicating that the close linkage between class I and class II genes was an ancestral character of vertebrate MHC. Further linkage analysis in nurse shark demonstrated that *tap1*, *tap2*, *psmb8*, *psmb8l*, and *psmb9* are linked to *mhc1a*, *mhc2a*, and *mhc2b* (Ohta *et al.*, 2002). The relationship between the *psmb8* and *psmb8l* is curious and complicated as was discussed previously. In addition, the complement *c4* and *cfb* genes were linked to the shark MHC (Terado *et al.*, 2003). Recently, the β-2-microglobulin (*b2m*) and *brd2* (*ring3*) genes were shown to be linked to the shark MHC (Ohta *et al.*, 2011). Unlike *mhc1a* genes that encode class I α chains, *b2m* is present outside of the MHC in all examined tetrapods and bony fish. However, the identification of *b2m* in the shark MHC region suggests that it is an original member of the MHC of the jawed vertebrate ancestor, and it moved to other genomic regions in early osteichthyian evolution. The bromodomain-containing protein 2 (BRD2), a putative nuclear transcriptional regulator, was thought to have no defined immune function until recently when it was proposed to influence inflammatory signal transduction (Wang *et al.*, 2010b). However, its gene (*brd2*) is linked to several

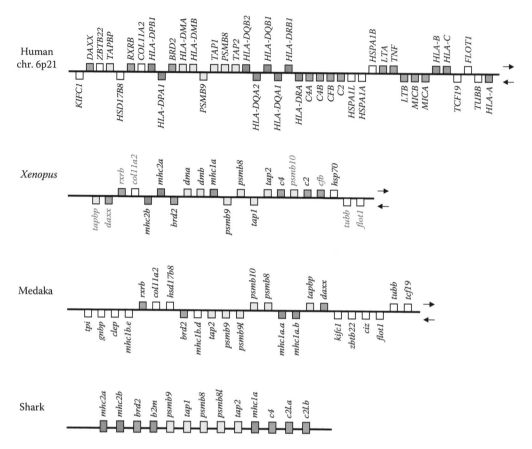

Figure 9.6 Comparison of the order and orientation of select genes within the human, *Xenopus*, medaka, and shark MHC regions. Actual distances between genes and sizes of loci are not represented. Classical class I genes, classical class II genes, and genes directly involved in antigen processing/transport are shown in red, blue, and yellow, respectively. Other genes with clear known or inferred immune functions are shown in green, whereas genes with no color are currently not associated with immune system function. The arrows at the right indicate the transcriptional orientation of the genes shown either above or below the line, respectively. For the shark, MHC gene order is still tentative, and the transcriptional orientation and distances between genes are not known. The gene for *b2m*, which encodes a single domain of the MHC class I protein, is shown in red. The two *cfb/c2*-like genes are shown as homologues *c2l.a* and *c2l.b*. Gene names are according to the HUGO Gene Nomenclature Committee (http://www.genenames.org/) and Xenbase (www.xenbase.org/). The convention of human gene names in upper case and other species names in lower case is observed.

MHC regions including humans, frog, and all examined bony fishes. The linkage of *brd2* to the shark MHC suggests that it is also an original member of the ancestral MHC, although functional or evolutionary significance is not yet clear (Figure 9.6).

Genome analysis of the elephant shark, *Callorhinchus milii*, was completed recently (Venkatesh *et al.*, 2014). However, the MHC region has eluded assembly (see Chapter 13), and therefore, it is still not clear if the MHC gene organization of holocephalans shows a significant difference from that of elasmobranchs. A genome project is in progress for an elasmobranch, the little skate, *Leucoraja erinacea* (Wang *et al.*, 2012; Yu *et al.*, 2008), although no useful linkage information of MHC-relevant genes can be obtained from the published sequence data so far. Thus, the presence of *b2m* and *brd2* on a single BAC clone of nurse shark is the only physical linkage data for these cartilaginous fish MHC genes (Ohta *et al.*, 2011). A tentative gene organization of the shark MHC is shown in Figure 9.6. However, the exact order and orientation of the genes as well as the distances between the genes are still not clear.

CONCLUSIONS

Analyses of the MHC and its related components in cartilaginous fishes find many similarities to other vertebrates but also some surprises. They provide useful information on ancestral features and evolutionary processes. But how extensive and similar are the antigen processing and presentation pathways? What other players may be found in sharks that are undiscovered in mammals because now they have different and possibly nonimmune roles? And what of the MHC region in cartilaginous fish? What is really missing from the shark MHC or, more specifically, what clearly maps elsewhere? How large is it and what genes will be found that have not yet been considered MHC linked?

We have yet to discover how multifunctional, malleable, and ancient MHC molecules are. Do MHC molecules function in shark mate selection? It would not be surprising if sharks use the powerful sense of smell to find MHC dissimilar mates. Will we find some nonclassical MHC molecules that are performing wonderfully unique and diverse functions in sharks and could these be a stepping stone to finally discover an MHC molecule in jawless fish and invertebrates? Our fascination with the MHC began 25 years ago when we first saw a single protein with a unique cleft and the capacity to hold a striking array of peptides. Twenty-five years hence we will know so much more, and hopefully, many of our answers will come from the study of cartilaginous fishes.

REFERENCES

Adams, E.J., and A.M. Luoma. The adaptable major histocompatibility complex (MHC) fold: Structure and function of nonclassical and MHC class I-like molecules. *Annu. Rev. Immunol.* 2012; 31:529–561.

Azumi, K., R. De Santis, A. De Tomaso, I. Rigoutsos, F. Yoshizaki, M.R. Pinto, R. Marino *et al.* Genomic analysis of immunity in a Urochordate and the emergence of the vertebrate immune system: "Waiting for Godot." *Immunogenetics* 2003; 55:570–581.

Bannai, H., and M. Nonaka. Comprehensive analysis of medaka major histocompatibility complex (MHC) class II genes: Implications for evolution in teleosts. *Immunogenetics* 2013; 65:883–895.

Bartl, S. What sharks can tell us about the evolution of MHC genes. *Immunol. Rev.* 1998; 166:317–331.

Bartl, S. New major histocompatibility complex class IIB genes from nurse shark. *Adv. Exp. Med. Biol.* 2001; 484:1–11.

Bartl, S., M.A. Baish, M.F. Flajnik, and Y. Ohta. Identification of class I genes in cartilaginous fish, the most ancient group of vertebrates displaying an adaptive immune response. *J. Immunol.* 1997; 159:6097–6104.

Bartl, S., and I.L. Weissman. Isolation and characterization of major histocompatibility complex class IIB genes from the nurse shark. *Proc. Natl. Acad. Sci. USA* 1994; 91:262–266.

Beckman, E., and M. Brenner. MHC dass I-like, class II-like and CD I molecules: Distinct roles in immunity. *Immunol. Today* 1995; 16:349–352.

Belov, K., J.E. Deakin, A.T. Papenfuss, M.L. Baker, S.D. Melman, H.V. Siddle, N. Gouin *et al.* Reconstructing an ancestral mammalian immune supercomplex from a marsupial major histocompatibility complex. *PLoS Biol.* 2006; 4:e46.

Bernatchez, L., and C. Landry. MHC studies in nonmodel vertebrates: What have we learned about natural selection in 15 years? *J. Evol. Biol.* 2003; 16:363–377.

Bevan, M. Cross-priming. *Nat. Immunol.* 2006; 7:363–365.

Bevan, M.J. Cross-priming for a secondary cytotoxic response to minor H antigens with H-2 congenic cells which do not cross-react in the cytotoxic assay. *J. Exp. Med.* 1976; 143:1283–1288.

Bingulac-Popovic, J., F. Figueroa, A. Sato, W.S. Talbot, S.L. Johnson, M. Gates, J.H. Postlethwait, and J. Klein. Mapping of mhc class I and class II regions to different linkage groups in the zebrafish, *Danio rerio.* *Immunogenetics* 1997; 46:129–134.

Bjorkman, P.J., M.A. Saper, B. Samraoui, W.S. Bennett, J.L. Strominger, and D.C. Wiley. Structure of the human class I histocompatibility antigen, HLA-A2. *Nature* 1987; 329:506–512.

Borysenko, M., and W.H. Hildemann. Reactions to skin allografts in the horn shark, *Heterodontis francisci*. *Transplantation* 1970; 10:545–551.

Bowes, J., K. Snyder, E. Segerdell, R. Gibb, C. Jarabek, N. Pollet, and P. Vize. Xenbase: A *Xenopus* biology and genomics resource. *Nucleic Acids Res.* 2008; 36:D761–D767.

Cadavid, L.F., A.L. Hughes, and D.I. Watkins. MHC class I-processed pseudogenes in New World primates provide evidence for rapid turnover of MHC class I genes. *J. Immunol.* 1996; 157:2403–2409.

Cannon, J.P., R.N. Haire, and G.W. Litman. Identification of diversified genes that contain immunoglobulin-like variable regions in a protochordate. *Nat. Immunol.* 2002; 3:1200–1207.

Carrington, M., G.W. Nelson, M.P. Martin, T. Kissner, D. Vlahov, J.J. Goedert, R. Kaslow, S. Buchbinder, K. Hoots, and S.J. O'Brien. HLA and HIV-1: Heterozygote advantage and B* 35-Cw* 04 disadvantage. *Science* 1999; 283:1748–1752.

Chen, H., S. Kshirsagar, I. Jensen, K. Lau, C. Simonson, and S.F. Schluter. Characterization of arrangement and expression of the beta-2 microglobulin locus in the sandbar and nurse shark. *Dev. Comp. Immunol.* 2010; 34:189–195.

Clark, M.S., L. Shaw, A. Kelly, P. Snell, and G. Elgar. Characterization of the MHC class I region of the Japanese pufferfish (*Fugu rubripes*). *Immunogenetics* 2001; 52:174–185.

Clarke, B., and D.R. Kirby. Maintenance of histocompatibility polymorphisms. *Nature* 1966; 211:999–1000.

Criscitiello, M.F., Y. Ohta, M. Saltis, E.C. McKinney, and M.F. Flajnik. Evolutionarily conserved TCR binding sites, identification of T cells in primary lymphoid tissues, and surprising trans-rearrangements in nurse shark. *J. Immunol.* 2010; 184:6950–6960.

Deakin, J.E., A.T. Papenfuss, K. Belov, J.G.R. Cross, P. Coggill, S. Palmer, S. Sims, T.P. Speed, S. Beck, and J.A. Marshall Graves. Evolution and comparative analysis of the MHC Class III inflammatory region. *BMC Genomics* 2006; 7:281–294.

Dishaw, L.J., S.L. Smith, and C.H. Bigger. Characterization of a C3-like cDNA in a coral: Phylogenetic implications. *Immunogenetics* 2005; 57:535–548.

Doherty, P.C., and R.M. Zinkernagel. Enhanced immunological surveillance in mice heterozygous at the H-2 gene complex. *Nature* 1975; 256:50–52.

Dohm, J.C., E. Tsend-Ayush, R. Reinhardt, F. Grutzner, and H. Himmelbauer. Disruption and pseudoautosomal localization of the major histocompatibility complex in monotremes. *Genome Biol.* 2007; 8:R175.

Escriva, H., L. Manzon, J. Youson, and V. Laudet. Analysis of lamprey and hagfish genes reveals a complex history of gene duplications during early vertebrate evolution. *Mol. Biol. Evol.* 2002; 19:1440–1450.

Esteves, P.J., D. Lanning, N. Ferrand, K.L. Knight, S.K. Zhai, and W. van der Loo. The evolution of the immunoglobulin heavy chain variable region (IgVH) in Leporids: An unusual case of transspecies polymorphism. *Immunogenetics* 2005; 57:874–882.

Ferguson, W., S. Dvora, J. Gallo, A. Orth, and S. Boissinot. Long-term balancing selection at the west nile virus resistance gene, Oas1b, maintains transspecific polymorphisms in the house mouse. *Mol. Biol. Evol.* 2008; 25:1609–1618.

Figueroa, F., E. Gunther, and J. Klein. MHC polymorphism pre-dating speciation. *Nature* 1988; 335:265–267.

Flajnik, M.F., and M. Kasahara. Origin and evolution of the adaptive immune system: Genetic events and selective pressures. *Nat. Rev. Genet.* 2010; 11:47–59.

Fujisawa, T., H. Ikegami, S. Noso, K. Yamaji, K. Nojima, N. Babaya, M. Itoi-Babaya *et al.* MHC-linked susceptibility to type 1 diabetes in the NOD mouse: Further localization of Idd16 by subcongenic analysis. *Ann. N.Y. Acad. Sci.* 2006; 1079:118–121.

Fujito, N.T., and M. Nonaka. Highly divergent dimorphic alleles of the proteasome subunit beta type-8 (PSMB8) gene of the bichir Polypterus senegalus: Implication for evolution of the PSMB8 gene of jawed vertebrates. *Immunogenetics* 2012; 64:447–453.

Garcia, K., L. Teyton, and I. Wilson. Structural basis of T cell recognition. *Annu. Rev. Immunol.* 1999; 17:369–397.

Gardiner, J.M., and J. Atema. The function of bilateral odor arrival time differences in olfactory orientation of sharks. *Curr. Biol.* 2010; 20:1187–1191.

Gersch, J.W. Microsatellite, mitochondrial, and major histocompatibility complex analyses of genetic structure in the nurse shark, *Ginglymostoma cirratum*, in the western Atlantic Ocean. Zoology. Southern Illinois University Carbondale, 2012, http://opensiuc.lib.siu.edu/theses/936/.

Goyos, A., J. Sowa, Y. Ohta, and J. Robert. Remarkable conservation of distinct nonclassical MHC class I lineages in divergent amphibian species. *J. Immunol.* 2011; 186:372–381.

Graham, M., Shin, D.-H., and S.L. Smith. Molecular and expression analysis of complement component C5 in the nurse shark (*Ginglymostoma cirratum*) and its predicted functional role. *Fish Shellfish Immunol.* 2009; 27(1):40–49.

Graser, R., C. O'HUigin, V. Vincek, A. Meyer, and J. Klein. Trans-species polymorphism of class II Mhc loci in danio fishes. *Immunogenetics* 1996; 44:36–48.

Gray, K., L. Daugherty, S. Gordon, R. Seal, M. Wright, and E. Bruford. Genenames.org: The HGNC resources in 2013. *Nucleic Acids Res.* 2013; 41:D545–D552.

Gruen, J.R., and S.M. Weissman. Evolving views of the major histocompatibility complex. *Blood* 1997; 90:4252–4265.

Hansen, J.D., P. Strassburger, G.H. Thorgaard, W.P. Young, and L. Du Pasquier. Expression, linkage, and polymorphism of MHC-related genes in rainbow trout, Oncorhynchus mykiss. *J. Immunol.* 1999; 163:774–786.

Hashimoto, K., T. Nakanishi, and Y. Kurosawa. Identification of a shark sequence resembling the major histocompatibility complex class I alpha 3 domain. *Proc. Natl. Acad. Sci. USA* 1992; 89:2209–2212.

Hedger, M. Immunologically privileged environments. *in* C. Halberstadt, and D. Emerich, eds. *Cellular Transplantation: From Laboratory to Clinic.* Elsevier, San Francisco, CA, 2007, pp. 567–590.

Hibino, T., M. Loza-Coll, C. Messier, A.J. Majeske, A.H. Cohen, D.P. Terwilliger, K.M. Buckley *et al.* The immune gene repertoire encoded in the purple sea urchin genome. *Dev. Biol.* 2006; 300:349–365.

Hildemann, W.H. Tissue transplantation immunity in Goldfish. *Immunology* 1958; 1:46–53.

Hill, A.V. HLA associations with malaria in Africa: Some implications for MHC evolution. *in Molecular Evolution of the Major Histocompatibility Complex.* Springer, Berlin, Heidelberg, Germany, 1991, pp. 403–420.

Hiscock, S.J., U. Kues, and H.G. Dickinson. Molecular mechanisms of self-incompatibility in flowering plants and fungi—different means to the same end. *Trends Cell Biol.* 1996; 6:421–428.

Holland, L.Z., R. Albalat, K. Azumi, E. Benito-Gutierrez, M.J. Blow, M. Bronner-Fraser, F. Brunet *et al.* The amphioxus genome illuminates vertebrate origins and cephalochordate biology. *Genome Res.* 2008; 18:1100–1111.

Horton, R., L. Wilming, V. Rand, R.C. Lovering, E.A. Bruford, V.K. Khodiyar, M.J. Lush *et al.* Gene map of the extended human MHC. *Nat. Rev. Genet.* 2004; 5:889–899.

Huang, A., A. Bruce, D. Pardoll, and H. Levitsky. in vivo cross-priming of MHC class I–restricted antigens requires the TAP transporter. *Immunity* 1996; 4:349–355.

Hughes, A. *Major Histocompatibility Complex (MHC) Genes: Evolution. Encyclopedia of Life Sciences (ELS).* Wiley, Chichester, UK, 2008.

Hughes, A.L., and M. Nei. Pattern of nucleotide substitution at major histocompatibility complex class I loci reveals overdominant selection. *Nature* 1988; 335:167–170.

Jeffery, K.J.M., and C.R.M. Bangham. Do infectious diseases drive MHC diversity? *Microbes Infect.* 2000; 2:1335–1341.

Kandil, E., C. Namikawa, M. Nonaka, A.S. Greenberg, M.F. Flajnik, T. Ishibashi, and M. Kasahara. Isolation of low molecular mass polypeptide complementary DNA clones from primitive vertebrates. Implications for the origin of MHC class I-restricted antigen presentation. *J. Immunol.* 1996; 156:4245–4253.

Kasahara, M. New insights into the genomic organization and origin of the major histocompatibility complex: Role of chromosomal (genome) duplication in the emergence of the adaptive immune system. *Hereditas* 1997; 127:59–65.

Kasahara, M., M. Hayashi, K. Tanaka, H. Inoko, K. Sugaya, T. Ikemura, and T. Ishibashi. Chromosomal localization of the proteasome Z subunit gene reveals an ancient chromosomal duplication involving the major histocompatibility complex. *Proc. Natl. Acad. Sci. USA* 1996a; 93:9096–9101.

Kasahara, M., E. Kandil, L. Salter-Cid, and M. Flajnik. Origin and evolution of the class I gene family: Why are some of the mammalian class I genes encoded outside the major histocompatibility complex? *Res. Immunol.* 1996b; 147:278–284.

Kasahara, M., E.C. McKinney, M.F. Flajnik, and T. Ishibashi. The evolutionary origin of the major histocompatibility complex: Polymorphism of class II alpha chain genes in the cartilaginous fish. *Eur. J. Immunol.* 1993; 23:2160–2165.

Kasahara, M., M. Vazquez, K. Sato, E.C. McKinney, and M.F. Flajnik. Evolution of the major histocompatibility complex: Isolation of class II A cDNA clones from the cartilaginous fish. *Proc. Natl. Acad. Sci. USA* 1992; 89:6688–6692.

Katsanis, N., J. Fitzgibbon, and E.M. Fisher. Paralogy mapping: Identification of a region in the human MHC triplicated onto human chromosomes 1 and 9 allows the prediction and isolation of novel PBX and NOTCH loci. *Genomics* 1996; 35:101–108.

Kaufman, J. Evolution and immunity. *Immunology* 2010; 130:459–462.

Kaufman, J., J. Jacob, I. Shaw, B. Walker, S. Milne, S. Beck, and J. Salomonsen. Gene organisation determines evolution of function in the chicken MHC. *Immunol. Rev.* 1999a; 167:101–117.

Kaufman, J., S. Milne, T. Göbel, B. Walker, J. Jacob, C. Auffray, R. Zoorob, and S. Beck. The chicken B locus is a minimal essential major histocompatibility complex. *Nature* 1999b; 401:923–925.

Kelley, J., L. Walter, and J. Trowsdale. Comparative genomics of major histocompatibility complexes. *Immunogenetics* 2005; 56:683–695.

Kimura, A., E. Sakaguchi, and M. Nonaka. Multi-component complement system of Cnidaria: C3, Bf, and MASP genes expressed in the endodermal tissues of a sea anemone, *Nematostella vectensis*. *Immunobiology* 2009; 214:165–178.

Klein, J., A. Sato, and N. Nikolaidis. MHC, TSP, and the origin of species: From immunogenetics to evolutionary genetics. *Annu. Rev. Genet.* 2007; 41:281–304.

Kosmrlj, A., E.L. Read, Y. Qi, T.M. Allen, M. Altfeld, S.G. Deeks, F. Pereyra, M. Carrington, B.D. Walker, and A.K. Chakraborty. Effects of thymic selection of the T-cell repertoire on HLA class[thinsp]I-associated control of HIV infection. *Nature* 2010; 465:350–354.

Kulski, J.K., T. Shiina, T. Anzai, S. Kohara, and H. Inoko. Comparative genomic analysis of the MHC: The evolution of class I duplication blocks, diversity and complexity from shark to man. *Immunol. Rev.* 2002; 190:95–122.

Kuraku, S., D. Hoshiyama, K. Katoh, H. Suga, and T. Miyata. Monophyly of lampreys and hagfishes supported by nuclear DNA-coded genes. *J. Mol. Evol.* 1999; 49:729–735.

Lawlor, D.A., F.E. Ward, P.D. Ennis, A.P. Jackson, and P. Parham. HLA-A and B polymorphisms predate the divergence of humans and chimpanzees. *Nature* 1988; 335:268–271.

Leinders-Zufall, T., P. Brennan, P. Widmayer, A. Maul-Pavicic, M. Jager, X.H. Li, H. Breer, F. Zufall, and T. Boehm. MHC class I peptides as chemosensory signals in the vomeronasal organ. *Science* 2004; 306:1033–1037.

Luer, C., C. Walsh, and A. Bodine. The immune system of sharks, skates, and rays. *in* J. Carrier, J. Musick, and M. Heithaus, eds. *Biology of Sharks and Their Relatives*. CRC Press, Boca Raton, FL, 2004, pp. 369–398.

Ma, Q., Y.Q. Su, J. Wang, Z.M. Zhuang, and Q.S. Tang. Molecular cloning and expression analysis of major histocompatibility complex class IIB gene of the Whitespotted bambooshark (*Chiloscyllium plagiosum*). *Fish Physiol. Biochem.* 2013; 39:131–142.

Madden, D. The three-dimensional structure of peptide-MHC complexes. *Annu. Rev. Immunol.* 1995; 13:587–622.

Matsuo, M.Y., S. Asakawa, N. Shimizu, H. Kimura, and M. Nonaka. Nucleotide sequence of the MHC class I genomic region of a teleost, the medaka (*Oryzias latipes*). *Immunogenetics* 2002; 53:930–940.

McConnell, T.J., W.S. Talbot, R.A. McIndoe, and E.K. Wakeland. The origin of MHC class II gene polymorphism within the genus Mus. *Nature* 1988; 332:651–654.

McCumber, L.J., M.M. Sigel, R.J. Trauger, and M.A. Cuchens. RES structure and function of the fishes. *in* N. Cohen, and M.M. Sigel, eds. *The Reticuloendothelial System*. Plenum, New York, 1982, pp. 393–422.

Michalova, V., B.W. Murray, H. Sultmann, and J. Klein. A contig map of the Mhc class I genomic region in the zebrafish reveals ancient synteny. *J. Immunol.* 2000; 164:5296–5305.

Milinski, M., S. Griffiths, K.M. Wegner, T.B. Reusch, A. Haas-Assenbaum, and T. Boehm. Mate choice decisions of stickleback females predictably modified by MHC peptide ligands. *Proc. Natl. Acad. Sci. USA* 2005; 102:4414–4418.

Miller, K.M., and R.E. Withler. Mhc diversity in Pacific salmon: Population structure and trans-species allelism. *Hereditas* 1997; 127:83–95.

Miller, M.M., C. Wang, E. Parisini, R.D. Coletta, R.M. Goto, S.Y. Lee, D.C. Barral *et al.* Characterization of two avian MHC-like genes reveals an ancient origin of the CD1 family. *Proc. Natl. Acad. Sci. USA* 2005; 102:8674–8679.

Miura, F., K. Tsukamoto, R.B. Mehta, K. Naruse, W. Magtoon, and M. Nonaka. Transspecies dimorphic allelic lineages of the proteasome subunit {beta}-type 8 gene (PSMB8) in the teleost genus Oryzias. *Proc. Natl. Acad. Sci. USA* 2010; 107:21599–21604.

Murphy, K. *Janeway's Immunobiology*. Garland Science, New York, 2011.

Nakao, M., J. Mutsuro, M. Nakahara, Y. Kato, and T. Yano. Expansion of genes encoding complement components in bony fish: Biological implications of the complement diversity. *Dev. Comp. Immunol.* 2003; 27:749–762.

Naruse, K., S. Fukamachi, H. Mitani, M. Kondo, T. Matsuoka, S. Kondo, N. Hanamura *et al.* A detailed linkage map of medaka, *Oryzias latipes*: Comparative genomics and genome evolution. *Genetics* 2000; 154:1773–1784.

Neefjes, J., M.L. Jongsma, P. Paul, and O. Bakke. Towards a systems understanding of MHC class I and MHC class II antigen presentation. *Nat. Rev. Immunol.* 2011; 11:823–836.

Nonaka, M., and A. Kimura. Genomic view of the evolution of the complement system. *Immunogenetics* 2006; 58:701–713.

Nonaka, M., C. Yamada-Namikawa, M.F. Flajnik, and L. Du Pasquier. Trans-species polymorphism of the major histocompatibility complex-encoded proteasome subunit LMP7 in an amphibian genus, Xenopus. *Immunogenetics* 2000; 51:186–192.

Nonaka, M.I., K. Aizawa, H. Mitani, H.P. Bannai, and M. Nonaka. Retained orthologous relationships of the MHC Class I genes during euteleost evolution. *Mol. Biol. Evol.* 2011; 28:3099–3112.

Ohta, Y., W. Goetz, M.Z. Hossain, M. Nonaka, and M.F. Flajnik. Ancestral organization of the MHC revealed in the amphibian *Xenopus*. *J. Immunol.* 2006; 176:3674–3685.

Ohta, Y., E.C. McKinney, M.F. Criscitiello, and M.F. Flajnik. Proteasome, transporter associated with antigen processing, and class I genes in the nurse shark *Ginglymostoma cirratum*: Evidence for a stable class I region and MHC haplotype lineages. *J. Immunol.* 2002; 168:771–781.

Ohta, Y., K. Okamura, E.C. McKinney, S. Bartl, K. Hashimoto, and M.F. Flajnik. Primitive synteny of vertebrate major histocompatibility complex class I and class II genes. *Proc. Natl. Acad. Sci. USA* 2000; 97:4712–4717.

Ohta, Y., T. Shiina, R.L. Lohr, K. Hosomichi, T.I. Pollin, E.J. Heist, S. Suzuki, H. Inoko, and M.F. Flajnik. Primordial linkage of beta2-microglobulin to the MHC. *J. Immunol.* 2011; 186:3563–3571.

Okamura, K., M. Ototake, T. Nakanishi, Y. Kurosawa, and K. Hashimoto. The most primitive vertebrates with jaws possess highly polymorphic MHC class I genes comparable to those of humans. *Immunity* 1997; 7:777–790.

Olsen, K.H., M. Grahn, and J. Lohm. Influence of MHC on sibling discrimination in Arctic char, *Salvelinus alpinus* (L.). *J. Chem. Ecol.* 2002; 28:783–795.

Ottova, E., A. Simkova, J.F. Martin, J.G. de Bellocq, M. Gelnar, J.F. Allienne, and S. Morand. Evolution and trans-species polymorphism of MHC class IIbeta genes in cyprinid fish. *Fish Shellfish Immunol.* 2005; 18:199–222.

Parham, P. The rise and fall of great dass I genes. *Semin. Immunol.* 1994; 6:373–382.

Penn, D.J., and P. Ilmonen. *Major Histocompatibility Complex (MHC)*. Wiley, Chichester, UK, 2005.

Piertney, S.B., and M.K. Oliver. The evolutionary ecology of the major histocompatibility complex. *Heredity (Edinb)* 2006; 96:7–21.

Potts, W., and E. Wakeland. Evolution of diversity at the major histocompatibility complex. *Trends Ecol. Evol.* 1990; 5:181–187.

Raghuraman, G., P.E. Lapinski, and M. Raghavan. Tapasin interacts with the membrane-spanning domains of both TAP subunits and enhances the structural stability of TAP1 x TAP2 complexes. *J. Biol. Chem.* 2002; 277:41786–41794.

Rajakaruna, R.S., J.A. Brown, K.H. Kaukinen, and K.M. Miller. Major histocompatibility complex and kin discrimination in Atlantic salmon and brook trout. *Mol. Ecol.* 2006; 15:4569–4575.

Robey, E., and B. Fowles. Selective events in T cell development. *Annu. Rev. Immunol.* 1994; 12:67S–70S.

Rock, K. A new foreign policy: MHC class I molecules monitor the outside world. *Immunol. Today* 1996; 17:131–137.

Rock, K.L., and A.L. Goldberg. Degradation of cell proteins and the generation of MHC class I-presented peptides. *Annu. Rev. Immunol.* 1999; 17:739–779.

Rock, K.L., I.A. York, and A.L. Goldberg. Post-proteasomal antigen processing for major histocompatibility complex class I presentation. *Nat. Immunol.* 2004; 5:670–677.

Ruff, J.S., A.C. Nelson, J.L. Kubinak, and W.K. Potts. MHC signaling during social communication. *in* C. López-Larrea, ed. *Self and Non-Self*. Landes Bioscience and Springer, Austin, TX, 2011, pp. 290–313.

Rumfelt, L.L., D. Avila, M. Diaz, S. Bartl, E.C. McKinney, and M.F. Flajnik. A shark antibody heavy chain encoded by a nonsomatically rearranged VDJ is preferentially expressed in early development and is convergent with mammalian IgG. *Proc. Natl. Acad. Sci. USA* 2001; 98:1775–1780.

Sato, A., F. Figueroa, B.W. Murray, E. Malaga-Trillo, Z. Zaleska-Rutczynska, H. Sultmann, S. Toyosawa, C. Wedekind, N. Steck, and J. Klein. Nonlinkage of major histocompatibility complex class I and class II loci in bony fishes. *Immunogenetics* 2000; 51:108–116.

Satta, Y., C. O'HUigin, N. Takahata, and J. Klein. The synonymous substitution rate of the major histocompatibility complex loci in primates. *Proc. Natl. Acad. Sci. USA* 1993; 90:7480–7484.

Shiina, T., C. Shimizu, A. Oka, Y. Teraoka, T. Imanishi, T. Gojobori, K. Hanzawa, S. Watanabe, and H. Inoko. Gene organization of the quail major histocompatibility complex (MhcCoja) class I gene region. *Immunogenetics* 1999; 49:384–394.

Shin, D.H., B. Webb, M. Nakao, and S.L. Smith. Molecular cloning, structural analysis and expression of complement component Bf/C2 genes in the nurse shark, *Ginglymostoma cirratum*. *Dev. Comp. Immunol.* 2007; 31:1168–1182.

Siddle, H.V., J.E. Deakin, P. Coggill, E. Hart, Y. Cheng, E.S. Wong, J. Harrow, S. Beck, and K. Belov. MHC-linked and un-linked class I genes in the wallaby. *BMC Genomics* 2009; 10:310.

Slade, R.W., and H.I. McCallum. Overdominant vs. frequency-dependent selection at MHC loci. *Genetics* 1992; 132:861–864.

Smith, J.J., S. Kuraku, C. Holt, T. Sauka-Spengler, N. Jiang, M.S. Campbell, M.D. Yandell *et al.* Sequencing of the sea lamprey (*Petromyzon marinus*) genome provides insights into vertebrate evolution. *Nat. Genet.* 2013; 45:415–421.

Smith, L.C., and E.H. Davidson. The echinoid immune system and the phylogenetic occurrence of immune mechanisms in deuterostomes. *Immunol. Today* 1992; 13:356–362.

Sommer, S. The importance of immune gene variability (MHC) in evolutionary ecology and conservation. *Front Zool.* 2005; 2:16.

Spehr, M., K.R. Kelliher, X.H. Li, T. Boehm, T. Leinders-Zufall, and F. Zufall. Essential role of the main olfactory system in social recognition of major histocompatibility complex peptide ligands. *J. Neurosci.* 2006; 26:1961–1970.

Spurgin, L.G., and D.S. Richardson. How pathogens drive genetic diversity: MHC, mechanisms and misunderstandings. *Proc. Biol. Sci.* 2010; 277:979–988.

Su, C., and M. Nei. Fifty-million-year-old polymorphism at an immunoglobulin variable region gene locus in the rabbit evolutionary lineage. *Proc. Natl. Acad. Sci. USA* 1999; 96:9710–9715.

Sutton, J.T., S. Nakagawa, B.C. Robertson, and I.G. Jamieson. Disentangling the roles of natural selection and genetic drift in shaping variation at MHC immunity genes. *Mol. Ecol.* 2011; 20:4408–4420.

Tam, M.R., A.L. Reddy, R.D. Karp, and W.H. Hildemann. Phylogeny of cellular immunity among vertebrates. *in* J.J. Marchalonis, ed. *Comparative Immunology*. Wiley, New York, 1976, pp. 98–119.

Tamura, K., D. Peterson, N. Peterson, G. Stecher, M. Nei, and S. Kumar. MEGA5: Molecular evolutionary genetics analysis using maximum likelihood, evolutionary distance, and maximum parsimony methods. *Mol. Biol. Evol.* 2011; 28:2731–2739.

Tan, Y.Y., R. Kodzius, B.H. Tay, A. Tay, S. Brenner, and B. Venkatesh. Sequencing and analysis of full-length cDNAs, 5'-ESTs and 3'-ESTs from a cartilaginous fish, the Elephant Shark (*Callorhinchus milii*). *PLoS One* 2012; 7:e47174.

Tanaka, K., and M. Kasahara. The MHC class I ligand-generating system: Roles of immunoproteasomes and the interferon-gamma-inducible proteasome activator PA28. *Immunol. Rev.* 1998; 163:161–176.

Terado, T., K. Okamura, Y. Ohta, D.H. Shin, S.L. Smith, K. Hashimoto, T. Takemoto *et al.* Molecular cloning of C4 gene and identification of the class III complement region in the shark MHC. *J. Immunol.* 2003; 171:2461–2466.

The MHC Sequencing Consortium. Complete sequence and gene map of a human major histocompatibility complex. The MHC sequencing consortium. *Nature* 1999; 401:921–923.

Trowsdale, J. The MHC, disease and selection. *Immunol. Lett.* 2011; 137:1–8.

Tsukamoto, K., F. Miura, N.T. Fujito, G. Yoshizaki, and M. Nonaka. Long-lived dichotomous lineages of the proteasome subunit beta type 8 (PSMB8) gene surviving more than 500 million years as alleles or paralogs. *Mol. Biol. Evol.* 2012; 29:3071–3079.

Tsukamoto, K., M. Sakaizumi, M. Hata, Y. Sawara, J. Eah, C.B. Kim, and M. Nonaka. Dichotomous haplotypic lineages of the immunoproteasome subunit genes, PSMB8 and PSMB10, in the MHC class I region of a Teleost Medaka, *Oryzias latipes*. *Mol. Biol. Evol.* 2009; 26:769–781.

Venkatesh, B., A.P. Lee, V. Ravi, A.K. Maurya, M.M. Lian, J.B. Swann, Y. Ohta *et al.* Elephant shark genome provides unique insights into gnathostome evolution. *Nature* 2014; 505(7482):174–179.

Venkatesh, B., Y. Ning, and S. Brenner. Late changes in spliceosomal introns define clades in vertebrate evolution. *Proc. Natl. Acad. Sci. USA* 1999; 96:10267–10271.

Villinger, J., and B. Waldman. Self-referent MHC type matching in frog tadpoles. *Proc. Biol. Sci.* 2008; 275:1225–1230.

Vyas, J.M., A.G. Van der Veen, and H.L. Ploegh. The known unknowns of antigen processing and presentation. *Nat. Rev. Immunol.* 2008; 8:607–618.

Wang, C., T.V. Perera, H.L. Ford, and C.C. Dascher. Characterization of a divergent non-classical MHC class I gene in sharks. *Immunogenetics* 2003; 55:57–61.

Wang, D., L. Zhong, Q. Wei, X. Gan, and S. He. Evolution of MHC class I genes in two ancient fish, paddlefish (*Polyodon spathula*) and Chinese sturgeon (*Acipenser sinensis*). *FEBS Lett.* 2010a; 584:3331–3339.

Wang, F., H. Liu, W.P. Blanton, A. Belkina, N.K. Lebrasseur, and G.V. Denis. Brd2 disruption in mice causes severe obesity without Type 2 diabetes. *Biochem. J.* 2010b; 425:71–83.

Wang, Q., C.N. Arighi, B.L. King, S.W. Polson, J. Vincent, C. Chen, H. Huang, *et al.* Community annotation and bioinformatics workforce development in concert—Little Skate Genome Annotation Workshops and Jamborees. *Database (Oxford)* 2012:bar064.

Yang, B.M., J. Harris, M. Gilpin, and A. Demaine. Isolation of major histocompatibility complex (MHC) class II genes from dogfish (*Scyliorhinus canicula*). *Immunol. Lett.* 1997; 56:230.

Yu, W.P., V. Rajasegaran, K. Yew, W.L. Loh, B.H. Tay, C.T. Amemiya, S. Brenner, and B. Venkatesh. Elephant shark sequence reveals unique insights into the evolutionary history of vertebrate genes: A comparative analysis of the protocadherin cluster. *Proc. Natl. Acad. Sci. USA* 2008; 105:3819–3824.

Zenewicz, L.A., C. Abraham, R.A. Flavell, and J.H. Cho. Unraveling the genetics of autoimmunity. *Cell* 2010; 140:791–797.

Zhong, G., F. Castellino, P. Romagnoli, and R.N. Germain. Evidence that binding site occupancy is necessary and sufficient for effective major histocompatibility complex (MHC) class II transport through the secretory pathway redefines the primary function of class II-associated invariant chain peptides (CLIP). *J. Exp. Med.* 1996; 184:2061–2066.

CHAPTER 10

Considering V(D)J Recombination in the Shark

Ellen Hsu

CONTENTS

SUMMARY

In order for B and T lymphocytes to respond specifically to immune challenge, only a single species of antigen receptor is expressed per cell. In cartilaginous fishes, such as sharks and skates, there can be >100 immunoglobulin (Ig) gene clusters, a situation in which regulatory processes permitting expression of only one allele of one cluster are not well understood. This chapter reviews the shark IgM heavy (H) chain gene organization, H chain rearrangement, and recombination events occurring in single B lymphocytes. In order to gain some insight into how the process may be controlled at multiple loci in B cells, somatic rearrangements at IgH genes in thymocytes and deduced recombination events in germ cells are examined. Studies in the shark demonstrate that certain mechanisms deemed crucial to V(D)J recombination in mammalian IgH systems—two-stage rearrangement, locus compaction—are processes that contribute to, but in themselves are not unconditionally required to bring about H chain exclusion.

INTRODUCTION

V(D)J rearrangement is the mechanism by which antigen receptor diversity is generated in lymphocytes of all jawed vertebrates. By the divergence of cartilaginous fishes >450 million years ago, specialized genes that somatically recombine were established for generating T-cell receptor (TCRα, TCRβ, TCRγ, TCRδ) in T lymphocytes and immunoglobulin (Ig) (heavy chain, light chain) in B lymphocytes (Flajnik and Du Pasquier, 2008). There has always been great interest in the immune systems of the earliest vertebrates, cyclostomes, and cartilaginous fishes and speculation on the origin of this specialized mechanism, called V(D)J rearrangement (Rast and Litman, 1998). Now that it is clear that lamprey and hagfish use entirely different antigen receptor gene systems (Hirano et al., 2011), the phylogenetically closest relatives to the ancestral vertebrates in which V(D)J recombination evolved are the cartilaginous fishes.

Adaptive immunity is mediated by lymphocytes, which possess two properties to achieve specificity and memory. First, the repertoire of antigen receptors is widely diverse, and second, after triggering by antigen, the few specifically activated lymphocytes will undergo extensive clonal expansion. For the expansion to be specific against the pathogen, each lymphocyte must express only one single type of antigen receptor, and in most diploid vertebrates, the expression of one antigen receptor gene is favored over its allele, primarily at the TCRβ and the IgH loci (Vetterman and Schlissel, 2010). This restriction is called allelic exclusion.

In one of the first publications on horned shark Ig genes, Hinds and Litman (1986) reported that its Ig multiple cluster organization is different from those established in tetrapods (Figure 10.1), demonstrating the uniqueness of the elasmobranch model for exploring fundamental immunology ideas developed in mammalian models. If the expression of a single Ig only per B cell is the underlying requirement for adaptive immune systems, it is not clear how this is managed in the shark system where there are estimated to be >100 IgH miniloci. Figure 10.1 illustrates the differences in organization as well as rearrangement steps between mouse and shark IgH systems: two-stage recombination process taking place over a >2 Mb locus contrasted to a single-stage event over <2 kb (Malecek et al., 2008; Zhu et al., 2011). Ig gene expression only takes place after successful V(D)J recombination, and in this chapter, we discuss how H chain exclusion is achieved through regulated rearrangement at the multiple autonomous IgH.

The review includes a brief section on the shark antibodies, followed by descriptions of their H chain gene organization, H chain rearrangement, and rearrangements in single B lymphocytes. Analysis of somatic IgH recombination events in shark thymocytes provides insights into how the process may be controlled in B cells. Lastly, parallels are drawn between rearrangement events in thymocytes and those deduced to occur in germ cells. The pre-rearranged or germline-joined Ig genes, also first discovered in horned shark (Kokubu et al., 1988), are inherited, and the fact of their existence implies that the recombinase effecting V(D)J rearrangement may be active in nonlymphocytes.

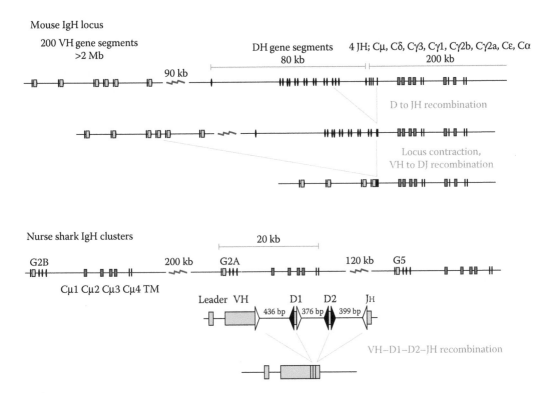

Figure 10.1 Organization of mouse IgH locus and shark IgH clusters. Top: murine IgH. The mouse IgH consists of multiple VH (blue boxes), DH and JH (narrow black boxes) gene segments, spread over >2 Mb. The C exons (brown boxes) of Cμ and Cδ are shown, followed by other Ig isotypes. Rearrangement takes place in two stages: D to JH, followed by VH to DJ. Bottom: nurse shark IgH miniloci. The orientation and order of the IgM H chain genes G2B, G2A, and G5 are shown, with the VH gene segments and C exons (TM, transmembrane), as labeled. The intragenic and intrasegmental distances are shown, although not to scale. (From Zhu, C. *et al.*, *J. Immunol.* 187, 2492–2501, 2011.) Recombination signal sequences (RSS) are shown as white (23 bp spacer RSS) and black (12 bp spacer RSS) triangles. Rearrangement occurs in one step, almost always as VHD1D2JH.

SHARK ANTIBODY RESPONSE

Shark antibodies were elicited to a wide variety of antigens: proteins (keyhole limpet hemocyanin, bovine serum albumin, hen egg lysozyme), bacteria (*Brucella abortus*, *Salmonella typhui*, Group A streptococcus), viruses (Ebolavirus), and hapten conjugates (dinitrophenyl, nitroiodophenyl, *p*-azobenzene arsonate, phenyloxazolone) (Goodchild *et al.*, 2011; Litman *et al.*, 1982; Nisonoff *et al.*, 1975). The smooth dogfish shark antibody made in response to hemocyanin was identified as IgM, based on high molecular weight (19S), H chain size, coupling with light (L) chain, and carbohydrate content (Marchalonis and Edelman, 1965, 1966). In nurse sharks, the 19S protein was isolated, showing a binding valence of five and efficacy in agglutinating and lytic abilities, characteristics of the polymeric Ig form (Voss *et al.*, 1969). Comparison of cloned horned shark C region sequences to Cμ of other species confirmed the homology (Kokubu *et al.*, 1988). Shark IgM also exists in a monomeric form (7S), which appears later in life (Small *et al.*, 1970). 19S antibodies are raised during the primary response; careful timing of boosts by monitoring the decline of the primary response elicited higher titers of mostly 7S antibodies with increased affinity. These experiments confirmed antibody maturation and immunological memory in shark antibody responses (Clem *et al.*, 1967; Dooley and Flajnik, 2000).

Besides IgM, there are two other classes of secreted Igs in shark serum, IgW and IgNAR. IgW is expressed as an Ig monomer and is believed to be an ortholog of IgD (Bernstein *et al.*, 1996; Greenberg *et al.*, 1996), although its function in elasmobranch fishes is not known. IgNAR does not require L chains and consists of a disulfide-bonded dimer, each chain by itself containing an independent combining site (Flajnik, 2002). IgW and/or IgNAR was found in all elasmobranchs studied: nurse shark, sandbar shark, little skate, spotted woebegong, shovelnose guitarfish, banded houndshark, and spiny dogfish (Anderson *et al.*, 1994; Honda *et al.*, 2010; Rumfelt *et al.*, 2004; Smith *et al.*, 2012). This review will focus on shark IgM, which has a clear structural and functional homology with IgM of other vertebrates, and this correlation makes it reasonable to extend mechanistic comparisons.

SHARK Igμ GENE CLUSTERS

Germline Elements and Organization

The IgM H chain gene organization in cartilaginous fishes is similar in two extant subclasses, Holocephali (ratfish, elephantfish) and Elasmobranchii. The latter subclass includes animals from two orders, Batoidea (long-nosed ray, little skate, clearnose skate) and Selachii (horned shark, nurse shark, sandbar shark). Because the subclasses diverged ~400 million years ago, the Ig gene organization is ancient and conserved; the main distinction among elasmobranch species is gene number. In horned shark, the IgH cluster number is estimated to be >100 (Hinds and Litman, 1986; Kokubu *et al.*, 1987), whereas in nurse sharks, between 9 and 12 functional IgH and three pseudogenes were described (Lee *et al.*, 2008).

The elasmobranch IgH cluster is considered the primordial organization, where the Ig gene unit consists of a few recombining gene segments and one set of C region exons (Figure 10.1). In nurse sharks, all Igμ genes contain one VH, two D, and one JH that need to be somatically rearranged; however, in all other elasmobranch species, some of the IgM clusters contain prejoined gene segments. That is, they exist in the germline as fully rearranged VDDJ and partially recombined VDD-J in horned sharks (Kokubu *et al.*, 1988) or VD-DJ in little skates (Harding *et al.*, 1990). These unusual configurations are discussed in detail in the last section of this chapter.

The J-C introns are 6.5–12.7 kb in length in horned shark IgH and 6.3–10 kb in nurse shark (Kokubu *et al.*, 1988; Lee *et al.*, 2008). Four exons encoding the C region follow, with two transmembrane (TM) exons. This brings the average size of IgH, from leader to the transmembrane exons, to ≥20 kb. Because older studies were carried out with bacteriophage vectors, and to date no elasmobranch genome was fully assembled, there is very little cluster linkage information available. In the clearnose skate, the many Igω and Igμ genes appear to be at multiple chromosomal sites, as observed by fluorescence *in situ* hybridization (Anderson *et al.*, 1994). From studies using a nurse shark BAC library, it appears that the Igμ is located far apart, and the most closely linked clusters are situated 120 kb apart (Figure 10.1; Lee *et al.*, 2008; Zhu *et al.*, 2011). The rearranging elements in any one Igμ cluster thus lie at a great distance from another, and possibly for this reason, intergenic Igμ rearrangement has not so far been observed in B cells (Hinds-Frey *et al.*, 1993; Lee *et al.*, 2008).

The gene segments, VH, D1, D2, and JH, of one cluster are located within <2 kb, each element separated by ~250–400 bp of intervening sequence. The motifs recognized by the recombinase, RAG (encoded by the recombination-activating genes RAG1 and RAG2), are the recombination signal sequences (RSS) that flanking the gene segments (Figures 10.1 and 10.2) (for a recent review, see Schatz and Swanson, 2011). They contain 12- or 23-bp spacer sequences (referred to as RSS-12 and RSS-23, respectively), so that rearrangement primarily occurs between VH and D1, D1 and D2, D2 and JH, following the 12/23 rule. Other rearrangement possibilities exist but occur rarely, and these are discussed in detail in a later section.

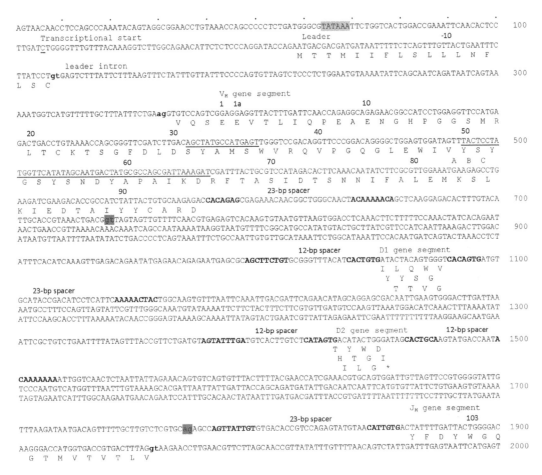

Figure 10.2 Organization of rearranging VH gene segments. The G2A gene with leader and VH gene segments (VH, D1, D2, JH) is shown with deduced amino acid sequence (single letter code). CDR1 and CDR2 are underlined. The flanking RSS have bolded heptamer and nonamer. A TATA motif is italicized. The 5′-most-observed transcriptional start site for G2A is underlined. The splice motifs at the leader intron and 3′ of JH are lower case and bolded; cryptic splice sites, discussed in the text, are in lower case and shaded dark gray.

Nurse shark and horned shark diverged 180 million years ago; sharks and skates/rays diverged 220 million years ago. Yet, the structure and configuration of the gene segments of the basic rearranging IgH unit are conserved as to the number of elements and the distances that separate them. The intersegmental sequences undergo change, and they differ sufficiently among the major nurse shark μ subfamilies (G1, G2, G3/G4, G5) such that they can be used as DNA probes to distinguish among the genes (Malecek *et al.*, 2008, see Figure S3). This is in contrast to a VH probe, which will cross-hybridize with all IgM VH in the animal. If the intersegmental sequences changed so extensively, why are they maintained at 250–400 bp? Perhaps this is the optimal distance allowing proper orientation of RSS/RAG pairs to synapse; certainly, it is possible for the RSS to be too close (Lewis and Hesse, 1991). Considering that a 400–800 bp difference seems insignificant when in mouse the distance from JH1 to Dh01b (DQ52) is 700 bp and to Dh16b is 50 kb (Riblet, 2004), it is probable that rearrangement to the adjacent gene segment in the cluster is also a preference based on the efficiency by which one RAG/RSS complex can capture a second RSS.

Except in one instance, all the VH gene segments within a species are classified as being in one family, which means that they share 80% or more sequence identity at the nucleotide level (Brodeur and Riblet, 1984). In horned shark, a second, pauci-copy IgM VH family was identified (Hinds-Frey *et al.*, 1993). Interestingly, in phylogenetic analyses, this VH (λ1113) clusters with VH sequences from IgW (Ota *et al.*, 2000), and it may be that this unique gene segment had been originally co-opted from an IgW gene. If so, this is an unusual instance of domain exchange between clusters, because the clusters by and large evolved independently (Lee *et al.*, 2008), and within each species, the genes evolved by continuous and gradual turnover (Ota *et al.*, 2000).

Among Igμ clusters, there is sequence variation in all coding regions, although the protein product is part of a classical IgM pentamer (Kokubu *et al.*, 1987, 1988; Rumfelt *et al.*, 2004; Smith *et al.*, 2012). The C exon differences have so far not been correlated with any effector functions or Fc receptor binding, but they are undoubtedly under selection pressure, as demonstrated by detailed analyses of substitution patterns (Lee *et al.*, 2008). Among the nine functional IgH in one nurse shark, the estimated substitution rates per nucleotide of the domains (VH, Cμ1, Cμ2, Cμ3, Cμ4) were not significantly different from each other; that is, no one domain is evolving faster than the others. However, it is only when analyzed with respect to synonymous versus non-synonymous changes that differences between the domains emerged. The nature of substitutions in Cμ4 shows that this exon is under strong stabilizing selection. In contrast, VH and Cμ2 both had the greatest replacement rates, demonstrating strong positive selection for increased sequence diversity. Given its known antigen-binding role, VH may be expected to diverge, but the surprising result for Cμ2 is suggestive of an as yet undetermined role affecting C region structure or effector function.

Transcription

IgH genes that have undergone somatic rearrangement in B cells are transcribed, whether or not the VDJ rearrangement encodes a functional protein (Zhu and Hsu, 2010). Pseudogenes with defects in the coding regions also can rearrange and are transcribed, probably as long as their cis-regulatory elements and RSS are not impaired (Zhu *et al.*, 2011). Considering that there are at least as many nonproductively rearranged as there are in-frame VDJ rearrangements in B cells (Malecek *et al.*, 2008), the apparent paucity of such sequences in cDNA libraries or amplified by RT-PCR must be attributed to faster degradation, probably by RNA surveillance mechanisms described in mammalian B cells (Tinguely *et al.*, 2011). Transcripts with out-of-frame VDJ can be preferentially amplified by placing the forward PCR primer within the leader intron (Zhu and Hsu, 2010), as it appears that leader intron splicing is less efficient in such cases. These sequences can contain many substitutions, demonstrating that somatic hypermutation (SHM) occurs at all active Ig loci. Partially rearranged genes (VD-DJ, VDD-J) are also transcribed and can be mutated (Malecek *et al.*, 2008), observations suggesting that once an IgH gene was activated, it remains active throughout B-cell development regardless of functionality.

Nonrearranged Ig genes (VH, D, and JH gene segments in germline (GL) configuration) are transcribed and easily detected in neonatal tissues (Miracle *et al.*, 2001; Rumfelt *et al.*, 2004) where there are few activated B cells secreting antibody. It is probable that such transcripts may derive from IgH poised for rearrangement because transcription is the culmination of activation processes rendering chromatin accessible to RAG (Krangel, 2003). The GL sequence transcripts are usually aberrantly spliced at cryptic splice sites, shaded dark gray in Figure 10.2, within the V-D1 and D2-JH intersegmental regions, with the JH spliced to C exons. It is not clear whether transcription of IgH in GL configuration continues into later B cells stages and, if so, this occurs in all B cells. What we have found is that they can be transcribed in (some) B cells undergoing SHM. Such transcripts can be detected if specifically amplified by RT-PCR where the primers target the leader and the intersegmental region between D2 and JH. We then screened for sequences where the leader intron was spliced out; in the mutants, multiple substitutions were present throughout VH and the intersegmental regions (Zhu *et al.*, 2011).

Cis-regulatory elements have not yet been identified with certainty in shark IgH genes. Although there is a TATA sequence upstream of the VH leader (Figure 10.2, in italics, line 1), other conventional Ig-related motifs such as the ubiquitous octamer are missing in IgH of both sharks and skates (Harding *et al.*, 1990; Kokubu *et al.*, 1988); However, it is present at elasmobranch L chain genes. There is unlikely to be another promoter in the J-C intron, as no one so far has reported germline Cμ-only transcripts in any cartilaginous fish. The other possibility is that Cμ-alone transcription occurs at such a low level that it has gone undetected. Northern blotting has not distinguished any transcripts identifiable as a germline Cμ-alone transcript. In the nurse shark, a probe to the C region detected only a single RNA species in the neonatal epigonal organ by northern blotting; this band, coinciding with one hybridizing to both VH and transmembrane μ probes, contains transcripts for the transmembrane μ form (Malecek and Hsu unpublished results). No other band, including the secreted form, was observed after prolonged exposure. Moreover, a probe to the region 5′ of the Cμ1 exon, one that cross-hybridizes to many IgH clusters, was negative when hybridized with RNA from lymphoid tissues of both neonate and adults (Zhu and Hsu unpublished results).

Junctional Diversity

All VH belong to one family and thus are of limited sequence diversity; rearrangement takes place within one Igμ gene cluster, which precludes combinatorial diversity in B cells (Hinds-Frey *et al.*, 1993; Lee *et al.*, 2008). What appears to be an exception is sandbar shark, from which highly variable VH cDNA sequences were cloned (Shen *et al.*, 1996). However, because genomic VH gene segments from that species were not investigated, the observed sequence diversity could have been due to somatic mutation.

The primary antibody repertoire in cartilaginous fishes thus depends almost entirely upon junctional diversity. Three rearrangements routinely take place to form the VDJ unit, one more than in most other animals. The extra rearrangement does not serve to increase overall CDR3 size; an average of 11 codons is found in nurse shark CDR3 (Rumfelt *et al.*, 2004), in comparison with other species such as the nine codons in *Xenopus* and mouse, 12 codons in rabbit, 13 codons in human and platypus (Johansson *et al.*, 2002). An additional rearrangement step undoubtedly increases junctional diversification, as illustrated in Figure 10.3, which shows CDR3 from genomic rearrangements of nurse shark IgH G2A. There is varying nucleotide deletion from the VH 3′ flank, the JH 5′ flank, both flanks of D1 and D2. No protein is translated from the nonproductive VDJ, so that these sequences (Figure 10.3, out-of-frame VDDJ) reflect the intrinsic activities of V(D)J rearrangement.

	VH	N/P	D1	N/P	D2	N/P	J gene segment
	TGTGCAAGAGAC		ATACTACAGTGGGT		ACATACTGGGATAG		ACTATTTTGATTACTGG
VDDJ							
In-frame							
JS4	TGTGCAAGAG--	GGG	ACTACAGTG	CGC	GGGATAG	CCCAC	------TTGATTACTGG
JS5	TGTGCAAGAG--	GCAT	ATACTACAG	GGGGATCTCGT	ATACTGGGA	CCGC	-------TGATTACTGG
JS5D	TGTGCAAGAGAC	GG	CT	CACCGGAG	TACT		---ATTTTGATTACTGG
JS6	TGTGCAAGAGAC	CC	ACAGTGGG	GCT	ATACTGGGAT	G	--------GATTACTGG
JS9D	TGTGCAA------	CCCCTCACGCC	GT	TCCGCGGGG	ACTGGG	TAT	-CTATTTTGATTACTGG
JS11	TGTGCAAG----	CCCCCGG	ACTACAGTGGGT	ATCATGGCTT	ACATACTG	CTC	--------GATTACTGG
Nonproductive							
JS3	TGTGCAAGAGAC	GTTCGGATT	TACAGTGG	ATGGACCGATCTCAAGCCCCCAT	TGGGATAG	CTAG	--------GATTACTGG
JS3D	TGTCCAAGAGA-	T	GGT	ATATCTTA	ACTGG	GGCC	------------ACTGG
JS4D	TGTGCAAGAGA-	GA	ACTACA	CG	ACAT	CCCG	-----TTTGATTACTGG
JS7	TGTGCAAGA---	CCACGGT	TACAGTGGGT	ACCA	ATAC	C	--------GATTACTGG
JS8	TGTGCAAGAGAC		TACAGTGGGT	ACCG	GGG	CGT	ACTATTTTGATTACTGG
JS10	TGTGCAAG----	GATGCAGCAA	ACAGTGGGT	ACGGGGTCAGCCAGGCGAAA	ACTGGGATAG	GCGA	-------TGATTACTGG
JS12D	TGTGCAAG----	CGGCCATACTCTG	ACAGTGGGT		ACATACT	ATCCCGGGGAAAG	---ATTTTGATTACTGG
JS20	TGTGCAAG----	GGCGA	ACT	CCATCCAC	CATA	GGCAA	---ATTTTGATTACTGG
JS26	TGTGCAAG----	CGG	CTACAGTGGG	G	ACTG	TGC	-----TTTGATTACTGG
JS28	TGTGCAAGA---	ATGGGT	CTACAGT	CGG	CT	C	---------------G

Figure 10.3 CDR3 junctions from genomic G2A VDJ. The reference sequence consists of the VH 3′ flank, the D1 and D2 genes, and the 5′ JH flank. The three rearrangements that make up shark H chain CDR3 involve highly variable deletion of six flank sequences as well as N addition and P nucleotides (underlined). The N region can be very extensive, as seen in a nonfunctional VDJ JS3 and JS10.

Typically, 0–5 bp are deleted from VH and 0–16 bp are deleted from JH relative to the initial break sites at the RSS. The deletion can range from 0 to 10–11 out of 14 bp from either end of D1 and D2.

Increased variability is created by nontemplated (N) insertions added by terminal deoxy-nucleotidyl transferase (TdT) at the three junctions (Kokubu et al., 1988; Malecek et al., 2005; Rumfelt et al., 2004). The N additions are GC-rich and can be very extensive, notably in clone JS3 (Figure 10.3, see N/P). Because the N regions in the nonfunctional VDJ can be overall longer than in-frame VDJ, it is presumed that the latter are receptors under selection, possibly through L chain pairing, to bear CDR3 within a certain size range. In adults, the average CDR3, at 11 codons, is three longer than in neonates (Rumfelt et al., 2004). This shift can be attributed to paucity of TdT action in neonatal cells.

N addition is equally extensive in L chain junctions of adult sharks and therefore more extensive than in L chain genes of any other animal (Fleurant et al., 2004). This combination of sequence and sequence length heterogeneity for both H and L chain CDR3 shows that the primary antibody reper-toire in sharks relies mainly on junctional diversity, which is the novel characteristic that RAG con-ferred to the immune system receptors (Lewis et al., 2004). The varying lengths of CDR3 loops are primarily responsible for the diverse topology of the antigen combining site (Chothia et al., 1989).

VDJ IN ONE STAGE

In the mouse model, V(D)J recombination was initially characterized in B-cell lines-mediated H chain rearrangement in vitro (Alt et al., 1981). Rearrangement occurred in two distinct steps and at both alleles. The order, D to J, followed by VH to the DJ, was strict and unvarying (see Figure 10.1, top) (Alt et al., 1984). This series of events is regulated at multiple levels including chromatin activation and recruitment of transcription factors. The initial activation of the IgH gene involves only the region including D and JH gene segments and the IgM C region exons (Chowdhury and Sen, 2001), during which D to JH rearrangement occurs. The ~200 upstream VH, which extends over 2 Mb, is activated separately. The nearest VH is ~90 kb from the D gene segments, and activation of the chromatin encompassing the many and distant VH is manifested by germline transcription of the gene segments (Yancopoulos and Alt, 1985) and by a looping pro-cess that physically draws the many VH into spatial proximity of the DJ. These three-dimensional changes cause a locus compaction that allows the many VH, unequally placed in linear terms, presumably to be available for the VH to DJ recombination step (Jhunjhunwala et al., 2008; for a recent critical review see Seitan et al., 2012).

In contrast, all the recombining elements in shark are within 2 kb; no locus contraction is needed for rearrangement over distances of 400 bp or less. Although it seemed unlikely that any portions of the 2 kb chromatin would be differentially activated, we undertook to see whether there existed a particular order in which the four gene segments recombined. In a defined nurse shark IgH gene subfamily, specific primers targeting its leader intron and the JH amplified the germline (GL) band from RBC DNA (Figure 10.4). At the left in Figure 10.4 are depicted the various configurations that may be expected from lymphocyte DNA. Because the four gene segments are equidistant gene within nurse shark IgH genes, three possible outcomes of one rearrangement event (1R) are roughly the same size, as are all the 2R outcomes. A ladder of VH-hybridizing bands was obtained from spleen DNA, which were cloned and shown to include every possible kind of configuration.

However, the three species of a configuration type (1R or 2R) differed in frequency. For the gene investigated, G2, VD-D-J was the most frequent species among 1R and VDD-J that among 2R (Figure 10.5) (Malecek et al., 2008). This result suggests that V to D1 is the first rearrangement event, followed by the VD to D2, and completed by VDD to the JH. Some 1R and 2R from adult and neonatal spleen DNA are shown (Figure 10.5, neonate shaded). The neonatal sequences have considerably less N region addition (Malecek et al., 2005; Rumfelt et al., 2004).

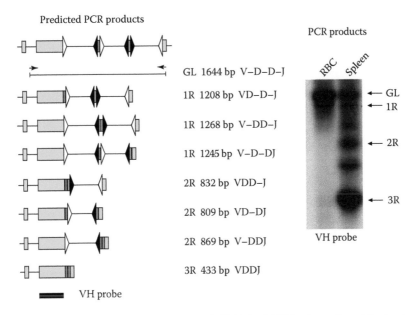

Predicted PCR products

PCR products

GL 1644 bp V–D–D–J

1R 1208 bp VD–D–J

1R 1268 bp V–DD–J

1R 1245 bp V–D–DJ

2R 832 bp VDD–J

2R 809 bp VD–DJ

2R 869 bp V–DDJ

3R 433 bp VDDJ

VH probe

VH probe

Figure 10.4 Amplifying rearrangements from genomic DNA. Left: PCR primers target the leader intron and JH sequence in Igμ gene G2. Two kinds of intermediates were expected (1R, 2R) in lymphoid tissue if cluster rearrangement is not completed (3R). All the possible configurations and their expected sizes are depicted; every type was isolated. Right: Electrophoresis of PCR products from erythrocyte DNA and spleen DNA. Filter was hybridized with VH probe (location depicted by black bar at left). Only the germline (GL) product appears in the RBC lane, whereas there is a ladder of hybridizing products for the lymphoid tissue. Arrows point to bands that proved to contain GL, 1R, 2R, or 3R sequences. Other bands could not be cloned, even when gel purified, and are assumed single-stranded PCR products.

Because every kind of configuration was in fact isolated, this series of events represents a preference rather than strict order and suggests that all the gene segments are simultaneously available for rearrangement. The basis for the preference is not clear because other IgH genes show a different order (Hua and Hsu unpublished results).

The gene configurations in individual B lymphocytes were examined by single-cell genomic PCR. It was observed that 1–3 IgH per cell had undergone recombination, and virtually all rearrangements were VDJ in configuration (Malecek *et al.*, 2008; Zhu *et al.*, 2011). The infrequently observed (1/15 B cells) incomplete 2R configuration led us to conclude that, once the gene is activated, rearrangement among the four gene segments takes place at once, without the complex intervening events of sequential chromatin activation and chromatin looping that occurs at the mouse IgH. Sharks thus display H chain exclusion (see subsection "H Chain Exclusion"), which suggests that the sequential chromatin activation and locus contraction in mammalian systems are adaptations to a 2–3-Mb locus and that these processes contribute to, but in themselves are not unconditionally required for the phenomenon of allelic exclusion.

REGULATION OF REARRANGEMENT

H Chain Exclusion

In two series of experiments involving single-cell genomic PCR, we found that not more than three VDJ existed per B cell, although the nurse shark carries up to 10 IgH genes (i.e., up to 20 alleles). The other genes were established to be in germline configuration. The linkage of three IgH

VH	N/P	D1	N/P	D2	N/P	J gene segment
TGTGCAAGAGAC		ATACTACAGTGGGT		ACATACTGGGATAG		ACTATTTTGATTACTGG
VD-D-J						
JS13 TGTGCAAGA---	AGTGACTACGTTAT	AGTGGGT		GL		GL
JS31 TGTGCAAGAG--	CCTCC	CAGTGGGT		GL		GL
JS53 TGTGCAAG----	GAT	CTACAGTGGGT		GL		GL
JS74 TGTGCAAGAGA-	TC	CAGTGGGT		GL		GL
JS22DD TGTGC-------	GAGTTG	TACAGTGGGT		GL		GL
JS24DD TGTGCAAGAGA-	CGTGG	ACAGTGGGT		GL		GL
JS26DD TGTGCAAGA---	AC	ACTACAGTGGGT		GL		GL
JSEDJ7 TGTGCAAGAGAC	CGCGGCT	CAGTGGGT		GL		GL
NA-45 TGTGCAAGAG--		TGGGT		GL		GL
V-DD-J						
JS66 GL		ATACTACAGT	CGGGATTGG	TGGGATAG		GL
JS6VD GL		ATACTACAGTGGGT	AGGACT	ATACTGGGATAG		GL
NA-28 GL		ATACTACAGTGGGT		ACTGGGATAG		GL
NA-64 GL		ATACTACAGT		ATACTGGGATAG		GL
NA-59 GL		ATACTACA		CTGGGATAG		GL
V-D-DJ						
JS23DD GL		GL		ACATACTG	ACCC	----TTTTGATTACTGG

VH	N/P	D1	N/P	D2	N/P	J gene segment
TGTGCAAGAGAC		ATACTACAGTGGGT		ACATACTGGGATAG		ACTATTTTGRTTACTGG
VDD-J						
JS18 TGTGCAAGA---	TT	GGGT	ACA	GGGATAG		GL
JS22 TGTGCAAGA---	ACG	ACTACAGTGGGT	ACAGT	ACATACTGGGATAG		GL
JS25 TGTGCAA-----	TTCGATC	GTGGGT	TC	ACTGGGATAG		GL
JS27 TGTGCAAG----	T	GTGGGT	AC	GGGATAG		GL
JS29 TGTGCAAG----		TACA	CTTC	TACTGGGATAG		GL
JS32 TGTGCAA-----	AGGGCCCT	CAGTGGGT	ACAAGGTTGT	GGATAG		GL
JS36 TGTGCAAGA---	TTGA	ATA	TGA	TGGGATAG		GL
JS43 TGTGCAA-----	CCATATAA	ATAC	TTCTCCCTA	TGGGATAG		GL
JS45 TGTGCAAG----	GTAGGG	ATACTACAGTGGGT	ACTTGGG	CATACTGGGATAG		GL
JS46 TGTGCAAG----	TAT	GGT	CGGAC	ACTGGGATAG		GL
NA-37 TGTGCAAGAGAC		AGTGGGT		CTGGGATAG		GL
NA-73 TGTGCAAGAGAC		AGTGGGT		ACTGGGATAG		GL
NA-74 TGTGCAA-----		ACTA		TACTGGGATAG		GL
NA-79 TGTGCAAGAGAC		TACAGTGGGT	ACGA	GATAG		GL
NA-84 TGTGCAAGAGAC		TACAGTGGG		ATAG		GL
NA-87 TGTGCAAG----				TAG		GL
NA-89 TGTGCAAGAG--	GGGG	CTACAGTGGG		ACTGGGATAG		GL
NA-103 TGTGCAA-----	A	TGGG	GCT	ATACTGGGATAG		GL
NA-106 TGTGCAAGAGA-		TACTACAGTGGG		ATAG		GL
NA-114 TGTGCAAGA---		AT	C	CATACTGGGATAG		GL
NA-115 TGTGCAAG----	T	TACAGTGGG		ATAG		GL
NA-120 TGTGCAAGAG--	GA	GGG		TACTGGGATAG		GL
VD-DJ						
JS8 TGTGCAAGAGAC	GGGGAT	GTGGGT		ACATACTGGGATAG	GGG	-----TTTGGTTACTGG
JS551 TGTGCAAGAG---	TCATATCC	CAGTGGGT		ACATAC	GGGACCCA	------TTGATTACTGG
NA-124 TGTGCAAGAG----		TGGGT		ACATACTGGGA		--TATTTTGGTTACTGG
V-DDJ						
JSEVD3 GL		ATACTACAGTGG	CCGCG	ACTGGGAT	CCCCGCTGAG	-----TTTGATTACTGG
NA-68 GL		ATACTACAGT		GGA		-----------TACTGG

Figure 10.5   Junctions from partially rearranged G2A and G2B genes. The reference sequence consists of the VH 3′ flank, the D1 and D2 genes, and the 5′ JH flank. Sequences were isolated from adult sIgM+ splenic DNA and neonatal spleen DNA (yellow). Only the junctions are shown, and rest are indicated as germline (GL). Top: One rearrangement configuration (1R). Rearrangement of V to D1; of D1 to D2; and of D2 to JH. Bottom: Two rearrangement configuration (2R). VDD-J, from V to DD-J or VD to D2; VD-DJ, from V to D-DJ or VD-D2 to JH; V-DDJ, from V-D to DJ or V-DD to JH. P nucleotides are underlined; N addition is sparse at neonatal junctions.

genes was determined (G2B-G2A-G5, Figure 10.1), and the pattern of activated genes showed no correlation with either allelic or linkage relationships. Therefore, activation occurred without any indication of cross-talk, and the IgH genes were apparently recombinationally autonomous. Taking into consideration the number of pseudogenes that are also capable of recombination and therefore could compete for nuclear factors that bring about chromatin activation, the data best fit with a probability model based on the hypothesis that V(D)J rearrangement occurs as a series of trials and with a maximum of four activated IgH per cell (Zhu et al., 2011).

Sequential IgH activation suggests the possibility for feedback inhibition, once a productive VDJ was generated. Moreover, the fact that we never observed more than three VDJ among the 18–24 functional IgH suggests that rearrangement takes place within a certain time frame, regardless of the number of available genes. This hypothesis is supported by a study involving single-cell RT-PCR, performed in clearnosed skate peripheral blood leukocytes (Eason et al., 2004), where the number of IgM H chain transcript species per B cell was also restricted to 1–3 despite the estimated hundreds of skate IgH genes. Whatever the basis for H chain exclusion in animals with multiple Ig clusters, it is independent of gene number, which varies greatly among elasmobranch species.

In conclusion, what the mammalian and shark IgH systems have in common is asynchronic rearrangement. Asynchrony at the VH to DJ step initiates allelic exclusion in mammals by allowing one gene to generate a VDJ before the other; asynchrony in the form of 1–3 successive IgH gene activations brings about H chain exclusion in sharks. The underlying causes are highly controversial, but one recent work proposes that, where two genes are autonomous and activation is random, they are unlikely to initiate simultaneously (Farcot et al., 2010). A time lag between two events, such as D to J and VH to DJ, allows the first productive VDJ to inhibit the second, rearranging gene. Because shark VDJ is formed in one stage, the critical time lag would occur between the activation of individual IgH clusters. If rearrangement initiation is random, this reduces the likelihood that more than one gene is rearranging at any time in a B cell. In shark, the rate-limiting step could be accessibility of the IgH to RAG, in other words, restricted activation of the IgH chromatin.

T-Cell/B-Cell Disparity

One unexpected finding in our studies was that IgH rearrangement also took place in thymocytes. The intermediates isolated from B-cell DNA by PCR were the same in relative proportion as those from thymocyte DNA, meaning that the rearranging preference order was the same in both cell types for a particular gene. The main difference is that in B cells, the vast majority of rearrangements were 3R, whereas in thymocytes, they were predominantly 1R and 2R. This discrepancy is so quantitatively significant that it can be observed by genomic Southern blotting of thymic DNA compared to DNA from sIgM-positive lymphocytes (Malecek et al., 2008).

A comparison of recombined IgH configurations and other characteristics in T and B cells allows us to make some deductions about regulation of rearrangement. We have not been able to detect μ transcripts in thymocytes by northern blotting or RT-PCR, but it is not known whether intergenic antisense transcription occurs (Bolland et al., 2004). No L chain rearrangement or transcription was observed. From these differences, we discerned some aspects of IgH regulation in shark pro-B cells. Table 10.1 compares data on Ig rearrangement and expression in thymocytes and sIg-positive cells (Malecek et al., 2008; Zhu et al., 2011). At the single-cell level, there were many more rearranged IgH in thymocytes than in B cells. The average number in single thymocytes, five rearranged genes, is an underestimate, because for technical reasons we were unable to screen for one IgH subfamily (G3). However, the majority of these rearrangements (82%) were not brought to completion, far higher than the 5% frequency of incomplete rearrangement in B cells.

Despite the fact that there are more than twice as many rearranging IgH genes present in thymocytes, no transcripts were detected by northern blotting. Parallel RT-PCR experiments performed to measure VDD-J transcripts from the gene G2 from thymocyte and spleen RNA samples were successful only in the latter tissue, even though G2 VDD-J was present in 10 out of 12 thymocytes. The VDD-J sequences that we isolated by RT-PCR from spleen carried substitutions. In this case, we believe that the mutants were from B lymphocytes, as bystanders in activated cells undergoing SHM. Although apparently mutated TCRγ were isolated from sandbar shark (Chen et al., 2009), not one of the 60 genomic rearrangements we isolated from thymocyte DNA contained the tandem substitutions that are the trademark of shark SHM (Lee et al., 2002). Thus, rearranged IgH genes with partial or completed VDJ are transcribed and subject to SHM in B cells. If rearranged IgH in thymocytes are transcribed, this was below the detection level for sense transcripts of a homogenous size.

Because of the great disparity of IgH configuration in the two cell types, it is clear that V(D)J rearrangement had not occurred in a common lymphoid progenitors (CLP). Therefore, IgH rearrangement in thymocytes probably took place in developing T cells during the time RAG was

Table 10.1   **Comparison of IgH Gene Rearrangement Configurations Scored in Single Shark T and B Cells**

	Thymocyte	sIgM+ cell
Single cell[a]	Up to 8 IgH recombined	Up to 3 IgH recombined
	Average 5	Average 2
	56 IgH in 12 cells	30 IgH in 16 cells
	13 1R	0 1R
	33 2R	1 2R
	10 3R	29 3R
	Average 9 events/cell	Average 6 events/cell
Genomic DNA[b]	IgH rearrangements	IgH rearrangements
	No IgL rearrangements	IgL rearrangements
RNA studies[c]	No H chain mRNA detected	H chain mRNA
	No L chain mRNA detected	L chain mRNA
	No VDD-J detected	VDD-J mutants cloned

Sources: [a] Single thymocyte data are from Malecek *et al.* (2008); the sIgM+ column is from Zhu *et al.* (2011). 1R is one rearrangement event: VD-D-J or V-DD-J or V-D-DJ. 2R is two events: VDD-J or V-DDJ or VD-DJ. 3R is three events or VDJ. The average number of rearrangement events is $(1 \times 1R\#) + (2 \times 2R\#) + (3 \times 3R\#)$ divided by # cells.
[b] Rearranged IgH in lymphocyte DNA, as detected by Southern blotting and PCR, was obtained from Malecek *et al.* (2008) and Zhu and Hsu (2010). Rearranged IgL in sIgM+ genomic DNA was obtained from Zhu and Hsu (2010); the absence of rearranged IgL in thymus DNA was reported in Malecek *et al.* (2008).
[c] Northern blotting and RT-PCR was obtained from Malecek *et al.* (2008).

expressed for TCR gene recombination. Considering the large number of IgH genes that had undergone recombination per thymocyte, most IgH genes were accessible to RAG, and yet, a minority completed all three rearrangements. There is no evidence of less recombinase activity in thymocytes (nine rearrangement events/thymocyte versus six events/B cell), but RAG appears not to be acting processively at most genes, perhaps because availability of the local gene segments is different in the two cell types.

Recent experiments show that RAG activity is regulated by the nature of chromatin structure or the epigenetic environment of the RSS (Liu *et al.*, 2007; Matthews *et al.*, 2007; for a review see Schatz and Ji, 2011). In activated chromatin, RAG1 protein binds RSS, and RAG2 protein is brought separately to the site by binding modified histone (Ji *et al.*, 2010). We suggest that in shark B cells, the activated IgH displays all epigenetic markings enabling it to recruit RAG efficiently. In thymocytes, many IgH are activated to some extent because so many undergo some recombination; tellingly, the absence of normal or even low levels of Ig transcription indicates that their chromatin is not fully modified. Possibly, it is the suboptimal epigenetic environment that does not permit efficient RAG recruitment/activity, and thymic IgH rearrangements usually are not completed to VDJ. What has not yet been investigated is if there exists a global opening of Ig chromatin, like that indicated by antisense transcription (for a review see Corcoran, 2010). In such case, the thymocyte IgH, because of close linkage to TCR genes or sharing common lymphocyte-initiating factors, may undergo a *first-pass* opening of the chromatin that is not completed by B-cell factors.

The IgH rearrangements in thymocytes demonstrated that B-cell lineage and stage-specific rearrangement factors exist in shark but B-cell-specific modifications are not required in order for IgH genes to be accessible to RAG. This leads us to speculate on changes in germ cell chromatin allowing germline rearrangements in the section so styled.

ATYPICAL REARRANGEMENT EVENTS

Although the vast majority of rearrangements take place between adjacent gene segments and are of the deletional type (Figures 10.4 and 10.5), we have on rare occasion found unusual configurations. The two cloned sequences shown in Figure 10.6 must have resulted from certain steps leading to an inverted intersegmental sequence and disrupted RSS. The lack of a pair of intact RSS prevented the final rearrangement to 3R and preserved the aberrant configuration in the cell.

Recombination by Inversion

In this G2A gene, the presence of the inverted D2-JH intersegmental sequence, inverted D2 sequence and its 3′ RSS joined to JH (Figure 10.6, bottom left), shows that a rearrangement between D2 and JH occurred by inversion. Both RSS flanking D2 contain 12 bp spacers (RSS-12);

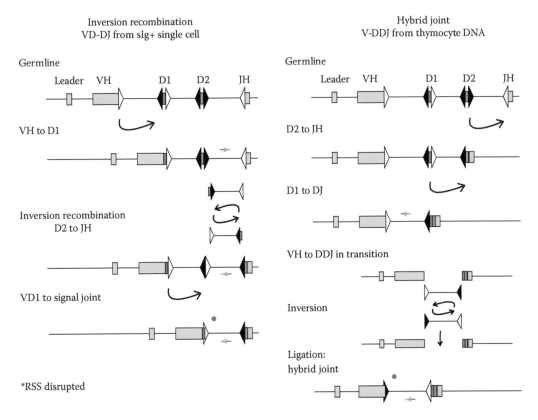

Figure 10.6 Atypical rearrangement events found at shark IgH. Left: Inversion recombination at G2A gene in B cell. A 2R configuration was cloned from sIgM+ genomic DNA, VD-DJ. The rearrangement events that led to this particular sequence are deduced as follows. The first step for G2A gene is most frequently VH to D1, as shown. The next event involves an inversion recombination that joins D2 and JH while inverting the intersegmental sequence (orientation shown with blue arrow). As this can only occur with an intact RSS 5′ of D2, the inversion is shown as step 2. The signal joint, joined RSS triangles, contains intact RSS so that they can be involved rearranging VD to RSS. The DSB at the joined RSS resolves with deletion at the broken ends; part of the RSS is affected, as shown (asterisk). Right: Hybrid joint in G5 gene in thymocyte. A 2R configuration was cloned from thymocyte DNA, V-DDJ. The first rearrangement steps for G5 gene tend to be D2 to JH, followed by D1 to DJ, as shown. It is the final event, VH to DDJ, where the freed intersegmental sequence becomes religated, although in inverted orientation. The rejoining disrupts one of the RSS (asterisk).

although recombination between the RSS-23 of JH and the 5′ RSS-12 of D2 is theoretically possible, it has not been definitively identified before. The second point of interest is that such an inversion must result in a signal joint within the gene as an intermediate state (Figure 10.6, left, VD1 to signal joint, which are the joined triangles). The availability of RSS in a signal joint for participation in rearrangement was documented (Lewis *et al.*, 1985; Marculescu *et al.*, 2003; Meier *et al.*, 1993). The subsequent recombination between VD and the RSS-12 component of the signal joint destroyed the abutting RSS-23 through the normal nucleolytic processing that occurs in sequence flanking RSS.

Hybrid Joint

This V-DDJ sequence (Figure 10.6, bottom, right), isolated from the G5 gene from thymus DNA (Hua and Hsu unpublished results), shows all the gene segments in the same transcriptional orientation, but the D1-D2 intersegmental sequence with its terminal RSS was inverted. Because the gene segments were not inverted, an inversional recombination involving them had not occurred; the state of the sequence suggests that deletional recombination was begun but was not completed. We suggest that rearrangement had gone as far as double-stranded DNA breaks at either RSS (Figure 10.6, right, VH to DDJ in transition), but the intersegmental DNA did not proceed to form a recombination circle. Instead, it was religated in inverted orientation to the coding ends. As with the G2 example, the normal end-processing events in the G5 genes deleted part of one RSS end before rejoining (Figure 10.6, asterisk).

Transrecombination

Although no recombination event between Igµ clusters has so far been isolated in B-cell cDNA, infrequent transcripts bearing transrecombination between Ig and TCR loci were reported (Criscitiello *et al.*, 2010). We have found genomic transrecombination between Igµ in thymus DNA (Hua and Hsu unpublished results). The rearrangements take place only between certain clusters and at low frequency. For example, as shown in Figure 10.7, we have detected a few G3 VH and D gene segments recombined with G1 D or JH. The reciprocal combinations were not detected. They may occur at a lower frequency or are only possible in certain relative chromosomal positions; the linkage between G1 and G3 is not known.

It is not clear why trans-rearrangements events occur at frequencies detectable in thymocytes, but perhaps this is a reflection that many antigen receptor genes are accessible, even if not optimally, to RAG at the same time. The RAG-mediated double-strand breaks at G3 and G1 are shown in Figure 10.7 as taking place independently, followed by joining of their separate broken ends during the NHEJ repair process like a translocation. However, it is also possible that the gene segments of G3 and G1 actively synapse as during normal V(D)J recombination. The strongest reason for the latter supposition is that Igµ trans-rearrangements follow the 12/23 rule, where the gene segment contributions can be identified—but on the other hand, where the components cannot be distinguished, it could well be proof of the opposite case. One can only speculate that G3 and G1 in Figure 10.7 are simultaneously available to RAG and were brought into sufficient proximity for recombination, but the circumstances that make this possible need to be elucidated.

Similarly, the varied trans-rearrangements (IgW or IgM V to TCRδ or TCRα sites) (Criscitiello *et al.*, 2010) must arise because so many RSS combinations are accessible to RAG protein. The shark TCR genes are probably organized like those in tetrapods, although there is little mapping information to date (Chen *et al.*, 2009; Criscitiello *et al.*, 2010). Gene compaction is not needed for V(D)J recombination in shark Ig genes, but it likely occurs to bring the many shark TCR V gene segments into DJ or J proximity. Perhaps certain IgW or IgM genes are bystanders in the TCR locus condensation process and become positioned close to TCR.

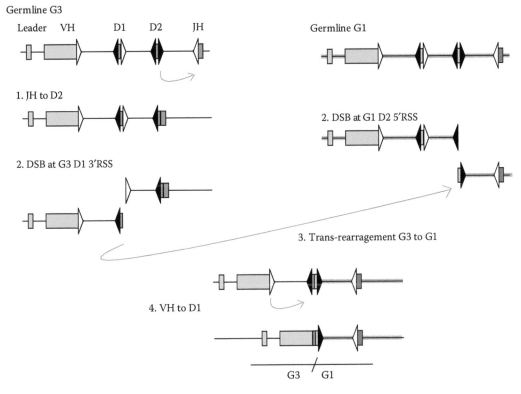

Figure 10.7   Trans-rearrangement event between genes G3 and G1. A 2R configuration was cloned from thymocyte DNA, VDD-J. The rearrangement events that led the formation of a chimeric sequence where VH and D1 belong to G3 and the intersegmental sequence and JH belong to G1 are hypothesized as follows. Because the first event in G3 is always D2 to JH, this is depicted. The rearrangement involves D1 and D2. Step 1 depicts DSB occurring at the 3′ RSS of D1. Step 2 depicts DSB at the 5′RSS of D2, which is part of the typical first event in G1 genes (D1 to D2). Step 3 shows that the broken end at G3 D1 ligates to the G1 end, forming a D1D2 joint. Step 4 shows VH to DD-J rearrangement.

GERMLINE REARRANGEMENT

In the course of characterizing the first isolated genomic Ig clones from horned shark, Litman and colleagues (Kokubu *et al.*, 1988) discovered that some IgH clusters were organized as V-D-D-J but the majority (8/13) contained already fused gene segments arrayed as VDD-J and VDJ. Moreover, one unique V-D-D-J clone carried an intersegmental sequence between D1 and D2 that appeared as if generated by inversion recombination, not only by the nature of the sequence but also the flipping of the RSS-12 and RSS-23, exchanging them at the 3′ flank of D1 and 5′ flank of D2, like a hybrid joint. Thereafter, both subclasses of cartilaginous fishes, ratfish as well as sharks/skates, were found to carry some form of pre-rearranged Ig gene segments (Table 10.2). Of the four L chain types, pre-joined VJ was described for type I/sigma-cart, type II/lambda, type III/kappa, but not so for sigma. Various forms of joined VDJ and partly joined (VD-DJ, VDD-J, V-D-DJ) H chains of Igμ and Igω clusters as well as of IgNAR were cloned from ratfish and various species of sharks and skates. Although some examples of germline rearrangement were found in a few other animal species (Table 10.2), elasmobranches are unique in possessing so many. Although this phenomenon may be detected mainly in genes organized as clusters and thus multiple copies, there is a surprising absence in the highly duplicated L chain clusters in teleost fishes.

Table 10.2   Germline-Rearranged Sequences in Various Species

	L Chain	H Chain (IgM, IgW)	Something Else
Hydrolagous colliei Spotted ratfish	Type II (VJ)[a]	V-D-D-J, VDJ (IgM)[b]	
Heterodontus franciscii Horn shark	Type I (V-J)[c] Type II (VJ)[e] Type III (V-J)[a]	V-D-D-J, VDJ, VD-J (IgM)[d]	
Carcharhinus plumbeus Sandbar shark	Type II (VJ)[f]		
Ginglyostoma cirratum Nurse shark	Type I (mixed)[g] Type II (VJ)[j] Type III (mixed)[m] Sigma (V-J)[n]	V-D-D-J, VD-D-J, VD-J (IgW)[h] V-D-D-J (IgM)[k]	V-DD-D-J (IgNAR)[i] VDJ (IgM1gj)[l]
Raja eglanteria Clearnose skate	Type I (VJ)[a]	V-D-D-J, V-D-DJ (IgW)[o]	
Raja erinacea Little skate	Type I (VJ)[a] Type II (VJ)[a]	V-D-D-J, VD-J, VD-DJ (IgM)[p]	
Catfish		VDJ[q]	
Chicken		VD Pseudogenes[r]	
Opossum			VD (VH3.1, TCRμ)[s]

Sources: [a] Rast *et al.* (1994). Type I is also referred to as sigma-cart or NS5, type II lambda or NS5, type III kappa
 or NS4. See Criscitiello and Flajnik (2007). VJ signifies pre-joined V to J; V-J indicates split gene
 segments.
 [b] Rast *et al.* (1998).
 [c] Shamblott and Litman (1989).
 [d] Kokubu *et al.* (1988).
 [e] Anderson *et al.* (1995).
 [f] Hohman *et al.* (1993).
 [g] Fleurant *et al.* (2004).
 [h] Hua, Zhang, and Hsu (unpublished results).
 [i] Diaz *et al.* (2002).
 [j] Lee *et al.* (2002).
 [k] Lee *et al.* (2008).
 [l] Rumfelt *et al.* (2001).
 [m] Lee *et al.* (2000).
 [n] Criscitiello and Flajnik (2007).
 [o] Anderson *et al.* (1994).
 [p] Harding *et al.* (1990).
 [q] Ghaffari and Lobb (1999).
 [r] Reynaud *et al.* (1989).
 [s] Wang *et al.* (2009).

As the leader intron was retained in the germline-rearranged IgH and IgL, they were not the result of retrotransposed mRNA. One suggestion was that these phenomena may have arisen from recombinase activity in germ cells, although V(D)J rearrangement was established as an activity crucial to lymphocyte development (Kokubu *et al.*, 1988). However, this apparently was the answer, as supported by studies that examined the nature of the VJ junctions demonstrating that they were generated from nonjoined gene segments (Lee *et al.*, 2000). On the other hand, the idea of RAG affecting germline DNA raises even more questions.

Why Would RAG Act on DNA in Germ Cells?

RAG protein was not demonstrated in elasmobranch germ cells due to various technical difficulties. If the recombinase were expressed in germ cells, its potential substrate must contain RSS recognition motifs that are part of accessible chromatin. Perhaps we can gain some insight into meiotic chromatin from studies on homologous recombination, which is initiated through double-strand DNA breakages mediated by the transesterase Spo11. Sites of Spo11 binding were mapped to gain understanding of circumstances favoring DSB (Pan *et al.*, 2011). There are many factors, including chromosomal position, chromatin structure, and sequence-specific DNA-binding proteins, but one consistent characteristic at these sites is low nucleosome occupancy. This suggests open chromatin in itself is not sufficient for efficient recruitment of RAG to its substrate (for a review see Schatz and Ji, 2011). However, we believe the chromatin does not have to be fully activated. Under as yet undefined conditions in shark thymocytes, there is unregulated RAG activity at IgH, where B-cell lineage-specific factors are absent and sense transcription cannot be detected (Malecek *et al.*, 2008). The partial rearrangements in thymocyte DNA resemble the various kinds of germline-joined IgH (VD-DJ, VD-J, etc.). We propose that RAG, when expressed in germ cells, can act on RSS at chromatin that is open but not fully activated, like in IgH in thymocytes. Because the multiple Ig cluster organization arose precisely as a result of gene duplication by unequal crossing over, there must exist open chromatin in their vicinity at some point in time.

What Is the Role of Germline-Rearranged Ig Genes?

The proportion of germline-joined Ig gene segments varies among elasmobranch species (Yoder and Litman, 2000). The type II/lambda L chains genes exist only as VJ clusters in sharks and skates (Table 10.2), which suggests that some such genes are functional, with their diversity primarily deriving from SHM (Anderson *et al.*, 1994; Lee *et al.*, 2002). However, there is no clear requirement for such genes; in horned shark, at least half of the >100 Igμ clusters are pre-rearranged to some extent but in nurse shark none of the 9–12 functional Igμ have done so (Kokubu *et al.*, 1988; Lee *et al.*, 2008). Lee *et al.* (2008) have argued for species-specific selection of such genes, where fortuitously generated germline rearrangement may fulfill a niche requirement.

Moreover, there probably is a difference between H and L chain genes. If one considers that most receptors generated in immature mouse B cells are potentially autoimmune (Retter and Nemazee, 1998), the chance-formed VDJ in the elasmobranch germline seems unlikely to have a positive impact on the immune system. In a model with relatively few Igμ genes, such as nurse shark, the single IgM-related germline VDJ we found cross-hybridizing to the VH probe was neither transcribed nor was linked to any C region exons (Lee *et al.*, 2008). Three pseudogene Igμ were also found to be available for rearrangement, suggesting perhaps that if the window of activated IgH is limited up to four genes then the nurse shark's 20%–25% pseudogene content is tolerable in the competition for B-cell activation factors. The large numbers of germline VDJ in horned shark or skate may be tolerable only as pseudogenes; and in fact, in the clearnose skate, Miracle *et al.* (2001) found 53 distinctive germline-joined IgM VDJ, 24 of which contain frame shift or stop codons between FR2 and JH. It is not known if the remaining 29 genes encode a viable protein because the complete sequences were not obtained. Several skate germline-joined VDJ, defective or not, were transcribed in embryo and hatchling; none was detected in the adult skate. To date, the extent to which these germline-joined VDJ sequences participate in the primary antibody repertoire remains to be elucidated.

What is interesting is that germline-joining in some other animals (Table 10.2)—the VDJ in catfish and the VD segments in chicken—has resulted in pseudogenes, and it is precisely as nonfunctional elements that they have their impact. The VD pseudogene segments at the chicken IgH locus are used as gene conversion templates in the course of generating the antibody repertoire. The VD cannot rearrange but their junctions are used to diversify the CDR3 of the rearranged template VDJ through gene conversion (Reynaud *et al.*, 1989). In catfish, there are tandem IgH clusters, one containing VH genes, a fused VDJ, and C exons and a second one, downstream, with multiple VH, D, and JH gene segments and C region exons (Bengtén *et al.*, 2005). The fused VDJ is not expressed (Edholm *et al.*, 2010), but the deletion of all D gene segments as a result of the germline-joining event effectually blocks any other somatic rearrangement at this site, obliging the other IgH gene to provide the B-cell antigen receptors. Yet, there is a protein product from the former gene: the C exons directly 3′ of the fused VDJ are expressed as secreted-only V-less proteins with specialized immune function (Edholm *et al.*, 2011).

GERMLINE REARRANGEMENT AS CATALYST IN EVOLUTION

The examples described earlier suggest that sometimes germline rearrangement of V gene segments does not affect the antibody repertoire so much as serve to *catalyze other events* shaping the fate of the Ig locus or cluster. In nurse shark, Rumfelt *et al.* (2001) discovered an Ig cluster with germline VDJ that was in-frame; this is the only pre-rearranged VDJ—so far—where a protein product was characterized. The H chain polypeptide can bind L chain and was a monomeric four-chain Ig unit prominent in neonatal sera. IgM1gj is probably expressed exclusively as a secreted molecule (Hsu *et al.*, 2006; Rumfelt *et al.*, 2001).

The loss of the transmembrane sequence is significant, as this suggests that IgM1gj-expressing cells do not proliferate as a direct result of this molecule-binding ligand. The IgM1gj combining site could have a limited spectrum, dependent on the array of kappa L chain it bound, but because there is no clonal expansion the specificity of any one particular IgM1gj is not critical. Because the VDJ coevolved with its unique 3-domain C region, its very invariance would seem to be the quality central to its as yet unknown function. If the IgM1gj gene cannot rearrange and is not a cell surface receptor, the absence of two essential features of adaptive immunity makes it not unlike the secreted V-less catfish molecule: an immune function molecule but no longer a B lymphocyte signaling receptor.

Now, let us rethink the implications of pre-rearranged VDJ. If an in-frame pre-rearranged VDJ could be expressed as protein, selection on the receptor would be fierce, and the absence of such transcripts in adult skates suggests that if there had been viable proteins made by B cells expressing any of the 29 in-frame germline-joined VDJ genes, those B cells are in fact not selected for. Why not? If the VDJ encodes a self-reactive combining site, the B cells expressing it would be selected against. If not autoimmune, its presence reduces antibody diversity. Moreover, a ligand whose specificity resides in that particular VDJ will be capable of activating very many lymphocytes at one time. Even if the IgH is one out of 100 clusters, clonal expansion involving 1% of the animal's B cells—as opposed to 0.001%–0.0001%—surely will bias its secondary repertoire in an undesirable direction. Logically, a germline IgH gene with an in-frame VDJ will very soon be eliminated; therefore, those that are yet transcribed and expressed as protein would be exceptional and potentially illuminate the sort of *evolution* that RAG can bring about. We need to look for more genes such as IgM1gj.

ACKNOWLEDGMENTS

I thank Susanna Lewis for her many helpful comments that improved this manuscript. This work was supported in part by funding from the National Institutes of Health R01 GM068095.

REFERENCES

Alt, F., Rosenberg, N., Lewis, S., Thomas, E., and Baltimore, D. Organization and reorganization of immunoglobulin genes in A-MULV-transformed cells: rearrangement of heavy but not light chain genes. *Cell* 1981; 27:381–390.

Alt, F.W., Yancopoulos, G.D., Blackwell, T.K., Wood, C., Thomas, E., Boss, M., Coffman, R., Rosenberg, N., Tonegawa, S., and Baltimore, D. Ordered rearrangement of immunoglobulin heavy chain variable region segments. *EMBO Journal* 1984; 3:1209–1219.

Anderson, M., Amemiya, C., Luer, C., Litman, R., Rast, J., Niimura, Y., and Litman, G. Complete genomic sequence and patterns of transcription of a member of an unusual family of closely related, chromosomally dispersed Ig gene clusters in Raja. *International Immunology* 1994; 6:1661–1670.

Anderson, M.K., Shamblott, M.J., Litman, R.T., and Litman, G.W. Generation of immunoglobulin light chain gene diversity in Raja erinacea is not associated with somatic rearrangement, an exception to a central paradigm of B cell immunity. *Journal of Experimental Medicine* 1995; 182:109–119.

Bengtén, E., Clem, L.W., Miller, N.W., Warr, G.W., and Wilson, M. Channel catfish immunoglobulins: Repertoire and expression. *Comparative and Developmental Immunology* 2005; 30:77–92.

Bernstein, R.M., Schluter, S.F., Shen, S., and Marchalonis, J.J. A new high molecular weight immunoglobulin class from the carcharhine shark: Implications for the properties of the primordial immunoglobulin. *Proceedings of the National Academy of Sciences of the United States of America* 1996; 93:3289–3293.

Bolland, D.J., Wood, A.L., Johnston, C.M., Bunting, S.F., Morgan, G., Chakalova, L., Fraser, P.J., and Corcoran, A.E. Antisense intergenic transcription in V(D)J recombination. Nature *Immunology*. 2004; 5:630–637.

Brodeur, P.H., and Riblet, R. The immunoglobulin heavy chain variable region (Igh-V) locus in the mouse. I. One hundred Igh-V genes comprise seven families of homologous genes. *European Journal of Immunology* 1984; 14:922–930.

Chen, H., Kshirsagar, S., Jensen, I., Lau, K., Covarrubias, R., Schluter, S.F., and Marchalonis, J.J. Characterization of arrangement and expression of the T cell receptor gamma locus in the sandbar shark. *Proceedings of the National Academy of Sciences of the United States of America* 2009; 106:8591–8596.

Chothia, C., Lesk, A.M., Tramontano, A., Levitt, M., Smith-Gill, S.J., Air, G., Sheriff, S. *et al.* Conformations of immunoglobulin hypervariable regions. *Nature* 1989; 342:877–883.

Chowdhury, D., and Sen, R. Stepwise activation of the immunoglobulin μ heavy chain gene locus. *EMBO Journal* 2001; 20:6394–6403.

Clem, L.W., De Boutaud, F., and Sigel, M.M. Phylogeny of immunoglobulin structure and function. II. Immunoglobulins of the nurse shark. *Journal of Immunology* 1967; 99:1226–1235.

Corcoran, A.E. The epigenetic role of non-coding RNA transcription and nuclear organization in immunoglobulin repertoire generation. *Seminars in Immunology* 2010; 22:353–361.

Criscitiello, M.F., and Flajnik, M.F. Four primordial immunoglobulin light chain isotypes, including lambda and kappa, identified in the most primitive living jawed vertebrates. *European Journal of Immunology* 2007; 37:2683–2694.

Criscitiello, M.F., Ohta, Y., Saltis, M., McKinney, E.C., and Flajnik, M.F. Evolutionarily conserved TCR binding sites, identification of T cells in primary lymphoid tissues, and surprising trans-rearrangements in nurse shark. *Journal of Immunology* 2010; 184:6950–6960.

Diaz, M., Stanfield, R.L., Greenberg, A.S., and Flajnik, M.F. Structural analysis, selection, and ontogeny of the shark new antigen receptor (IgNAR): Identification of a new locus preferentially expressed in early development. *Immunogenetics* 2002; 54:501–512.

Dooley, H., and Flajnik, M.F. Shark immunity bites back: Affinity maturation and memory response in the nurse shark, *Ginglymostoma cirratum. European Journal of Immunology* 2005; 35:935–945.

Eason, D.D., Litman, R.T., Luer, C.A., Kerr, W., and Litman, G.W. Expression of individual immunoglobulin genes occurs in an unusual system consisting of multiple independent loci. *European Journal of Immunology* 2004; 14: 2551–2558.

Edholm, E.S., Bengtén, E., Stafford, J.L., Sahoo, M., Taylor, E.B., Miller, N.W., and Wilson, M. Identification of two IgD⁺ B cell populations in channel catfish, *Ictalurus punctatus. Journal of Immunology* 2010; 185:4082–4094.

Edholm, E.S., Bengtén, E., and Wilson, M. Insights into the function of IgD. *Developmental and Comparative Immunology* 2011; 35:1309–1316.

Farcot, E., Bonnet, M., Jaeger, S., Spicuglia, S., Fernandez, B., and Ferrier, P. TCR beta allelic exclusion in dynamical models of V(D)J recombination based on allele independence. *Journal of Immunology* 2010; 185:1622–1632.

Flajnik, M.F. Comparative analyses of immunoglobulin genes: Surprises and portents. *Nature Reviews in Immunology* 2002; 2:688–698.

Flajnik, M.F., and Du Pasquier, L. Evolution of the immune system. In: Paul, W.E., editor. *Fundamental Immunology*. Philadelphia, PA: Lippincott Williams & Wilkins, 2008, pp. 57–124.

Fleurant, M., Changchien, L., Chen, C.-T., Flajnik, M.F., and Hsu, E. Shark immunoglobulin light chain junctions are as diverse as in heavy chains. *Journal of Immunology* 2004; 173:5574–5582.

Ghaffari, S.H., and Lobb, C.J. Structure and genomic organization of a second cluster of immunoglobulin heavy chain gene segments in the channel catfish. *Journal of Immunology* 1999; 162:1519–1529.

Goodchild, S.A., Dooley, H., Schoepp, R.J., Flajnik, M., and Lonsdale, S.G. Isolation and characterisation of Ebolavirus-specific recombinant antibody fragments from murine and shark immune libraries. *Molecular Immunology* 2011; 48:2027–2037.

Greenberg, A.S., Hughes, A.L., Guo, J., Avila, D., McKinney, E.C., and Flajnik, M.F. A novel "chimeric" antibody class in cartilaginous fish: IgM may not be the primordial immunoglobulin. *Journal of Immunology* 1996; 26:1123–1129.

Harding, F.A., Cohen, N., and Litman, G.W. Immunoglobulin heavy chain gene organization and complexity in the skate, *Raja erinacea. Nucleic Acids Research* 1990; 18:1015–1020.

Hinds, K.R., and Litman, G.W. Major reorganization of immunoglobulin V_H segmental elements during vertebrate evolution. *Nature* 1986; 320:546–549.

Hinds-Frey, K.R., Nishikata, H., Litman, R.T., and Litman, G.W. Somatic variation precedes extensive diversification of germline sequences and combinatorial joining in the evolution of immunoglobulin heavy chain diversity. *Journal of Experimental Medicine* 1993; 178:815–824.

Hirano, M., Das, S., Guo, P., and Cooper, M.D. The evolution of adaptive immunity in vertebrates. *Advances in Immunology* 2011; 109:125–157.

Hohman, V.S., Schuchman, D.B., Schluter, S.F., and Marchalonis, J.J. Genomic clone for sandbar shark lambda light chain: Generation of diversity in the absence of gene rearrangement. *Proceedings of the National Academy of Sciences of the United States of America* 1993; 90:9882–9886.

Honda, Y., Kondo, H., Caipang, C.M., Hirono, I., and Aoki, T. cDNA cloning of the immunoglobulin heavy chain genes in banded houndshark *Triakis scyllium. Fish Shellfish Immunology* 2010; 29:854–861.

Hsu, E., Pulham, N., Rumfelt, L.L., and Flajnik, M.F. The plasticity of immunoglobulin gene systems in evolution. *Immunological Reviews* 2006; 210:8–26.

Jhunjhunwala, S., van Zelm, M.C., Peak, M.M., Cutchin, S., Riblet, R., van Dongen, J.J., Grosveld, F.G., Knoch, T.A., and Murre, C. The 3D structure of the immunoglobulin heavy-chain locus: implications for long-range genomic interactions. *Cell* 2008; 133:265–279.

Ji, Y., Resch, W., Corbett, E., Yamane, A., Casellas, R., and Schatz, D.G. The in vivo pattern of binding of RAG1 and RAG2 to antigen receptor loci. *Cell* 2010; 141:419–431.

Johansson, J., Aveskogh, M., Munday, B., and Hellman, L. Heavy chain V region diversity in the duck-billed platypus (*Ornithorhinchus anatinus*): Long and highly variable complementarity-determining region 3 compensates for limited germline diversity. *Journal of Immunology* 2002; 168:5155–5162.

Kokubu, F., Hinds, K., Litman, R., Shamblott, M.J., and Litman, G.W. Extensive families of constant region genes in a phylogenetically primitive vertebrate indicate an additional level of immunoglobulin complexity. *Proceedings of the National Academy of Sciences of the United States of America* 1987; 84:5868–5872.

Kokubu, F., Litman, R., Shamblott, M.J., Hinds, K., and Litman, G.W. Diverse organization of immunoglobulin VH gene loci in a primitive vertebrate. *EMBO Journal* 1988; 7:3413–3422.

Krangel, M.S. Gene segment selection in V(D)J recombination: Accessibility and beyond. *Nature Immunology* 2003; 4:624–630.

Lee, S.S., Fitch, D., Flajnik, M.F., and Hsu, E. Rearrangement of immunoglobulin genes in shark germ cells. *Journal of Experimental Medicine* 2000; 191:1637–1648.

Lee, S.S., Tranchina, D.S., Ohta, Y., Flajnik, M.F., and Hsu, E. Hypermutation in shark immunoglobulin light chain genes results in contiguous substitutions. *Immunity* 2002; 16:571–582.

Lee, V., Huang, J.L., Lui, M.F., Malecek, K., Ohta, Y., Mooers, A., and Hsu, E. The evolution of multiple isotypic IgM heavy chains in the shark. *Journal of Immunology* 2008; 180:7461–7470.

Lewis, S., Gifford, A., and Baltimore, D. DNA elements are asymmetrically joined during the site-specific recombination of kappa immunoglobulin genes. *Science* 1985; 228:677–685.

Lewis, S.M., and Hesse, J.E. Cutting and closing without recombination in V(D)J joining. *EMBO Journal* 1991; 10:3631–3639.

Lewis, S.M., Wu, G.E., and Hsu, E. The origin of V(D)J diversification. In: Honjo, T., Alt, F.W., and Neuberger, M.S., editors. *Molecular Biology of B Cells*. London: Elsevier Academic Press, 2004, pp. 473–489.

Litman, G.W., Erickson, B.W., Lederman, L., and Mäkelä, O. Antibody response in *Heterodontus. Molecular and Cellular Biochemistry* 1982; 45:49–57.

Liu, Y., Subrahmanyam, R., Chakraborty, T., Sen, R., and Desiderio, S. A plant homeodomain in RAG-2 that binds Hypermethylated lysine 4 of histone H3 is necessary for efficient antigen-receptor-gene rearrangement. *Immunity* 2007; 27:561–571.

Malecek, K., Brandman, J., Brodsky, J.E., Ohta, Y., Flajnik, M.F., and Hsu, E. Somatic hypermutation and junctional diversification at immunoglobulin heavy chain loci in the nurse shark. *Journal of Immunology* 2005; 175:8105–8115.

Malecek, K., Lee, V., Feng, W., Huang, J.L., Flajnik, M.F., Ohta, Y., and Hsu, E. Immunoglobulin heavy chain exclusion in the shark. *PLoS Biology* 2008; 6:e157.

Marchalonis, J., and Edelman, G.M. Phylogenetic origins of antibody structure. I. Multichain structure of immunoglobulins in the smooth dogfish (*Mustelus canis*). *Journal of Experimental Medicine* 1965; 122:601–618.

Marchalonis, J., and Edelman, G.M. Polypeptide chains of immunoglobulins from the smooth dogfish (*Mustelus canis*). *Science* 1966; 154:1567–1568.

Marculescu, R., Vanura, K., Le, T., Simon, P., Jager, U., and Nadel, B. Distinct t(7;9)(q34;q32) breakpoints in healthy individuals and individuals with T-ALL. *Nature Genetics* 2003; 33: 342–344.

Matthews, A.G., Kuo, A.J., Ramón-Maiques, S., Han, S., Champagne, K.S., Ivanov, D., Gallardo, M. *et al.* RAG2 PHD finger couples histone H3 lysine 4 trimethylation with V(D)J recombination. *Nature* 2007; 450:1106–1110.

Meier, J.T., and Lewis, S.M. P nucleotides in V(D)J recombination: A fine-structure analysis. *Molecular and Cellular Biology* 1993; 13:1078–1092.

Miracle, A.L., Anderson, M.K., Litman, R.T., Walsh, C.J., Luer, C.A., Rothenberg, E.V., and Litman, G.W. Complex expression patterns of lymphocyte-specific genes during the development of cartilaginous fish implicate unique lymphoid tissues in generating an immune repertoire. *International Immunology* 2001; 13:567–580.

Nisonoff, A., Hopper, J.E., and Spring, S.B. *The Antibody Molecule*. New York: Academic Press, 1975.

Ota, T., Sitnikova, T., and Nei, M. Evolution of vertebrate immunoglobulin variable gene segments. *Current Topics in Microbiology Immunology* 2000; 248:221–245.

Pan, J., Sasaki, M., Kniewel, R., Murakami, H., Blitzblau, H.G., Tischfield, S.E., Zhu, X. *et al.* Hierarchical combination of factors shapes the genome-wide topography of yeast meiotic recombination initiation. *Cell* 2011; 4:719–731.

Rast, J.P., Amemiya, C.T., Litman, R.T., Strong, S.J., and Litman, G.W. Distinct patterns of IgH structure and organization in a divergent lineage of chondrichthyan fishes. *Immunogenetics* 1998; 47:234–245.

Rast, J.P., Anderson, M.K., Ota, T., Litman, R.T., Margittai, M., Shamblott, M.J., and Litman, G.W. Immunoglobulin light chain class multiplicity and alternative organizational forms in early vertebrate phylogeny. *Immunogenetics* 1994; 40:83–99.

Rast, J.P., and Litman, G.W. Towards understanding evolutionary origins and early diversification of rearranging antigen receptors. *Immunological Reviews* 1998; 166:79–86.

Retter, M.W., and Nemazee, D. Receptor editing occurs frequently during normal B cell development. *Journal of Experimental Medicine* 1998; 188:1231–1238.

Reynaud, C.A., Dahan, A., Anquez, V., and Weill, J.C. Somatic hyperconversion diversifies the single Vh gene of the chicken with a high incidence in the D region. *Cell* 1989; 59:171–183.

Riblet, R. Immunoglobulin heavy chain genes of mouse. In: Honjo, T., Alt, F.W., and Neuberger, M.S., editors. *Molecular Biology of B Cells*. London: Elsevier Academic Press, 2004, pp. 19–26.

Rumfelt, L.L., Avila, D., Diaz, M., Bartl, S., McKinney, E.C., and Flajnik, M.F. A shark antibody heavy chain encoded by a nonsomatically rearranged VDJ is preferentially expressed in early development and is convergent with mammalian IgG. *Proceedings of the National Academy of Sciences of the United States of America* 2001; 98:1775–1780.

Rumfelt, L.L., Diaz, M., Lohr, R.L., Mochon, E., and Flajnik, M.F. Unprecedented multiplicity of Ig transmembrane and secretory mRNA forms in the cartilaginous fish. *Journal of Immunology* 2004; 173:1129–1139.

Rumfelt, L.L., Lohr, R.L., Dooley, H., and Flajnik, M.F. Diversity and repertoire of IgW and IgM VH families in the newborn nurse shark. *BMC Immunology* 2004; 5:8.

Schatz, D.G., and Ji, Y. Recombination centres and the orchestration of V(D)J recombination. *Nature Reviews Immunology* 2011; 11:251–263.

Schatz, D.G., and Swanson, P.C. V(D)J recombination: Mechanisms of initiation. *Annual Reviews in Genetics* 2011; 45:167–202.

Seitan, V.C., Krangel, M.S., and Merkenschlager, M. Cohesin, CTCF and lymphocyte antigen receptor locus rearrangement. *Trends in Immunology* 2012; 33:153–159.

Shamblott, M.J., and Litman, G.W. Genomic organization and sequences of immunoglobulin light chain genes in a primitive vertebrate suggest coevolution of immunoglobulin gene organization. *EMBO Journal* 1989; 8:3733–3739.

Shen, S.X., Bernstein, R.M., Schluter, S.F., and Marchalonis, J.J. Heavy-chain variable regions in carcharhine sharks: development of a comprehensive model for the evolution of VH domains among the gnathanstomes. *Immunology Cellular Biology* 1996; 74:357–364.

Small, P.A., Jr., Klapper, D.G., and Clem, L.W. Half-lives, body distribution and lack of interconversion of serum 19S and 7S IgM of sharks. *Journal of Immunology* 1970; 105:29–37.

Smith, L.E., Crouch, K., Cao, W., Müller, M.R., Wu, L., Steven, J., Lee, M. *et al.* Characterization of the immunoglobulin repertoire of the spiny dogfish (*Squalus acanthias*). *Developmental and Comparative Immunology* 2012; 36:2349–2355.

Tinguely, A., Chemin, G., Péron, S., Sirac, C., Reynaud, S., Cogné, M., and Delpy, L. Crosstalk between IgH transcription and RNA surveillance during B cell development. *Molecular and Cellular Biology* 2011; 32: 107–117.

Vettermann, C., and Schlissel, M.S. Allelic exclusion of immunoglobulin genes: Models and mechanisms. *Immunological Reviews* 2010; 237:22–42.

Voss, E.W., Jr., Russell, W.J., and Sigel, M.M. Purification and binding properties of nurse shark antibody. *Biochemistry* 1969; 12:4866–4872.

Wang, X., Olp, J.J., and Miller, R.D. On the genomics of immunoglobulins in the gray, short-tailed opossum, *Monodelphis domestica*. *Immunogenetics* 2009; 61:581–596

Yancopoulos, G.D., and Alt, F.W. Developmentally controlled and tissue-specific expression of unrearranged VH gene segments. *Cell* 1985; 40:271–281.

Yoder, J.A., and Litman, G.W. Immune-type diversity in the absence of somatic rearrangement. *Current Topics in Microbiology and Immunology* 2000; 248:271–282.

Zhu, C., Feng, W., Weedon, J., Hua, P., Stefanov, D., Ohta, Y., Flajnik, M.F., and Hsu E. The multiple shark immunoglobulin heavy chain genes rearrange and hypermutate autonomously. *Journal of Immunology* 2011; 187:2492–2501.

Zhu, C., and Hsu E. Error-prone DNA repair activity during somatic hypermutation in shark B lymphocytes. *Journal of Immunology* 2010; 185:5336–5359.

CHAPTER 11

Shark Immunoglobulin Light Chains

Michael F. Criscitiello

CONTENTS

SUMMARY

Shark immunoglobulin heavy chains do not always heterodimerize with light chains, but when they do four isotypes are available. In addition to IgLλ and IgLκ, sharks have IgLσ and IgLσ-cart. The IgLλ loci of sharks are always germline-joined, which may shed light on the origins of the V(D)J rearrangement.

INTRODUCTION

The humoral adaptive immune system of sharks is partly based on the same H_2L_2 heterodimeric antibody structure seen in all jawed vertebrates (reviewed in Flajnik, 2002). The constituent vertebrate immunoglobulin (Ig) heavy (H) and light (L) chains provide obligate defense against pathogens and their toxins through neutralization, opsonization, and activation of the classical pathway of the complement cascade. The two IgL are generally identical to one another in any antibody, as are the two IgH, and this antibody is identical to all others made for plasma membrane expression or secretion by a particular B cell at a point in time. However, receptor editing and somatic hypermutation can genetically alter the IgL (and IgH in the case of hypermutation) during the life of a B cell.

The IgL chain is always comprised of two Ig superfamily domains: one variable (V) and one constant (C) (Figure 11.1). The diverse V domains at the amino terminus of both the IgH and IgL together form the antigen recognition site (paratope) that recognizes the epitope of the antigen. Each V domain has three hypervariable loops called complementarity-determining regions (CDR) connecting the β-strands that form the Ig superfamily domain fold. Both the H and the L chains contribute three CDRs each to each of the two paratopes of the antibody, the region that interacts with the epitope on an antigen. Disulfide bonds link a cysteine in the constant domain of the IgL chain to a cysteine in the first constant domain of the IgH chain. The gross structure of an IgL chain resembles a T-cell receptor (TCR), without the transmembrane or cytoplasmic portions.

In most vertebrates, the genes encoding the V domain of IgL chains (and IgH and TCR chains) are not functional until recombination-activating gene (RAG)-mediated somatic recombination mechanisms act on lymphocyte DNA to assemble the V gene from variable (V) and joining (J) segments (and diversity (D) segments in the case of IgH, TCRβ and TCRδ). Although CDR1 and CDR2 of IgL chains are encoded within the V gene segment, CDR3 is encoded by the juxtaposition and modified ligation of the V and J segments (Figure 11.1). The most common of these modifications at the V-J juncture is N (nontemplate encoded) additions catalyzed by the terminal deoxynucleotidyl transferase (TdT), more common in IgH rearrangements of most species but also contributing to IgL CDR3 diversity. Artemis and the DNA-dependent protein kinase resolve the DNA hairpins at the ends of V(D)J coding segments randomly, which donate P (palindromic) nucleotides. Finally, the exonuclease activity of several DNA repair enzymes removes nucleotides, even while TdT is adding them. Thus, several mechanisms contribute to CDR3 diversity in the IgL protein with immunogenetics at the V-J recombination site.

Although most shark antibodies are of the common H_2L_2 quaternary structure, sharks also use antibodies that contain a pair of H chains without associated L chains, dubbed NAR (Greenberg et al., 1995) for new or nurse shark antigen receptor. IgNAR of cartilaginous fish is not unique in its abandonment of L chains, as mammalian camels, alpacas, and llamas (camelids) have evolved a distinct IgG subclass that functions as IgH dimers (Conrath et al., 2003). There is strong evidence that affinity maturation occurs in the IgNAR isotype (Dooley et al., 2006). Yet other IgH chain isotypes along with their paired IgL somatically hypermutate in sharks as well, albeit possibly with

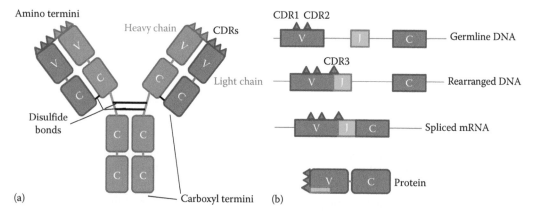

Figure 11.1   Schematic of IgL chains in a heterodimeric antibody and the genes that encode them. (a) Protein structure: heavy chains are blue and light chains are red. CDR3 is depicted as a larger triangle than CDR1 and CDR2. (b) Corresponding gene structure.

different mechanisms (see the following text for more details) (Hinds-Frey *et al.*, 1993). A doubly rearranging form of TCRδ of sharks uses an additional V domain very similar to that of the IgL chain-less IgNAR, called NARTCR (Criscitiello *et al.*, 2006). Nonplacental mammals convergently make TCRs with a similar fundamental structure (Miller, 2010), another example of evolution finding similar immune innovations in both cartilaginous fish and more recent vertebrates. More on NARTCR can be found in Chapter 12.

Mammalian IgL chains were originally identified as Bence-Jones proteins in the serum and urine of lymphoma and myeloma patients (Edelman and Gally, 1962). Mammals express two IgL isotypes from distinct genomic loci, λ and κ (Wu and Kabat, 1970). Yet four isotypes of IgL were now recognized in sharks (Criscitiello and Flajnik, 2007), this identification being eased by the wealth of molecular sequence data from cartilaginous fish and other vertebrates that became available in the early 2000s.

In this chapter, an older body of formative biochemical work will be reviewed first to set the stage for the molecular cloning to come. Genomic functional work that sheds light on exclusion and hypermutation in sharks, as well as the multiple cluster Ig organization found in bony fish light chain and cartilaginous fish heavy chain genes (reviewed in Dooley and Flajnik, 2006; Litman *et al.*, 1999), will be discussed. Finally, discussion will turn to the evolution of IgL chains, building on what we know from these oldest vertebrates that employ them in adaptive immunity. Focus will remain on sharks proper, but knowledge gleaned from other cartilaginous Chondrichthyes will be included when it is likely to have bearing on shark immunobiology (species and references in which they are studied are summarized in Table 11.1).

PIONEERING BIOCHEMICAL STUDIES

Immunological competence was shown in elasmobranchs as early as 1963 by immunizing lemon sharks, *Negaprion brevirostris,* with human influenza virus (Sigel and Clem, 1963). Anamnesis (as defined by an accelerated antibody response upon secondary immunization) was recognized in horned sharks, *Heterodontus francisci,* stimulated with hemocyanin (Papermaster *et al.*, 1964), but it was the work of Marchalonis and Edelman in 1965 that showed that IgH/IgL antibody structure existed in sharks (Marchalonis and Edelman, 1965). They used urea to reduce immunized smooth dogfish *Mustelus canis* antiserum for resolution on starch gels. Peptide maps indicated that the IgL had different primary structure than the IgH. Continued work in the dogfish model resolved the molecular weight of the IgL to 20 kDa and showed the amino acid composition of IgL associated with the pentameric 17S and monomeric 7S forms of shark IgM to be nearly identical (Marchalonis and Edelman, 1966). Similarly, biochemical work on lemon shark sera following bovine serum albumin immunizations showed the established H_2L_2 structure and electrophoretic heterogeneity of the IgL chains (as well as the IgH chains) (Clem *et al.*, 1967). Additionally, various reducing experiments with antibodies of horned shark immunized with *Brucella abortus* showed the necessity of inter-chain disulfide bonds to antibody activity (Frommel *et al.*, 1971). These data suggest that in sharks, like mammals, both chains of Ig are contributing to the diversity of the antigen-binding paratope.

The first amino-terminal protein sequencing of shark IgL was performed by Edman degradation on antibodies of the leopard shark, *Triakis semifasciata* (Suran and Papermaster, 1967). The sequence of the first five positions of this shark IgL (D/E-I-V-L/V/G-T) suggested Igκ, even though nucleic acid work in this species still has neither confirmed nor refuted the original biochemistry. More extensive amino-terminal IgL sequencing of nurse shark *Ginglymosotma cirratum* identified IgL more related to Igκ than Igλ as well in that species (Sledge *et al.*, 1974), and light chains were later resolved to 23 kDa by reducing polyacrylamide gel electrophoresis (Fuller *et al.*, 1978). Similar to what was seen by Sledge *et al.* in nurse shark, IgL from the tiger shark, *Galeocerdo cuvieri,* and Galapagos shark,

Table 11.1 Summary of Experimental Investigations in Different Species

Name/type	Species	Chain isotype	Immunization	Biochemistry	cDNA	Genomic	Expression
Lemon shark	*Negaprion brevirostris*		Sigel (1963)	Clem (1967)			
Smooth dogfish	*Mustelis canis*		Marchalonis (1965)	Marchalonis (1965), Marchalonis (1966)			
Leopard shark	*Triakis semifasciata*			Suran (1967)			
Horned shark	*Heterodontus francisci*	σ-cart Shamblott (1989a, b), κ λ Rast (1994), σ Criscitiello (2007)	Papermaster (1964)	Frommel (1971), Kehoe (1978)	Shamblott (1989a), Rast (1994), σ Criscitiello σ Greenberg (1993), Criscitiello (2007)	Shamblott (1989b) (σ-cart), Rast (1994)	
Nurse shark	*Ginglymostoma cirratum*	κ, σ-cart, λ Greenberg (1993), σ Criscitiello (2007)		Sledge (1974), Fuller (1978)	Greenberg (1993), Criscitiello (2007)	Greenberg (1993), Lee (2000), Fleurant (2004)	Criscitiello (2007)
Tiger shark	*Galeocerdo cuvieri*			Marchalonis (1988), Schluter (1990) Schluter (1987)			
Galapagos shark	*Carcharhinus galapagensis*						
Sandbar shark	*Carcharhinus plumbeus*	λ Schluter (1989), Hohman (1992)		Marchalonis (1988), Schluter (1990)	Schluter (1989) Hohman (1992)	Hohman (1993)	
Little skate	*Leucoraja erinacea*	λ Rast (1994)			Rast (1994)	Rast (1994)	
Spotted ratfish	*Hydrolagus colliei*	λ Rast (1994)			Rast (1994)	Rast (1994)	
Ratfish	*Callorhinchus Callorhinchus*			De Ioannes (1989)			
Clearnose skate	*Raja eglanteria*						Miracle (2001)
Little skate	*Leucoraja erinacea*	σ-cart Anderson (1995)			Anderson (1995), Criscitiello (2007)	Anderson (1995)	
Spiny dogfish	*Squalus acanthus*	σ Criscitiello (2007)		Smith et al. (2012)	Criscitiello (2007), Smith et al. (2012)	Smith et al. (2012)	Smith (2012)

Note: For brevity, only the first author name is given in the table.

Carcharhinus galapagensis, showed high identity with Igκ of mammals (Marchalonis *et al.*, 1988). This amino-terminal sequencing work was accompanied by isoelectric focusing that showed heterogeneous bands for the sandbar shark, *Carcharhinus plumbeus*, as well as the other two carcharhine sharks. A second 0-iodosobenzoic acid-liberated peptide was also sequenced from the tiger shark that corroborates the Igκ homology. Binding of rabbit antisera raised to a synthetic human TCRβJ segment cross-reacted to IgL of Galapagos shark, suggesting conservation of the F-G-x-G-T-R-L motif broadly in vertebrate TCR and IgL (Schluter and Marchalonis, 1986; Schluter *et al.*, 1987). The amino-terminal sequencing in sandbar and tiger sharks was later complemented by tandem mass spectrometry (Schluter *et al.*, 1990). Amino terminal sequence from the Holocephalian ratfish was analyzed with distant homology noted to κ (De Ioannes and Aguila, 1989). Although ratfish are not sharks, the Holocephali they represent is one extant subclass of Chondrichthyes, with the Elasmobranchii being the other (which includes the modern sharks and rays).

MOLECULAR WORK YIELDING cDNA SEQUENCES

Work in the horned shark yielded the first nucleic acid sequence of a shark IgL chain and confirmed the framework and CDR characteristic of vertebrate Ig and TCR proteins (Shamblott and Litman, 1989a, b), as was suggested by V domain amino acid sequencing of this same isotype (Kehoe *et al.*, 1978). The predicted amino acid sequence showed homology to mammalian Igλ in the more limited database of the day, and the nucleic acid sequence showed over 50% identity with a mammalian TCRβ. Work in the sandbar shark revealed partial cDNAs of an Igλ homologue in that species (Schluter *et al.*, 1989).

The next cDNA publication from Marchalonis' sandbar shark team described a complete Igλ sequence (Hohman *et al.*, 1992). This work by Hohman *et al.* is notable for two reasons, one to be discussed in the section "Genomic Organization." It is the first study of shark IgL chains to employ tree-building phylogenetic analysis based on molecular sequence data, using just V sequences to correctly place this sandbar shark IgL with the Igλ of mammals (no small feat, as there were no other poikilothermic or "cold-blooded" sequences in the analysis).

The first IgL cloned from the nurse shark was Igκ (Greenberg *et al.*, 1993), which confirmed both earlier amino-terminal sequencing data of pooled IgL (Sledge *et al.*, 1974) and the emergence of multiple IgL chain isotypes early in vertebrate evolution. This work by Greenberg *et al.* also mentions cloning of two other IgL isotypes from *Ginglymostoma*, and for the first time suggests three or more isotypes in the shark. Evidence for IgL isotypes other than λ and κ had previously only been suggested in the Anuran amphibian *Xenopus* (Hsu *et al.*, 1991).

Rast *et al.* extended the study of elasmobranch IgL to other cartilaginous fish (Rast *et al.*, 1994). They analyzed the more ancient out-group to the living elasmobranchs, the Holocephali, with the spotted ratfish, *Hydrolagus colliei*. They also investigated the other major extant radiation of elasmobranchs besides the sharks (Selachii), the skates and rays (Batoidea), by studying the little skate, *Leucoraja erinacea*. An anchored polymerase chain reaction (PCR) strategy on spotted ratfish cDNA resulted in the cloning of Igλ from this ancient group of fishes. Rast *et al.* used the ratfish Igλ to successfully probe cDNA libraries of the little skate and used a nurse shark κ probe to identify that isotype in the horned shark as well.

GENOMIC ORGANIZATION

The Litman laboratory provided the first evidence that shark IgL genes were organized in a similar multiple cluster organization (Shamblott and Litman, 1989b), as that seen in the shark IgH (Hinds and Litman, 1986) and some bony fish IgL (Bengten *et al.*, 2000; Daggfeldt *et al.*, 1993;

Hsu and Criscitiello, 2006). They sequenced a cluster containing a single V gene (with intron "split leader" typical of antigen receptors), single J gene, and single C gene occupying a comparatively small 2.7 kb. The conserved heptamer and nonamer recognized by the RAG recombinase were found 3' of the V gene and were spaced by 12 nucleotides. These motifs 5' of the J gene were spaced by 23 nucleotides, abiding the rule of 12-bp-spaced recombination signal sequences (RSS) only rearranging with 23-bp-spaced RSS observed in mammals. This orientation (V12/23J) of the RSS is curious, known only at Igκ loci among mammalian antigen receptors. Igλ, all known TCR, and all known Ig V gene segments are flanked by 3' RSS spaced by 23 nucleotides (Criscitiello and Flajnik, 2007). Restriction mapping of genomic clones and genomic Southern blotting with V and C region probes all suggested multiple light chain clusters of very limited complexity (likely 1V-1J-1C). This seminal IgL work did not find evidence for germline-joined V-J genes, a hallmark of many IgH clusters in shark. As discussed in Chapter 10 (Hsu, IgH VDJ), V, multiple D, and J gene segments are often partially or completely germline-joined for many IgH loci in cartilaginous fish (Kokubu et al., 1988).

The description of the complete sandbar shark IgLλ cDNA (Hohman et al., 1992) included a most interesting note added in proof. The addendum shares the sequencing of two sandbar shark IgLλ genomic clones, both of which have the V and J segments fused in the germline. This is the first suggestion of germline-joined V domains encoding genes in IgL chains. Subsequent work by the Marchalonis group confirmed more germline-joined IgLλ in the horned shark (Hohman et al., 1993). In fact, every IgLλ genomic clone or V-J PCR amplicon analyzed in this shark showed fusion. Although IgLκ was originally thought to always be unjoined in cartilaginous fish (Rast et al., 1994), Lee et al. (2000) showed that in the nurse shark, some IgLκ loci are split and some are joined. The joining of V and J segments in the germline is made possible by the expression of RAG in shark germ cells. Genomic Southern blotting of IgLκ in the nurse shark showed the cluster organization now accepted for shark immunoglobulin genes for the first time for this isotype (Greenberg et al., 1993). The RSS orientation of this shark IgLκ is V12/23J, consistent with Igκ in other vertebrates. Genomic V-J PCR suggested that at least one nurse shark IgLκ locus was germline-joined. Later work on *Ginglymostoma* showed that of around 60 IgLκ loci, six are germline-joined but the rest can rearrange (Lee et al., 2000). Phylogenetic dendrograms had been drawn with shark IgL data before (Hohman et al., 1992; Zezza et al., 1992), but Greenberg et al. for the first time showed the distinct patterns C domain data yield compared to that from V domains. Both methods are confounded by different problems: V domains diversify for a broad repertoire (especially problematic if CDR columns of the alignment are included), and C domains tend to cluster as much or more by taxonomic group as by isotype, presumably for heterodimerization of C1 domain peculiarities of the particular taxon. The different physiology of these functional domains of L chains would be expected to exert different evolutionary pressures on the exons encoding them.

Genomic analysis by Southern blotting suggests multiple clusters in *Hydrolagus*. In the same study, PCR and limited sequencing found two V families and no unjoined IgLλ in the spotted ratfish (Rast et al., 1994), consistent with sharks. Rast et al. used the spotted ratfish IgLλ to probe genomic libraries of horned shark and little skate, finding germline-joined IgLλ in both *Heterodontus* and *Leucoraja*.

Genomic library screening and PCR from genomic DNA of the little skate *L. erinacea* failed to find any of the isotype called Type I (σ-cart, see the following text) unjoined (Anderson et al., 1995). Some of the exact same sequences (including the CDR3 region) were cloned from cDNA and genomic DNA, proving that these joined loci are at least expressed at the mRNA level. This gives credence to the idea that loci of a particular IgL isotypes may be all joined, all split by intron, or some combination of the two. Additionally, this arrangement may not be consistent among loci of the same isotype when they are examined in different cartilaginous fish species.

CLASSIFICATION

In hindsight, the classification and nomenclature of IgL was quite confusing across vertebrates, and nowhere more so than in the cartilaginous fish (although bony fish come close). Some comparative studies strove to classify newly discovered IgL as either mammalian IgLλ or IgLκ, some accepted the notion that there could be IgLλ, IgLκ, or other isotypes in these vertebrates, whereas others designed new independent systems of nomenclature for the isotypes of the particular species or taxonomic group being studied. Recognizing that three cartilaginous fish sequences did not all belong to IgLλ or IgLκ, a system of NS3, NS4, and NS5 was adopted in the nurse shark (Greenberg *et al.*, 1993) simply based on the original monikers of bands in a particularly fruitful PCR experiment (dissertation of Andrew Greenberg, University of Miami, 1994). This system was later replaced with type I (NS5), type II (NS3), and type III (NS4) after standardization with other cartilaginous fish (Rast *et al.*, 1994). Several authors recognized that type I were difficult to classify as either IgLκ or IgLλ, whereas type II were more like mammalian IgLλ, and type III were more like IgLκ.

The discovery of a fourth isotype in the nurse shark allowed new connections to be made between isotypes in cartilaginous fish, bony fish, and tetrapods (Criscitiello and Flajnik, 2007). A mini-expressed sequence tag (EST) library from spleen and pancreas produced a sequence of an isotype not previously identified in cartilaginous fish, but that shared homology with a teleost isotype called IgL2 and the amphibian isotype IgLσ. The "new" shark sequence was named for the original discovery of its orthologue in frog (Schwager *et al.*, 1991). IgLσ in nurse shark was corroborated by clones from horn shark, dogfish shark, and the little skate (Criscitiello and Flajnik, 2007). The cartilaginous fish IgLσ is most similar to the type I/NS5 IgL, yet is distinct. Phylogenetic trees made from V domain, C domains, and RSS orientations suggest that all vertebrate IgL can be placed in one of the four (once more renamed) groups present in sharks: IgLσ, IgLσ-cart (type I/NS5), IgLλ (type II/NS3), and IgLκ (type III/NS4). Because IgLσ, IgLλ, and IgLκ were discovered first and are found in other major vertebrate radiations, these names are used now in shark as well. IgLσ-cart (type I/NS5), however, has now been found outside of the cartilaginous fish in the coelacanth genome (Amemiya *et al.*, 2013).

Recently, studies in the spiny dogfish, *Squalus acanthias*, demonstrated the existence of these four light chain isotypes in that species as well (Smith *et al.*, 2012). In this molecular and biochemical work on the dogfish immunoglobulins, immunoprecipitations with a monoclonal raised to nurse shark IgLκ preferentially bound to polymeric IgM, whereas monomeric IgM tended to heterodimerize with the other IgL isotypes. Figure 11.2 shows these four isotypes in sharks and their representation in other vertebrate groups.

This system of four ancient clades of IgL present in sharks, three of which are shared with most poikilothermic vertebrates (IgLλ, IgLκ, and IgLσ), two with mammals (IgLλ and IgLκ) and one with birds (IgLλ), has withstood recent discoveries and reevaluations (Das *et al.*, 2008; Edholm *et al.*, 2011) including those taking advantage of recent genomic projects in reptiles and nonplacental mammals (Figure 11.3 adapted from (Smith *et al.*, 2012). However, the next decade will bring high-throughput transcriptomic and genomic analysis within the reach of many more model systems, and this flood of data will determine if other IgL isotypes remain and if nomenclature again may require reanalysis. Table 11.1 organizes the available IgL literature by species.

DIVERSITY GENERATION

The Hsu laboratory took advantage of the relatively few (four) functional IgLλ loci in the nurse shark to rigorously explore IgL somatic hypermutation in this species (Lee *et al.*, 2002). In these V-J fused genes, they found a preponderance of contiguous (sometimes called tandem) mutations

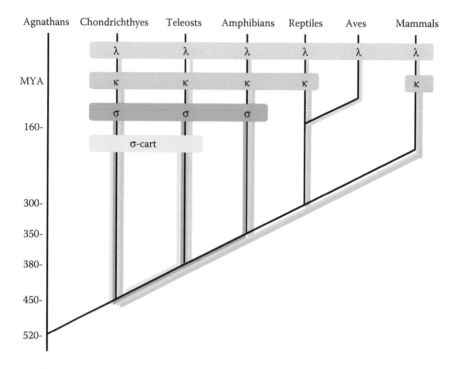

Figure 11.2   Phylogenetic representation of the four IgL isotypes of sharks. These four isotypes were given other names in other species. These other monikers include for IgLσ-cart (yellow) Type I and NS5 in cartilaginous fish; for IgLσ (magenta) L2 and Type 2 in bony fish; for IgLκ (green) Type III and NS4 in cartilaginous fish, Type 1, Type 3, L1 (L1A and L1B), L3, F and G in bony fish, as well as ρ in amphibians; and for IgLλ (blue) Type II and NS3 in cartilaginous fish, L2 in bony fish, and Type III in amphibians. (Adapted from Criscitiello, M.F., and Flajnik, M.F., *Eur. J. Immunol.*, 37, 2683–2694, 2007.)

occurring in 2–4 bp stretches. The tandem mutations bring to three the count of general mutation patterns that were seen in cartilaginous fish: point mutations with a GC bias (Du Pasquier *et al.*, 1998), mammalian-like point mutations without the bias (Diaz *et al.*, 1999), and these contiguous or tandem mutations. Lee *et al.* found no evidence of donor sequences, discounting the possibility of gene conversion.

Junctional diversity was investigated using IgLσ-cart (type 1/NS5) in the nurse shark (Fleurant *et al.*, 2004). The genomic loci encoding this isotype were characterized to enable this comparison of germline sequence with that of somatically joined genes. Fleurant *et al.* (2004) found only three (two functional) unjoined and one joined loci, a situation this group again exploited to definitively analyze rearrangement at these two functional loci. They found N (nontemplate) extensions at CDR3 to be unusually common and long compared to tetrapod IgL chains; such activity levels are more often associated with IgH chains than with IgL. The authors speculate that in sharks (employing multiple IgH and IgL clusters), simultaneous rearrangement of both IgH and IgL may make the genes encoding both chains accessible to the same processing factors. To date, there is no evidence for a surrogate light chain in sharks, perhaps it is not needed if IgH and IgL are rearranging and being expressed simultaneously. IgLσ in nurse shark shows evidence of N additions and exonuclease activity as well, even in young animals (Criscitiello and Flajnik, 2007). TdT is expressed in shark primary lymphoid organs along with RAG and is likely responsible for the N additions (Criscitiello *et al.*, 2010; Rumfelt *et al.*, 2001). Hence, germline-joined genes and 1V-1J-1C clusters do not appear to damn shark IgL genes to limited diversity.

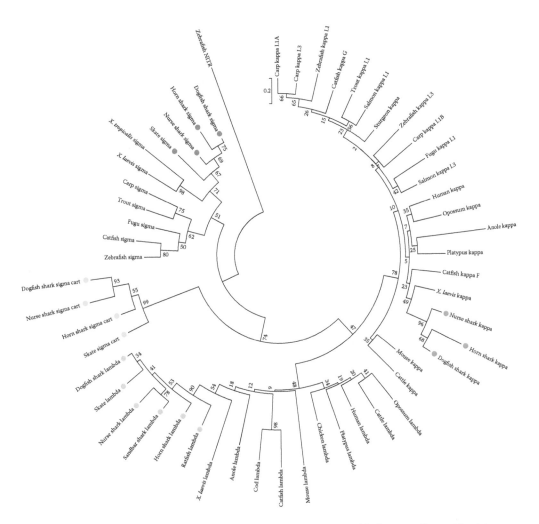

Figure 11.3 Neighbor-joining tree based on IgL variable domain amino acid alignment. Percent bootstrap support after 1000 samplings is shown at each node. A novel immune type receptor (NITR) sequence (GenBank accession: NM001005577) was employed as an out-group. Clusters of sequences in both trees are marked with blocks showing the four ancestral vertebrate isotypes. Spiny dogfish sequences are shaded. Accession numbers of sequences used are as follows: spiny dogfish (*Squalus acanthias*) kappa [JN419107], lambda [JN419108], sigma cart [JN419096], sigma [JN419086]; horn shark (*Heterodontos francisci*) kappa [L25561], lambda [L25560], sigma cart [X15315], sigma [EF114760]; sandbar shark (*Carcharhinus plumbeus*) lambda [M81314]; nurse shark (*Ginglymostoma cirratum*) kappa [GSU15144], lambda (see note), sigma cart [AAV34678.1], sigma [EF114759]; little skate (*Leucoraja erinacea*) lambda [L25566], sigma cart [L25568]; sturgeon (*Acipenser baerii*) kappa [CAB44624]; zebrafish (*Danio rerio*) kappa L1 [AF246185], kappa L3 [AB246193], sigma [AF246183]; spotted ratfish (*Hydrolagus colliei*) lambda [L25549]; pufferfish (*Takifugu rubripes*) kappa L1 [AB126061], sigma [DQ471453]; carp (*Cyprinus carpio*) kappa L1A [AB073328], kappa L1B [AB073332], kappa L3 [AB073335], sigma [AB091113]; salmon (*Salmo salar*) kappa L1 [AF273012], kappa L3 [AF406956]; trout (*Oncorhynchus mykiss*) kappa L1 [X65260], sigma [AAB41310]; cod (*Gadus morhua*) lambda [CAC03754.1]; catfish (*Ictalurus punctatus*) kappa F [U25705], kappa G [L25533], lambda [ACG70845], sigma [CK403931]; African clawed frog (*Xenopus laevis*) kappa [L15570], lambda [L76575], sigma [AAB34863]; western clawed frog (*Xenopus tropicalis*) sigma [from JGI genome project v4.1 scaffold 289]; anole (*Anolis carolinensis*) kappa [ACB45836], lambda [ADB22722]; chicken (*Gallus gallus*) lambda [M24403]; platypus (*Ornithorhynchus anatinus*) kappa [AF116942], lambda [AAO16074]; opossum (*Monodelphis domestica*) kappa [AAF25575], lambda [AAC98667]; cattle (*Bos taurus*) kappa [AA151501], lambda [AAC48564]; mouse (*Mus musculus*) kappa [CAA81787], lambda [AA053422]; human (*Homo sapiens*) kappa [S46371], lambda [AAA59013]. Cartilaginous fish IgLσ sequences are denoted with magenta dots, IgLσ-cart yellow, IgLκ green, and IgLλ blue.

EXPRESSION AND FUNCTION

Hohman *et al.* (1995) found diversity within the noncoding regions of sandbar shark IgLλ genes when many genomic clones were sequenced. Promoter elements in different positions and orientations suggest that there may be differential regulation of antibodies produced by different clusters of the same isotype in this shark. The first application of quantitative real-time PCR to elasmobranchs light chain use was done in the clearnose skate, *Raja eglanteria* (Miracle *et al.*, 2001). Developmentally, IgLσ-cart showed moderate expression throughout the first 11 weeks of embryonic life with a peak at week eight, whereas IgLλ was nearly absent save for a peak at week eight. Differential expression of IgL σ-cart and λ were seen in individual tissues in this skate as well, IgLλ being more abundant in Leydig organ (a B-cell primary lymphoid tissue of some elasmobranchs) whereas IgLσ-cart is higher in the gonad and liver; both isotypes are high in spleen. In all tissues, IgLσ-cart expression dominated IgLλ at the eight-week embryo and hatchling stages, a pattern reversed in the adult skate.

Different mammalian species have different ratios of isotype expression in their mature repertoires. In mice, the average κ:λ ratio is 20:1, in humans it is 2:1, yet in cattle it is 1:20 (Du *et al.*, 2012). Not surprisingly, limited analysis of cartilaginous species suggests diversity in relative IgL isotype use as well. IgLκ was found to be dominant in nurse shark (Greenberg *et al.*, 1993) and accounted for 79% of IgL clones in a secondary lymphoid tissue EST library (Criscitiello and Flajnik unpublished observations). In contrast, IgLσ-cart and IgLλ predominated in cDNA libraries from horn shark (Rast *et al.*, 1994). IgLλ also appears to predominate in the sandbar shark (Hohman *et al.*, 1995).

COMPARATIVE GENOMICS

Although no elasmobranch genomes are available at the time this chapter is being composed, that of the holocephalan elephant shark, *Callorhinchus milii*, was reported by Venkatesh *et al.* (2014). In this more primordial cartilaginous fish lineage, about 20 IgLλ, two IgLκ, and one partial Igσ-cart loci were found in the assembled scaffolds yet no IgLσ. Additionally, although evidence for IgNAR was not found, one unconventional IgM locus was found without canonical residues needed for association with light chains and a duplicated CH2 domain in lieu of CH1. This putative single chain IgM suggests an alternative single-chain antibody alternative to IgNAR in the earliest jawed vertebrates.

EVOLUTION

The genesis of the heterodimeric H_2L_2 antibody system occurred in an ancestor we share with the jawed cartilaginous fishes (Figure 11.2). Both the IgLλ and IgLκ isotypes common in mammals emerged in sharks and their kin, but IgLσ and IgLσ-cart also arose in the cartilaginous fish and have had different evolutionary fates. IgLσ-cart appears to have not been passed to other bony vertebrates, yet IgLσ is found in all non-amniotic jawed vertebrates (cartilaginous fish, bony fish, and amphibians). The orientations of RSS flanking the 3′ of the V gene segment and 5′ of the J segment are consistent in these isotypes across vertebrate groups, as IgLσ, IgLσ-cart, and IgLκ all have V12-23J-spaced RSS, whereas IgLλ is always in the V23-12J orientation (Criscitiello and Flajnik, 2007).

The unusual germline joining of shark Ig genes due to gonadal RAG expression may have had several ramifications in IgL evolution. Joined antigen receptor genes occur at least in one other place in vertebrate evolution: the Ig-like supporting V domain of marsupial TCRμ is germline-joined

(Parra *et al.*, 2007), although the monotreme orthologue still rearranges (Vandesompele *et al.*, 2002). Also in the opossum, a IgHV exists (IGHV3.1) with a germline-joined VD (yet not joined D to J) (Livak and Schmittgen, 2001). So far, publications with shark IgL have found both joined and unjoined σ-cart and κ, yet only unjoined σ and only joined λ.

Several models for the evolution of vertebrate antigen receptor loci taking shark IgL character states such as RSS orientation and germline-joining into account are presented in Figure 11.4. RAG can act in the germline as a recombinase that will swap the 12-bp- and 23-bp-spaced RSS between two segments (Lewis and Wu, 2000; Lewis *et al.*, 1988), a process first recognized in shark Ig (Kokubu *et al.*, 1988). The conservation of RSS orientations not only for IgL but for IgH and TCR as well suggests that this swapping is relatively rare in evolutionary terms, and thus, RSS orientation may be of some value in reconstructing the natural history of antigen receptors. We previously suggested that four possibilities are most parsimonious (Criscitiello and Flajnik, 2007). Model *A* suggests the IgLλ isotype evolved from the ancestral common antigen receptor with a V23-12J orientation, with RSS inversion birthing the other IgL chains. This model only invokes one RSS swap and the subsequent fusing of IgLλ and other loci is a shared derived character. Model *B* requires an additional RSS swap but allows the ancestral IgL to be σ/κ, which was supported in some phylogenies (Criscitiello and Flajnik, 2007). A related model that would only require one RSS flip is shown as model *C*. This scheme has the paths of IgH and TCR splitting from the σ/κ-like IgL after the RSS flip along with IgLλ. This model would suggest ancestral IgL function without other chains, perhaps as membrane homodimers with a gross structure similar to TCR. Lastly, perhaps IgLλ was

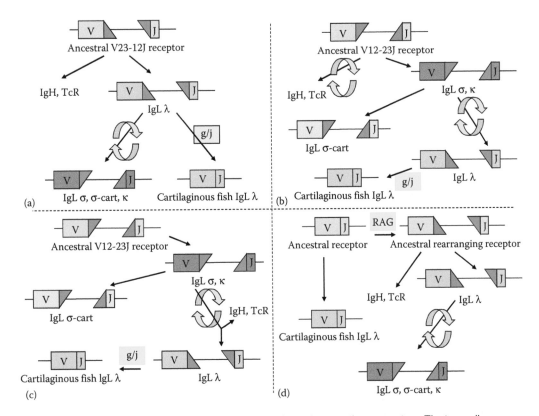

Figure 11.4 Models of shark IgL in the natural history of vertebrate antigen receptors. The two yellow arrows represent RSS swaps (presumable rare and evolutionarily informative). *RAG* symbolizes ancestral RAG transposon insertion event. Yellow *g/j* indicates germline joining of RSS. (Adapted from Criscitiello, M.F., and Flajnik, M.F., *Eur. J. Immunol.*, 37, 2683–2694, 2007.)

not rejoined in the germline as this was the ancestral state of the antigen receptor. If the RAG insertion event occurred in one site of a multicopy IgLλ locus system, the extant cartilaginous fish IgLλ could be a holdover from the pre-RAG adaptive immune system (model *D*). We look forward to cartilaginous fish genomic projects elucidating the likelihood of these and other models of antigen receptor locus history (Venkatesh *et al.*, 2007).

The maintenance of four distinct isotypes for nearly 500 million years in sharks and apparently at least two isotypes in all vertebrates that do not employ gene conversion (birds) suggests a differential functional basis for their emergence and maintenance. Ample evidence is lacking for distinct physiology between mammalian IgLλ and IgLκ (Stanfield *et al.*, 2006; Sverremark *et al.*, 2000), though it exists (Hershberg and Shlomchik, 2006). We have suggested that the distinct CDR length patterns between different IgL isotypes may be a clue (Criscitiello and Flajnik, 2007). IgLσ and IgLσ-cart have shorter CDR1 and longer CDR2, whereas IgLκ and IgLλ have the opposite (longer CDR1 and shorter CDR2). These different IgL isotypes may have paired differentially with different VH domains, accommodating different paratopes that are not sterically hindered by the very different CDR loops of the IgL. Long CDR2 such as that found in IgLσ would be expected to pair with IgHV with a short CDR3 (Padlan, 1994). Alternatively, different IgL isotypes may have evolved for different IgH isotype heterodimerization. IgLσ in the frog, *Xenopus laevis*, was found to associate with only two of the three IgH (IgM and IgX, but not IgY) (Hsu *et al.*, 1991). Additionally, work in the cartilaginous skate showed a large disparity in relative IgLσ-cart and IgLλ expression from intestine (Miracle *et al.*, 2001). These two hypotheses are not mutually exclusive, different IgL isotypes could have evolved for both binding paratope structure in the V domain and distinct IgH isotype IgHC1 domain pairing with the IgL C domain.

Much work is left to do in shark IgL biology. It is anticipated that the new accessibility to genomics will direct novel hypothesis testing back in the animal, where basic science should be able to reveal fundamental properties of the antibodies of sharks and all vertebrates.

ACKNOWLEDGMENTS

The author thanks Ellen Hsu and Valerie Hohman for critical reading of the manuscript, Jeannine O. Eubanks for expert phylogenetic figure preparation, Thaddeus C. Deiss and Natalie Jacobs for helpful discussions, and the National Science Foundation (IOS1257829) for their support.

REFERENCES

Amemiya, C.T., Alfoldi, J., Lee, A.P., Fan, S., Philippe, H., Maccallum, I., Braasch, I. *et al.* The African coelacanth genome provides insights into tetrapod evolution. *Nature* 2013; 496:311–316.

Anderson, M.K., Shamblott, M.J., Litman, R.T., and Litman, G.W. Generation of immunoglobulin light chain gene diversity in *Raja erinacea* is not associated with somatic rearrangement, an exception to a central paradigm of B cell immunity. *J. Exp. Med.* 1995; 182:109–119.

Bengten, E., Wilson, M., Miller, N., Clem, L.W., Pilstrom, L., and Warr, G.W. Immunoglobulin isotypes: Structure, function, and genetics. *Curr. Top. Microbiol. Immunol.* 2000; 248:189–219.

Clem, I.W., De Boutaud, F., and Sigel, M.M. Phylogeny of immunoglobulin structure and function. II. Immunoglobulins of the nurse shark. *J. Immunol.* 1967; 99:1226–1235.

Conrath, K.E., Wernery, U., Muyldermans, S., and Nguyen, V.K. Emergence and evolution of functional heavy-chain antibodies in Camelidae. *Dev. Comp. Immunol.* 2003; 27:87–103.

Criscitiello, M.F., and Flajnik, M.F. Four primordial immunoglobulin light chain isotypes, including lambda and kappa, identified in the most primitive living jawed vertebrates. *Eur. J. Immunol.* 2007; 37:2683–2694.

Criscitiello, M.F., Ohta, Y., Saltis, M., McKinney, E.C., and Flajnik, M.F. Evolutionarily conserved TCR binding sites, identification of T cells in primary lymphoid tissues, and surprising trans-rearrangements in nurse shark. *J. Immunol.* 2010; 184:6950–6960.

Criscitiello, M.F., Saltis, M., and Flajnik, M.F. An evolutionarily mobile antigen receptor variable region gene: Doubly rearranging NAR-TcR genes in sharks. *Proc. Natl. Acad. Sci. USA* 2006; 103:5036–5041.

Daggfeldt, A., Bengten, E., and Pilstrom, L. A cluster type organization of the loci of the immunoglobulin light chain in Atlantic cod (*Gadus morhua* L.) and rainbow trout (*Oncorhynchus mykiss* Walbaum) indicated by nucleotide sequences of cDNAs and hybridization analysis. *Immunogenetics* 1993; 38:199–209.

Das, S., Nikolaidis, N., Klein, J., and Nei, M. Evolutionary redefinition of immunoglobulin light chain isotypes in tetrapods using molecular markers. *Proc. Natl. Acad. Sci. USA* 2008; 105:16647–16652.

De Ioannes, A.E., and Aguila, H.L. Amino terminal sequence of heavy and light chains from ratfish immunoglobulin. *Immunogenetics* 1989; 30:175–180.

Diaz, M., Velez, J., Singh, M., Cerny, J., and Flajnik, M.F. Mutational pattern of the nurse shark antigen receptor gene (NAR) is similar to that of mammalian Ig genes and to spontaneous mutations in evolution: The translesion synthesis model of somatic hypermutation. *Int. Immunol.* 1999; 11:825–833.

Dooley, H., and Flajnik, M.F. Antibody repertoire development in cartilaginous fish. *Dev. Comp. Immunol.* 2006; 30:43–56.

Dooley, H., Stanfield, R.L., Brady, R.A., and Flajnik, M.F. First molecular and biochemical analysis of in vivo affinity maturation in an ectothermic vertebrate. *Proc. Natl. Acad. Sci. USA* 2006; 103:1846–1851.

Du, C.C., Mashoof, S.M., and Criscitiello, M.F. Oral immunization of the African clawed frog (*Xenopus laevis*) upregulates the mucosal immunoglobulin IgX. *Vet. Immunol. Immunopathol.* 2012; 145:493–498.

Du Pasquier, L., Wilson, M., Greenberg, A.S., and Flajnik, M.F. Somatic mutation in ectothermic vertebrates: Musings on selection and origins. *Curr. Top. Microbiol. Immunol.* 1998; 229:199–216.

Edelman, G.M., and Gally, J.A. The nature of Bence-Jones proteins. Chemical similarities to polypetide chains of myeloma globulins and normal gamma-globulins. *J. Exp. Med.* 1962; 116:207–227.

Edholm, E.S., Wilson, M., and Bengten, E. Immunoglobulin light (IgL) chains in ectothermic vertebrates. *Dev. Comp. Immunol.* 2011; 35:906–915.

Flajnik, M.F. Comparative analyses of immunoglobulin genes: Surprises and portents. *Nat. Rev. Immunol.* 2002; 2:688–698.

Fleurant, M., Changchien, L., Chen, C.T., Flajnik, M.F., and Hsu, E. Shark Ig light chain junctions are as diverse as in heavy chains. *J. Immunol.* 2004; 173:5574–5582.

Frommel, D., Litman, G.W., Finstad, J., and Good, R.A. The evolution of the immune response. XI. The immunoglobulins of the horned shark, *Heterodontus francisci*: Purification, characterization and structural requirement for antibody activity. *J. Immunol.* 1971; 106:1234–1243.

Fuller, L., Murray, J., and Jensen, J.A. Isolation from nurse shark serum of immune 7S antibodies with two different molecular weight H-chains. *Immunochemistry* 1978; 15:251–259.

Greenberg, A.S., Avila, D., Hughes, M., Hughes, A., McKinney, E.C., and Flajnik, M.F. A new antigen receptor gene family that undergoes rearrangement and extensive somatic diversification in sharks. *Nature* 1995; 374:168–173.

Greenberg, A.S., Steiner, L., Kasahara, M., and Flajnik, M.F. Isolation of a shark immunoglobulin light chain cDNA clone encoding a protein resembling mammalian kappa light chains: Implications for the evolution of light chains. *Proc. Natl. Acad. Sci. USA* 1993; 90:10603–10607.

Hershberg, U., and Shlomchik, M.J. Differences in potential for amino acid change after mutation reveals distinct strategies for kappa and lambda light-chain variation. *Proc. Natl. Acad. Sci. USA* 2006; 103:15963–15968.

Hinds, K.R., and Litman, G.W. Major reorganization of immunoglobulin VH segmental elements during vertebrate evolution. *Nature* 1986; 320:546–549.

Hinds-Frey, K.R., Nishikata, H., Litman, R.T., and Litman, G.W. Somatic variation precedes extensive diversification of germline sequences and combinatorial joining in the evolution of immunoglobulin heavy chain diversity. *J. Exp. Med.* 1993; 178:815–824.

Hohman, V.S., Schluter, S.F., and Marchalonis, J.J. Complete sequence of a cDNA clone specifying sandbar shark immunoglobulin light chain: gene organization and implications for the evolution of light chains. *Proc. Natl. Acad. Sci. USA* 1992; 89:276–280.

Hohman, V.S., Schluter, S.F., and Marchalonis, J.J. Diversity of Ig light chain clusters in the sandbar shark (*Carcharhinus plumbeus*). *J. Immunol.* 1995; 155:3922–3928.

Hohman, V.S., Schuchman, D.B., Schluter, S.F., and Marchalonis, J.J. Genomic clone for sandbar shark lambda light chain: Generation of diversity in the absence of gene rearrangement. *Proc. Natl. Acad. Sci. USA* 1993; 90:9882–9886.

Hsu, E., and Criscitiello, M.F. Diverse immunoglobulin light chain organizations in fish retain potential to revise B cell receptor specificities. *J. Immunol.* 2006; 177:2452–2462.

Hsu, E., Lefkovits, I., Flajnik, M., and Du, P.L. Light chain heterogeneity in the amphibian Xenopus. *Mol. Immunol.* 1991; 28:985–994.

Kehoe, J.M., Sharon, J., Gerber-Jenson, B., and Litman, G.W. The structure of immunoglobulin variable regions in the horned shark, *Heterodontus francisci*. *Immunogenetics* 1978; 7:35–40.

Kokubu, F., Litman, R., Shamblott, M.J., Hinds, K., and Litman, G.W. Diverse organization of immunoglobulin VH gene loci in a primitive vertebrate. *EMBO J.* 1988; 7:3413–3422.

Lee, S.S., Fitch, D., Flajnik, M.F., and Hsu, E. Rearrangement of immunoglobulin genes in shark germ cells. *J. Exp. Med.* 2000; 191:1637–1648.

Lee, S.S., Tranchina, D., Ohta, Y., Flajnik, M.F., and Hsu, E. Hypermutation in shark immunoglobulin light chain genes results in contiguous substitutions. *Immunity* 2002; 16:571–582.

Lewis, S.M., Hesse, J.E., Mizuuchi, K., and Gellert, M. Novel strand exchanges in V(D)J recombination. *Cell* 1988; 55:1099–1107.

Lewis, S.M., and Wu, G.E. The old and the restless. *J. Exp. Med.* 2000; 191:1631–1636.

Litman, G.W., Anderson, M.K., and Rast, J.P. Evolution of antigen binding receptors. *Annu. Rev. Immunol.* 1999; 17:109–147.

Livak, K.J., and Schmittgen, T.D. Analysis of relative gene expression data using real-time quantitative PCR and the 2(-Delta Delta C(T)) Method. *Methods* 2001; 25:402–408.

Marchalonis, J., and Edelman, G.M. Phylogenetic origins of antibody structure. I. Multichain structure of immunoglobulins in the smooth dogfish (*Mustelus canis*). *J. Exp. Med.* 1965; 122:601–618.

Marchalonis, J., and Edelman, G.M. Polypeptide chains of immunoglobulins from the smooth dogfish (*Mustelus canis*). *Science* 1966; 154:1567–1568.

Marchalonis, J.J., Schluter, S.F., Rosenshein, I.L., and Wang, A.C. Partial characterization of immunoglobulin light chains of carcharhine sharks: Evidence for phylogenetic conservation of variable region and divergence of constant region structure. *Dev. Comp. Immunol.* 1988; 12:65–74.

Miller, R.D. Those other mammals: The immunoglobulins and T cell receptors of marsupials and monotremes. *Semin. Immunol.* 2010; 22:3–9.

Miracle, A.L., Anderson, M.K., Litman, R.T., Walsh, C.J., Luer, C.A., Rothenberg, E.V., and Litman, G.W. Complex expression patterns of lymphocyte-specific genes during the development of cartilaginous fish implicate unique lymphoid tissues in generating an immune repertoire. *Int. Immunol.* 2001; 13:567–580.

Padlan, E.A. Anatomy of the antibody molecule. *Mol. Immunol.* 1994; 31:169–217.

Papermaster, B.W., Condie, R.M., Finstad, J., and Good, R.A. Evolution of the immune response. I. The phylogenetic development of adaptive immunologic responsiveness in vertebrates. *J. Exp. Med.* 1964; 119:105–130.

Parra, Z.E., Baker, M.L., Schwarz, R.S., Deakin, J.E., Lindblad-Toh, K., and Miller, R.D. A unique T cell receptor discovered in marsupials. *Proc. Natl. Acad. Sci. USA* 2007; 104:9776–9781.

Rast, J.P., Anderson, M.K., Ota, T., Litman, R.T., Margittai, M., Shamblott, M.J., and Litman, G.W. Immunoglobulin light chain class multiplicity and alternative organizational forms in early vertebrate phylogeny. *Immunogenetics* 1994; 40:83–99.

Rumfelt, L.L., Avila, D., Diaz, M., Bartl, S., McKinney, E.C., and Flajnik, M.F. A shark antibody heavy chain encoded by a nonsomatically rearranged VDJ is preferentially expressed in early development and is convergent with mammalian IgG. *Proc. Natl. Acad. Sci. USA* 2001; 98:1775–1780.

Schluter, S.F., Beischel, C.J., Martin, S.A., and Marchalonis, J.J. Sequence analysis of homogeneous peptides of shark immunoglobulin light chains by tandem mass spectrometry: Correlation with gene sequence and homologies among variable and constant region peptides of sharks and mammals. *Mol. Immunol.* 1990; 27:17–23.

Schluter, S.F., Hohman, V.S., Edmundson, A.B., and Marchalonis, J.J. Evolution of immunoglobulin light chains: cDNA clones specifying sandbar shark constant regions. *Proc. Natl. Acad. Sci. USA* 1989; 86:9961–9965.

Schluter, S.F., and Marchalonis, J.J. Antibodies to synthetic joining segment peptide of the T-cell receptor beta-chain: serological cross-reaction between products of T-cell receptor genes, antigen binding T-cell receptors, and immunoglobulins. *Proc. Natl. Acad. Sci. USA* 1986; 83:1872–1876.

Schluter, S.F., Rosenshein, I.L., Hubbard, R.A., and Marchalonis, J.J. Conservation among vertebrate immunoglobulin chains detected by antibodies to a synthetic joining segment peptide. *Biochem. Biophys. Res. Commun.* 1987; 145:699–705.

Schwager, J., Burckert, N., Schwager, M., and Wilson, M. Evolution of immunoglobulin light chain genes: Analysis of *Xenopus* IgL isotypes and their contribution to antibody diversity. *EMBO J.* 1991; 10:505–511.

Shamblott, M.J., and Litman, G.W. Complete nucleotide sequence of primitive vertebrate immunoglobulin light chain genes. *Proc. Natl. Acad. Sci. USA* 1989a; 86:4684–4688.

Shamblott, M.J., and Litman, G.W. Genomic organization and sequences of immunoglobulin light chain genes in a primitive vertebrate suggest coevolution of immunoglobulin gene organization. *EMBO J.* 1989b; 8:3733–3739.

Sigel, M.M., and Clem, L.W. Immunological response of an elasmobrance to human influenza virus. *Nature* 1963; 197:315–316.

Sledge, C., Clem, L.W., and Hood, L. Antibody structure: Amino terminal sequences of nurse shark light and heavy chains. *J. Immunol.* 1974; 112:941–948.

Smith, L.E., Crouch, K., Cao, W., Muller, M.R., Wu, L., Steven, J., Lee, M., Liang, M., Flajnik, M.F., Shih, H.H., Barelle, C.J., Paulsen, J., Gill, D.S., and Dooley, H. Characterization of the immunoglobulin repertoire of the spiny dogfish (*Squalus acanthias*). *Dev Comp Immunol.* 2012; 36:665–679.

Stanfield, R.L., Zemla, A., Wilson, I.A., and Rupp, B. Antibody elbow angles are influenced by their light chain class. *J. Mol. Biol.* 2006; 357:1566–1574.

Suran, A.A., and Papermaster, B.W. N-terminal sequences of heavy and light chains of leopard shark immunoglobulins: Evolutionary implications. *Proc. Natl. Acad. Sci. USA* 1967; 58:1619–1623.

Sverremark, E., Rietz, C., and Fernandez, C. Kappa-deficient mice are non-responders to dextran B512: Is this unresponsiveness due to specialization of the kappa and lambda Ig repertoires? *Int. Immunol.* 2000; 12:431–438.

Vandesompele, J., De Preter, K., Pattyn, F., Poppe, B., Van Roy, N., De Paepe, A., and Speleman, F. Accurate normalization of real-time quantitative RT-PCR data by geometric averaging of multiple internal control genes. *Genome Biol.* 2002; 3:RESEARCH0034.

Venkatesh, B., Kirkness, E.F., Loh, Y.H., Halpern, A.L., Lee, A.P., Johnson, J., Dandona, N. *et al.* Survey sequencing and comparative analysis of the elephant shark (*Callorhinchus milii*) genome. *PLoS Biol.* 2007; 5:e101.

Venkatesh, B., Lee, A.P., Ravi, V., Maurya, A.K., Lian, M.M., Swann, J.B., Ohta, Y. *et al.* Elephant shark genome provides unique insights into gnathostome evolution. *Nature* 2014; 505:174–179.

Wu, T.T., and Kabat, E.A. An analysis of the sequences of the variable regions of Bence Jones proteins and myeloma light chains and their implications for antibody complementarity. *J. Exp. Med.* 1970; 132:211–250.

Zezza, D.J., Stewart, S.E., and Steiner, L.A. Genes encoding *Xenopus laevis* Ig L chains. Implications for the evolution of kappa and lambda chains. *J. Immunol.* 1992; 149:3968–3977.

CHAPTER 12

Shark T-Cell Receptors

Michael F. Criscitiello

CONTENTS

SUMMARY

Sharks possess the four canonical T-cell receptor (TCR) chains known from other vertebrates: α, β, γ, and δ. The loci encoding these chains employ recombination-activating gene (RAG)-mediated somatic cell V(D)J rearrangement mechanisms for diverse repertoires. Sharks have some additional immunogenetic TCR capacity, including the doubly rearranging NARTCR δ, somatic hypermutation, and trans-rearrangements that use immunoglobulin (Ig) variable segments.

T-CELL RECEPTORS

The cellular arm of the adaptive immune system of sharks depends upon lymphocytes derived from the thymus with hallmark T-cell receptors (TCRs) of the immunoglobulin superfamily. Lymphocytes of the only older vertebrates, the jawless lampreys and hagfish, use a very different receptor system based on leucine-rich repeat domain molecules more akin to toll-like receptors than immunoglobulins (reviewed in Boehm *et al.*, 2012). Thus, together with antibodies and the major histocompatibility complex (MHC) appearing in cartilaginous fish and all more recent vertebrates, but not in the jawless fishes, the existence of TCRs in sharks suggests that the adaptive immune system evolved in cartilaginous fish with the same fundamental major components that exist in warm-blooded vertebrates such as humans (Flajnik and Rumfelt, 2000).

Canonical TCRs are heterodimers of proteins each consisting a variable (V) and constant (C) region. Each of the two protein chains of the TCR contains an amino-terminal immunoglobulin superfamily V domain and an immunoglobulin superfamily C domain (Figure 12.1). TCRs are type I transmembrane glycoproteins, with extracellular V and C domains and a short cytoplasmic tail. Unlike the immunoglobulin that is secreted as antibody and serves as the B-cell receptor, TCRs are always membrane bound. It is with the V domains of both protein chains that interact with ligand. Similar to the B-cell receptor, the short cytoplasmic tails of TCR chains are incapable of transducing a signal, so TCR heterodimers cluster with a complex of accessory molecules called CD3 that recruit cytoplasmic kinases to signal receptor ligation. There are usually only two TCR chain heterodimerization partnerships, αβ and γδ, and individual T cells bear receptors of one or the other form. The ligand of most αβ TCR is peptide antigen in the context of MHC. The γδ TCR binds a more diverse set of poorly characterized ligands and is not restricted to classical antigen presentation by MHC.

A great deal is known about the physiology of mammalian T cells bearing the αβ receptor. These subfunctionalize into CD8+ cytotoxic T cells, CD4+ regulatory T cells, and CD4+ helper (T_H) cells. These T_H can be further classified as inflammatory T_H1 that activate macrophages, T_H2 that stimulate B cells to produce antibodies, and T_H17 that recruit neutrophils to sites of infection (to name but a few). Much less is known about γδ T cells. They develop from a distinct fate decision in the thymus and sometimes recognize antigen directly, in a manner more akin to antibody than the αβ TCR (Allison *et al.*, 2001). Use of the comparative method to query the function of these cells in sharks may provide a clearer picture of the fundamental role of these cells in broader vertebrate immunity.

As we shall see, TCR genes of cartilaginous fish are organized much like TCR genes of other vertebrates. This is not the case with the Ig heavy and light chain gene loci that encode shark

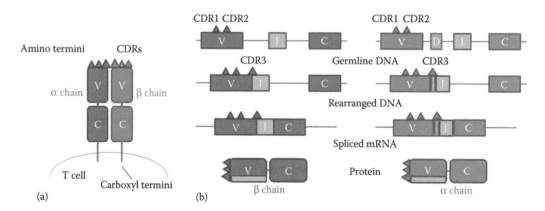

Figure 12.1   Schematic of protein chains in a heterodimeric TCR and the genes that encode them. (a) Domain structure of TCR heterodimer: rounded rectangles represent immunoglobulin superfamily domains and triangles represent complementary determining regions (CDR) that are most variable and articulate most with MHC/peptide antigen. (b) Simplified schematic of genetic elements that combine to form mature TCR gene and cartoons showing what those elements encode in the translated proteins. TCRδ rearrangement is similar to that of TCRβ (including the use of D segment), and TCRγ is similar to that of TCRα.

antibodies. The Ig gene multiple cluster organization (Hinds and Litman, 1986) and frequent germline joining (Kokubu *et al.*, 1988) are deviations from the vertebrate norm. Much of what is known of shark TCR is consistent with their form and function in "higher" vertebrates, but shark TCR has provided some surprising deviations whose functional consequences for their immune system are just starting to be explored.

PIONEERING FUNCTIONAL STUDIES

Early work studying adaptive immunity in sharks yielded neither robust responses nor conclusive results. The first work by Robert Good's group found evidence for memory to allografts in Chondrichthyes (Perey *et al.*, 1968) but a study in the horned shark, *Heterodontus francisci*, by Bill Hildemann's team determined that the response to skin allografts was chronic (Borysenko and Hildemann, 1970; Hildemann, 1970). It has since been suggested that maintaining the animals at 22°C may have marred these results described in the nurse shark, *Ginglymostoma cirratum*. Although the adherent, cold-loving effectors were likely macrophages (Pettey and McKinney, 1983), they may have been regulated by T cells (although MHC restriction was lacking) (Haynes and McKinney, 1991). Now that we are better prepared with immunogenetic data and experimental tools, T-cell function in shark can and should be reexamined.

MOLECULAR WORK YIELDING CANONICAL TCR cDNA SEQUENCES

A series of publications by Gary Litman's group in the 1990s elucidated the expressed TCR of cartilaginous fishes. Then, a student in the lab, Jonathan Rast, devised short, minimally degenerate PCR primers that exploited the conserved WYRQ and YYCA motifs of the framework 2 and framework 3 regions of immunoglobulin light chains and TCR. This approach yielded an amplicon from horned shark genomic DNA that was used as a probe to screen a spleen cDNA library, resulting in the discovery of mature TCRβ homologue transcripts (Rast and Litman, 1994). Four complete and one partial V sequences showed diversity in this domain and the C region sequence clearly showed highest homology to tetrapod TCRβ versus other antigen receptor chains.

A second, more extensive cDNA study of horned shark TCRβ characterized 55 clones from the spleen (Hawke *et al.*, 1996). This work, first-authored by Noel Hawke (yes, brother of actor Ethan), found seven diverse V families, a consensus D sequence, and 18 J sequences used in the TCRβ repertoire of this species. The dataset showed V family multiplicity that was more suggestive of a translocon arrangement of TCR loci for the horned shark, in contrast to the multicluster organization described for shark antibody genes. Diverse length was found in the CDR3 regions formed by the juxtaposition of V, D, and J segments, indicative of both complex shark TCR loci and complex resolution of the coding joints in elasmobranch RAG-mediated recombination (specifically, the existence of nontemplate additions by terminal deoxynucleotidyl transferase (TdT) and removal of bases by exonuclease activity).

This same PCR approach was successful in identifying chicken TCRγ, *Xenopus* TCRα, pufferfish TCRα, and horned shark TCRδ from genomic and cDNA libraries (Rast *et al.*, 1995). At the time, the authors were conservative in assigning the horned shark sequences to α or δ and left open the possibility that the two loci had yet to diverge to form separate chains in the cartilaginous fish. Time would tell, however, that their suggestion that these sequences encoded TCRδ was indeed correct.

Rast's last publication while in Litman's group definitively showed that cartilaginous fish share all four TCR chains found in other vertebrates. In the clearnose skate, *Raja eglanteria*, he refined his minimally degenerate framework 2 and 3 PCR primers to clone α, β, γ, and δ from this batoid elasmobranch (Rast et al., 1997), in which lymphoid tissue architecture and development had already been defined by Carl Luer's group (Luer et al., 1995). The sequences described in this work placed α, β, γ, and δ TCR at the dawn of jawed vertebrate adaptive immunity, suggesting chains of similar structure to their mammalian counterparts and similar repertoire diversity. This manuscript described complete cDNAs for each of the four chains as well as representative V and J gene segment and junctional diversity.

Thirteen years passed after Rast's *Immunity* paper before the characterization of TCR α, β, γ, and δ was completed in a shark. All four chains were identified in the nurse shark (by four different approaches) detailing shark sequences and repertoire diversity that are fundamentally similar to that found in the skate, and with a couple of additional surprises (Criscitiello et al., 2010) that are described in section "Genomic Organization".

GENOMIC ORGANIZATION

In Rast and Litman's first horned shark TCR study, genomic Southern blots with V and C probes showed much concordance in hybridizing bands, though not absolutely (Rast and Litman, 1994). These data were seen as consistent with the existence of multiple V-C loci in the shark genome, an organization akin to that found for the immunoglobulin genes. Screening of a horned shark genomic phage library with these V and C probes identified nearly 300 hybridizing clones. Twelve unique V-containing clones were restriction mapped, eight contained two Vs, and 11 hybridized to a J probe (Rast and Litman, 1994). Twelve C clones were similarly mapped and 10 found to be unique, five bearing J sequences. This same publication described genomic V genes lacking the intron-split leader peptide typical of vertebrate antigen receptor genes but found in chicken TCRα (Gobel et al., 1994).

Deeper cDNA analysis of horned shark TCRβ made the multicluster organization less obvious. Multiple, diverse V families and many J sequences with relative uniformity in C sequences suggested the combinatorial mechanisms of the translocon genomic organization (Hawke et al., 1996). This led to the conclusion that TCR genes of sharks may be more similar between sharks and men than immunoglobulin genes. This latter idea was corroborated by work in the skate and nurse shark, where genomic Southern blotting with constant region probes yielded at most two bands and often one for any TCR chain, again suggesting translocon organization (Criscitiello et al., 2010; Rast et al., 1997).

The translocon genomic organization was definitively confirmed for the sandbar shark, *Carcharhinus plumbeus*, TCRγ locus by the group of Jack Marchalonis (Chen et al., 2009). The 32 kb locus containing five Vs, three Js, and a C gene is about one-fifth the size of the human locus, which contains more Vs, Js, and nonfunctional V pseudogenes. The locus was assembled from a combination of PCR products and chromosomal walking fragments. The careful assembly of this TCRγ locus in the sandbar shark set the stage for a major immunogenetic discovery that would depend upon a well-defined germline genomic sequence.

GENERATION OF DIVERSITY

The trailblazing horned shark TCR work identified canonical recombination signal sequences (RSS) flanking germline genomic elements used in the construction of mature variable domain encoding exons (Rast and Litman, 1994). V segments had 23-bp-spaced RSS 3′ of their coding sequences and J genes had 12-bp-spaced RSS 5′ of their coding sequence. It is predicated that

D elements flanked by 5′ 12-bp-spaced and 3′ 23-bp-spaced RSS could be contributing to the horned shark TCRβ diversity at CDR3.

The clonal selection theory requires a single antigen receptor specificity for a given lymphocyte, though we know this is not a steadfast rule (Brady *et al.*, 2010). Much work was performed by Ellen Hsu's group elucidating isotypic and allelic exclusion at B-cell receptor loci in shark (Malecek *et al.*, 2005, 2008; Zhu *et al.*, 2011), yet comparatively little is understood of these mechanisms at the shark TCR loci. At the antibody genes of shark (as at those genes of chicken, rabbit, and other vertebrates), there does not appear to be the need for the rigid, sequential, stepwise process for maintenance of isotypic and allelic exclusion at the cellular level yet a diverse repertoire in the organism (Hsu, 2009).

The group of Jack Marchalonis identified both recombination-activating genes RAG1 and RAG2 in the sandbar shark. Interestingly, in sharks, as in humans, the RAG lack introns and are closely linked (Bernstein *et al.*, 1994, 1996; Schluter and Marchalonis, 2003). This is consistent with their origin in the vertebrate adaptive immune system through horizontal transfer as the transposase system of a prokaryotic transposon, with RSS evolving from the terminal repeats of the ancestral transposon.

Nontemplate nucleotides are found at the V(D)J junctures of rearranged B- and T-cell receptor CDR3 regions. TdT, the family X polymerase responsible for these additions, was identified in both shark and skate (Bartl *et al.*, 2003). In elasmobranchs, as in bony vertebrates, the predicted structure of TdT suggests both a lack of substrate nucleotide specificity and template independence. Phylogenetically, the TdT of cartilaginous fish appears to be more akin to ancient polymerases than to polymerases lambda and beta.

The outcome of RAG-mediated V(D)J recombination with CDR3-extending TdT action is a diverse shark repertoire for all four TCR chains (Criscitiello *et al.*, 2010). The range of CDR3 lengths are larger in the limited shark sampling than those seen for mouse and man, suggesting great diversity in this most important loop for antigen recognition. Significant diversity is also added to the TCR paratope at CDR1 and CDR2 by diverse V genes, and these V genes display trans-species evolutionary maintenance (Criscitiello *et al.*, 2010), as was noted for MHC alleles (Klein, 1987).

Our understanding of shark TCR repertoire development, diversity, and dynamics is still very much in its infancy, but two distinct themes are beginning to emerge. As detailed thus far in this chapter, the first is that sharks are capable of making αβ and γδ TCR similar to "higher" vertebrates with just as much diversity and employing similar immunogenetic mechanisms of rearrangement. The second theme, to be described in the next three sections, is that shark TCR loci perform some extraordinary feats very much at odds with mouse-centric immune dogma.

NARTCR

Nurse shark TCR characterization work in the lab of Martin Flajnik revealed a subset of unheralded, longer TCR δ cDNA sequences by 5′ RACE (rapid amplification of cDNA ends) PCR (Criscitiello *et al.*, 2006). These sequences encoded two V domains and a TCRδ C domain, which at the time was the first occurrence of a lymphocyte antigen receptor chain containing more than one V domain. Further inquiry determined that about 20% of nurse shark TCRδ rearrangements held to this longer, two-V form. Some TCRδV gene families were always found to have a canonical leader peptide and existed in mature cDNAs with only one V and TCRδ C, whereas other TCRδV gene families were exclusively found as the membrane proximal V domain in the larger TCRδ products, supporting an additional membrane distal V domain. The additional membrane distal V domain encoding 5′ of the TCRδV was always closely related to the V domains of IgNAR, an immunoglobulin heavy chain peculiar to the cartilaginous fish that does not heterodimerize with light chain (Greenberg *et al.*, 1995). Thus, the special TCRδ chains employing the additional V domain were given the moniker of NARTCR. This was satisfying as the hypothesized NARTCR δ is expected to heterodimerize with a TCRγ chain that lacks the additional domain, and therefore, both the IgNAR V domain and the NARTCR δ V domain would be expected to lack a pairing partner.

Both V domains in the NARTCR δ chain are the products of RAG-mediated V(D)J recombination. Genomic DNA sequencing by long-range PCR identified several blocks of NARTCRV-NARTCRD-NARTCRJ-TCRδV separated by short introns of ~300 bp (Criscitiello *et al.*, 2006). Importantly, there is no split leader peptide encoded between the NARTCRJ and supporting TCRδV, yet standard GT/AG intron splicing does connect the J encoding one V domain to the V segment encoding the other at the RNA level. Mature NARTCR would always be constructed from the NARTCR V-D-J and supporting TCRδV all from one of these genomic blocks, whereas the J used in the supporting TCR V domain could be from any of the ~30 that also are used in rearrangements to canonical (non-NARTCR supporting) TCRδV. Adding further diversity, the CDR3 regions of both the NARTCRV and the supporting TCRV were diversified with palindromic and N nucleotide additions. Thus, a model emerged where V(D)J rearrangement of the membrane and C domain proximal domain encoding V is instructive for the generation of a canonical (one V) TCRδ or a NARTCRδ (with two V). If this initial rearrangement selects a canonical TCRδV (with a leader peptide 5′ of it) a canonical TCRδ will result. If the initial rearrangement selects a supporting TCRδV (with a NARTCRV-D-J block 5′ of it) that block will subsequently rearrange (or has concurrently rearranged) and will result in the NARTCRδ.

Data from the elephant shark, *Callorhinchus milli*, genome project suggests that NARTCR has similar diversity and genomic organization in that more ancient Holocephalian fish (Venkatesh *et al.*, 2007), which suggests that NARTCR may be widespread in the cartilaginous fish. Curiously, TCR with two V domains is now known not to be restricted to cartilaginous fish. Nonplacental mammals have a fifth TCR chain called TCRμ that appears to have converged on a similar, two V domain structure as shark NARTCRδ (reviewed in Miller, 2010). Rob Miller's group has elucidated the evolution of this fifth TCR chain in early mammals that is most closely related to TCRδ and uses V domains more akin to those of the IgH locus. In marsupials, the supporting V is germline-joined yet in the monotreme platypus it diversifies as in shark NARTCR (Parra *et al.*, 2007, 2012).

TRANS-REARRANGEMENTS

The Flajnik group's characterization of nurse shark TCR also unearthed a second unusual subset of TCRδ chain rearrangements. These use Ig heavy chain V segments RAG-rearranged with TCRδ D and J, and spliced to C of TCR δ (Criscitiello *et al.*, 2010). The first of these cDNAs was isolated by Rebecca Lohr in an IgWV library screen (IgW is the shark orthologue of IgD) (Ohta and Flajnik, 2006), when she found TCRδ with a clone of IgWV-TCRδD-J-C from neonatal shark. These trans-rearrangements between immunoglobulin and TCR antigen receptor locus elements employ only a few V genes from IgM and IgW, never the IgNAR that shun Ig light chains. Since that publication, we have characterized many such chimeric rearrangements from thymus, spleen, and spiral valve of adult sharks. The CDR3 encoding regions of these sequences are almost always in frame, at a much higher frequency than if they were not being selected for functional protein products on the surface of T cells. Quantitative real-time PCR suggests that these chimeric Ig-TCR mRNAs are being expressed with a tissue distribution consistent with T-cell expression (most signal in thymus) and at a level less than but on the same order of magnitude as the canonical TCRδ rearrangements using TCRδV.

At the same time the trans-rearrangements were discovered in shark, the Miller group at the University of New Mexico described the amphibian *Xenopus* TCRα/δ locus, finding that it contains IgH V genes that are used in TCRδ rearrangements (Parra *et al.*, 2010). Subsequent studies have unearthed seemingly related combinations of IgHV and the TCRδ loci in birds (Parra and Miller, 2012; Parra *et al.*, 2012), platypus (Parra *et al.*, 2012), and coelacanth (Amemiya *et al.*, 2013). Although more shark genomic antigen receptor locus characterization is required to understand if the Ig/TCR trans-rearrangements in shark are orthologous or convergent with the IgHV use in the TCRδ loci of Sarcopterygii (lobe-finned fish and tetrapods), it is intriguing to think that vertebrates old and (relatively) new employ antibody V gene segments in their TCRδ repertoires.

SOMATIC HYPERMUTATION

Shark B cells also use the activation-induced cytidine deaminase (AID) (Conticello *et al.*, 2005) in what Tonegawa called the "fourth somatic diversifier" after gene segment combinatorial diversity, imprecise coding ends, and nontemplate additions: somatic hypermutation (SHM) (Tonegawa, 1983). In addition to SHM in all jawed vertebrate immunoglobulins, AID was now implicated in the processes of heavy chain class switch recombination in sharks (Zhu *et al.*, 2012) as well as tetrapods, and immunoglobulin gene conversion in birds and mammals as well (Barreto and Magor, 2011).

The notion of SHM at TCR loci was recently given credence by work in the sandbar shark driven by Sam Schluter. Sequencing of the entire TCRγ translocon in *Carcharhinus plumbeus* allowed definitive determination that SHM is occurring at that locus (Chen *et al.*, 2009). The shark TCRγ SHM occurs in two distinct patterns: point mutations and tandem mutations characteristic of SHM in cartilaginous fish (Anderson *et al.*, 1995; Lee *et al.*, 2002; Zhu and Hsu, 2010), possibly suggesting two different mechanisms (Chen *et al.*, 2012). The sandbar shark analysis found targeted nucleotide motifs of AID activity at the TCRγ locus and evidence for SHM being used for repertoire diversification rather than affinity maturation. Our study of TCR expression in the nurse shark showed some evidence for SHM in that species, in both the γ and α chains (Criscitiello *et al.*, 2010), although it was not recognized as such before Schluter's seminal discovery. This preliminary nurse shark mutation data mandated a more rigorous analysis of the TCRα expression in our nurse shark model.

Additional cloning and sequencing of TCR from multiple sharks found evidence for somatic hypermutation not only in the TCRα sequences isolated from peripheral lymphoid tissues such as spleen and spiral valve, but in the thymus as well. In the absence of a fully assembled TCRα locus, somatic hypermutation was confirmed by mutated sequences with the same CDR3 rearrangement. Interestingly for the nurse shark TCRα set, more mutations are found in the framework regions compared to the CDR regions, yet more tandem mutations are found in the CDRs. There are more nonsynonymous mutations that change the amino acid coded for by a codon than synonymous ones, and more of these nonsynonymous mutations in the CDR compared to the framework regions. We found no transition, transversion, or particular nucleotide bias in the mutations and nearly no evidence of mutation at TCRβ, TCRδ nor in the constant region exons of these clones.

We were curious if this SHM was due to AID activity in the shark thymus and found evidence for just that through quantitative PCR and *in situ* hybridization (nurse shark AID sequence kind gift of Ellen Hsu). Intriguingly, Niels Jerne suggested over 40 years ago that the thymus was a "mutant-breeding organ," in which mutation was used to modify overly self-reactive cells to generate both self-tolerance and diversity in repertoire specificity (Jerne, 1971). Much work remains to determine definitively whether TCR SHM is used for repertoire diversification, rescue for passage of thymic selection, affinity maturation in the periphery, or something entirely novel altogether.

THYMUS AND T-CELL DEVELOPMENT

In the clearnose skate, real-time PCR showed expression of all four TCR chains in the thymus of eight-week-old embryos, then an apparent shut down of that thymic expression after hatching (Miracle *et al.*, 2001). Northern blotting of adult nurse shark with TCR C region probes gave a more expected robust signal (Criscitiello *et al.*, 2010). *In situ* hybridization experiments in nurse shark thymus generally showed great conservation of thymic architecture and gene expression between shark and mammal. RAG1 and TdT were strongest in the subcapsular region and expressed throughout the cortex, indicating active V(D)J recombination and junctional diversification there. MHC class I and II were generally expressed in the medulla yet more sparsely in the cortex. TCRα and β were brightest in the central cortex but weaker in the subcapsular region and medulla, γ and δ were also in the cortex but brighter in the medulla, with delta having the greatest medullary expression of the four. There is evidence for greater γ and δ expression relative to α and β in shark thymus versus that of mammals (Criscitiello *et al.*, 2010).

PERIPHERAL EXPRESSION AND FUNCTION

In the clearnose skate, both northern blotting and real-time PCR were used to track antigen receptor, TdT, and RAG expression in different tissues and developmental stages (Miracle *et al.*, 2001). At eight weeks of development, all four TCR genes appear to be expressed, β perhaps a week earlier and Δ in a bell curve of expression over several weeks. This expression is exclusively in the thymus in the eight-week-old embryo but showed diverse, chain-specific expression profiles in the spleen, intestine, and liver of hatchling skates. In addition to thymus and spleen, TCR was expressed heavily in rectal gland, intestine, and liver of adult skate.

Adult nurse shark relative TCR expression in peripheral tissues by northern blotting was generally consistent among the four chains: spleen > spiral valve > gill > peripheral blood leukocytes > pancreas > liver. Epigonal organ, ovary, and brain were negative (Criscitiello *et al.*, 2010).

No data are available describing actual function of these receptors in shark cell systems or mammalian transfections. It is of note that a glutamine (position 136 in human) in the TCRβ-connecting peptide, thought to be crucial for signaling efficiency, is present in shark, skate, and mammal but generally not in bony fish, amphibians, reptiles, and birds (Backstrom *et al.*, 1996).

COMPARATIVE GENOMICS

The recent publication of the elephant shark, *Callorhinchus milii*, genome gives much insight into the evolution of TCR loci and T-cell biology in the holocephalans and broader cartilaginous fish as well (Venkatesh *et al.*, 2014). Major findings of the elephant shark genome include the close linkage of Ig and TCR genes, NARTCR yet no IgW or IgNAR, and no canonical CD4 and evidence for a primitive, very limited helper T-cell capacity. These results are consistent with the trans-rearrangements seen in nurse shark being enabled by the close proximity of immunoglobulin V genes and TCR D segments. Additionally, they suggest that the NAR V domain was originally a TCR component that was secondarily co-opted for B-cell receptor use in evolution. However, the distinction between T- and B-cell receptor immunogenetics appears to have been weaker 450 million years ago and in extant cartilaginous fish.

EVOLUTION

When all four TCR chains were found in the skate without obvious multicluster or germline-joined genes, the authors suggested the possibility of a longer or more stable evolutionary history of TCR genes compared to those of Ig (Rast *et al.*, 1997). I like this notion, as vertebrates have evolved myriad Ig isotypes (I think it is likely that at least IgM, IgD, IgT, IgA, IgY, IgG, and IgE are distinct classes) and divergent organs for B-cell development (Leydig organ, epigonal, bursa, bone marrow, fetal liver, ileal Peyers patch, anterior kidney, and appendix are all primary B-cell tissues in some species), yet it appears most have held fast to α, β, γ, and δ TCR developing in the thymus (save some recent notable developments). That should be one take-home message for the evolution of vertebrate TCR, and it is anchored by shark data as the oldest group with TCR. However, sharks have also introduced a second theme in TCR evolution that may be able to employ some antigen receptor immunogenetic tricks thought only to be associated with B cells. The doubly rearranging NARTCR with its presumably partnerless V domain paratope, SHM at TCR loci, and the use of IgH V in TCRδ are all exciting findings that will help us reevaluate the physiological boundaries of TCR and T-cell physiology in mammals (Figure 12.2).

AID is a member of the APOBEC family of nucleic acid mutators, some of which were found to diversify the variable lymphocyte receptor (VLR) system in the more ancient vertebrate lineages of lamprey and hagfish (Guo *et al.*, 2009). There is an emerging connection between the use of AID as a diversifying agent for adaptive immunity now in shark B- and T-cell repertoires with APOBEC family members being used to diversify the older VLR system in lamprey and hagfish (Rogozin *et al.*, 2007). More functional work in shark may shed light on the relative importance of RAG versus AID-mediated diversification mechanisms for the genesis of the gnathastome adaptive immune system.

We have every reason to believe that the V genes of αβ T cells have evolved with MHC for 450 million years (Yin *et al.*, 2012), as sharks have polygenic, polymorphic MHC (Kasahara *et al.*, 1992). Shark MHC/TCR physiology may help answer the chicken-egg conundrum of how the system began (Kurosawa and Hashimoto, 1997). Although a smoking gun, nonrearranging receptor gene that was clearly the recipient of a RAG transposon invasion one half billion years ago has yet to be found (Rast and Litman, 1998), studies of TCR in sharks and other lower chordates will continue to elucidate the natural history of man's immune system and expose new potential paths of clinical intervention for its future.

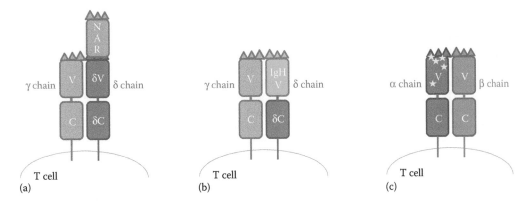

Figure 12.2 Three immunogenetic tricks available to shark T-cell receptors that depart from the mouse/human norm. (a) NARTCR extends the TCRδ chain with a second somatically rearranged V domain supported by special TCRδV domains. (b) Trans-rearrangements allow TCRδ to draw from Ig heavy chain V segments to create V domains on T cells that are largely the product of antibody genes (the D and J are contributed by δ gene segments). Mounting evidence suggests that many vertebrate groups have a similar capability. (c) Somatic hypermutation, known as a B-cell phenomenon, acts upon the light chains of in sharks TCR, α and γ.

ACKNOWLEDGMENTS

The author thanks Jonathan Rast and Sam Schluter for review of the manuscript, Thaddeus C. Deiss and Natalie Jacobs for helpful discussions, and the National Science Foundation (IOS1257829) for their support.

REFERENCES

Allison, T.J., Winter, C.C., Fournie, J.J., Bonneville, M., and Garboczi, D.N. Structure of a human gammadelta T-cell antigen receptor. *Nature* 2001; 411:820–824.

Amemiya, C.T., Alfoldi, J., Lee, A.P., Fan, S.H., Philippe, H., MacCallum, I., Braasch, I. *et al.* The African coelacanth genome provides insights into tetrapod evolution. *Nature* 2013; 496:311–316.

Anderson, M.K., Shamblott, M.J., Litman, R.T., and Litman, G.W. Generation of immunoglobulin light chain gene diversity in *Raja erinacea* is not associated with somatic rearrangement, an exception to a central paradigm of B cell immunity. *J. Exp. Med.* 1995; 182:109–119.

Backstrom, B.T., Milia, E., Peter, A., Jaureguiberry, B., Baldari, C.T., and Palmer, E. A motif within the T cell receptor alpha chain constant region connecting peptide domain controls antigen responsiveness. *Immunity* 1996; 5:437–447.

Barreto, V.M., and Magor, B.G. Activation-induced cytidine deaminase structure and functions: A species comparative view. *Dev. Comp. Immunol.* 2011; 35:991–1007.

Bartl, S., Miracle, A.L., Rumfelt, L.L., Kepler, T.B., Mochon, E., Litman, G.W., and Flajnik, M.F. Terminal deoxynucleotidyl transferases from elasmobranchs reveal structural conservation within vertebrates. *Immunogenetics* 2003; 55:594–604.

Bernstein, R.M., Schluter, S.F., Bernstein, H., and Marchalonis, J.J. Primordial emergence of the recombination activating gene 1 (RAG1): Sequence of the complete shark gene indicates homology to microbial integrases. *Proc. Natl. Acad. Sci. USA* 1996; 93:9454–9459.

Bernstein, R.M., Schluter, S.F., Lake, D.F., and Marchalonis, J.J. Evolutionary conservation and molecular cloning of the recombinase activating gene 1. *Biochem. Biophys. Res. Commun.* 1994; 205:687–692.

Boehm, T., McCurley, N., Sutoh, Y., Schorpp, M., Kasahara, M., and Cooper, M.D. VLR-based adaptive immunity. *Annu. Rev. Immunol.* 2012; 30:203–220.

Borysenko, M., and Hildemann, W.H. Reactions to skin allografts in the horn shark, *Heterodontis francisci*. *Transplantation* 1970; 10:545–551.

Brady, B.L., Steinel, N.C., and Bassing, C.H. Antigen receptor allelic exclusion: An update and reappraisal. *J. Immunol.* 2010; 185:3801–3808.

Chen, H., Bernstein, H., Ranganathan, P., and Schluter, S.F. Somatic hypermutation of TCR gamma V genes in the sandbar shark. *Dev. Comp. Immunol.* 2012; 37:176–183.

Chen, H., Kshirsagar, S., Jensen, I., Lau, K., Covarrubias, R., Schluter, S.F., and Marchalonis, J.J. Characterization of arrangement and expression of the T cell receptor gamma locus in the sandbar shark. *Proc. Natl. Acad. Sci. USA* 2009; 106:8591–8596.

Conticello, S.G., Thomas, C.J., Petersen-Mahrt, S.K., and Neuberger, M.S. Evolution of the AID/APOBEC family of polynucleotide (deoxy)cytidine deaminases. *Mol. Biol. Evol.* 2005; 22:367–377.

Criscitiello, M.F., Ohta, Y., Saltis, M., McKinney, E.C., and Flajnik, M.F. Evolutionarily conserved TCR binding sites, identification of T cells in primary lymphoid tissues, and surprising trans-rearrangements in nurse shark. *J. Immunol.* 2010; 184:6950–6960.

Criscitiello, M.F., Saltis, M., and Flajnik, M.F. An evolutionarily mobile antigen receptor variable region gene: Doubly rearranging NAR-TcR genes in sharks. *Proc. Natl. Acad. Sci. USA* 2006; 103:5036–5041.

Flajnik, M.F., and Rumfelt, L.L. The immune system of cartilaginous fish. *Curr. Top. Microbiol. Immunol.* 2000; 248:249–270.

Gobel, T.W., Chen, C.L., Lahti, J., Kubota, T., Kuo, C.L., Aebersold, R., Hood, L., and Cooper, M.D. Identification of T-cell receptor alpha-chain genes in the chicken. *Proc. Natl. Acad. Sci. USA* 1994; 91:1094–1098.

Greenberg, A.S., Avila, D., Hughes, M., Hughes, A., McKinney, E.C., and Flajnik, M.F. A new antigen receptor gene family that undergoes rearrangement and extensive somatic diversification in sharks. *Nature* 1995; 374:168–173.

Guo, P., Hirano, M., Herrin, B.R., Li, J., Yu, C., Sadlonova, A., and Cooper, M.D. Dual nature of the adaptive immune system in lampreys. *Nature* 2009; 459:796–801.

Hawke, N.A., Rast, J.P., and Litman, G.W. Extensive diversity of transcribed TCR-beta in phylogenetically primitive vertebrate. *J. Immunol.* 1996; 156:2458–2464.

Haynes, L., and McKinney, E.C. Shark spontaneous cytotoxicity: Characterization of the regulatory cell. *Dev. Comp. Immunol.* 1991; 15:123–134.

Hildemann, W.H. Transplantation immunity in fishes: Agnatha, Chondrichthyes and Osteichthyes. *Transplant. Proc.* 1970; 2:253–259.

Hinds, K.R., and Litman, G.W. Major reorganization of immunoglobulin VH segmental elements during vertebrate evolution. *Nature* 1986; 320:546–549.

Hsu, E. V(D)J Recombination: Of mice and sharks. *Adv. Exp. Med. Biol.* 2009; 650:166–179.

Jerne, N.K. The somatic generation of immune recognition. *Eur. J. Immunol.* 1971; 1:1–9.

Kasahara, M., Vazquez, M., Sato, K., McKinney, E.C., and Flajnik, M.F. Evolution of the major histocompatibility complex: Isolation of class II A cDNA clones from the cartilaginous fish. *Proc. Natl. Acad. Sci. USA* 1992; 89:6688–6692.

Klein, J. Origin of major histocompatibility complex polymorphism: The trans-species hypothesis. *Hum. Immunol.* 1987; 19:155–162.

Kokubu, F., Litman, R., Shamblott, M.J., Hinds, K., and Litman, G.W. Diverse organization of immunoglobulin VH gene loci in a primitive vertebrate. *EMBO J.* 1988; 7:3413–3422.

Kurosawa, Y., and Hashimoto, K. How did the primordial T cell receptor and MHC molecules function initially? *Immunol. Cell Biol.* 1997; 75:193–196.

Lee, S.S., Tranchina, D., Ohta, Y., Flajnik, M.F., and Hsu, E. Hypermutation in shark immunoglobulin light chain genes results in contiguous substitutions. *Immunity* 2002; 16:571–582.

Luer, C., Walsh, C.J., Bodine, A.B., Wyffels, J.T., and Scott, T.R. The Elasmobranch Thymus: Anatomical, histological, and preliminary functional characterization. *J. Exp. Zool.* 1995; 273:342–354.

Malecek, K., Brandman, J., Brodsky, J.E., Ohta, Y., Flajnik, M.F., and Hsu, E. Somatic hypermutation and junctional diversification at Ig heavy chain loci in the nurse shark. *J. Immunol.* 2005; 175:8105–8115.

Malecek, K., Lee, V., Feng, W., Huang, J.L., Flajnik, M.F., Ohta, Y., and Hsu, E. Immunoglobulin heavy chain exclusion in the shark. *PLoS Biol.* 2008; 6:e157.

Miller, R.D. Those other mammals: The immunoglobulins and T cell receptors of marsupials and monotremes. *Semin. Immunol.* 2010; 22:3–9.

Miracle, A.L., Anderson, M.K., Litman, R.T., Walsh, C.J., Luer, C.A., Rothenberg, E.V., and Litman, G.W. Complex expression patterns of lymphocyte-specific genes during the development of cartilaginous fish implicate unique lymphoid tissues in generating an immune repertoire. *Int. Immunol.* 2001; 13:567–580.

Ohta, Y., and Flajnik, M. IgD, like IgM, is a primordial immunoglobulin class perpetuated in most jawed vertebrates. *Proc. Natl. Acad. Sci. USA* 2006; 103:10723–10728.

Parra, Z.E., Baker, M.L., Schwarz, R.S., Deakin, J.E., Lindblad-Toh, K., and Miller, R.D. A unique T cell receptor discovered in marsupials. *Proc. Natl. Acad. Sci. USA* 2007; 104:9776–9781.

Parra, Z.E., Lillie, M., and Miller, R.D. A model for the evolution of the Mammalian T-cell receptor alpha/delta and mu Loci based on evidence from the Duckbill Platypus. *Mol. Biol. Evol.* 2012; 29:3205–3214.

Parra, Z.E., and Miller, R.D. Comparative analysis of the chicken TCR alpha/delta locus. *Immunogenetics* 2012; 64:641–645.

Parra, Z.E., Mitchell, K., Dalloul, R.A., and Miller, R.D. A second TCRdelta locus in Galliformes uses antibody-like V domains: Insight into the evolution of TCRdelta and TCRmu genes in tetrapods. *J. Immunol.* 2012; 188:3912–3919.

Parra, Z.E., Ohta, Y., Criscitiello, M.F., Flajnik, M.F., and Miller, R.D. The dynamic TCRdelta: TCRdelta chains in the amphibian Xenopus tropicalis utilize antibody-like V genes. *Eur. J. Immunol.* 2010; 40:2319–2329.

Perey, D.Y., Finstad, J., Pollara, B., and Good, R.A. Evolution of the immune response. VI. First and second set skin homograft rejections in primitive fishes. *Lab. Invest.* 1968; 19:591–597.

Pettey, C.L., and McKinney, E.C. Temperature and cellular regulation of spontaneous cytotoxicity in the shark. *Eur. J. Immunol.* 1983; 13:133–138.

Rast, J.P., Anderson, M.K., Strong, S.J., Luer, C., Litman, R.T., and Litman, G.W. Alpha, beta, gamma, and delta T cell antigen receptor genes arose early in vertebrate phylogeny. *Immunity* 1997; 6:1–11.

Rast, J.P., Haire, R.N., Litman, R.T., Pross, S., and Litman, G.W. Identification and characterization of T-cell antigen receptor-related genes in phylogenetically diverse vertebrate species. *Immunogenetics* 1995; 42:204–212.

Rast, J.P., and Litman, G.W. T-cell receptor gene homologs are present in the most primitive jawed vertebrates. *Proc. Natl. Acad. Sci. USA* 1994; 91:9248–9252.

Rast, J.P., and Litman, G.W. Towards understanding the evolutionary origins and early diversification of rear-ranging antigen receptors. *Immunol. Rev.* 1998; 166:79–86.

Rogozin, I.B., Iyer, L.M., Liang, L., Glazko, G.V., Liston, V.G., Pavlov, Y.I., Aravind, L., and Pancer, Z. Evolution and diversification of lamprey antigen receptors: Evidence for involvement of an AID-APOBEC family cytosine deaminase. *Nat. Immunol.* 2007; 8:647–656.

Schluter, S.F., and Marchalonis, J.J. Cloning of shark RAG2 and characterization of the RAG1/RAG2 gene locus. *FASEB J.* 2003; 17:470–472.

Tonegawa, S. Somatic generation of antibody diversity. *Nature* 1983; 302:575–581.

Venkatesh, B., Kirkness, E.F., Loh, Y.H., Halpern, A.L., Lee, A.P., Johnson, J., Dandona, N. *et al.* Survey sequencing and comparative analysis of the elephant shark (*Callorhinchus milii*) genome. *PLoS Biol.* 2007; 5:e101.

Venkatesh, B., Lee, A.P., Ravi, V., Maurya, A.K., Lian, M.M., Swann, J.B., Ohta, Y. *et al.* Elephant shark genome provides unique insights into gnathostome evolution. *Nature* 2014; 505:174–179.

Yin, L., Scott-Browne, J., Kappler, J.W., Gapin, L., and Marrack, P. T cells and their eons-old obsession with MHC. *Immunol. Rev.* 2012; 250:49–60.

Zhu, C., Feng, W., Weedon, J., Hua, P., Stefanov, D., Ohta, Y., Flajnik, M.F., and Hsu, E. The multiple shark Ig H chain genes rearrange and hypermutate autonomously. *J. Immunol.* 2011; 187:2492–2501.

Zhu, C., and Hsu, E. Error-prone DNA repair activity during somatic hypermutation in shark B lymphocytes. *J. Immunol.* 2010; 185:5336–5347.

Zhu, C., Lee, V., Finn, A., Senger, K., Zarrin, A.A., Du Pasquier, L., and Hsu, E. Origin of immunoglobulin isotype switching. *Curr. Biol.* 2012; 22(10):872–880.

CHAPTER 13

The *Shark-Family* (Cartilaginous Fish) Immunogenome

Byrappa Venkatesh and Yuko Ohta

CONTENTS

SUMMARY

In this chapter, we propose that cartilaginous fishes, by virtue of their unique phylogenetic position, are valuable models for understanding the origin and evolution of adaptive immune system of jawed vertebrates. Analyses of immune system genes in cartilaginous fishes such as elephant shark, nurse shark, and little skate have suggested that the cartilaginous fishes lack certain *bona fide* components of the adaptive immune system of bony vertebrates. In particular, the paucity of some cytokines and their receptors that define the wide range of helper T-cell lineages in bony vertebrates has indicated that the adaptive immune system of jawed vertebrates evolved in a stepwise manner rather than in a single *big-bang* event, and TH1-restricted responses may have dominated as primordial adaptive immune T cells.

INTRODUCTION

Cartilaginous fishes (the *shark-family*) are the oldest group of living jawed vertebrates (gnathostomes) and a sister group of bony vertebrates. Thus, they are a critical out-group for understanding the evolution and diversity of bony vertebrates including humans. Cartilaginous fishes are divided into two broad groups, the Holocephalans (chimeras) and Elasmobranchs (sharks and rays), which diverged ~420 million years ago (MYA) (Inoue *et al.*, 2010). Elasmobranchs comprise two major groups, the Selachii (sharks) and Batoidea (rays), which diverged ~280 MYA (Inoue *et al.*, 2010).

GENOMICS

Genome sizes of majority of cartilaginous fishes range from 2.9×10^9 bp to 6.8×10^9 bp/haploid genome, which is 1~2 times the size of human's (3.0×10^9 bp/haploid genome), with the exception of some Squaliformes that contain sharks (e.g., angular rough shark) with a much larger genome size (16.6×10^9 bp/haploid genome). On the contrary, Holocelphalan species (e.g., elephant shark and spotted ratfish) generally possess smaller genomes (1.0 and 1.5×10^9 bp/haploid genome, respectively) (data calculated from DNA content in *Animal Genome Size Database* (www .genomesize.com). The chromosome number (karyotype) was analyzed for several cartilaginous fish species and shown to range from $2n = 62$ to $2n = 90$. Interestingly, the number of chromosomes seems to correlate inversely with genome size in Elasmobranch group: species with larger genomes (i.e., Scyliorhinidae) have fewer chromosomes (e.g., $2n = 62$ for *Scyliorhinus canicula*; $5.5 – 7.4 \times 10^9$ bp/haploid genome), whereas species with a smaller genome have more chromosomes (e.g., $2n = 86$ for *Carcharhinus limbatus*; $3.6–4 \times 10^9$ bp/haploid genome) (data compiled in the reference: Asahida *et al.*, 1995). Among holocephalans, the chromosome number was analyzed only in the ratfish, *Hydrolagus colliei*, that contains 29 pairs ($2n = 58$) of dot-like chromosomes resembling microchromosomes in birds (Ohno *et al.*, 1969). It is expected that other holocephalans, such as the elephant shark, contain a similar karyotype.

The human genome project revealed that homologous genes are often found as sets in a similar order on four different chromosomal regions (Kasahara *et al.*, 1996). This finding led to revival of the old theory proposing that, early in vertebrate history, the genome experienced two rounds (1R and 2R) of whole-genome duplication (WGD) (Ohno, 1970). Hox clusters (Pendleton *et al.*, 1993) and MHC (Kasahara *et al.*, 1996) provided the earliest clear examples of WGD. Originally, it was proposed that the first round (1R) of duplication had occurred before the emergence of jawless vertebrates and the second round (2R) before the emergence of jawed vertebrates (Pendleton *et al.*, 1993). However, analysis of the somatic genome of the sea lamprey suggested that both 1R and 2R occurred before the divergence of jawless vertebrates and jawed vertebrates (Smith *et al.*, 2013). The timings of 1R and 2R remains unresolved with the recent findings that the Japanese lamprey and the sea lamprey contain more than six Hox clusters (Mehta *et al.*, 2013), suggesting that the lamprey lineage has undergone three rounds of WGD, and not two as previously thought. In any case, because cartilaginous fishes like the elephant shark which contains four Hox clusters (Ravi *et al.*, 2009), the two rounds of duplications clearly occurred before the divergence of cartilaginous fishes and bony vertebrates. Thus, we can expect, overall, a similar genome organization and genes in cartilaginous fishes and bony vertebrates.

Many teleost fish species have evolved rapidly, and thus, their genomes are highly derived (Ravi and Venkatesh, 2008), at least partially attributable to a third WGD that occurred in a teleost fish ancestor (Meyer and Málaga-Trillo, 1999). Synteny of MHC genes is not well conserved in most teleost fishes, as class I, class II, and class III genes are dispersed onto many chromosomal regions (Flajnik *et al.*, 1999), suggestive of a highly diverged and modified genome. In contrast, synteny among MHC class I, class II, and class III genes (e.g., *C4* and *Bf*) is highly conserved in nurse shark, *Ginglymostoma cirratum*, and banded houndshark, *Triakis scyllium* (Ohta *et al.*, 2000; Terado *et al.*, 2003). Thus, cartilaginous fish appear to be much more appropriate as model species than teleost fish to infer the ancestral vertebrate genome. Furthermore, because genomes of cartilaginous fishes seem to be evolving slowly (Renz *et al.*, 2013), one could expect that these species still possess some of the primordial features that can be used to infer the ancestral state of the vertebrate genome. One such example is the linkage of beta-2 microglobulin (B2M) to the shark MHC (Ohta *et al.*, 2011). B2M is a member of the immunoglobulin superfamily, which is shared with MHC class I, and it associates with MHC class I. In addition, the *B2M* gene demonstrates inconsistent synteny in various species, and it was speculated that *B2M* and MHC class I were derived from a common ancestor by tandem duplication and translocated multiple times or independently during vertebrate evolution

(Flajnik and Kasahara, 2010). Therefore, mapping of *B2M* to the nurse shark MHC resolved this long-lasting conjecture and proved that shark genome maintained this primordial feature. These examples highlight the cartilaginous fish as a useful model for comparative analysis of gnathostome genomes.

In contrast to the completed genome sequences of several bony vertebrates, the genome sequence of only one cartilaginous fish, the elephant shark, *Callorhinchus milii*, was completed (Venkatesh *et al.*, 2014). In addition, a few other cartilaginous fish genomes such as the little skate, *Leucoraja erinacea*, are in the pipeline and expected to be completed in the near future (Bernardi *et al.*, 2012; Wang *et al.*, 2012). In section "Immunogenome," we will discuss the immunogenome based on the elephant shark genome in conjunction with new and previous work using various other cartilaginous fish species.

IMMUNOGENOME

As mentioned earlier, the elephant shark belongs to the subclass Holocephali (chimaera), the sister group of Elasmobranchii. From genome sequencing, it was estimated to contain 910 Mbp haploid genome, which is approximately one-third the size of the human genome (Venkatesh *et al.*, 2014). It has a GC content of 42.3% and a repeat content of 30%. A total of 17, 449 protein-coding genes were predicted in the assembly. A major finding of the whole-genome analysis of the elephant shark is that it is the slowest evolving genome among all the sequenced vertebrate genomes, including that of the coelacanth, which is commonly known as a *living fossil* because of its slow evolutionary rate (Amemiya *et al.*, 2013).

The genome assembly of the elephant shark (see http://esharkgenome.imcb.a-star.edu.sg/) is of a high quality with an N50 scaffold size of 4.5 Mbp. In fact, it includes 186 scaffolds that are larger than 1 Mbp and 13 scaffolds that are larger than 10 Mbp. The long contiguity of the assembly has allowed comparison of long-range synteny of genes in elephant shark and other sequenced vertebrate genomes. In general, the gene synteny is highly conserved with that of the chicken and human genomes. By contrast, the synteny of teleost fish genomes such as zebrafish and medaka shows a lower degree of conservation with elephant shark and human genomes (Venkatesh *et al.*, 2014). Among them, a close linkage of immunoglobulin (Ig) and T-cell receptor (TCR) genes was found in one scaffold. This is compatible with the hypotheses that the antigen receptor (AgR) (i.e., Ig and TCR) genes evolved from a common ancestor by tandem duplications (Flajnik and Kasahara, 2010). Moreover, some AgR genes are mapped within an MHC paralogous region, supporting the coevolutionary origin of genes involved in antigen recognition and presentation in a primordial immune complex.

Detailed analysis of the elephant shark genome revealed insights into the origin of the immune system of jawed vertebrates. There is a wide array of immune genes involved in both innate and adaptive immunity, consistent with our existing knowledge of cartilaginous fish immunology. Most innate immune genes are present including all complement components, pattern recognition receptors (e.g., toll-like receptors [TLR], Nod-like receptors [NLR]), and interferons. Like in other species such as the sea urchin (Sea Urchin Genome Sequencing Consortium 2006) and lamprey (Smith *et al.*, 2013), there is an expansion of certain pattern recognition receptors (e.g., NLRP3) in the cartilaginous fish genome. One exception is TLR4, which seems to have been lost; only a fragment of TLR4 is present within the conserved synteny around TLR4 in other vertebrates, and most adaptor molecules that associate with TLR4 seem to have been lost. Lack of TLR4 in shark is consistent with the notion of weak immune stimulation induced by lipopolysaccharide (LPS) (Dooley and Flajnik, unpublished).

From previous studies, MHC/AgR-based adaptive immunity is present in sharks (Flajnik, 1998): most of the genes that play roles in adaptive immune system were identified and analyzed. For example, IgM loci were extensively analyzed for many cartilaginous fish species, revealing the number of genes and their genomic organization. Unlike all other jawed vertebrate species that demonstrate the translocon-type organization, shark Ig loci are organized as mini clusters (Hinds and Litman, 1986). The number of clusters differs depending on the species; for example, ~15 for nurse shark (Malecek *et al.*, 2005), but over 100 for horn shark (Kokubu *et al.*, 1988) and predicted to contain a similar number of genes in the little skate (Harding *et al.*, 1990). We found ~10 complete clusters in the elephant shark genome with many more pseudogenes and orphan Vh fragments. The presence of pseudo Vh fragments was also observed in another Holocephalan, the spotted ratfish (Rast *et al.*, 1998). In contrast, TCR organization in cartilaginous fish is similar to that of mammals, having the translocon organization (Rast *et al.*, 1997) and TCRα and δ in the same locus. Interestingly, IgM is the only Ig isotype found in the Holocephalan species: IgW and NAR appeared to have been lost or only emerged in elasmobranchs (mentioned previously in Venkatesh *et al.*, 2007).

Likewise, both MHC class I and class II genes were previously identified from many cartilaginous fishes and have the typical features of MHC, such as domain structure, high level of polymorphism, and evolutionarily conserved amino acid residues (Hashimoto *et al.*, 1999). Although we could not reconstruct the MHC region in the elephant shark genome, both MHC class I and class II genes were present in the genome and transcriptomes. Therefore, we expected both CD4 helper- and CD8 cytotoxic-type T-cell subsets. However, the genes involved in the development, maintenance, and function of many helper T (TH) lineage cells (e.g., TH2, Tfh TH17, Treg) seemed either absent or atypical (transcription factors FoxP3, ThPok, RORγt; cytokines and cytokine receptors IL2, IL2RA, IL4, IL9, IL13, IL17E, IL21, IL25, IL31, IL23, IL23RA; canonical CD4 coreceptor). Furthermore, the expression levels of atypical TH genes were either undetectable or low, confirmed by searching EST databases (i.e., nurse shark [*Ginglymostoma cirratum*], spiny dogfish [*Squalus acanthias*], and small-spotted catshark [*Scyliorhinus canicula*]), suggesting that this feature is not limited to the elephant shark. These data strongly suggest that the shark TH system is not *full-blown* as we know it from tetrapods, and especially, TH2, TH17, and Treg systems are primordial. Thus, the mechanism for selection of helper T-cell lineages seems relatively simple compared to tetrapods and further suggests that the adaptive immune system evolved in a stepwise fashion, that is, cartilaginous fish represent a transitional stage of development of adaptive immunity. One puzzling finding is how MHC class II presents antigen to T cells in the absence of a *bona fide* CD4 coreceptor: Lack of a conventional CD4 gene and CD4-binding residues in shark class II proteins (Dijkstra *et al.*, 2013) suggests that the mechanisms for positive selection and T-cell activation in sharks differ from all other jawed vertebrates. Sharks have a variety of costimulation molecules, both in the tumor necrosis factor (TNF) and B7 families, and thus, costimulation might compensate for the lack of CD4; alternatively, CD8 might bind to both MHC class I and class II molecules or TCR of class II-recognizing T cells might not require a coreceptor. Understanding the mechanism of the MHC class II-antigen presentation system in shark will likely reveal insights into the mammalian TH lineages. Other major genes missing from shark genome are lymphotoxin, LTα, LTβ, and lymphotoxin receptor, LTRβ; the lack of these genes in sharks is consistent with the absence of lymph nodes and germinal centers in cartilaginous fish (Flajnik, 2002).

Restricted helper T-cell functions (exemplified by perhaps a primordial T-follicular helper lineage) in cartilaginous fish is consistent with the requirement for the long response time for antibody responses, especially in affinity maturation (Dooley *et al.*, 2005). In contrast, all genes involved in the mammalian TH1 lineage are present, and it may be the predominant type of immunity in cartilaginous fishes. Thus, the primordial adaptive immune system may have been focused on cytotoxicity and macrophage activation, as well as a primordial *helper* system perhaps geared toward a TH1-type response. So far, there are no obvious shark NK receptors (NKR) of either the C2-type or lectin-type. However, spontaneous cytotoxic cells were observed (Pettey and McKinney, 1983), and a cluster of genes encoding an activating NKR, NKp30, was identified in various sharks (Flajnik *et al.*, 2012). It will be of great interest whether NKp30 is a marker for shark NK cells.

ACKNOWLEDGMENT

We thank Dr. Martin F. Flajnik for reading the draft and giving us insightful advice on this chapter.

REFERENCES

Amemiya, C.T., Alföldi, J., Lee, A.P., Fan, S., Philippe, H., Maccallum, I., Braasch, I. *et al.* The African coelacanth genome provides insights into tetrapod evolution. *Nature* 2013; 496(7445):311–316.

Asahida, T., Ida, H., and Hayashizaki, K. Karyotypes and cellular DNA contents of some Sharks in the order Carcharhiniformes. *Jpn. J. Ichthyol.* 1995; 42(1):21–26.

Bernardi, G., Wiley, E.O., Mansour, H., Miller, M.R., Orti, G., Haussler, D., O'Brien, S.J., Ryder, O.A., and Venkatesh, B. The fishes of Genome 10K. *Mar. Genomics* 2012; 7:3–6. doi:10.1016/j.margen.2012.02.002.

Dijkstra, J.M., Grimholt, U., Leong, J., Koop, B.F., and Hashimoto, K. Comprehensive analysis of MHC class II genes in teleost fish genomes reveals dispensability of the peptide-loading DM system in a large part of vertebrates. *BMC Evol. Biol.* 2013; 13:260.

Dooley, H., and Flajnik, M.F. Shark immunity bites back: Affinity maturation and memory response in the nurse shark, *Ginglymostoma cirratum. Eur. J. Immunol.* 2005; 35(3):936–945.

Flajnik, M.F. Churchill and the immune system of ectothermic vertebrates. *Immunol. Rev.* 1998; 166:5–14. Review.

Flajnik, M.F. Comparative analyses of immunoglobulin genes: Surprises and portents. *Nat. Rev. Immunol.* 2002; 2(9):688–698. Review.

Flajnik, M.F., and Kasahara, M. Origin and evolution of the adaptive immune system: Genetic events and selective pressures. *Nat. Rev. Genet.* 2010; 11:47–59. Review.

Flajnik, M.F., Ohta, Y., Namikawa-Yamada, C., and Nonaka, M. Insight into the primordial MHC from studies in ectothermic vertebrates. *Immunol. Rev.* 1999; 167:59–67. Review.

Flajnik, M.F., Tlapakova, T., Criscitiello, M.F., Krylov, V., and Ohta, Y. Evolution of the B7 family: Co-evolution of B7H6 and NKp30, identification of a new B7 family member, B7H7, and of B7's historical relationship with the MHC. *Immunogenetics* 2012; 64(8):571–590. doi:10.1007/s00251-012-0616-2. Epub 2012 Apr 11. Erratum in: *Immunogenetics* 2013; 65(7):559.

Harding, F.A., Cohen, N., and Litman, G.W. Immunoglobulin heavy chain gene organization and complexity in the skate, *Raja erinacea. Nucleic Acids Res.* 1990; 18(4):1015–1020.

Hashimoto, K., Okamura, K., Yamaguchi, H., Ototake, M., Nakanishi, T., and Kurosawa, Y. Conservation and diversification of MHC class I and its related molecules invertebrates. *Immunol. Rev.* 1999; 167:81–100. Review.

Hinds, K.R., and Litman, G.W. Major reorganization of immunoglobulin VH segmental elements during vertebrate evolution. *Nature* 1986; 320(6062):546–549.

Inoue, J.G., Miya, M., Lam, K., Tay, B.-H., Danks, J.A., Bell, J., Walker, T.I., and Venkatesh, B. Evolutionary origin and phylogeny of the modern holocephalans (Chondrichthyes: Chimaeriformes): A mitogenome perspective. *Mol. Biol. Evol.* 2010; 27:2576–2586.

Kasahara, M., Hayashi, M., Tanaka, K., Inoko, H., Sugaya, K., Ikemura, T., and Ishibashi, T. Chromosomal localization of the proteasome Z subunit gene reveals an ancient chromosomal duplication involving the major histocompatibility complex. *Proc. Natl. Acad. Sci. USA* 1996; 93(17):9096–9101.

Kokubu, F., Litman, R., Shamblott, M.J., Hinds, K., and Litman, G.W. Diverse organization of immunoglobulin VH gene loci in a primitive vertebrate. *EMBO J.* 1988; 7(11):3413–3422.

Malecek, K., Brandman, J., Brodsky, J.E., Ohta, Y., Flajnik, M.F., and Hsu, E. Somatic hypermutation and junctional diversification at Ig heavy chain loci in the nurse shark. *J. Immunol.* 2005; 175(12):8105–8115.

Mehta, T.K., Ravi, V., Yamasaki, S., Lee, A.P., Lian, M.M., Tay, B., Tohari, S. *et al.* Evidence for at least six Hox clusters in the Japanese lamprey (*Lethenteron japonicum*). *Proc. Natl. Acad. Sci. USA* 2013; 110:16044–16049.

Meyer, A., and Málaga-Trillo, E. Vertebrate genomics: More fishy tales about Hoxgenes. *Curr. Biol.* 1999; 9(6):R210–R213. Review.

Ohno, S. *Evolution by Gene Duplication.* New York: Springer-Verlag, 1970.

Ohno, S., Muramoto, J., Stenius, C., Christian, L., Kittrell, W.A., and Atkin, N.B. Microchromosomes in holo-cephalian, chondrostean and holostean fishes. *Chromosoma* 1969; 26(1):35–40.

Ohta, Y., Okamura, K., McKinney, E.C., Bartl, S., Hashimoto, K., and Flajnik, M.F. Primitive synteny of ver-tebrate major histocompatibility complex class I and class II genes. *Proc. Natl. Acad. Sci. USA* 2000; 97(9):4712–4717.

Ohta, Y., Shiina, T., Lohr, R.L., Hosomichi, K., Pollin, T.I., Heist, E.J., Suzuki, S., Inoko, H., and Flajnik, M.F. Primordial linkage of β2-microglobulin to the MHC. *J. Immunol.* 2011; 186(6):3563–3571.

Pendleton, J.W., Nagai, B.K., Murtha, M.T., and Ruddle, F.H. Expansion of the Hox gene family and the evolu-tion of chordates. *Proc. Natl. Acad. Sci. USA* 1993; 90(13):6300–6304.

Pettey, C.L., and McKinney, E.C. Temperature and cellular regulation of spontaneous cytotoxicity in the shark. *Eur. J. Immunol.* 1983; 13(2):133–138.

Rast, J.P., Amemiya, C.T., Litman, R.T., Strong, S.J., and Litman, G.W. Distinct patterns of IgH structure and organization in a divergent lineage of chrondrichthyan fishes. *Immunogenetics* 1998; 47(3):234–245. Erratum in: *Immunogenetics* 1998 April;47(5):415.

Rast, J.P., Anderson, M.K., Strong, S.J., Luer, C., Litman, R.T., and Litman, G.W. Alpha, beta, gamma, and delta T cell antigen receptor genes arose early in vertebrate phylogeny. *Immunity* 1997; 6(1):1–11.

Ravi, V., Lam, K., Tay, B.H., Tay, A., Brenner, S., and Venkatesh, B. Elephant shark (*Callorhinchus milii*) pro-vides insights into the evolution of Hox gene clusters in gnathostomes. *Proc. Natl. Acad. Sci. USA* 2009; 106(38):16327–16332.

Ravi, V., and Venkatesh, B. Rapidly evolving fish genomes and teleost diversity. *Curr. Opin. Genet Dev.* 2008; 18(6):544–550. Review.

Renz, A.J., Meyer, A., and Kuraku, S. Revealing less derived nature of cartilaginous fish genomes with their evolutionary time scale inferred with nuclear genes. *PLoS One* 2013; 8(6):e66400. doi:10.1371/journal. pone.0066400.

Sea Urchin Genome Sequencing Consortium. Sodergren, E., Weinstock, G.M., Davidson, E.H., Cameron, R.A., Gibbs, R.A., Angerer, R.C. *et al.* The genome of the sea urchin *Strongylocentrotus purpuratus.* *Science* 2006; 314(5801):941–952. Erratum in: *Science* 2007 February 9;315(5813):766.

Smith, J.J., Kuraku, S., Holt, C., Sauka-Spengler, T., Jiang, N., Campbell, M.S., Yandell, M.D. *et al.* Sequencing of the sea lamprey (*Petromyzon marinus*) genome provides insights into vertebrate evolution. *Nat. Genet.* 2013; 45(4):415–421, 421e1–421e2.

Terado, T., Okamura, K., Ohta, Y., Shin, D-H., Smith, S.L., Hashimoto, K., Takemoto, T. *et al.* Molecular clon-ing of C4 gene and identification of the class III complement region in the shark MHC. *J. Immunol.* 2003; 171(5):2461–2466.

Venkatesh, B., Kirkness, E.F., Loh, Y.H., Halpern, A.L., Lee, A.P., Johnson, J., Dandona, N. *et al.* Survey sequencing and comparative analysis of the elephant shark (*Callorhinchus milii*) genome. *PLoS Biol.* 2007; 5(4):e101.

Venkatesh, B., Lee, A.P., Ravi, V., Maurya, A.K., Lian, M.M., Swann, J.B., Ohta, Y. *et al.* Elephant shark genome provides unique insights into gnathostome evolution. *Nature* 2014; 505(7482):174–179.

Wang, Q., Arighi, C.N., King, B.L., Polson, S.W., Vincent, J., Chen, C., Huang, H. *et al.* Community annotation and bioinformatics workforce development in concert—Little Skate Genome Annotation Workshops and Jamborees. *Database (Oxford).* 2012 March 20; 2012:bar064. doi:10.1093/database/bar064.

CHAPTER 14

In Vitro Culture of Elasmobranch Cells

Catherine J. Walsh and Carl A. Luer

CONTENTS

SUMMARY

Cell culture has become one of the major tools in contemporary studies of life sciences and has supported research in many areas of cell and molecular biology; it is a complex process in which cells are grown under controlled laboratory conditions outside their natural environment. Culture conditions vary widely for each species and for each cell type, with the culture medium the most important factor for cell growth. Cell culture medium contains amino acids, vitamins, minerals, and carbohydrates and is involved in regulating pH and osmolality. Although conditions for mammalian and teleost cell culture were well established, conditions for elasmobranch cell culture were challenging to establish, mostly as a result of the unique physiological properties of the elasmobranch system. This chapter reviews the history of elasmobranch cell culture, outlines various media formulations, describes functional characteristics of cultured elasmobranch cells, addresses the challenges associated with cultures of somatic and embryonic cell lines, and highlights the prospects for future success.

INTRODUCTION

Elasmobranch fishes (sharks, skates, and rays) appeared in the fossil record more than 400 million years ago, with extant elasmobranch species serving as useful models for comparative vertebrate immunology, cell biology, physiology, and genomics. Contributing to their value as biomedical models is the unique phylogenetic niche occupied by elasmobranch species, where they currently represent the earliest jawed vertebrate animal group to possess all the components necessary for an adaptive immune system (Criscitiello *et al.*, 2010; Luer *et al.*, 1995; Rast *et al.*, 1997).

The successful culture of elasmobranch immune cells in the laboratory is critical for the advancement of comparative immunology research. Appropriate media, reagents, techniques for isolation of cells, and effective methods for culture and maintenance of elasmobranch cells are currently quite limited and represent a significant factor that has hindered progress in understanding elasmobranch immune function. In addition, identification of cell surface markers to confirm efficacy of cell separation techniques has yet to be developed.

Comparative approaches using elasmobranch fishes as animal models in immunology, cell biology, and genomics were limited by availability of *in vitro* models of cell culture (Bewley *et al.*, 2006). Although genomic approaches are definitely informative, cell culture approaches provide experimental verification and functional validation of genomic sequences and thus are being used in greater frequency to complement hypotheses developed through molecular biology (Forest *et al.*, 2007). The ability to assign functional properties to cells in culture has presented significant roadblocks to understanding elasmobranch immune cellular function. Elasmobranch fishes have fallen behind mammalian model systems in development of species-specific cell cultures and cell lines. Currently, there are two reported cell lines from elasmobranch fish that are of somatic origin (Parton *et al.*, 2007), which are described in the following section. No cell lines derived from immune cells were developed for elasmobranch fishes.

REVIEW OF PUBLISHED EFFORTS TO CULTURE ELASMOBRANCH CELLS

Several studies have described early efforts at culturing elasmobranch cells, with varying success. Approaches to isolation and separation of leukocyte populations, a process greatly complicated by nucleated erythrocytes and high osmolarity of blood and tissues of elasmobranch species, are listed in Table 14.1. A number of different media formulations were tested, with adjustment of osmotic conditions a common feature among them. Some of the various media preparations that were used for culturing elasmobranch cells are listed in Table 14.2.

Leukocyte Separation

A major difficulty associated with culturing elasmobranch immune cells was lack of available and effective techniques for separating leukocytes and separation of cell populations from elasmobranch whole blood. Several methods to achieve these separations were attempted, with varying degrees of success. Although methods for separating cell populations are desirable, the properties of elasmobranch peripheral blood cells, including nucleated erythrocytes and high osmolarity of blood, complicate isolation and eliminate the effectiveness of traditional cell separation media. Challenges associated with separating cell types have led to most cell culture efforts with elasmobranch peripheral blood cells utilizing primarily mixed cell populations, except for a few reports of cell culture separating adherent from nonadherent cell populations (McKinney, 1992; Pettey and McKinney, 1983).

Table 14.1   **Published Methods Used to Separate Leukocytes from Elasmobranch Whole Blood**

Species	Cell Types Isolated	Method	References
Ginglymostoma cirratum	Leukocytes from whole blood	Nylon wool filtration	Pettey and McKinney (1983)
Scyliorhinus canicula	Thrombocytes and granulocytes from whole blood	Differential adhesion/ adherence to glass	Mainwaring and Rowley (1985)
Carcharhinus plumbeus Rhizoprionodon porosus Dasyatis americana	Leukocytes from whole blood	Ficoll-Hypaque density gradient followed by nylon wool	Grogan and Lund (1990)
Ginglymostoma cirratum Raja eglanteria	Leukocytes from whole blood and lymphomyeloid tissues	Slow speed centrifugation	Walsh and Luer (1998)
Leucoraja erinacea	Leukocytes from whole blood	Discontinuous gradient separation	Tomana *et al.* (2008)

Table 14.2   **Cell Culture Media Formulations Used for Culture and Functional Assay of Elasmobranch Cells**

Species	Cell Type	Culture Medium	Functional Assay	References
Ginglymostoma cirratum	Peripheral blood leukocytes	Modified RPMI 1640	Cytotoxicity	Pettey and McKinney (1983)
Rhizoprionodon porosus Squalus acanthias	Brain explant	Leibovitz L-15	Cell line development	Garner (1988)
Caracharhinus plumbeus Rhizoprionodon porosus Dasyatis americana	Leukocytes	Modified RPMI 1640	Growth assays, cell proliferation, conditioned media-enhanced growth	Grogan and Lund (1990)
Carcharhinus falciformis	Embryonic heart and brain explants	Modified Opti-MEM I	Cell line development	Poyer and Hartmann (1992)
Ginglymostoma cirratum Raja eglanteria	Leukocytes from peripheral blood and lymphomyeloid tissue	Modified RPMI 1640	Phagocytosis, apoptosis, nitric oxide production	Walsh and Luer (1998); Walsh *et al.* (2002, 2006b)
Squalus acanthias	Embryonic cell line	LDF	MRP3 expression	Kobayashi *et al.* (2007)
Squalus acanthias	Embryonic mesenchymal cells	LDF	Cell line development	Parton *et al.* (2007)

A summary of published approaches to the separation of cell populations from elasmobranch whole blood is listed in Table 14.1. Some success with density gradient centrifugation followed by differential adhesion to separate leukocytes was reported from small-spotted catshark, *Scyliorhinus canicula* (Mainwaring and Rowley, 1985). Using this method, thrombocytes and several types of granulocytes were separated using single-step methods. Separation of other cell types, however, required an additional step, such as adherence to glass, for effective purification. A method to achieve separation of leukocytes from erythrocytes in elasmobranch whole blood using repetitive slow-speed centrifugation was described, in which denser, nucleated erythrocytes fall to the bottom of the tube, whereas the lighter leukocyte population remains suspended in the plasma

(Walsh and Luer, 1998). Other methods include a Ficoll-Hypaque density gradient followed by nylon wool to isolate leukocytes from whole blood of shark and stingrays (Garner, 1988; Grogan and Lund, 1990). Whole blood layered onto a mixture of two parts lymphocyte separation medium (LSM; Bionetics) mixed with one part 0.85% saline, followed by separation by adherence using glass bead columns and solutions adjusted to contain 0.2 M NaCl and 0.35 M urea, and washing nonadherent cells through with EDTA and agitation was somewhat successful in separating cell populations (Nonaka and Smith, 2000; Pettey and McKinney, 1983). In other studies, phagocytic cells were removed from nylon wool columns equilibrated with 0.85% NaCl using iron carbonyl (Greenberg et al., 1975). Adherence to fibronectin was also used for elasmobranch leukocyte separation (McKinney et al., 1986) in a protocol, in which cells were separated with LSM (two parts LSM mixed with 1 part 0.85% saline), adhered to glass, further separated by Percoll density gradient centrifugation, with the final step being adherence to fibronectin. Cells were then fractionated on glass bead columns, with nonadherent cells being washed through and adherent cells eluted with EDTA buffer and agitation. Leukocytes separated by Percoll density gradient centrifugation were further fractioned by adding to fibronectin-coated culture dishes. The population of resulting cells was determined to be comprised of 87%–100% macrophages.

One of the more successful elasmobranch leukocyte separation techniques utilized a shark relative (little skate, *Leucoraja erinacea*). Cells were separated by modifying an existing density gradient separation technique (through addition of sodium chloride to a final osmolarity of 1000 mOsm) to establish a purification procedure based on discontinuous, two-gradient separation (Tomana et al., 2008). This procedure excluded erythrocytes as well as dead cells and debris and yielded leukocytes of high purity and viability, with flow cytometric analysis confirming a 10-fold purification of viable leukocytes. The authors suggest these techniques are amenable to refinement to facilitate further separation of leukocytes into lymphocytes, macrophages, granulocytes, or other subpopulations.

Media Formulations

Attempts at elasmobranch cell culture have involved a number of different media formulations and additives. Nearly all media formulations utilized with elasmobranch cells have adjusted osmolarity of commercially available cell culture media to more closely mimic the physiological osmolarity found in the blood and tissues of elasmobranch species, primarily involving addition of urea, sodium chloride, and trimethylamine oxidase (TMAO) in various combinations. The normal osmolarity of elasmobranch peripheral blood is approximately three times higher than that of mammalian blood, ranging from 935 to 1092 mOsm for those elasmobranch fishes measured (Griffith, 1981; Maren, 1967) compared to 310 mOsm for mammals, most falling within the range of 290 to 305 mOsm (Waymouth, 1970). Some media preparations have included addition of TMAO that occurs in relatively high concentrations in the extracellular fluids of elasmobranchs, ranging from 22 to 99 mM in those species measured (Griffith, 1981; Grogan and Lund 1990).

Elasmobranch immune cells were successfully cultured using elasmobranch-modified RPMI-1640 (E-RPMI) and elasmobranch-phosphate buffered saline (E-PBS) for tissue culture and cell isolation procedures (Luer et al., 1995; Walsh and Luer, 1998; Walsh et al., 2002). E-RPMI is prepared by adding 360 mM urea and 188 mM NaCl to mammalian-formulated RPMI-1640 (Gibco, Grand Island, NY). Penicillin (50 units/ml), streptomycin (50 μg/ml), neomycin (0.1 mg/ml), and amphotericin B (0.25 μg/ml) can be added to control bacterial and fungal growth. Sodium bicarbonate (238 mM) is added for pH balance, and the pH of the medium adjusted to 7.2–7.4. E-PBS is prepared by increasing the salt concentration of normal phosphate-buffered saline to achieve elasmobranch osmolarity. Modifications to other commercially available media have included DMEM to culture immunocytes (Grogan and Lund, 1990) and Leibovitz L-15 to culture shark brain tissue

(Garner, 1988). In addition, numerous media supplements were tried, including fetal bovine serum (FBS) (Walsh *et al.*, 2006b), mixtures of FBS and horse serum (McKinney, 1992), shark serum (McKinney, 1992), nonessential amino acids (Parton *et al.*, 2007), or chemically defined lipids (Parton *et al.*, 2007).

In addition to peripheral blood leukocytes (PBL), successful culture of immune cells from elasmobranch lymphomyeloid tissues was achieved by mincing fresh tissue and culturing as described for PBL using serum-free E-RPMI (Walsh *et al.*, 2006a). Temperatures range, depending on species, with nurse shark cells typically incubated at 25°C (Walsh and Luer, 1998) and cells from more temperate skates incubated at 18°C (little skate, *L. erinacea*) (Parton *et al.*, 2010; Tomana *et al.*, 2008) or 21°C (clearnose skate, *Raja eglanteria*) (Walsh and Luer, 1998).

Functional Assays

Establishing conditions for *in vitro* culture of elasmobranch immune cells has allowed characterization of numerous innate immune functions, including phagocytosis and pinocytosis (Walsh and Luer, 1998), nitric oxide production (Walsh *et al.*, 2006b), nonspecific cytotoxicity (McKinney *et al.*, 1986; Pettey and McKinney, 1981, 1988), complement activation (Nonaka and Smith, 2000), and dexamethasone-induced apoptosis (Walsh *et al.*, 2002).

Cytotoxicity was measured in nurse shark, *Ginglymostoma cirratum*, leukocytes using modification of RPMI to contain 0.2 M NaCl and 0.35 M urea in a solution isotonic for shark cells (Pettey and McKinney, 1983). This medium included antibiotics (100 U/ml penicillin, 100 µg/ml streptomycin, and 0.25 µg/ml fungizone) and 10% heat-inactivated FBS, as well as HEPES buffer (25 mM) to maintain pH. *In vitro* proliferation of shark leukocytes stimulated with phorbol myristate acetate and calcium ionophore (A12187) was described by McKinney (1992), with glass nonadherent populations shown to contain the majority of proliferating cells. Addition of $MgCl_2$ was believed to be the most critical addition to the growth medium.

Short-term cultures of cells from epigonal tissue are shown in Figure 14.1. These cells were isolated by finely mincing tissue and allowing epigonal cells to migrate out of the tissue and were cultured in E-RPMI (Walsh *et al.*, 2006a; Figure 14.1). Media produced by these cells were demonstrated to produce cytotoxic factors (Walsh *et al.*, 2006a). Mixed populations of peripheral blood leukocytes from nurse shark, *G. cirratum*, in short-term (48 h) culture are shown in Figure 14.2a and b, with viability demonstrated using the conversion by live cells of the fluorescent vital dye, carboxyfluorescein diacetate succinimidyl ester (CFDA-SE), to carboxyfluorescein diacetate (CFDA) (Figure 14.2c and d). Phagocytosis of Congo red-stained yeast is shown in Figure 14.3. Figure 14.3a and b shows nonadherent peripheral blood leukocytes from nurse shark, whereas Figure 14.3c and d shows an adherent population of peripheral blood leukocytes from the same animal. Cultured epigonal cells with and without yeast are shown in Figure 14.3e and f.

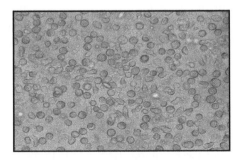

Figure 14.1   Short-term culture (72 h) of immune cells isolated from bonnethead shark, *Sphyrna tiburo*, epigonal tissue.

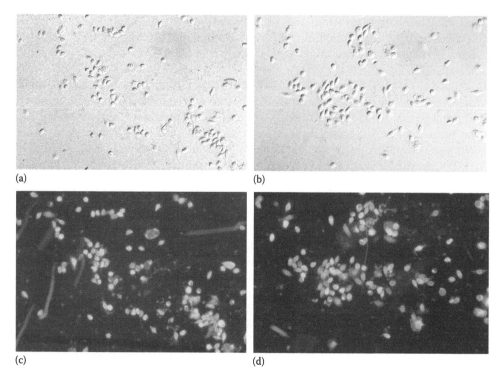

Figure 14.2   Viability of nurse shark, *G. cirratum* peripheral blood leukocytes after 24 h in culture. (a, b) Unstained peripheral blood leukocytes; (c, d) Same fields of peripheral blood leukocytes as in (a, b), respectively, stained with the vital dye, CFDA. Viable cells are able to convert the nonfluorescent ester form of the dye to its fluorescent acetate form.

SOMATIC CELL LINES

More success was achieved with the culture of elasmobranch somatic cells than with the cells of immune origin. Although not immune in origin, development of cell culture techniques for somatic cells represents a major step forward in the available cell culture tools for elasmobranch species. As summarized by Barnes *et al.* (2008), successful derivation of cell lines from cartilaginous fish has relied on culture medium formulations based on knowledge of the physiology of the cell type to be cultured, with the addition of only a minimum of heterologous components such as FBS, rather than adapting cells to standard medium formulations (e.g., basal nutrient medium supplemented only with 10% FBS). To develop additional tools for cellular, molecular, and genomic studies of cartilaginous fish, cell lines were derived from embryos of spiny dogfish, *Squalus acanthias*, and little skate, *L. erinacea* (Forest *et al.*, 2007; Hwang *et al.*, 2008; Kobayashi *et al.*, 2007; Parton *et al.*, 2007).

Short-Term Explant Cultures

Short-term cultures of elasmobranch somatic tissue were reported and include *in vitro* brain explant cultures from Caribbean sharpnose shark, *Rhizoprionodon porosus*, and spiny dogfish, *S. acanthias* (Garner, 1988). For these cultures, Leibovitz L-15 medium was used after adjusting to 1000 mOsm by adding 350 mM urea and 20 mM NaCl. Medium was supplemented with either 2% FBS, 2% bull shark (*Carcharhinus leucus*) serum, or 2% spiny dogfish serum. Cultures were

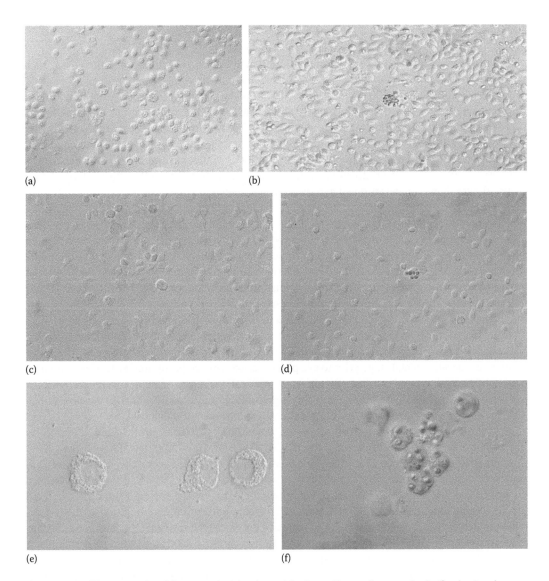

Figure 14.3 Phagocytosis of Congo red-stained yeast by the cultures of nurse shark, *G. cirratum* immune cells. 1-ml cell suspensions containing 2–5 × 10⁶ cells/ml were incubated for 24 h with stained yeast cells at a yeast:leukocyte ratio of 20:1. (a) Nonadherent peripheral blood leukocytes, no yeast, 40X magnification; (b) nonadherent peripheral blood leukocytes, Congo red-stained yeast, 40X magnification; (c) adherent peripheral blood leukocyte population, no yeast, 40X magnification; (d) adherent peripheral blood leukocyte population, Congo red-stained yeast, 40X magnification; (e) epigonal cells, no yeast, 100X; (f) epigonal cells, Congo red-stained yeast, 100X.

maintained at 22°C. Cell outgrowth from explants appeared after one month and was comprised of nonneuronal cells exhibiting cell division. Cells were subcultured once with 20% survival and were viable at the end of six months. Culture of ciliated shark cells in Eagle's Minimum Essential Medium supplemented with 350 mM urea resulted in observations of metabolism, but no cell division (Wolf and Quimby, 1969). Jones *et al.* (1983) cultured explants of shark rectal gland, kidneys, intestine, and stomach, but reported no outgrowth.

Embryonic Elasmobranch Cell Lines

The first report of a cell line from an elasmobranch species is thought to be the passage and maintenance of an embryonic silky shark, *Carcharhinus falciformis* cell line through 45 passages (Hartmann *et al.*, 1992). For this cell line, cells of embryonic brain and heart were cultured in elasmobranch-modified medium prepared from Opti-MEM I that was adjusted to elasmobranch physiological conditions through the addition of NaCl, TMAO, and urea. The morphology of silky shark brain (SSB) cells remained constant through 45 passages and was epithelial-like in appearance. SSB cells were found to grow from 20°C to 37°C, although optimal growth occurred at 29°C–30°C with a population doubling time of approximately 67 h. Cell cultures derived from cardiac tissues, however, deteriorated and died during the eighth passage.

Many challenges were associated with the development of conditions for continuously proliferating elasmobranch cells *in vitro*. Successful derivation of cell lines from shark and skate was based on the concept that conventional heterologous bovine serum supplementation to culture medium should be minimized and replaced with defined peptide growth factors and other growth-promoting components more specifically suited to the cells to be cultured (Barnes *et al.*, 2008). Two cell lines derived from elasmobranch tissue, the spiny dogfish shark, *S. acanthias*, and the little skate, *L. erinacea*, were developed as continuously proliferating cell lines (Forest *et al.*, 2007; Mattingly *et al.*, 2004; Parton *et al.*, 2007). The cell line derived from spiny dogfish embryo, referred to as SAE, is mesenchymal in origin and represents the first multipassage, continuously proliferating cell line from a cartilaginous fish (Barnes *et al.*, 2008). The SAE cell line was derived from a pooled group of embryos at a developmental stage prior to eye pigmentation, with cells cultured at 18°C in a modified basal nutrient medium as described (Parton *et al.*, 2007). Population doubling time of the SAE line was approximately two weeks. This relatively long doubling time is consistent with the slow growth of these colder water animals and low temperature at which cells were cultured.

A continuously proliferating cell line was later developed from little skate, *L. erinacea* (Kobayashi *et al.*, 2007), and referred to as LEE-1. The LEE-1 cell line was initiated from a single stage 28 embryo (Hwang *et al.*, 2008). Cells for LEE-1 cell line were derived from an early stage skate embryo at a developmental point at which gill filaments and eye pigmentation were not yet observed. Cultures were initiated and grown for 7–10 passages on collagen-coated cell culture vessels and then adapted to growth without collagen. Population doubling time of LEE-1 was approximately 10 days. Cold-water organisms exhibit reduced rates of metabolism, ion transport, and oxygen consumption compared with many animal models, leading to increased stability of cells in culture compared to cells from mammals or other animals preferring warmer temperatures (Mattingly *et al.*, 2004). The medium developed for LEE-1 cell line represents a complicated formulation of nutrients, growth factors, and other growth-promoting molecules.

In general, these embryonic cell lines utilized LDF medium (Helmrich and Barnes, 1999), a mixture of 50% Dulbecco's modified medium, 35% L-15, 15% Ham's F-12, 2 mM sodium bicarbonate, and 15 mM HEPES buffer, pH 7.2, supplemented with 10 μg/ml insulin, 10 μg/ml transferrin, 10 nM selenous acid, 50 ng/ml human recombinant epidermal growth factor (EGF), 50 ng/ml basic fibroblast growth factor (FGF), 2 mM sodium bicarbonate, 2 mM GlutaMAX, 0.1X minimum essential medium (MEM) nonessential amino acids, 0.05X essential amino acids, 1:1000 chemically defined lipids, and 2% heat-inactivated FBS. Cells were grown at 18°C and passaged using 2 mg/ml collagenase in LDF medium and 0.2% trypsin with 1 mM ethylenediaminetetraacetic acid (Parton *et al.*, 2007). Osmolarity of the medium was 250 mOsm, an osmolarity lower than in adult cartilaginous fish, suggesting that the embryos had not yet developed the high-osmolarity circulatory system seen in adult animals. This may be because early stage embryos are protected from full-strength sea water in the internal environment of the egg. Although elasmobranchs maintain a high urea concentration as an osmoregulator, it should be noted that the embryonic cells did not require urea in the medium.

ADVANTAGES OF ELASMOBRANCH CELL CULTURE IN BIOMEDICAL AND IMMUNOLOGICAL RESEARCH

Elasmobranch immune cells provide a model for studying the evolution of the immune system. Many novel components of the elasmobranch immune system were identified from peripheral cells or from cultured cells. In previous studies, cells with genes for T-cell receptor were identified (Criscitiello *et al.*, 2010; Rast *et al.*, 1997), but demonstration of functional activities typical of an adaptive cell-mediated immune system was hampered by lack of reliable cell culture systems.

Additionally, fish cell lines are becoming increasingly useful in various areas of marine toxicology (Schirmer, 2006). Cell culture approaches facilitate reliable control of the extracellular environment and provide an opportunity to study a number of transport phenomena at cellular and subcellular levels (Barnes *et al.*, 2008). Although molecular and physiological studies were critical in elucidating the nature of elasmobranch anion/xenobiotic transport, cell culture was a valuable tool that was lacking. For example, the SAE cell line was used to study multidrug resistance-associated protein 3, a member of the ATP-binding cassette (ABC) protein family of membrane transporters (Kobayashi *et al.*, 2007). The skate embryonic cell line, LEE-1, was used to characterize Ost (Hwang *et al.*, 2008), a unique membrane transport protein involved in transport of conjugated steroid and prostaglandin E2 across plasma membranes and first identified in cartilaginous fish. Other transport genes, such as TAP genes, were identified in horn shark, *Heterodontus francisci* (Ohta *et al.*, 1999). Development of cell culture techniques may allow further functional characterization of genes that were identified in the immune system, such as the role of CD83-expressing cells in sharks (Ohta *et al.*, 2004).

Genomic experiments are also gaining momentum with the development of elasmobranch cell culture techniques. Recently, EST libraries from elasmobranch cell lines (SAE and LEE-1) have provided information for assembly of genomic sequences and were useful in revealing gene diversity, new genes, and molecular markers, as well as providing means for elucidation of full-length cDNAs and probes for gene array analyses (Forest *et al.*, 2007). Using ESTs defining mRNAs derived from the SAE cell line, Forest *et al.* (2007) identified highly conserved gene-specific nucleotide sequences in the noncoding 3' UTRs of eight genes involved in the regulation of cell growth and proliferation. With the development of elasmobranch cell lines, transfected cell lines can provide a useful tool to examine effectors of transcriptional and translational control, providing the opportunity to test molecular hypotheses that are difficult to test through other means.

CONCLUSIONS

In summary, culture of elasmobranch cells has presented a number of challenging hurdles. Among these are techniques for the effective separation of cell populations, methods to confirm identity of separated cells, and development of appropriate media formulations that consistently support cell growth and function. Although considerable effort and progress was made, optimization of cell culture methodology, as it applies to elasmobranch cells, is still critical, especially for our continued understanding of immune function in this important subclass of fishes.

REFERENCES

Barnes, D.W., Parton, A., Tomana, M., Hwang, J.H., Czechanski, A., Fan, L., and Collodi, P. Stem cells from cartilaginous and bony fish. *Methods Cell Biol.* 2008; 86:343–367.

Bewley, M.S., Pena, J.T., Plesch, F.N., Decker, S.E., Weber, G.J., and Forrest, J.N., Jr. Shark rectal gland vasoactive intestinal peptide receptor: Cloning, functional expression, and regulation of CFTR chloride channels. *Am. J. Physiol. Regul. Integr. Comp. Physiol.* 2006; 291(4):R1157–R1164.

Criscitiello, M.F., Ohta, Y., Saltis, M., McKinney, E.C., and Flajnik, M.F. Evolutionarily conserved TCR binding sites, identification of T cells in primary lymphoid tissues, and surprising trans-rearrangements in nurse shark. *J. Immunol.* 2010; 184(12):6950–6960.

Forest, D., Nishikawa, R., Kobayashi, H., Parton, A., Bayne, C.J., and Barnes, D.W. RNA expression in a cartilaginous fish cell line reveals ancient 3′ noncoding regions highly conserved in vertebrates. *Proc. Natl. Acad. Sci. USA* 2007; 104(4):1224–1229.

Garner, W.D. Elasmobranch tissue culture: In vitro growth of brain explants from a shark (*Rhizoprionodon*) and dogfish (*Squalus*). *Tissue Cell* 1988; 20(5):759–761.

Greenberg, A.H., Shen, L., and Medley, G. Characteristics of the effector cells mediating cytotoxicity against antibody-coated target cells. I. Phagocytic and non-phagocytic effector cell activity against erythrocyte and tumour target cells in a 51Cr release cytotoxicity assay and [125I]IUdR growth inhibition assay. *Immunology* 1975; 29(4):719–729.

Griffith, R.W. Composition of the blood serum of deep-sea fishes. *Biol. Bull.* 1981; 160:250–264.

Grogan, E.D., and Lund, R. A culture system for the maintenance and proliferation and shark and sting ray immunocytes. *J. Fish Biol.* 1990; 36(5):633–642.

Hartmann, J.X., Bissoon, L.M., and Poyer, J.C. Routine establishment of primary elasmobranch cell cultures. In Vitro *Cell. Dev. Biol.* 1992; 28A(2):77–79.

Helmrich, A., and Barnes, D. Zebrafish embryonal cell culture. *Methods Cell Biol.* 1999; 59:29–37.

Hwang, J.H., Parton, A., Czechanski, A., Ballatori, N., and Barnes, D. Arachidonic acid-induced expression of the organic solute and steroid transporter-beta (Ost-beta) in a cartilaginous fish cell line. *Comp. Biochem. Physiol. C Toxicol. Pharmacol.* 2008; 148(1):39–47.

Jones, R.T., Hudson, E.A., and Sato, T. Explant cultures of shark tissues. *In Vitro* 1983; 19:258.

Kobayashi, H., Parton, A., Czechanski, A., Durkin, C., Kong, C.C., and Barnes, D. Multidrug resistance-associated protein 3 (Mrp3/Abcc3/Moat-D) is expressed in the SAE *Squalus acanthias* shark embryo-derived cell line. *Zebrafish* 2007; 4(4):261–275.

Luer, C.A., Walsh, C.J., Bodine, A.B., Wyffels, J.T., and Scott, T.R. The elasmobranch thymus: Anatomical, histological, and preliminary functional characterization. *J. Exp. Zool.* 1995; 273(4):342–354.

Mainwaring, G., and Rowley, A.F. Separation of leucocytes in the dogfish (*Scyliorhinus canicula*) using density gradient centrifugation and differential adhesion to glass coverslips. *Cell Tissue Res.* 1985; 241(2):283–290.

Maren, T.H. Special body fluids of the elasmobranch. In: Gilbert, P.W., Mathewson, R.F., and Rall, D.P., eds. *Sharks, Skates, and Rays.* John Hopkins Press, Baltimore, MD; 1967. pp. 287–292.

Mattingly, C., Parton, A., Dowell, L., Rafferty, J., and Barnes, D. Cell and molecular biology of marine elasmobranchs: *Squalus acanthias* and *Raja erinacea*. *Zebrafish* 2004; 1(2):111–120.

McKinney, E.C. Proliferation of shark leukocytes. In Vitro *Cell. Dev. Biol.* 1992; 28A(5):303–305.

McKinney, E.C., Haynes, L., and Droese, A.L. Macrophage-like effector of spontaneous cytotoxicity from the shark. *Dev. Comp. Immunol.* 1986; 10(4):497–508.

Nonaka, M., and Smith, S.L. Complement system of bony and cartilaginous fish. *Fish Shellfish Immunol.* 2000; 10(3):215–228.

Ohta, Y., Haliniewski, D.E., Hansen, J., and Flajnik, M.F. Isolation of transporter associated with antigen processing genes, TAP1 and TAP2, from the horned shark *Heterodontus francisci*. *Immunogenetics* 1999; 49(11–12):981–986.

Ohta, Y., Landis, E., Boulay, T., Phillips, R.B., Collet, B., Secombes, C.J., Flajnik, M.F., and Hansen, J.D. Homologs of CD83 from elasmobranch and teleost fish. *J. Immunol.* 2004; 173(7):4553–4560.

Parton, A., Bayne, C.J., and Barnes, D.W. Analysis and functional annotation of expressed sequence tags from in vitro cell lines of elasmobranchs: Spiny dogfish shark (*Squalus acanthias*) and little skate (*Leucoraja erinacea*). *Comp. Biochem. Physiol. Part D Genomics Proteomics* 2010; 5(3):199–206.

Parton, A., Forest, D., Kobayashi, H., Dowell, L., Bayne, C., and Barnes, D. Cell and molecular biology of SAE, a cell line from the spiny dogfish shark, *Squalus acanthias*. *Comp. Biochem. Physiol. C Toxicol. Pharmacol.* 2007; 145(1):111–119.

Pettey, C.L., and McKinney, E.C. Mitogen induced cytotoxicity in the nurse shark. *Dev. Comp. Immunol.* 1981; 5(1):53–64.

Pettey, C.L., and McKinney, E.C. Temperature and cellular regulation of spontaneous cytotoxicity in the shark. *Eur. J. Immunol.* 1983; 13(2):133–138.

Pettey, C.L., and McKinney, E.C. Induction of cell-mediated cytotoxicity by shark 19S IgM. *Cell. Immunol.* 1988; 111(1):28–38.

Poyer, J.C., and Hartmann, J.X. Establishment of a cell line from brain tissue of the silky shark, *Carcharhinus falciformis*. In Vitro *Cell Dev. Biol.* 1992; 28(11/12):682–684.

Rast, J.P., Anderson, M.K., Strong, S.J., Luer, C., Litman, R.T., and Litman, G.W. Alpha, beta, gamma, and delta T cell antigen receptor genes arose early in vertebrate phylogeny. *Immunity* 1997; 6(1):1–11.

Schirmer, K. Proposal to improve vertebrate cell cultures to establish them as substitutes for the regulatory testing of chemicals and effluents using fish. *Toxicology* 2006; 224(3):163–183.

Tomana, M., Parton, A., and Barnes, D.W. An improved method for separation of leucocytes from peripheral blood of the little skate (*Leucoraja erinacea*). *Fish Shellfish Immunol.* 2008; 25(1–2):188–190.

Walsh, C.J., and Luer, C.A. Comparative phagocytic and pinocytic activities of leucocytes from the peripheral blood and lymphomyeloid tissues of the nurse shark (*Ginglymostoma cirratum* Bonaterre) and the clearnose skate (*Raja eglanteria* Bosc). *Fish Shellfish Immunol.* 1998; 8:197–215.

Walsh, C.J., Luer, C.A., Bodine, A.B., Smith, C.A., Cox, H.L., Noyes, D.R., and Maura, G. Elasmobranch immune cells as a source of novel tumor cell inhibitors: Implications for public health. *Integr. Comp. Biol.* 2006a; 46(6):1072–1081.

Walsh, C.J., Toranto, J.D., Gilliland, C.T., Noyes, D.R., Bodine, A.B., and Luer, C.A. Nitric oxide production by nurse shark (*Ginglymostoma cirratum*) and clearnose skate (*Raja eglanteria*) peripheral blood leucocytes. *Fish Shellfish Immunol.* 2006b; 20(1):40–46.

Walsh, C.J., Wyffels, J.T., Bodine, A.B., and Luer, C.A. Dexamethasone-induced apoptosis in immune cells from peripheral circulation and lymphomyeloid tissues of juvenile clearnose skates, *Raja eglanteria*. *Dev. Comp. Immunol.* 2002; 26(7):623–633.

Waymouth, C. Osmolality of mammalian blood and of media for culture of mammalian cells. *In Vitro* 1970; 6(2):109–127.

Wolf, K., and Quimby, M.C. Progress report on the in vitro culture of cyclostome and elasmobranch cells and tissues. *In Vitro* 1969; 4:125.

CHAPTER 15

Antimicrobial Molecules of Sharks and Other Elasmobranchs
A Review

Nichole Hinds Vaughan and Sylvia L. Smith

CONTENTS

SUMMARY

This review will discuss the many types of antimicrobial molecules that were conserved across phyla as an innate mechanism of defense/resistance against microorganisms. Innate immune mechanisms of vertebrates and invertebrates include a variety of antimicrobial molecules that show a remarkable degree of structural and functional conservation across species representing diverse taxa. Structurally and in their mode of action, these molecules vary considerably and include proteins (e.g., lysozyme, alpha-2-macroglobulin [α2M], chitinase, granulysin), peptides (e.g., defensins), bioactive fragments of larger proteins (e.g., histone fragments and cathelicidins), enzyme systems (such as those responsible for intracellular oxidative killing of microorganisms), and nonprotein antimicrobials such as squalamine. These molecules vary in their antimicrobial activity from those that use enzymatic digestion to lyse bacterial cell walls to molecules that kill by using nonenzymatic chelation. Herein the antimicrobial molecules that were reported in elasmobranch species are described and a comparison with similar molecules found in other taxa is provided.

INTRODUCTION

There is a plethora of information on antimicrobial molecules that are part of the innate defense system of diverse organisms, both vertebrates and invertebrates (Bulet *et al.*, 2004). Although much is known of such factors in mammals, birds, reptiles, amphibians, and bony fish (teleosts), relatively little information is available on innate immune defense against microorganisms in elasmobranchs (cartilaginous fish), particularly the antimicrobial proteins/peptides and related molecules. In contrast, there is an impressive body of work on molecules that play a key role in the adaptive immune response of the shark against microorganisms and other foreign targets for further details refer to chapters 9, 10, 11, and 12 in this book (Crouch *et al.*, 2013; Dooley and Flajnik, 2005; Flajnik, 1996; Laing and Hansen, 2011; Litman, 1996; Ohta *et al.*, 2002, 2011; Pettey and McKinney, 1983; Rijkers, 1982).

Antimicrobial proteins and peptides are a diverse group of molecules that play an essential role in the initial and immediate response of the innate immune system against microbes (Boman *et al.*, 1985; Cole *et al.*, 1997; Moore *et al.*, 1996; Noga and Silphaduang, 2003; Relf *et al.*, 1999; Schnapp *et al.*, 1996; Selsted *et al.*, 1985; Zanetti *et al.*, 1995; Zasloff, 1987). Such molecules were found in both prokaryotic and eukaryotic cells and are ubiquitous (Boman, 1998, 2003; Rinehart *et al.*, 1990; Sang and Blecha, 2008). One or several examples of these molecules are present in most organisms studied (Kindt *et al.*, 2007). A significant interspecies variation in expression was noted although most are constitutively expressed (Boman, 1998, 2003). Antimicrobials comprise a large heterogeneous group of molecules that can be found endogenously in phagocytic granulocytes and macrophages, and/or extracellularly in body fluids, such as plasma, mucus, and tears, or secreted into the adjacent external environment, which is the case of certain peptide antibiotics and enzymes (Befus *et al.*, 1999; Diamond *et al.*, 1991; Mallow *et al.*, 1996; Petit and Jolles, 1963; Tsai and Bobek, 1998). Sharks, skates, and rays (fish with cartilaginous skeletons) are no exception and several antimicrobials were described for this diverse group of fish.

Recognition of nonself molecules is crucial in innate defense. Germline-encoded pattern recognition receptors (PRRs), present on and produced by a variety of cells, identify pathogen-associated molecular patterns (PAMPs) characteristic of microorganisms (Akira *et al.*, 2000; Boman, 2003; Magnadóttir, 2006). The target molecules are usually associated with microorganisms and can be polysaccharides, lipopolysaccharides (LPS), peptidoglycan, bacterial DNA, and double-stranded viral RNA (Fearon, 1997; Janeway, 1989; Magnadóttir, 2006). The reaction of each PRR is specific for the molecular pattern/structure it recognizes. *Toll-Like Receptors* (TLRs) are examples of PRRs and were described for many organisms (Arancibia *et al.*, 2007; Assavalapsakul *et al.*, 2012; Cuperus *et al.*, 2013; Fluhr and Kaplan-Levy, 2002; Martinelli and Reichhart, 2005; Medzhitov *et al.*, 1997; Rakoff-Nahoum and Medzhitov, 2006; Wiens *et al.*, 2007), including cartilaginous fish. They play an essential role in the initial, immediate, and specific response of the innate immune system against microbes expressing recognizable molecular patterns (Boman, 1998; Claeys *et al.*, 2003; Medzhitov *et al.*, 1997). In addition to TLRs and similar receptors (e.g., natural killer cell scavenger receptors), organisms produce innate defense molecules (i.e., antimicrobials) that are found intra- and extracellularly. These include proteins with enzymatic antimicrobial activity such as those that mediate the production of reactive oxygen and reactive nitrogen intermediates (e.g., myeloperoxidase, NADPH oxidase, and nitric oxide synthetase [NOS] [Kindt, 2007; Rieger and Barreda, 2011]), the granulysins, intra- and extracellular proteases, lysozyme, LPS-binding peptide, and α2M (Bourazopoulou *et al.*, 2009; Odeberg and Olsson, 1975; Senior *et al.*, 1982; Spitznagel, 1984, 1990; Tang *et al.*, 1999; Ursic-Bedoya *et al.*, 2008). In addition to enzymes, there are antimicrobial peptides (AMPs) that are usually 6–10 kDa in size (typically less than 100 amino acids), although larger forms up to 30 kDa were described (Bevins, 1994; Boman, 1995, 1998; Dennis 1997; Elsbach and Weiss, 1993; Gallo and

Huttner, 1998; Jolles and Jolles, 1984; Lehrer *et al.*, 1991; Zanetti *et al.*, 1995; Zasloff, 1992). Of the large number of antimicrobial peptides that were described, the best studied are the defensins, catheli-cidins, histatins (histine-rich peptides), and peptides generated by cleavage of larger protein precursors with different function, such as lactoferricin from lactoferrin and buforin, a histone fragment derived from histone H2A (reviewed in Steinstraesser *et al.*, 2009). Antimicrobial histone-derived fragments were described for a variety of species (Boman *et al.*, 1985; Cole *et al.*, 1997; Destoumieux *et al.*, 1997; Lee *et al.*, 1997; Moore *et al.*, 1996; Noga *et al.*, 2003; Relf *et al.*, 1999; Schnapp *et al.*, 1996; Selsted *et al.*, 1985; Zanetti *et al.*, 1995; Zasloff, 1987). Occasionally, antimicrobial peptides were referred to as host defense molecules (Steinstraesser *et al.*, 2009). Not all antimicrobial molecules are protein in nature, several have been described which are complex lipids or carbohydrates (Bergsson *et al.*, 2001; Desbois and Smith, 2010; Kabara *et al.*, 1972; Ohta *et al.*, 1994; Schröder *et al.*, 2003). Squalamine, structurally lipid in nature, is an example of an antimicrobial aminosterol (Moore *et al.*, 1996).

Unlike the comprehensive studies carried out on the antimicrobials of teleosts such as trout, salmon, and carp (Cole *et al.*, 1997; Fernandes *et al.*, 2002; Noga *et al.*, 2003; Patrzykat *et al.*, 2001) and certain invertebrate species such as crabs, shrimp, and oysters (Defer *et al.*, 2013; Destoumieux *et al.*, 1997; Itoh *et al.*, 2010; Relf *et al.*, 1999; Schnapp *et al.*, 1996), the study of shark antimicrobial factors has not been as extensive. However, as a result of major conservation efforts directed to the preservation of species, there is a growing interest to further our understanding of innate immune mechanisms of survival and a comprehensive picture is gradually emerging of these molecules, their distribution, function, and role in innate immunity, in sharks and other ecologically and com-mercially important species. In this chapter, we review our current knowledge of antimicrobial molecules in elasmobranchs, with the exception of the humoral complement system and cytokines, which are discussed separately elsewhere in this book (for further details refer to Smith and Nonaka, Chapter 8 and Secombes *et al.*, Chapter 7, respectively; Fox, 2013).

Specific, specialized cells of the immune system, such as leukocytes and secretory epithelial cells (Paneth) of the intestine, secrete soluble proteins with antimicrobial properties. Some of these proteins are enzymes and have been characterized in a variety of different animals, including some shark species, and from a variety of body sites. The mode of antimicrobial action of endogenous enzymes, such as the oxidative antimicrobial activity of NADPH oxidase and myeloperoxidase, the pore-forming property of granulysin, the lytic activity of lysozyme, and the binding of essential nutrients by transferrin and lactoferrin were well documented (Davies *et al.*, 1987; Got *et al.*, 1967; Hinds Vaughan and Smith, 2013; Ingram, 1980). Granulysin is constitutively expressed in activated human cytotoxic T cells (CTLs) and natural killer (NK) cells (Krensky and Clayberger, 2009) as a 15-kDa precursor molecule. This molecule is then cleaved to produce a 9-kDa molecule that is sequestered in cytolytic granules (Hanson *et al.*, 1999; Krensky and Clayberger, 2009). Granulysin belongs to the family of saposin-like lipid-binding proteins and is found within the granules along with granzymes and perforin (Kumar *et al.*, 2001; Tewary *et al.*, 2010). Granulysin is a highly cationic peptide that is folded in a 5-helix bundle, which is stabilized by two highly conserved intramolecular disulfide bonds. This structure allows it to lyse bacterial mem-branes thus facilitating its broad bactericidal activity (Tewary *et al.*, 2010; Zitvogel and Kroemer, 2010). Granulysins may also induce apoptotic mechanisms by permeabilizing the mitochondria of human cells once they are attacked by CTL or NK cells. Granulysins were also found to be chemoattractants for T lymphocytes, monocytes, and other inflammatory cells, stimulating release of several cytokines includ-ing RANTES, MCP-1, MCP-3, IL-1, IL-6, IL-10, and IFNα (Krensky and Clayberger, 2009) and have since been shown to be chemotactic for human monocyte-derived dendritic cells by Tewary *et al.* (2010).

Cathepsin G is a 26-kDa serine protease, found in neutrophils and some monocytes, where it is sequestered in azurophilic granules (Kudo *et al.*, 2009; Odeberg and Olsson, 1975; Senior *et al.*, 1982; Spitznagel, 1990). A neutral enzyme that plays an important role in inflammation when released by activated neutrophils in humans, cathepsin G, hydrolyzes a variety of proteins, including proteins of the extracellular matrix and some cytokines such as TNF-α and IL-8 (Meyer-Hoffert, 2009; Pham, 2006; Wiedow and Meyer-Hoffert, 2005). Cathepsin G takes part in intracellular killing of bacteria contained within phagolysosomes along with myeloperoxidase and reactive oxygen species generated by the NADPH oxidase complex (Kobayashi *et al.*, 2005). Cathepsin G, along with other neutrophil serine proteases, can directly bind to negatively charged bacterial membranes, and may inhibit bacterial protein synthesis and lead to the depolarization and disruption of the membrane. The recent discovery of a cathepsin L orthologue on the invariant chain of the nurse shark, and a link to MHC class II, suggests a role for cathepsins in antigen presentation. In mammals, cathepsin L is responsible for degrading lysosomal contents for antigen loading and is crucial in antigen processing (Criscitiello *et al.*, 2012).

ANTIMICROBIAL ENZYMES

Lysozyme

Lysozyme is a bacteriolytic enzyme that was first described by Fleming and Allison (1922). Lysozymes are basic, cationic proteins that induce bacterial cell lysis by hydrolyzing the β-1-4-linked glycosidic bonds of the peptidoglycan cell wall between *N*-acetylmuramic acid and *N*-acetylglucosamine (Imoto *et al.*, 1972). They are produced mainly by monocytes and granulocytes in mammals and are present in intracellular lysosomes and body fluids (Jolles and Jolles, 1966; Pruzanski and Saito, 1969). Three types of lysozyme have been described: c-type (e.g., hen egg white lysozyme [HEWL]), g-type (e.g., goose egg white lysozyme), and i-type, which is primarily found in invertebrates. Other forms of lysozyme have been found in bacteria and bacteriophages (Fange *et al.*, 1976; Jolles, 1969, 1996; Jolles and Jolles, 1975, 1984; Tsugita and Inouye, 1968). Although lysozymes may differ in molecular size, enzymatic properties, and amino acid sequence, sequence homology, however, is conserved across the active site of the enzyme responsible for its specific hydrolytic activity (Jolles and Jolles, 1975). Interestingly, in mammals, lysozyme was found to have an indirect effect on the complement system by inhibiting the polymorphonuclear response toward anaphylatoxins C5a, C3a, and C4a (Ogundele, 1998), thus behaving as an anti-inflammatory component. In bony fish, c- and g-type lysozymes were detected in serum, mucus, and in certain other tissues of carp, trout, salmon, plaice, and wolffish (Fange *et al.*, 1976; Fletcher and White, 1973). Lundblad *et al.* (1979) reported the presence of lysozyme in lymphomyeloid tissues of several chondrichthyan species (sharks, ray, and *Chaemera monstrasa*), whereas Fange *et al.* (1980) examined the lymphoid tissue of several elasmobranch species for glycosidases and found β-*N*-acetyl glucosaminidase activity present in all species studied. They reported high lysozyme and chitinase activity in Leydig's organ of *Etmopterus spinax* (velvet belly lantern shark), *Sominiosus microcephalus* (Greenland shark), and *Torpedo nobiliana* (Atlantic torpedo or dark electric ray). High lysozyme activity levels were also found in the Leydig's and epigonal organs of *Raja radiate* (thorny skate) and *Rhinoptera bonasus* (cownose ray) and in the spleen of the former. In contrast, little or no lysozyme activity was detected in the epigonal organ of *Ginglymostoma cirratum* (nurse shark), *Heterodontus francisci* (hornshark), and *Scyliorhinus canicula* (small-spotted cat shark). A later study reported weak lysozyme activity in a single houndshark species *Galiorhinus galeus* (Lie *et al.*, 1989). Lysozyme activity in nurse shark serum was, however, reported by Hedayati (1992). Although shark serum gave a positive reaction in dot-blot assays using sheep antiserum to human lysozyme, the nonspecific reaction of shark immunoglobulin with sheep serum proteins could not be ruled out because a control using nonimmune sheep serum was not included. The presence of lysozyme activity in leukocytes of the nurse shark was shown by Hinds (2000) and recently confirmed by isolation and characterization of the enzyme from peripheral blood leukocytes (Hinds Vaughan and Smith, 2013). Shark lysozyme is a 14-kDa protein that hydrolyzes *Micrococcus lysodeikticus* cell walls and inhibits the growth of Gram-positive bacterium (e.g., *Planococcus citreus*). It is a c-type lysozyme based on sequence analysis, and the enzyme bears sufficient structural similarity to HEWL to permit reactivity (weak) with heterologous anti-HEWL antibody in Western blots. The complete primary structure of the enzyme has not been established. There is also evidence that suggests the presence in the nurse shark of a second molecule with muramidase activity that differs in ionic properties. To date, no g-type or i-type lysozyme (the latter is usually found in insects) was reported for any elasmobranch species. However, a lysozyme variant, based on cDNA sequence analysis, was described for another cartilaginous fish, the elephant shark *Callorhinchus milii*, a chimera (Tan *et al.*, 2012).

Chitinase

Chitinase is a hydrolytic enzyme that hydrolyzes the β-1,4 glucosamide linkage between repeating units of *N*-acetyl-D-glucosamine that constitute the chitin monomer (Ingram, 1980). However, unlike lysozyme, chitinase lacks muramidase activity. Chitinases have a molecular size of about 30 kDa and occur as extracellular enzymes usually found in fungi and bacteria, and also in the gut of various invertebrates and in secretions of insectivorous plants (Ingram, 1980). Chitinase activity was detected in lymphomyeloid tissue in certain cartilaginous fish species including *Chimaera monstrosa*, *Etmopterus spinax*, and *Raja radiata* and certain bony fish, *Melanogrammus aeglefinus*, *Gadus morhua*, and *Pollachius pollachius* (Fang *et al.*, 1976). The spleen, plasma, and lymph showed the highest chitinase activity in bony fish, whereas the spleen of certain cartilaginous fish showed relatively weaker reactions (Ingram, 1980). High chitinase activity was found in gastric mucosa of three different species of elasmobranchs (Fange *et al.*, 1976). As stated earlier, chitinase activity was reported in the Leydig's and epigonal organs of several elasmobranch species including *Etmopterus spinax*, *Sominiosus microcephalus*, *Torpedo nobiliana*, *Raja radiate*, *Rhinoptera bonasus*, *G. cirratum*, *Heterodontus francisci*, and *Scyliorhinus canicula* (Fange *et al.*, 1980). Chitinase has shown a protective ability in fish as it is an effective defense against fungi and invertebrate parasites (Alexander and Ingram, 1992; Brunt *et al.*, 2007; Leiro *et al.*, 1997; Manson *et al.*, 1992). Its presence was reported in the elephant shark by sequence analysis of cDNAs and ESTs (Tan *et al.*, 2012).

Elastase

Elastase is another serine protease (usually colocalized with cathepsin G) present in neutrophil azurophilic granules and approximately 26 kDa in size (Kobayashi *et al.*, 2005). Elastases cleave a wide range of substrates, including elastin, proteoglycans, collagen, and fibronectin. When released during inflammation, elastase is rapidly bound by α1-proteinase inhibitor and α2-macroglobulin to form an inhibitor-elastase complex (Döring, 1994; Unanue *et al.*, 1976). Elastase also modifies the function of NK cells and monocytes and inhibits C5a-dependent neutrophil enzyme release and chemotaxis (Döring, 1994; Tralau *et al.*, 2004). Two elastase isoforms were isolated from activated pancreatic extract of dogfish, *Scyliorhinus canicula*. These two isoforms have similar physicochemical properties and are similar in molecular size and amino acid composition showing 85% sequence homology at their amino-terminal sequences with human and rat elastase 2 (Smine and Le Gal 1995).

Miscellaneous Antimicrobial Enzymes

Phospholipase A2 selectively hydrolyzes phospholipids in bacterial membranes and is a member of a large family of *secretory* phospholipases that are produced by vertebrates and invertebrates (Dennis, 1997). The phospholipase A2 family shares several highly conserved features, including size (14–16 kDa), high disulfide bond content, a Ca^{2+}-binding loop, and closely similar catalytic machinery and secondary and tertiary structure (Dennis, 1997; Ganz and Weiss, 1997). Phospholipase A2 was isolated from the intestine of the common stingray, *Dasyatis pastinaia*, in keeping with murine phospholipase A2 secreted by the Paneth cells of the intestines, and like its murine counterpart, it is highly effective against Gram-positive bacteria (Bacha *et al.*, 2012). Other properties of elasmobranch phospholipase A2 are comparable with murine phospholipase, such as the chelation of divalent cations, like calcium, which significantly decreases its antimicrobial activity and that high levels of nutrition (such as bovine serum albumin) and salt do not significantly affect its bactericidal ability (Bacha *et al.*, 2012).

In addition to enzymes that have a direct antimicrobial effect on microorganisms (e.g., lysozymes), there are enzymes that indirectly affect microbial viability by the action of antimicrobial metabolites that are released from their enzymatic activity. Examples of the latter include the oxygen-dependent mechanisms of killing carried out by phagocytic cells, which result in the production of a number of reactive oxygen intermediates (ROI) and reactive nitrogen intermediates. The activation of membrane-bound oxidases will catalyze the reduction of oxygen to a superoxide anion, an intermediate that is toxic to ingested microbes. The superoxide anion formation is catalyzed by NADPH-oxidase, a membrane-bound multicomponent complex that assembles on the inner surface of the cell membrane as it folds to engulf the invading microbe (Briggs *et al.*, 1975; Rieger and Barreda, 2011). During the engulfing process, myeloperoxidase is crucial in the production of hypochlorite from hydrogen peroxide and chloride ions. Other ROI products include hydroxyl radicals and hydrogen peroxide. Oxidative stress indicators (catalase, superoxide dismutase) were found in the liver and kidney of the blue shark, but have not been attributed an immune function (Barrera-García *et al.*, 2013).

Several NADPH-oxidases were identified in teleost fish based on cDNA analysis (Boltana *et al.*, 2009; Inoue *et al.*, 2004; Kawahara *et al.*, 2007; Olavarria *et al.*, 2010; Rieger and Barreda, 2011). Nitric oxide is another important oxygen-dependent molecule involved in effective antimicrobial defenses. Molecular patterns such as LPS, a component of Gram-negative bacterial cell wall, or muramyl dipeptide (MDP), a component of mycobacterial cell wall, will activate macrophages, in conjunction with interferon-γ (IFN-γ), to express high levels of nitric oxide synthetase (NOS). NOS catalyzes the breakdown of L-arginine to produce L-citrulline and nitric oxide (NO). NO, although a potent antimicrobial, can also combine with superoxide anion to yield even more potent antimicrobial product (Kindt *et al.*, 2007). Inducible NOS (iNOS) is the only one of three variants that has been shown to be involved in immune defense. It is a nonspecific defense molecule and its production is up regulated following LPS stimulation *in vitro*. iNOS was identified in goldfish, carp, rainbow trout, and zebrafish (Laing *et al.*, 1996; Lepiller *et al.*, 2009; Rieger and Barreda, 2011; Saeij *et al.*, 2000; Wang *et al.*, 2001). iNOS was first identified in a cartilaginous species by cloning of the iNOS gene from the small-spotted catshark, *Scyliorhinus canicula* (Reddick *et al.*, 2006). This cDNA was coded for a protein of approximately 1125 amino acid residues with a predicted molecular weight of 127.8 kDa. The deduced amino acid sequence was shown to be 70%–73% and 77% similar to known iNOS molecules from bony fish and chickens, respectively. However, the protein has not been isolated and functionally characterized from an elasmobranch species.

ANTIMICROBIAL PROTEASE INHIBITORS

α2M, a relatively nonspecific humoral factor in plasma, is considered an immune defense protein because of its ability to limit the activity of proteases, particularly those of bacterial origin. α2M exhibits broad specificity and limits the damage caused by a variety of exogenous proteases secreted by invading microorganisms. This is accomplished by cleavage of α2M by the protease, followed by refolding of the α2M to form a cage round the protease, which restricts its access to protein substrates (Alexander and Ingram, 1992; Armstrong and Quigley, 1999; Magnadóttir, 2006). α2M is a major representative of a group of plasma proteins that includes complement components C3, C4, and C5 as well as pregnancy zone protein (PZP) that are collectively called the thiolester family of proteins (TEPs) because of a unique internal cyclic thiolester bond. Even though C5 molecules lack the characteristic thiolester bond, they are included in the TEP family due to considerable conserved structural similarities. In some vertebrates, α2M is found in several forms. In human plasma, α2M can occur as a monomer, dimer but mainly a tetramer composed of identical subunits of 180 kDa. Multiple reactive sites conserved within the molecule suggest diversified and complex function. Although α2M has not been isolated from shark serum, its presence is typically and consistently seen as a large protein band in analytical gels of nonreduced shark serum. The presence of a thiolester bond in the molecule can be confirmed by autoradiography. Thiolester bonds react with small amines; consequently, shark serum when treated with C^{14}-methylamine incorporates the label into all serum TEPs, including α2M and complement proteins C3 and C4. Because of its size, α2M is easily distinguishable from other TEPs as a large radioactive protein band in autoradiographs. Functional data to confirm its activity in sharks are lacking, however (Acton *et al.*, 1972; Nonaka and Kimura, 2006; Smith, 1998; Starkey and Barret, 1982). Recently, Shin and Smith (unpublished data) have cloned and sequenced the complete α2M cDNA from the nurse shark (Gcα2M GenBank Accession # KJ406706), and the predicted amino acid sequence is compatible with the presence of the internal thiolester bond.

NONENZYMATIC IRON-BINDING ANTIMICROBIAL PROTEINS

Transferrins are nonheme, iron-binding glycoproteins that are globular and between 80–90 kDa in size. They are found in the serum of most vertebrates and are also present in avian egg white and in mammalian milk (Ingram, 1980; Putnam, 1975). Some transferrins are bacteriostatic when not fully saturated with iron: they work by chelating any free available iron present, making it unavailable to the invading organism (Ingram, 1980; Weinberg, 1974). Transferrins have a high degree of polymorphism and can be found in several different genotypic variants in animals (Fine *et al.*, 1967; Utter *et al.*, 1970). Elasmobranchs also exhibit multiple forms of transferrin. Dogfish sharks synthesize 2–4 forms that display many similarities in physicochemical characteristics to human transferrin (Boffa *et al.*, 1965, 1966; Davies *et al.*, 1987; Got *et al.*, 1967). Recently, transferrin cDNA was cloned and sequenced from the elephant shark, *Callorhinchus melii* (Tan *et al.*, 2012). Studies carried out on salmon suggest that the greater avidity for iron binding by certain genotypes is directly related to an organism's resistance to infection (Suzomoto *et al.*, 1977). Lactoferrin, a member of the transferrin family, is a high-affinity iron-chelator in milk and displays bacteriostatic properties (Bellamy *et al.*, 1993; Bullen and Armstrong, 1979; Ellison and Giehl, 1991). The antimicrobial activity of lactoferrin is often reinforced by lysozyme activity (Ellison and Giehl, 1991). In humans, both of these enzymes are in high concentration in mucosal secretions (Bullen and Armstrong, 1979; Ellison and Giehl, 1991; Masson *et al.*, 1966). Lactoferrin contains an antimicrobial sequence near its N-terminus that appears to function by a mechanism distinct from iron chelation (Tomita *et al.*, 1994). Lactoferricin is the antimicrobial peptide derived from the digested N-terminal portion of lactoferin (Bellamy *et al.*, 1993).

ANTIMICROBIAL PEPTIDES

Antimicrobial peptides have been classified based on their physicochemical characteristics such as molecular size and primary or secondary structure and further divided into five subcategories on the basis of their amino acid composition and primary structure: (1) anionic peptides, (2) linear amphiphatic α-helical peptides, (3) cationic peptides enriched for specific amino acids, (4) peptides with cysteines that form intramolecular bonds, and (5) peptide fragments derived from larger precursor proteins (Boman, 1995, 2000, 2003; Brogden, 2005; Diamond *et al.*, 2009).

Anionic peptides were identified from surfactant extracts, bronchoalveolar lavage fluid, and airway epithelial cells. They require zinc as a cofactor for activity and are active against both

Gram-positive and Gram-negative bacteria. Examples include dermcidin in humans and maximin H5 from amphibians (Brogden, 2005). An anionic antibacterial peptide was not reported for sharks. Linear amphiphatic α-helical peptides are diverse structurally and in their mechanism of bacteriocidal activity and are usually less than 40 amino acid residues in length. They adopt an α-helical secondary structure when in contact with membranes, but are disordered in aqueous solution (Diamond *et al.*, 2009). Examples of these peptides are the cecropins, first isolated from the Cecropia moth (Steiner *et al.*, 1981), magainin from the African clawed tree frog *Xenopus laevis* (Zasloff, 1987), pleurocidin (Cole *et al.*, 1997), melittin (Fennell *et al.*, 1968), and penaeidins from penaeid shrimp and horseshoe crab (Destoumieux *et al.*, 1997; Schnapp *et al.*, 1996; Tassanakajon *et al.*, 2010). Antimicrobial peptides of this class that have been isolated from fish include pardaxins, isolated from the skin glands of the Red Sea Moses sole, *Parachirus marmoratus*; pleurocidin, isolated from the skin mucus of the white flounder, *Pleuronectes americanus*; and piscidin (22 residues), first isolated from the skin and gills of the striped bass (Noga and Silphaduang, 2003). Other members of this family are dicentracin, isolated from the European bass, *Dicentrarchus labrax*; chrysophsins from red sea bream, *Chrysophrys major*; and epinecidin from the orange-spotted grouper, *Epinephelus coioides* (Iijima *et al.*, 2003; Salerno *et al.*, 2007; Smith *et al.*, 2010; Yin *et al.*, 2006). To date, similar defense molecules have not been described for elasmobranchs; although given the diverse distribution of these antimicrobials in species across the phylogenetic spectrum, it would be reasonable to assume that similar and related molecules are likely present in cartilaginous fish.

Many peptides are positively charged (cationic) molecules and are known to bind negatively charged microbial membranes and thus disrupt the integrity of the microbial membrane causing peptide-mediated permeabilization (Oren and Shai, 1998). Cationic peptides that are enriched for specific amino acids and lack cysteines residues are the bactenecins and proline-arginine-rich peptide (PR-39) (Raj *et al.*, 1996), both of which are rich in proline and arginine. Examples are the proline-rich abaecin from bees, and indolicidin from cattle, which is rich in tryptophan (Brogden, 2005). There are no known fish antimicrobial peptides that fall into this category; however, several new fish antimicrobial peptide structures were reported but not fully characterized and classified.

Another group of antimicrobial peptides are the charged peptides that are fragments of larger proteins with different function. Similar to lactoferricin, mentioned earlier, cathelicidins are peptides, which are found at the C-terminus of precursors, that share N-termini homology with the porcine serine protease known as cathelin (Diamond *et al.*, 2009; Zanetti *et al.*, 1995). This family includes human LL37, Bac5, and Bac7 (ovine, bovine, and porcine cathelicidins), and porcine protegrins (Bucki *et al.*, 2010; Zanetti *et al.*, 1995). Some cathelicidins undergo extracellular proteolytic cleavage that releases the active C-terminal peptide from the precursor molecule (Zanetti *et al.*, 1995). Also in this group are peptides derived from oxygen-binding proteins, such as hemacyanin- and hemoglobin-derived peptides (Vizioli and Salzet, 2002). Putative antimicrobial hemoglobin fragments were reported in sharks (see Table 15.1); however, their antimicrobial activity has not been defined (Hinds Vaughan unpublished data).

Defensins are peptides with six conserved cysteine motifs as a characteristic feature. The cysteine residues form three intramolecular disulfide bridges between the amino and carboxyl terminal regions of the peptide. This creates a cyclic, triple-stranded, amphiphatic β-sheet structure. The disulfide bridges serve to stabilize its β-sheet structure and increase its resistance to proteolysis (Brogden, 2005; Campopiano *et al.*, 2004; Diamond *et al.*, 2009). Defensins are grouped into α- and β-defensins that are similar in three-dimensional (3D) structure and in antimicrobial activity; however, the two groups differ in molecular size with the β-defensins tending to be larger molecules; some defensins may have modified termini and show differences in the linkage pattern of the cysteine residues (Boman, 2003). The α-defensins are produced constitutively, whereas most β-defensins are inducible peptides (Boman, 2003; Diamond *et al.*, 1991). Defensins were identified

Table 15.1   Amino Acid Sequence of Putative Antibacterial Peptides Identified in Elasmobranchs. Comparison with Peptides of Other Taxa Including Buforin I from Toad, *Bufu bufu*

Source	Protein/Peptide Sequence	Identification
Skate *Raja kenojei*	PKLRRHTKAALPAKGSLRGGVKPFYQKG	Kenojeinin I: 3 kDa cationic peptide
Nurse shark *Ginglymostoma cirratum*	VHWTEGERAEIEGLWKKLDLVSIVKEA	β Hemoglobin fragments
	VHWTEGERAEIEGQ	
Stingray *Potamotrygon* cf. *henlei*	GKLTDSQEDYIRHVWDDVNRKLITAKALERVN// LVAEALSSNYN	β Hemoglobin fragment
Bovine	QADFQKVVAGVANALAHRYH	β Hemoglobin peptide
Nurse shark	MQIFVKTLTGKTITLEV	Ubiquitin
	LIFAGK	
Nurse shark	TITLQVEPSDTIENVK	Polyubiquitin (3 unique peptides)
	IQDKEGIPPDQQR	
	KSTLHLVLR	
Nurse shark	SRSE**RAGLQF**QK	H2A-1 Histone fragment
	SIAISRSY**RAGLQF**	H2A-2
Toad (*Bufu bufu*)	AGRGKQGGKVRAKAKTRSS**RAGLQF**PVG	Buforin I

Notes: // = inserted gap.
Bold text indicates the common sequence (RAGLQF) shared between shark and toad fragments.

in diverse marine species from bivalve mollusks and decapod crustaceans to teleost fish (Smith *et al.*, 2010). Most fish defensins and cathelicidins were identified from sequence analysis of ESTs and complete genome data (of zebrafish and pufferfish), which show that fish defensins are similar to β-defensins of birds and mammals (Zou *et al.*, 2007). β-defensins protect the mucosa both in the lungs and along the digestive tract in mammals (Bevins *et al.*, 1999). The α-defensins seem to have evolved to work inside the phagosome, where their propeptides are processed into the active peptide. These defensins are sequestered in storage granules as their natural cytotoxicity could harm the host. Another group that belongs to the defensin family is the θ-defensins, which are the product of two separate genes; each produces two separate molecules whose peptide backbones are linked covalently to form small circular molecules (Tang *et al.*, 1999). The θ-defensins differ from the α- and β-defensins in that they have a double-stranded β-sheet and have only been found in primates (Boman, 2003; Tang *et al.*, 1999). Similar to α-defensins is the *liver-expressed anti-microbial* peptide (LEAP), which is cysteine-rich and forms two β-sheets (Krause *et al.*, 2000; Smith and Fernandes, 2009). Peptides that belong to the LEAP family include hepcidin, which has been isolated from several species of fish and exhibits both antimicrobial activity and iron-regulatory activity (Douglas *et al.*, 2003; Ganz, 2006; Hirono *et al.*, 2005; Krause *et al.*, 2000; Liang *et al.*, 2013; Smith *et al.*, 2010). Defensin-like peptides have not been described for any elasmobranch species.

Historically, histones have been associated with the chromatin and the packaging of DNA in the cell nucleus. There are several types of histones, the core histones that form the nucleosome consisting of H2A, H2B, H3, and H4 and the linker histones, H1 and H5. However, fragments of these histones have now emerged as potent antimicrobial agents, as investigators have been able to attribute antimicrobial activity specifically to histone-derived peptides, working independently or synergistically with each other and other peptides (Agerberth *et al.*, 2000; Kawasaki and Iwamura, 2008; Kim *et al.*, 2000; Patrzykat *et al.*, 2001; Rose *et al.*, 1998). Histone peptides with broad spectrum antimicrobial activity have been found in shrimp (Patat *et al.*, 2004); teleost fish (Bergsson *et al.*, 2005; De Zoysa *et al.*, 2009; Fernandes *et al.*, 2002; Nam *et al.*, 2012; Noga *et al.*, 2001; Richards *et al.*, 2001; Robinette *et al.*, 1998); frog (*Rhacophorus schlegelii*)

(Kawasaki *et al.*, 2003); chicken (Li *et al.*, 2007; Silphaduang *et al.*, 2006); and mammals (reviewed in Hiemstra *et al.*, 1993; Howell *et al.*, 2003; Jacobsen *et al.*, 2005; Kawasaki and Iwamuro, 2008; Kim *et al.*, 2002; Rose *et al.*, 1998). In flounder, there is evidence that synthesized histone H1 may work synergistically with flounder pleurocidin to potentiate and enhance its antibiotic effects against fish pathogens (Patrzykat *et al.*, 2001). In the chicken, full length histone H1, as well as H2B, was purified from extracts of the ovary and oviduct (Silphaduang *et al.*, 2006). A histone H1 fragment has been isolated from skin secretions of rainbow trout, termed *oncorhyncin II*, with broad spectrum antimicrobial activity against not only bacteria but also fungi (Fernandes *et al.*, 2004). Histone H2A is the best studied member of this family of proteins, and the full length protein has been isolated from human placenta (Kim *et al.*, 2002), chicken liver (Li *et al.*, 2007), and rainbow trout (Fernandes *et al.*, 2002). An antimicrobial fragment of histone H2A was isolated from gastric tissue of the Asian toad, *Bufo bufo*, designated Buforin I, which is a potent 39 amino acid residue antimicrobial peptide (Park *et al.*, 1996). Further studies have shown that Histone H2A is cleaved by the pepsin C isozymes, Ca and Cb, specifically between the Tyr39-Ala40 bond, confirmed by structural and immunological analyses of the processed peptide (Park *et al.*, 2000). By removing several amino acids from Buforin I by cleavage with endoproteinase Lys-C, a H2A-derived antimicrobial peptide (21 amino acid residues) was generated that showed a broad-spectrum of activity against several microorganisms, but did not kill microbes by cell lysis. This modified antimicrobial agent, designated Buforin II, has a strong affinity for DNA and RNA has been shown to penetrate through the cell membrane and to accumulate inside bacterial cells suggesting that it may inhibit cellular function by binding to nucleic acid (Park *et al.*, 2000). A novel histone antimicrobial peptide, synthesized from Buforin II, shows homology to histone H3 with broad spectrum antimicrobial activity (Tsao *et al.*, 2009). Two other forms of histone H2A-related peptides (fragments) have been found in the skin mucus of fish, Parasin I, from the skin mucus of a wounded catfish (Park *et al.*, 1998) and hipposin, from the Atlantic halibut (Birkemo *et al.*, 2003). Histone H2B was first isolated from the cytoplasm of mouse (murine) macrophage cell lines, RAW264.7 and J774A.1 (Hiemstra *et al.*, 1993). H2B has also been isolated from colonic epithelial cells in tandem with H1 (Howell *et al.*, 2003). Calf intestinal H2B was found to have anti-parasitic activity, similar to the histone H1-like protein isolated from the skin, gill, and spleen extracts of rainbow trout and sunshine bass (Noga *et al.*, 2001). Histone H2B fragments have been isolated from extracts of human blister wound fluid, which displayed very high activity against Gram-positive bacteria (Frohm *et al.*, 1996). Lee *et al.* (2009) have reported on the antimicrobial activity of histone H4.

Histone fragments with putative antibacterial activity were first identified in lysates from peripheral blood leukocytes from the nurse shark, *G. cirratum* (Hinds, 2000). SDS-PAGE analysis of partially purified antimicrobial lysate fractions revealed small, <14 kDa, peptides. N-terminal amino acid sequence analysis of the peptides identified them as fragments of histones H2A, H2B, and H4 (Hinds Vaughan and Smith, unpublished data). Sequence comparison with known antimicrobial peptides showed sequence similarity to antimicrobial histone fragments described for several species of bony fish as well as several invertebrates (Bergsson *et al.*, 2005; De Zoysa *et al.*, 2009; Fernandes *et al.*, 2002; Frohm *et al.*, 1996; Park *et al.*, 1996, 1998; Patat *et al.*, 2004; Robinette *et al.*, 1998; Seo *et al.*, 2011). Interestingly, the nurse shark H2A fragment (Cirratin) showed higher sequence homology with Buforin I than with Buforin II and likely functions as an antimicrobial in a similar manner to both (Table 15.1).

Although not much is known of antimicrobial peptides in elasmobranchs, a 3-kDa cationic antimicrobial peptide was described for the skate, *Raja kenojei*, isolated from fermented skin extracts (Cho *et al.*, 2005) (Table 15.1). In the stingray, *Potamotrygon* cf. *henlei*, a protein similar in structure to the β-chain of hemoglobin was isolated from its epidermal mucus. The stingray protein has a molecular size of 16 kDa and was shown to be active against Gram-positive and Gram-negative bacteria, as well as yeast, without displaying hemolytic activity against RBCs (Conceição *et al.*, 2012). Amino acid sequence analysis of nurse shark peptides present in leukocyte lysates generated some unexpected findings. Ubiquitin and two protein fragments very similar in sequence to hemoglobin fragments were isolated from fractionated lysates. Hemoglobin's primary function in erythrocytes is the transport of oxygen; however, it has been shown to be antibacterial in its intact as well as fragmented form in various species (Altman and Katz, 1961; Nakajima *et al.*, 2003; Parish *et al.*, 2001). The putative antimicrobial hemoglobin-derived fragments identified in the nurse shark do not, however, align with the stingray sequence or bovine β hemoglobin fragment (Table 15.1). Ubiquitin's main function is to covalently bind to proteins to act as a tag and to shuttle the tagged protein to the proteasome for degradation, thereby playing a role in MHC Class I antigen presentation, and it has a role in the vertebrate immune response through CTLs (Kindt *et al.*, 2007). Ubiquitin on its own has not been shown to be antibacterial; however, it has been associated, that is, covalently bound, with histones (e.g., histone H1), which were antibacterial (Wang *et al.*, 2002). In addition, ubiquicidin, shown to be antibacterial (Hiemstra *et al.*, 1999), is a protein with high homology to the ribosomal protein S30 (which itself has shown high homology to ubiquitin). Ubiquicidin was purified from the cytosol of IFN-γ activated murine macrophages and has been theorized to be a precursor molecule of ubiquitin (Hiemstra *et al.*, 1999). The presence of ubiquicidin in the cytosol of activated mouse macrophages may enable it to inhibit the growth of intracellular pathogens such as *Listeria monocytogenes*, *Ricketsia* spp., or *Shigella* spp., which can survive in the cytosol. Indeed, ubiquicidin was shown to be as effective against *Listeria* as NP-1, a rabbit defensin (Hiemstra *et al.*, 1999), and may have clinical applications, as it has proven to be effective at killing methichillin-resistant *Staphylococcus aureus* in mouse infection models (Brouwer *et al.*, 2006). Synthetic peptides derived from ubiquicidin have been used in a small clinical trial to test their efficacy as markers for imaging sites of infection in bone and soft tissue, by identifying sites of inflammation (Gandomkar *et al.*, 2009).

NONPROTEIN ANTIMICROBIAL MOLECULES

Squalamine is a sterol that was originally isolated from the liver of the dogfish shark (Moore *et al.*, 1993). The compound, a natural cholesterol-like molecule, has a net positive charge. Initially, interest in the molecule was to determine how squalamine acted as an immune agent in sharks. Squalamine was found to enter only certain cells, such as those in blood vessels, capillaries, and liver. Studies showed that when squalamine enters cells, it displaces positively charged proteins that are bound to the negatively charged surface of the cell's inner membrane. Interestingly, some of the displaced proteins are used by viruses to replicate, and without these proteins, a virus's life cycle is disrupted rendering the virus inert and allowing subsequent destruction of the infected cell. This suggests that squalamine has broad antiviral properties. Fortunately, squalamine can be synthesized in the laboratory, a process that does not involve the use of any harvested shark tissues; thus, sharks do not have to be sacrificed to obtain this compound. Squalamine's anti-angiogenic and antitumor activity has since been investigated (Alhanout *et al.*, 2010; Cho and Kim, 2002) as well as its effectiveness as a topical treatment for dermatophyte infections and an antitubercular agent (Coulibaly *et al.*, 2013; Walker and Houston, 2013). Molecules related to squalamine have not been reported for other shark species.

CONCLUSION

From sporadic studies carried out over several decades, it is apparent that antimicrobial molecules identified in many vertebrate and invertebrate species are also present in several elasmobranch species as a form of innate defense against microorganisms. Many of the antimicrobial molecules reported in cartilaginous fish are known to be essential elements of mammalian and higher vertebrate immune systems indicating high evolutionary conservation of structure and function. Because of a growing interest in the protection and preservation of elasmobranch species, recent years have seen a more concerted effort on the part of investigators to examine cartilaginous fish specifically for antimicrobial molecules.

REFERENCES

Acton, R.T., Niedermeier, W., Weinheimer, P.F., Clem, L.W., Leslie, G.A., and Bennett, J.C. The carbohydrate composition of immunoglobulins from diverse species of vertebrates. *J. Immunol.* 1972; 09(2):371–381.

Agerberth, B., Charo, J., Werr, J., Olsson, B., Idali, F., Lindbom, L., Kiessling, R., Jornvall, H., Wigzell, H., and Gudmundsson, G.H. The human antimicrobial and chemotactic peptides LL-37 and alpha-defensins are expressed by specific lymphocyte and monocyte populations. *Blood* 2000; 96(9):3086–3093.

Akira, S., Hoshino, K., and Kaisho T. The role of Toll-like receptors and MyD88 in innate immune responses. *J. Endotoxin. Res.* 2000; 6(5):383–387.

Alexander, J., and Ingram G. Noncellular nonspecific defense mechanisms of fish. *Annu. Rev. Fish Dis.* 1992; 2:249–279.

Alhanout, K., Rolain, J.M., and Brunel, J.M. Squalamine as an example of a new potent antimicrobial agents class: A critical review. *Curr. Med. Chem.* 2010; 17(32):3909–3917.

Altman, P.L., and Katz, D.D. Blood and other body fluids. *Pediatrics* 1961; 28(3):361.

Arancibia, S.A., Beltrán, C.J., Aguirre, I.M., Silva, P., Peralta, A.L., Malinarich, F., and Hermoso, M. Toll-like receptors are key participants in innate immune responses. *Biol. Res.* 2007; 40(2):97–112.

Armstrong, P., and Quigley, J. A2-macroglobulin: An evolutionarily conserved arm of the innate immune system. *Dev. Comp. Immunol.* 1999; 23:375–390.

Assavalapsakul, W., and Panyim, S. Molecular cloning and tissue distribution of the Toll receptor in the black tiger shrimp, *Penaeus monodon. Genet. Mol. Res.* 2012; 11(1):484–493.

Bacha, B., Abid, I., and Horchani, H. Antibacterial properties of intestinal phospholipase A2 from the common stingray *Dasyatis pastinaca. Appl. Biochem. Biotechnol.* 2012; 168(5):1277–1287.

Barrera-García, A., O'Hara, T., Galván-Magaña, F., Méndez-Rodríguez, L.C., Castellini, J.M., and Zenteno-Savín, T. Trace elements and oxidative stress indicators in the liver and kidney of the blue shark (*Prionace glauca*). *Comp. Biochem. Physiol. A Mol. Integr. Physiol.* 2013; 165(4):483–490.

Befus, A.D., Mowat, C., Gilchrist, M., Hu, J., Solomon, S., and Bateman, A. Neutrophil defensins induce histamine secretion from mast cells: Mechanisms of action. *J. Immunol.* 1999; 163(2):947–953.

Bellamy, W., Wakabayashi, H., Takase, M., Kawase, K., Shimamura, S., and Tomita, M. Killing of *Candida albicans* by lactoferricin B, a potent antimicrobial peptide derived from the N-terminal region of bovine lactoferrin. *Med. Microbiol. Immunol.* 1993; 182(2):97–105.

Bergsson, G., Agerberth, B., Jornvall, H., and Gudmundsson, G.H. Isolation and identification of antimicrobial components from the epidermal mucus of Atlantic cod (*Gadus morhua*). *FEBS J.* 2005; 272(19):4960–4969.

Bergsson, G., Arnfinnsson, J., Steingrímsson, O., and Thormar, H. Killing of Gram-positive cocci by fatty acids and monoglycerides. *APMIS* 2001; 109(10):670–678.

Bevins, C.L. Antimicrobial peptides as agents of mucosal immunity. In: Boman, H.G., March, J., Goode, J., eds. *Antimicrobial Peptides*. Ciba Foundation Symposium, Chichester, UK: Wiley, 1994; 186, 250–269.

Bevins, C.L., Martin-Porter, E., and Ganz, T. Defensins and innate host defence of the gastrointestinal tract. *Gut* 1999; 45(6):911–915.

Birkemo, G.A., Luders, T., Andersen, O., Nes, I.F., and Nissen-Meyer, J. Hipposin, a histone-derived antimicrobial peptide in Atlantic halibut (*Hippoglossus hippoglossus* L.). *Biochim. Biophys. Acta* 2003; 1646(1–2):207–215.

Boffa, G.A., Faure, A., Got, R., Drilhon, A., and Fine, J.M. Sur les caracteres physico-chimiques des transferrines serique de la lamproie (cyclostome) et de la roussette (selacien). In: Peeters, D.H., ed. *Protides of Biological Fluids*. Amsterdam, the Netherlands: Elsevier Publishing Co, 1966; 14:97–102.

Boffa, G.A., Zakin, M.M., Drilhon, A., Jacquot-Armand, Y., Amouch, P., and Fine, J.M. Existence et caracteres de la transferrine chez la roussette (selacien du g. *Scyllium*). *C. r. Seanc. Soc. Biol.* 1965; 159:2317–2322.

Boltana, S., Donate, C., Goetz, F.W., MacKenzie, S., and Balasch, J.C. Characterization and expression of NADPH oxidase in LPS-, poly(I:C)- and zymosan-stimulated trout (*Oncorhynchus mykiss* W.) macrophages. *Fish Shellfish Immunol.* 2009; 26:651–661.

Boman, H.G. Peptide antibiotics and their role in innate immunity. *Annu. Rev. Immunol.* 1995; 13:61–92.

Boman, H.G. Gene-encoded peptide antibiotics and the concept of innate immunity: An update review. *Scand. J. Immunol.* 1998; 48(1):15–25.

Boman, H.G. Innate immunity and the normal microflora. *Immunol. Rev.* 2000; 173:5–16.

Boman, H.G. Antibacterial peptides: Basic facts and emerging concepts. *J. Intern. Med.* 2003; 254(3):197–215.

Boman, H.G., Faye, I., von Hofsten, P., Kockum, K., Lee, J.Y., Xanthopoulos, K.G., Bennich, H., Engstrom, A., Merrifield, R.B., and Andreu, D. On the primary structures of lysozyme, cecropins and attacins from *Hyalophora cecropia. Dev. Comp. Immunol.* 1985; 9(3):551–558.

Bourazopoulou, E., Kapsogeorgou, E.K., Routsias, J.G., Manoussakis, M.N., Moutsopoulos, H.M., and Tzioufas, A.G. Functional expression of the alpha 2-macroglobulin receptor CD91 in salivary gland epithelial cells. *J. Autoimmun.* 2009; 33(2):141–146.

Briggs, R.T., Drath, D.B., Karnovsky, M.L., and Karnovsky, M.J. Localization of NADH oxidase on the surface of human polymorphonuclear leukocytes by a new cytochemical method. *J. Cell. Biol.* 1975; 67:566–586.

Brogden, K.A. Antimicrobial peptides: Pore formers or metabolic inhibitors in bacteria? *Nat. Rev. Microbiol.* 2005; 3(3):238–250.

Brouwer, C.P., Bogaards, S.J., Wulferink, M., Velders, M.P., and Welling, M.M. Synthetic peptides derived from human antimicrobial peptide ubiquicidin accumulate at sites of infections and eradicate (multi-drug resistant) *Staphylococcus aureus* in mice. *Peptides* 2006; 27:2585–2591.

Brunt, J., Newaj-Fyzul, A., and Austin, B. The development of probiotics for the control of multiple bacterial diseases of rainbow trout, *Oncorhynchus mykiss* (Walbaum). *J. Fish Dis.* 2007; 30(10):573–579.

Bucki, R., Leszczynska, K., Namiot, A., and Sokolowski, W. Cathelicidin LL-37: A multitask antimicrobial peptide. *Arch. Immunol. Ther. Exp. (Warsz)* 2010; 58(1):15–25.

Bulet, P., Stöcklin, R., and Menin, L. Anti-microbial peptides: From invertebrates to vertebrates. *Immunol. Rev.* 2004; 198:169–184.

Bullen, J.J., and Armstrong, J.A. The role of lactoferrin in the bactericidal function of polymorphonuclear leucocytes. *Immunology* 1979; 36(4):781–791.

Campopiano, D.J., Clarke, D.J., Polfer, N.C., Barran, P.E., Langley, R.J., Govan, J.R., Maxwell, A., and Dorin, J.R. Structure-activity relationships in defensin dimers: A novel link between beta-defensin tertiary structure and antimicrobial activity. *J. Biol. Chem.* 2004; 279(47):48671–48679.

Cho, J., and Kim, Y. Sharks: A potential source of antiangiogenic factors and tumor treatments. *Mar. Biotechnol. (N.Y.)* 2002; 4(6):521–525.

Cho, S.H., Lee, B.D., An, H., and Eun, J.B. Kenojeinin I, antimicrobial peptide isolated from the skin of the fermented skate, *Raja kenojei. Peptides* 2005; 26(4):581–587.

Claeys, S., de Belder, T., Holtappels, G., Gevaert, P., Verhasselt, B., van Cauwenberge, P., and Bachert, C. Human beta-defensins and toll-like receptors in the upper airway. *Allergy* 2003; 58(8):748–753.

Cole, A.M., Weis, P., and Diamond, G. Isolation and characterization of pleurocidin, an antimicrobial peptide in the skin secretions of winter flounder. *J. Biol. Chem.* 1997; 272(18):12008–120013.

Conceição, K., Monteiro-dos-Santos, J., Seibert, C.S., Silva, P.I., Jr., Marques, E.E., Richardson, M., and Lopes-Ferreira, M. Potamotrygon cf. henlei stingray mucus: biochemical features of a novel antimicrobial protein. *Toxicon* 2012; 60(5):821–829.

Coulibaly, O., Alhanout, K., L'Ollivier, C., Brunel, J.M., Thera, M.A., Djimdé, A.A., Doumbo, O.K., Piarroux, R., and Ranque, S. in vitro activity of aminosterols against dermatophytes. *Med. Mycol.* 2013; 51(3):309–312.

Criscitiello, M.F., Ohta, Y., Graham, M.D., Eubanks, J.O., Chen, P.L., and Flajnik M.F. Shark class II invariant chain reveals ancient conserved relationships with cathepsins and MHC class II. *Dev. Comp. Immunol.* 2012; 36(3):521–533.

Crouch, K., Smith, L.E., Williams, R., Cao, W., Lee, M., Jensen, A., and Dooley, H. Humoral immune response of the small-spotted catshark, *Scyliorhinus canicula. Fish Shellfish Immunol.* 2013; 34(5):1158–1169.

Cuperus, T., Coorens, M., van Dijk, A., and Haagsman, H.P. Avian host defense peptides. *Dev. Comp. Immunol.* 2013; 41(3):352–369.

Davies, D., Lowson, R., Burch, S., and Hanson, J. Evolutionary relationships of a Primitive shark (*Heterodontus*) assessed by micro-complement fixation of serum transferrin. *J. Mol. Evol.* 1987; 25(1):74–80.

Defer, D., Desriac, F., Henry, J., Bourgougnon, N., Baudy-Floc'h, M., Brillet, B., Le Chevalier, P., and Fleury, Y. Antimicrobial peptides in oyster hemolymph: The bacterial connection. *Fish Shellfish Immunol.* 2013; 34(6):1439–1447.

Dennis, E.A. The growing phospholipase A2 superfamily of signal transduction enzymes. *Trends Biochem. Sci.* 1997; 22(1):1–2.

Desbois, A.P., and Smith, V.J. Antibacterial free fatty acids: Activities, mechanisms of action and biotechnological potential. *Appl. Microbiol. Biotechnol.* 2010; 85:1629–1642.

Destoumieux, D., Bulet, P., Loew, D., Van Dorsselaer, A., Rodriguez, J., and Bachere, E. Penaeidins, a new family of antimicrobial peptides isolated from the shrimp *Penaeus vannamei* (Decapoda). *J. Biol. Chem.* 1997; 272(45):28398–28406.

De Zoysa, M., Nikapitiya, C., Whang, I., Lee, J.S., and Lee, J. Abhisin: A potential antimicrobial peptide derived from histone H2A of disk abalone (*Haliotis discus discus*). *Fish Shellfish Immunol.* 2009; 27(5):639–646.

Diamond, G., Beckloff, N., Weinberg, A., and Kisich, K.O. The roles of antimicrobial peptides in innate host defense. *Curr. Pharm. Des.* 2009; 15(21):2377–2392.

Diamond, G., Zasloff, M., Eck, H., Brasseur, M., Maloy, W.L., and Bevins, C.L. Tracheal antimicrobial peptide, a cysteine-rich peptide from mammalian tracheal mucosa: Peptide isolation and cloning of a cDNA. *Proc. Natl. Acad. Sci. USA* 1991; 88(9):3952–3956.

Dooley, H., and Flajnik, M.F. Shark immunity bites back: Affinity maturation and memory response in the nurse shark, *Ginglymostoma cirratum*. *Eur. J. Immunol.* 2005; 35(3):936–945.

Döring, G. The role of neutrophil elastase in chronic inflammation. *Am. J. Respir. Crit. Care Med.* 1994; 150(6):S114–S117.

Douglas, S.E., Gallant, J.W., Liebscher, R.S., Dacanay, A., and Tsoi, S.C. Identification and expression analysis of hepcidin-like antimicrobial peptides in bony fish. *Dev. Comp. Immunol.* 2003; 27(6–7):589–601.

Ellison, R.T., III, and Giehl, T.J. Killing of gram-negative bacteria by lactoferrin and lysozyme. *J. Clin. Invest.* 1991; 88(4):1080–1091.

Elsbach, P., and Weiss, J. The bactericidal/permeability-increasing protein (BPI), a potent element in host-defense against gram-negative bacteria and lipopolysaccharide. *Immunobiology* 1993; 187(3–5):417–429.

Fange, R., Lundblad, G., and Lind, J. Lysozyme and chitinase in blood and lymphomyeloid tissues of marine fish. *Mar. Biol.* 1976; 36:277–282.

Fange, R., Lundblad, G., Slettengren, K., and Lind, J. Glycosidases in lymphomyeloid (hematopoietic) tissues of elsmobranch fish. *Comp. Biochem. Physiol. B: Comp. Biochem.* 1980; 67(4):527–532.

Fearon, D.T. Seeking wisdom in innate immunity. *Nature* 1997; 388:323–324.

Fennell, J.F., Shipman, W.H., and Cole, L.J. Antibacterial action of melittin, a polypeptide from bee venom. *Proc. Soc. Exp. Biol. Med.* 1968; 127(3):707–710.

Fernandes, J.M., Kemp, G.D., Molle, M.G., and Smith, V.J. Anti-microbial properties of histone H2A from skin secretions of rainbow trout, *Oncorhynchus mykiss*. *Biochem. J.* 2002; 368(Pt 2):611–620.

Fernandes, J.M., Kemp, G.D., and Smith, V.J. Two novel muramidases from skin mucosa of rainbow trout (*Oncorhynchus mykiss*). *Comp. Biochem. Physiol. B Biochem. Mol. Biol.* 2004; 138(1):53–64.

Fine, J.M., Drilhon, A., Ridgway, C.J., Amouch, P., and Boffa, G. Groups of transferrins in the *Anguilla* species. Differences in the phenotypic frequencies of transferrins in *Anguilla anguilla* and *Anguilla rostrata*. *C. R. Acad. Sci. Hebd. Seances Acad. Sci. D.* 1967; 265(1):58–60.

Flajnik, M.F. The immune system of ectothermic vertebrates. *Vet. Immunol. Immunopathol.* 1996; 54(1–4):145–150.

Fleming, A., and Allison, V.D. Observations on a bacteriolytic substance ("lysozyme") found in secretions and tissues. *Brit. J. Exp. Path.* 1922; 13:252–260.

Fletcher, T.C., and White, A. Lysozyme activity of the plaice (*Pleuronectes platessa* L.). *Experientia* 1973; 29:1283–1285.

Fluhr, R., and Kaplan-Levy, R.N. Plant disease resistance: Commonality and novelty in multicellular innate immunity. *Curr. Top. Microbiol. Immunol.* 2002; 270:23–46.

Fox, J.L. Antimicrobial peptides stage a comeback. *Nat. Biotechnol.* 2013; 31(5):379–382.

Frohm, M., Gunne, H., Bergman, A.C., Agerberth, B., Bergman, T., Boman, A., Liden, S., Jornvall, H., and Boman, H.G. Biochemical and antibacterial analysis of human wound and blister fluid. *Eur. J. Biochem.* 1996; 237(1):86–92.

Gallo, R.L., and Huttner, K.M. Antimicrobial peptides: An emerging concept in cutaneous biology. *J. Invest. Derm.* 1998; 3:739–743.

Gandomkar, M., Najafi, R., Shafiei, M., Mazidi, M., Goudarzi, M., Mirfallah, S.H., Ebrahimi, F., Heydarpor, H.R., and Abdie, N. Clinical evaluation of antimicrobial peptide [(99 m)Tc/Tricine/HYNIC(0)] ubiquicidin 29–41 as a human-specific infection imaging agent. *Nucl. Med. Biol.* 2009; 36:199–205.

Ganz, T. Hepcidin and its role in regulating systemic iron metabolism. *Hematology Am. Soc. Hematol. Educ. Program* 2006; 29–35:507.

Ganz, T., and Weiss, J. Antimicrobial peptides of phagocytes and epithelia. *Ser. Hematol.* 1997; 34:343–354.

Got, R., Font, J., and Goussault, Y. Study on transferrin of a Selachii, the great spotted dogfish (*Scyllium stellare*). *Comp. Biochem. Physiol.* 1967; 23(2):317–327.

Hanson, D.A., Kaspar, A.A., Poulain, F.R., and Krensky, A.M. Biosynthesis of granulysin, a novel cytolytic molecule. *Mol. Immunol.* 1999; 36(7):413–422.

Hedayati, A. Studies of lysozyme-like activity in nurse shark serum. Medical Laboratory Science, Masters Thesis. 1992. Florida International University, Miami, FL.

Hiemstra, P.S., Eisenhauer, P.B., Harwig, S.S., van den Barselaar, M.T., van Furth, R., and Lehrer, R.I. Antimicrobial proteins of murine macrophages. *Infect. Immun.* 1993; 61(7):3038–3046.

Hiemstra, P.S., van den Barselaar, M.T., Roest, M., Nibbering, P.H., and van Furth, R. Ubiquicidin, a novel murine microbicidal protein present in the cytosolic fraction of macrophages. *J. Leukoc. Biol.* 1999; 66(3):423–428.

Hinds, N. Assessment of antibacterial activity of leukocyte lysates of the nurse shark, (*Ginglymostoma cirratum*) Medical Laboratory Sciences, Masters Thesis. 2000. Florida International University, Miami, FL.

Hirono, I., Hwang, J.Y., Ono, Y., Kurobe, T., Ohira, T., Nozaki, R., and Aoki, T. Two different types of hepcidins from the Japanese flounder *Paralichthys olivaceus. FEBS J.* 2005; 272(20):5257–5264.

Howell, S.J., Wilk, D., Yadav, S.P., and Bevins, C.L. Antimicrobial polypeptides of the human colonic epithelium. *Peptides* 2003; 24(11):1763–1770.

Iijima, N., Tanimoto, N., Emoto, Y., Morita, Y., Uematsu, K., Murakami, T., and Nakai, T. Purification and characterization of three isoforms of chrysophsin, a novel antimicrobial peptide in the gills of the red sea bream, *Chrysophrys major. Eur. J. Biochem.* 2003; 270:675–686.

Imoto, T., Johnson, L.N., North, A.C.T., Phillips, D.C., and Rupley, J.A. Vertebrate lysozyme. In: Boyer, P.D., ed. *The Enzymes.* New York: Academic Press, 1972; 1:665–836.

Ingram, G.A. Substances involved in natural resistance of fish to infection—A review. *J. Fish Biol.* 1980; 16:23–60.

Inoue, Y., Suenaga, Y., Yoshiura, Y., Moritomo, T., Ototake, M., and Nakanishi, T. Molecular cloning and sequencing of Japanese pufferfish (*Takifugu rubripes*) NADPH oxidase cDNAs. *Dev. Comp. Immunol.* 2004; 28:911–925.

Itoh, N., Okada, Y., Takahashi, K.G., and Osada, M. Presence and characterization of multiple mantle lysozymes in the Pacific oyster, *Crassostrea gigas. Fish Shellfish Immunol.* 2010; 29(1):126–135.

Jacobsen, F., Baraniskin, A., Mertens, J., Mittler, D., Mohammadi-Tabrisi, A., Schubert, S., Soltau, M. *et al.* Activity of histone H1.2 in infected burn wounds. *J. Antimicrob. Chemother.* 2005; 55(5):735–741.

Janeway, C. Immunogenicity signals 1, 2, 3, and 0. *Immunol. Today* 1989; 10:283–286.

Jolles, P. Lysozymes: A chapter of molecular biology. *Angew. Chem. Int. Ed.* 1969; 8(4):227–294.

Jolles, P. From the discovery of lysozyme to the characterization of several lysozyme families. *EXS* 1996; 75:3–5.

Jolles, J., and Jolles, P. Isolation and purification of the biologically active des-lysylvalyl phenylalanine-lysozyme. *Biochem. Biophys. Res. Commun.* 1966; 22(1):22–25.

Jolles, J., and Jolles, P. The lysozyme from *Asterias rubens. Eur. J. Biochem.* 1975; 54(1):19–23.

Jolles, P., and Jolles, J. What's new in lysozyme research? Always a model system, today as yesterday. *Mol. Cell. Biochem.* 1984; 63(2):165–189.

Kabara, J.J., Swieczkowski, D.M., Conley, A.J., and Truant, J.P. Fatty acids and derivatives as antimicrobial agents. *Antimicrob. Agents Chemother.* 1972; 2:23–28.

Kawahara, T., Quinn, M.T., and Lambeth, J.D. Molecular evolution of the reactive oxygen-generating NADPH oxidase (Nox/Duox) family of enzymes. *BMC Evol. Biol.* 2007; 7:109.

Kawasaki, H., Isaacson, T., Iwamuro, S., and Conlon, J.M. A protein with antimicrobial activity in the skin of Schlegel's green tree frog *Rhacophorus schlegelii* (Rhacophoridae) identified as histone H2B. *Biochem. Biophys. Res. Commun.* 2003; 312(4):1082–1086.

Kawasaki, H., and Iwamuro, S. Potential roles of histones in host defense as antimicrobial agents. *Infect. Disord. Drug Targets* 2008; 8(3):195–205.

Kim, H.S., Cho, J.H., Park, H.W., Yoon, H., Kim, M.S., and Kim, S.C. Endotoxin-neutralizing antimicrobial proteins of the human placenta. *J. Immunol.* 2002; 168(5):2356–2364.

Kim, H.S., Yoon, H., Minn, I., Park, C.B., Lee, W.T., Zasloff, M., and Kim, S.C. Pepsin-mediated processing of the cytoplasmic histone H2A to strong antimicrobial peptide buforin I. *J. Immunol.* 2000; 165(6):3268–3274.

Kindt, T.J., Goldsby, R.A., Osborne, B.A., and Kuby, J. *Kuby Immunology.* New York: W.H. Freeman, 2007.

Kobayashi, S.D., Voyich, J.M., Burlak, C., and DeLeo, F.R. Neutrophils in the innate immune response. *Arch. Immunol. Ther. Exp. (Warsz)* 2005; 53:505–517.

Krause, A., Neitz, S., Mägert, H.J., Schulz, A., Forssmann, W.G., Schulz-Knappe, P., and Adermann, K. LEAP-1, a novel highly disulfide-bonded human peptide, exhibits antimicrobial activity. *FEBS Lett.* 2000; 480(2–3):147–150.

Krensky, A.M., and Clayberger, C. Biology and clinical relevance of granulysin. *Tissue Antigens* 2009; 73(3):193–198.

Kudo, T., Kigoshi, H., Hagiwara, T., Takino, T., Yamazaki, M., and Yui, S. Cathepsin G, a neutrophil protease, induces compact cell-cell adhesion in MCF-7 human breast cancer cells. *Mediators Inflamm.* 2009; 2009:850940.

Kumar, J., Okada, S., Clayberger, C., and Krensky, A.M. Granulysin: A novel antimicrobial. *Expert. Opin. Investig. Drugs* 2001; 10(2):321–329.

Laing, K.J., Grabowski, P.S., Belosevic, M., and Secombes, C.J. A partial sequence for nitric oxide synthase from a goldfish (*Carassius auratus*) macrophage cell line. *Immunol. Cell. Biol.* 1996; 74:374–379.

Laing, K.J., and Hansen, J.D. Fish T cells: Recent advances through genomics. *Dev. Comp. Immunol.* 2011; 35(12):1282–1295.

Lee, D.Y., Huang, C.M., Nakatsuji, T., Thiboutot, D., Kang, S.A., Monestier, M., and Gallo, R.L. Histone H4 is a major component of the antimicrobial action of human sebocytes. *J. Invest. Dermatol.* 2009; 129(10):2489–2496.

Lee, I.H., Cho, Y., and Lehrer, R.I. Styelins, broad-spectrum antimicrobial peptides from the solitary tunicate, *Styela clava. Comp. Biochem. Physiol. B Biochem. Mol. Biol.* 1997; 118(3):515–521.

Lehrer, R.I., Rosenman, M., Harwig, S.S., Jackson, R., and Eisenhauer, P. Ultrasensitive assays for endogenous antimicrobial polypeptides. *J. Immunol. Methods* 1991; 137(2):167–173.

Leiro, J., Ortega, M., Siso, M.I., Sanmartín M.L., and Ubeira F.M. Effects of chitinolytic and proteolytic enzymes on in vitro phagocytosis of microsporidians by spleen macrophages of turbot, *Scophthalmus maximus* L. *Vet. Immunol. Immunopathol.* 1997; 59(1–2):171–180.

Lepiller, S., Franche, N., Solary, E., Chluba, J., and Laurens, V. Comparative analysis of zebrafish nos2a and nos2b genes. *Gene* 2009; 445:58–65.

Li, G.H., Mine, Y., Hincke, M.T., and Nys, Y. Isolation and characterization of antimicrobial proteins and peptide from chicken liver. *J. Pept. Sci.* 2007; 13(6):368–378.

Liang, T., Ji, W., Zhang, G.R., Wei, K.J., Feng, K., Wang, W.M., and Zou, G.W. Molecular cloning and expression analysis of liver-expressed antimicrobial peptide 1 (LEAP-1) and LEAP-2 genes in the blunt snout bream (*Megalobrama amblycephala*). *Fish Shellfish Immunol.* 2013; 35(2):553–563.

Lie, Ø., Evensen Ø., Sørensen, A., and Frøysadal, E. Study on lysozyme activity in some fish species. *Dis. Aquat. Org.* 1989; 6:1–5.

Litman, G.W. Sharks and the origins of vertebrate immunity. *Sci. Am.* 1996; 275(5):67–71.

Lundblad, G., Fang, R., Slettengren, K., and Lind, J. Lysozyme, chintinase and exo-N-acetyl-beta-D-glucosaminidase (NAGase) in lymphomyeloid tissue of marine fishes. *Mar. Biol.* 1979; 53:311–315.

Magnadóttir, B. Innate immunity of fish (overview). *Fish Shellfish Immunol.* 2006; 20(2):137–151.

Mallow, E.B., Harris, A., Salzman, N., Russell, J.P., DeBerardinis, R.J., Ruchelli, E., and Bevins, C.L. Human enteric defensins. Gene structure and developmental expression. *J. Biol. Chem.* 1996; 271(8):4038–4045.

Manson, F.D.C., Fletcher, T.C., and Gooday, G.W. Localization of chitinolytic enzymes in blood of turbot, *Scophthalmus maximus*, and their possible roles in defence. *J. Fish Biol.* 1992; 40:919–927.

Martinelli, C., and Reichhart, J.M. Evolution and integration of innate immune systems from fruit flies to man: Lessons and questions. *J. Endotoxin. Res.* 2005; 11(4):243–248.

Masson, P.L., Heremans, J.F., Prignot, J.J., and Wauters, G. Immunohistochemical localization and bacteriostatic properties of an iron-binding protein from bronchial mucus. *Thorax* 1966; 21(6):538–544.

Medzhitov, R., Preston-hurlburt, P., and Janeway, C.A., Jr. A human homologue of the *Drosophila* Toll protein signals activation of adaptive immunity. *Nature* 1997; 388:394–397.

Meyer-Hoffert, U. Neutrophil-derived serine proteases modulate innate immune responses. *Front. Biosci.* 2009; 14:3409–3418.

Moore, A.J., Beazley, W.D., Bibby, M.C., and Devine, D.A. Antimicrobial activity of cecropins. *J. Antimicrob. Chemother.* 1996; 37(6):1077–1089.

Moore, K.S., Wehrli, S., Roder, H., Rogers, M., Forrest, J.N., Jr., McCrimmon, D., and Zasloff, M. Squalamine: An aminosterol antibiotic from the shark. *Proc. Natl. Acad. Sci. USA* 1993; 90(4):1354–1358.

Nakajima, Y., Ogihara, K., Taylor, D., and Yamakawa, M. Antibacterial hemoglobin fragments from the midgut of the soft tick, *Ornithodoros moubata* (Acari: Argasidae). *J. Med. Entomol.* 2003; 40(1):78–81.

Nam, B.H., Seo, J.K., Go, H.J., Lee, M.J., Kim, Y.O., Kim, D.G., Lee, S.J., and Park, N.G. Purification and characterization of an antimicrobial histone H1-like protein and its gene from the testes of olive flounder, *Paralichthys olivaceus. Fish Shellfish Immunol.* 2012; 33(1):92–98.

Noga, E.J., Fan, Z., and Silphaduang, U. Histone-like proteins from fish are lethal to the parasitic dinoflagellate *Amyloodinium ocellatum. Parasitology* 2001; 123(Pt 1):57–65.

Noga, E.J., and Silphaduang, U. Piscidins, a novel family of peptide antibiotics from fish. *Drug News Perspect.* 2003; 16:87–92.

Nonaka, M., and Kimura, A. Genomic view of the evolution of the complement system. *Immunogenetics* 2006; 58(9):701–713.

Odeberg, H., and Olsson, I. Antibacterial activity of cationic proteins from human granulocytes. *J. Clin. Invest.* 1975; 56(5):1118–1124.

Ogundele, M.O. A novel anti-inflammatory activity of lysozyme: Modulation of serum complement activation. *Mediators Inflamm.* 1998; 7(5):363–365.

Ohta, S., Chang, T., Kawashima, A., Nagate, T., Murase, M., Nakanishi, H., Miyata, H., and Kondo, M. Methicillin-resistant *Staphylococcus aureus* (MRSA) activity by linolenic acid isolated from the marine microalga *Chlorococcum* HS-101. *Bull. Environ. Contam. Toxicol.* 1994; 52:673–680.

Ohta, Y., McKinney, E.C., Criscitiello, M.F., and Flajnik, M.F. Proteasome, transporter associated with antigen processing, and class I genes in the nurse shark, *Ginglymostoma cirratum*: Evidence for a stable class I region and MHC haplotype lineages. *J. Immunol.* 2002; 168(2):771–781.

Ohta, Y., Shiina, T., Lohr, R.L., Hosomichi, K., Pollin, T.I., Heist, E.J., Suzuki, S., Inoko, H., and Flajnik, M.F. Primordial linkage of β2-microglobulin to the MHC. *J. Immunol.* 2011; 186(6):3563–3571.

Olavarria, V.H., Gallardo, L., Figueroa, J.E., and Mulero, V. Lipopolysaccharide primes the respiratory burst of Atlantic salmon SHK-1 cells through protein kinase C-mediated phosphorylation of p47phox. *Dev. Comp. Immunol.* 2010; 34(12):1242–1253.

Oren, Z., and Shai, Y. Mode of action of linear amphipathic alpha-helical antimicrobial peptides. *Biopolymers* 1998; 47(6):451–463.

Parish, C.A., Jiang, H., Tokiwa, Y., Berova, N., Nakanishi, K., McCabe, D., Zuckerman, W., Xia, M.M., and Gabay, J.E. Broad-spectrum antimicrobial activity of hemoglobin. *Bioorg. Med. Chem.* 2001; 9(2):377–382.

Park, C.B., Kim, M.S., and Kim, S.C. A novel antimicrobial peptide from *Bufo bufo gargarizans. Biochem. Biophys. Res. Commun.* 1996; 218(1):408–413.

Park, C.B., Yi, K.S., Matsuzaki, K., Kim, M.S., and Kim, S.C. Structure-activity analysis of buforin II, a histone H2A-derived antimicrobial peptide: The proline hinge is responsible for the cell-penetrating ability of buforin II. *Proc. Natl. Acad. Sci. USA* 2000; 97(15):8245–8250.

Park, I.Y., Park, C.B., Kim, M.S., and Kim, S.C. Parasin I, an antimicrobial peptide derived from histone H2A in the catfish, *Parasilurus asotus. FEBS Lett.* 1998; 437(3):258–262.

Patat, S.A., Carnegie, R.B., Kingsbury, C., Gross, P.S., Chapman, R., and Schey, K.L. Antimicrobial activity of histones from hemocytes of the Pacific white shrimp. *Eur. J. Biochem.* 2004;271(23–24):4825–4833.

Patrzykat, A., Zhang, L., Mendoza, V., Iwama, G.K., and Hancock, R.E. Synergy of histone-derived peptides of coho salmon with lysozyme and flounder pleurocidin. *Antimicrob. Agents Chemother.* 2001; 45(5):1337–1342.

Petit, J.F., and Jolles, P. Purification and analysis of human saliva lysozyme. *Nature* 1993; 200: 168–169.

Pettey, C.L., and McKinney, E.C. Temperature and cellular regulation of spontaneous cytotoxicity in the shark. *Eur. J. Immunol.* 1983; 13(2):133–138.

Pham, C.T. Neutrophil serine proteases: Specific regulators of inflammation. *Nat. Rev. Immunol.* 2006; 6:541–550.

Pruzanski, W., and Saito, S.G. The diagnostic value of lysozyme (muramidase) estimation in biological fluids. *Am. J. Med. Sci.* 1969; 258(6):405–415.

Putnam, F.W., ed. Transferrin. In: *The Plasma Proteins*. London: Academic Press, 1975; 1:265–316.

Raj, P.A., Marcus, E., and Edgerton, M. Delineation of an active fragment and poly(L-proline) II conformation for candidacidal activity of bactenecin 5. *Biochemistry* 1996; 35(14):4314–4325.

Rakoff-Nahoum, S., and Medzhitov, R. Role of the innate immune system and host-commensal mutualism. *Curr. Top. Microbiol. Immunol.* 2006; 308:1–18.

Reddick, J.I., Goostrey, A., and Secombes, C.J. Cloning of iNOS in the small spotted catshark (*Scyliorhinus canicula*). *Dev. Comp. Immunol.* 2006; 30:1009–1022.

Relf, J.M., Chisholm, J.R., Kemp, G.D., and Smith, V.J. Purification and characterization of a cysteine-rich 11.5-kDa antibacterial protein from the granular haemocytes of the shore crab, *Carcinus maenas*. *Eur. J. Biochem.* 1999; 264(2):350–357.

Richards, R.C., O'Neil, D.B., Thibault, P., and Ewart, K.V. Histone H1: An antimicrobial protein of Atlantic salmon (*Salmo salar*). *Biochem. Biophys. Res. Commun.* 2001; 284(3):549–555.

Rieger, A.M., and Barreda, D.R. Antimicrobial mechanisms of fish leukocytes. *Dev. Comp. Immunol.* 2011; 35(12):1238–1245.

Rijkers, G.T. Non-lymphoid defense mechanisms in fish. *Dev. Comp. Immunol.* 1982; 6(1):1–13.

Rinehart, K.L., Holt, T.G., Fregeau, N.L., Keifer, P.A., Wilson, G.R., Perun, T.J., Jr., Sakai, R. *et al.* Bioactive compounds from aquatic and terrestrial sources. *J. Nat. Prod.* 1990; 53(4):771–792.

Robinette, D., Wada, S., Arroll, T., Levy, M.G., Miller, W.L., and Noga, E.J. Antimicrobial activity in the skin of the channel catfish *Ictalurus punctatus*: Characterization of broad-spectrum histone-like antimicrobial proteins. *Cell. Mol. Life Sci.* 1998; 54(5):467–475.

Rose, F.R., Bailey, K., Keyte, J.W., Chan, W.C., Greenwood, D., and Mahida, Y.R. Potential role of epithelial cell-derived histone H1 proteins in innate antimicrobial defense in the human gastrointestinal tract. *Infect. Immun.* 1998; 66(7):3255–3263.

Saeij, J.P., Stet, R.J., Groeneveld, A., Verburg-van Kemenade, L.B., van Muiswinkel, W.B., and Wiegertjes, G.F. Molecular and functional characterization of a fish inducible-type nitric oxide synthase. *Immunogenetics* 2000; 51(4–5):339–346.

Salerno, G., Parrinello, N., Roch, P., and Cammarata, M. cDNA sequence and tissue expression of an antimicrobial peptide, dicentracin; a new component of the moronecidin family isolated from head kidney leukocytes of sea bass, *Dicentrarchus labrax*. *Comp. Biochem. Physiol.* 2007; 146B:521–529.

Sang, Y., and Blecha, F. Antimicrobial peptides and bacteriocins: Alternatives to traditional antibiotics. *Anim. Health Res. Rev.* 2008; 9(2):227–235.

Schnapp, D., Kemp, G.D., and Smith, V.J. Purification and characterization of a proline-rich antibacterial peptide, with sequence similarity to bactenecin-7, from the haemocytes of the shore crab, *Carcinus maenas*. *Eur. J. Biochem.* 1996; 240(3):532–539.

Schröder, H.C., Ushijima, H., Krasko, A., Gamulin, V., Thakur, N.L., Diehl-Seifert, B., Müller, I.M., and Müller, W.E.G. Emergence and disappearance of an immune molecule, an antimicrobial lectin, in basal metazoan. *J. Biol. Chem.* 2003; 278:32810–32817.

Selsted, M.E., Harwig, S.S., Ganz, T., Schilling, J.W., and Lehrer, R.I. Primary structures of three human neutrophil defensins. *J. Clin. Invest.* 1985; 76(4):1436–1439.

Senior, R.M., Campbell, E.J., Landis, J.A., Cox, F.R., Kuhn, C., and Koren, H.S. Elastase of U-937 monocytelike cells. Comparisons with elastases derived from human monocytes and neutrophils and murine macrophagelike cells. *J. Clin. Invest.* 1982; 9(2):384–393.

Seo, J.K., Stephenson, J., and Noga, E.J. Multiple antibacterial histone H2B proteins are expressed in tissues of American oyster. *Comp. Biochem. Physiol. B Biochem. Mol. Biol.* 2011; 158(3):223–229.

Silphaduang, U., Hincke, M.T., Nys, Y., and Mine, Y. Antimicrobial proteins in chicken reproductive system. *Biochem. Biophys. Res. Commun.* 2006; 340(2):648–655.

Smine, A., and Le Gal, Y. Purification and characterization of two pancreatic elastase isoforms from dogfish (*Scyliorhinus canicula*). *Mol. Mar. Biol. Biotechnol.* 1995; 4(4):295–303.

Smith, S.L. Shark complement: An assessment. *Immunol. Rev.* 1998; 166:67–78.

Smith, V.J., Desbois, A.P., and Dyrynda, E.A. Conventional and unconventional antimicrobials from fish, marine invertebrates and micro-algae. *Mar. Drugs* 2010; 8(4):1213–1262.

Smith, V.J., and Fernandes, J.M.O. Non-specific antimicrobial proteins of the innate system. In: Zaccone, G., Masseguer, J., García-Ayala, A., Kapoor, B.G., eds. *Fish Defences*. Enfield, NH: Science Publishers, 2009; 1:241–275.

Spitznagel, J.K. Nonoxidative antimicrobial reactions of leukocytes. *Contemp. Top. Immunobiol.* 1984; 14:283–343.

Spitznagel, J.K. Antibiotic proteins of human neutrophils. *J. Clin. Invest.* 1990; 86(5):1381–1386.

Starkey, P.M., and Barrett, A.J. Evolution of alpha 2-macroglobulin. The demonstration in a variety of vertebrate species of a protein resembling human alpha 2-macroglobulin. *Biochem. J.* 1982; 205(1):91–95.

Steiner, H., Hultmark, D., Engstrom, A., Bennich, H., and Boman, H.G. Sequence and specificity of two antibacterial proteins involved in insect immunity. *Nature* 1981; 292(5820):246–248.

Steinstraesser, L., Kraneburg, U.M., Hirsch, T., Kesting, M., Steinau, H.U., Jacobsen, F., and Al-Benna, S. Host defense peptides as effector molecules of the innate immune response a sledgehammer for drug resistance? *Int. J. Mol. Sci.* 2009; 10(9):3951–3970.

Suzomoto, B.K., Schreck, C.B., and McIntyre, J.D. Relative resistance of three transferring genotypes of coho salmon (*Oncorhyncus Kisutch)* and their hematological responses to bacterial kidney disease. *J. Fish Res. Bd. Can.* 1977; 34:1–8.

Tan, Y.Y., Kodzius, R., Tay, B.H., Tay, A., Brenner, S., and Venkatesh, B. Sequencing and analysis of full length cDNAs, 5′-ESTs and 3′-ESTs from a cartilaginous fish, the Elephant shark (*Callorhinchus millii*). *PLoS One* 2012; 7(10):E47174.

Tang, Y.Q., Yuan, J., Miller, C.J., and Selsted, M.E. Isolation, characterization, cDNA cloning, and antimicrobial properties of two distinct subfamilies of alpha-defensins from rhesus macaque leukocytes. *Infect. Immun.* 1999; 67(11):6139–6144.

Tassanakajon, A., Amparyup, P., Somboonwiwat, K., and Supungul, P. Cationic antimicrobial peptides in Penaeid Shrimp. *Mar. Biotechnol. (NY)* 2010; 12(5):487–505.

Tewary, P., Yang, D., de la Rosa, G., Li, Y., Finn, M.W., Krensky, A.M., Clayberger, C., and Oppenheim, J.J. Granulysin activates antigen-presenting cells through TLR4 and acts as an immune alarmin. *Blood* 2010; 116(18):3465–3474.

Tomita, M., Takase, M., Bellamy, W., and Shimamura, S. A review: The active peptide of lactoferrin. *Acta Paediatr. Jpn.* 1994; 36(5):585–591.

Tralau, T., Meyer-Hoffert, U., Schroder, J.M., and Wiedow, O. Human leukocyte elastase and cathepsin G are specific inhibitors of C5a-dependent neutrophil enzyme release and chemotaxis. *Exp. Dermatol.* 2004; 13:316–325.

Tsai, H., and Bobek, L.A. Human salivary histatins: Promising anti-fungal therapeutic agents. *Crit. Rev. Oral. Biol. Med.* 1998; 9(4):480–497.

Tsao, H.S., Spinella, S.A., Lee, A.T., and Elmore, D.E. Design of novel histone-derived antimicrobial peptides. *Peptides* 2009; 30(12):2168–2173.

Tsugita, A., and Inouye, M. Complete primary structure of phage lysozyme from *Escherichia coli* T4. *J. Mol. Biol.* 1968; 37(1):201–212.

Unanue, E.R., Beller, D.I., Calderon, J., Kiely, J.M., and Stadecker, M.J. Regulation of immunity and inflammation by mediators from macrophages. *Am. J. Pathol.* 1976; 5(2):465–478.

Ursic-Bedoya, R.J., Nazzari, H., Cooper, D., Triana, O., Wolff, M., and Lowenberger, C. Identification and characterization of two novel lysozymes from *Rhodnius prolixus*, a vector of Chagas disease. *J. Insect. Physiol.* 2008; 54(3):593–603.

Utter, F.M., Ames, W.E., and Hodgin, H.O. Transferrin polymorphism in coho salmon (*Oncorhynchus kisutch*). *J. Fish Res. Bd. Can.* 1970; 27:2371–2373.

Vaughan, N.H., and Smith, S.L. Isolation and characterization of a c type lysozyme from the nurse shark. *Fish Shellfish Immunol.* 2013; 35:1824–1828.

Vizioli, J., and Salzet, M. Antimicrobial peptides from animals: Focus on invertebrates. *Trends Pharmacol. Sci.* 2002; 23(11):494–496.

Walker, B.T., and Houston, T.A. Squalamine and its derivatives as potential antitubercular compounds. *Tuberculosis (Edinb)* 2013; 93(1):102–103.

Wang, T., Ward, M., Grabowski, P., and Secombes, C.J. Molecular cloning, gene organization and expression of rainbow trout (Oncorhynchus mykiss) inducible nitric oxide synthase (iNOS) gene. *Biochem. J.* 2001; 358:747–755.

Wang, Y., Griffiths, W.J., Jornvall, H., Agerberth, B., and Johansson, J. Antibacterial peptides in stimulated human granulocytes: Characterization of ubiquitinated histone H1A. *Eur. J. Biochem.* 2002; 269(2):512–518.

Weinberg, E.D. Iron and susceptibility to infectious disease. *Science* 1974; 184(140):952–956.

Wiedow, O., and Meyer-Hoffert, U. Neutrophil serine proteases: Potential key regulators of cell signalling during inflammation. *J. Intern. Med.* 2005; 257(4):319–328.

Wiens, M., Korzhev, M., Perovic-Ottstadt, S., Luthringer, B., Brandt, D., Klein, S., and Müller, W.E. Toll-like receptors are part of the innate immune defense system of sponges (demospongiae: Porifera). *Mol. Biol. Evol.* 2007; 24(3):792–804.

Yin, Z.-X., Chen, W.-J., Yan, J.-H., Yang, J.-N., Chan, S.-M., and He, J.-G. Cloning,expression and antimicrobial activity of an antimicrobial peptide, epinecidin-1, from the orange-spotted grouper, *Epinephelus coioides*. *Aquaculture* 2006; 253:204–211.

Zanetti, M., Gennaro, R., and Romeo, D. Cathelicidins: A novel protein family with a common proregion and a variable C-terminal antimicrobial domain. *FEBS Lett.* 1995; 374(1):1–5.

Zasloff, M. Magainins, a class of antimicrobial peptides from *Xenopus* skin: Isolation, characterization of two active forms, and partial cDNA sequence of a precursor. *Proc. Natl. Acad. Sci. USA* 1987; 84(15):5449–5453.

Zasloff, M. Antibiotic peptides as mediators of innate immunity. *Curr. Opin. Immunol.* 1992; 4:3–7.

Zasloff, M. Antimicrobial peptides of multicellular organisms. *Nature* 2002; 415:389–395.

Zitvogel, L., and Kroemer, G. The multifaceted granulysin. *Blood* 2010; 116(18):3379–3380.

Zou, J., Mercier, C., Koussounadis, A., and Secombes, C. Discovery of multiple beta-defensin like homologues in teleost fish. *Mol. Immunol.* 2007; 44:638–647.

CHAPTER 16

Shark-Derived Immunomodulators

Liza Merly and Sylvia L. Smith

CONTENTS

SUMMARY

The aim of this chapter is to review and highlight our current understanding of bioactive compounds derived from the shark, particularly those with immunomodulating properties. Historically, shark-derived products have been used in traditional medicine in certain countries for centuries. However, the last several decades have seen a worldwide increase in the intake of shark products as dietary supplements and in their use as natural remedies. Sharks possess robust immune capacity, exhibiting all of the major innate and adaptive immune mechanisms present in higher vertebrates such as man. Their relatively high degree of immunological resilience along with other physical and behavioral aspects of shark biology has fueled the notion that consuming shark products could be beneficial to human health. In this chapter, we review the various shark-derived components that are believed to be effective anticancer agents because of their anti-angiogenic properties and those that exhibit a range of bioactive properties including antimicrobial activity and the ability to modulate immune function, that is, immunomodulatory agents. For example, squalamine exhibits a variety of distinct bioactive properties ranging from anti-angiogenic to antiviral effects. Some products have made their way through early stages of the clinical pipeline and entered clinical trials with varying degrees of success. Our present knowledge of the immunomodulatory properties of shark liver oil and shark cartilage-derived products will be reviewed along with their use as dietary supplements and as therapeutic and prophylactic agents against disease. In case of the latter, the impact on public health will be examined, particularly because there is an absence of strong scientific corroborative evidence in support of their purported effectiveness. In addition, the impact of shark harvests and of the trade in shark products on environmental health will be addressed, taking into account the role that sharks play in the health of marine ecosystems.

INTRODUCTION

Immunomodulators either enhance or suppress immune function or can do both as a function of dose. The immunomodulatory capability of shark-derived compounds and their potential application in human medicine have been the focus of several different lines of scientific inquiry and public curiosity. Sharks were harvested for both food and medicinal purposes by various peoples for centuries (Feretti *et al.*, 2010). The mystique surrounding their physical strength and agility as well as their relatively long survival record has fueled the increased exploitation of shark resources. Sharks represent an extant group of vertebrates that have survived for millions of years and their evolutionary success may at least, in part, be explained by a competent innate and adaptive immune system consisting components similar to those found in mammals. Sharks are capable of maintaining effective defenses within their microbe-rich marine environment. The immunological resilience noted in sharks has lead, unfortunately, to erroneous and misleading conclusions to be drawn and unsubstantiated claims to be made that a variety of shark products enhance human immunity unequivocally. The purpose here is to examine the scientific evidence for immunomodulatory properties of shark-derived compounds with a focus on two specific substances, shark liver oil and shark cartilage, both of which are popular substances usually taken as dietary supplements. Their historical use as well as the current trends the natural product industry is engaged to market them will be reviewed. In addition, the immunomodulatory properties of various shark compounds, the consequences of their unrestrictive use on public health, and the effects on global shark populations as trade in shark resources increases worldwide will be discussed.

SHARKS AS A RESOURCE

The life history of sharks has not evolved under high rates of natural mortality because they are top-level predators; this makes them particularly vulnerable to high levels of exploitation. One can argue that the very biological features of sharks that make them immunologically resilient and have allowed them to survive millions of years are being exploited and the result is that survival of some shark species is now threatened. Sharks were used as food by coastal communities for over 5000 years, and many shark products have been a component of traditional Chinese medicine for centuries (Clarke *et al.*, 2007). Besides being consumed as food, sharks are harvested for their fins and oil in the global market in which many nations participate. Pelagic sharks are generally characterized by their strength and speed as well as their elusive life history. This, combined with the folk belief in the beneficial properties of shark products, has led to the practice of consuming shark, particularly in the form of shark fin soup in many Asian countries, including China and Japan. It is now accepted that the consumption of shark fin soup is a major driver in the trade of shark resources worldwide.

The status of shark species is characterized as threatened in many parts of the world primarily due to over exploitation (Lotze *et al.*, 2010). Sharks exhibit slow growth, late sexual maturity, and low rates of reproduction so sustainability under fishing pressure is very difficult to achieve. Until very recently, most shark fisheries were not managed in an effective way to ensure sustainability of shark stocks. Stock assessments for oceanic sharks are not readily available. Although there has been a call for plans of action by nations regarding the management and conservation of sharks, only a few have been put in place, particularly with open ocean species. The harvesting of sharks for the shark fin industry is also a potential source of other shark products such as liver oil and shark cartilage, in order to take advantage of the whole animal rather than discarding carcasses or live animals after finning takes place. Unfortunately, the consumption of these natural products may be creating additional markets that provide significant incentive for the shark fin trade.

SHARK-DERIVED NATURAL PRODUCTS

Immunomodulatory compounds derived from sharks contain putative bioactive components with diverse chemical structures and mechanisms of action, many of which remain to be isolated and fully characterized. Bioactivity may include properties of that of an antioxidant, antibacterial, antiviral, anti-angiogenic, antitumor, immunoregulatory, and immunostimulatory agent. The potential use of these compounds and/or synthetic alternatives in clinical medicine remains unclear as does the safety and effectiveness of commercial natural products currently available to the public. In some cases, the nature of the immunomodulatory activity is well understood and studies are underway to determine how best to develop therapeutics based on either natural products or similar synthetic compounds. The spectrum of immunomodulatory properties of a variety of shark cartilage products still remain to be defined and verified, by both *in vivo* and *in vitro* studies, before their usefulness can be defined with confidence. Presently, their practical use remains highly debatable.

A few of the shark-derived immunomodulating compounds that have been described are proteins or glycoproteins, but others, obtained primarily from shark livers, are lipid in nature. Elasmobranchs have relatively large livers, which serve along with a cartilaginous skeleton to provide them with the necessary buoyancy for their aquatic habitat (Baldridge, 1970). Shark livers contain a high abundance of unsaturated fats and oils, many of which were used traditionally in natural remedies due to their purported healing abilities. More recently, compounds isolated from shark liver oil have been shown to contain various bioactive properties (Belo *et al.*, 2010). One such compound that has exhibited various activities is squalamine, a compound first isolated from the stomach and various other organs from dogfish shark, *Squalus acanthias*. It is antibacterial and kills a broad range of microorganisms including *Candida*, an opportunistic fungus, and is as effective as ampicillin against commonly encountered clinically relevant microorganisms (Moore *et al.*, 1993). This aminosterol compound also exhibits anti-proliferative, antioxidant, and anti-angiogenic properties in endothelial cells, and most recently, it has been shown to contain antiviral activity against several human pathogens (Zasloff, 2011). Structurally, the molecule is formed of two ubiquitous constituents: a steroid similar to that occurring in the cholesterol synthetic pathway in animals, and spermadine, an amine that interacts with nucleic acids and has a role in stabilizing membranes. Fortunately, squalamine can be synthetically produced in the laboratory and exhibits comparable bioactivity to the natural product and has been shown to be safe for human consumption. Consequently, because it can be readily prepared synthetically, its potential application in clinical medicine does not pose a threat to shark populations.

Another sterol derived from shark is squalene, a precursor for cholesterol synthesis in most animals. It is present in significant quantities in the liver of most sharks and has been attributed to both anticancer and antioxidant properties (Bhilwade *et al.*, 2010). Squalene is readily available from two major sources, olive oils and liver oil of deep sea sharks. However, because the purity and yield of squalene from sharks is substantially better than from olive oil with much reduced processing time (Camin *et al.*, 2010), it is preferred. Shark squalene has also been used to prepare oil-in-water emulsions to boost the immunogenicity of antigens when used as adjuvants. Recently, a plant-derived squalene of high purity has been shown to be equally effective as an adjuvant for immunization purposes (Brito *et al.*, 2011). These examples illustrate the importance of identifying and developing alternative sources for natural compounds, preferably by synthetic manufacture or as a byproduct of other industrial processes with less negative impact on environmental resources which presently are under siege for many natural products. It is clear that this can only be achieved when bioactive components of natural products are thoroughly investigated and analyzed biochemically and their biological properties unequivocally established.

Shark Liver Oil

Shark liver oil has been used by northern European fishing nations and Japan as an alternative treatment for cancer for many years (Denaiu, 2010). It has been a component of traditional medicine in these countries for centuries, particularly for immune-related diseases (Lewkowicz *et al.*, 2006). In Scandinavia, it is believed that shark liver oil upregulates the immune response and its use to treat cancer stems from traditional medicinal practices. The main components of shark liver oil are primarily 1-*O*-alkylglycerols and omega-3 fatty acids. Alkylglycerols are ether lipids that serve as precursors to phospholipids, which make up the structural and functional components of cellular membranes (Hallgren and Larsson, 1962). It is believed that one of the putative mechanisms of action for these lipids in cancer is their involvement as precursors in phospholipid metabolism. Because phospholipids are critical structural and functional components of cells, their metabolism is tightly linked to the regulation of cell proliferation. Alteration in lipid metabolism is one of the first major changes that occur in transformed cells (Hilvo and Oresic, 2012).

Shark liver oil has been used commercially for over 50 years as an adjunctive treatment for cancer, particularly to protect against tissue damage following radiation therapy (Pugliese *et al.*, 1998). In early experiments, alkylglycerols derived from shark liver oil were shown to regress tumor growth when administered before radiation therapy used to treat uterine cervical cancer (Brohult *et al.*, 1978). This regression was quantified as a change in the quotient between early and more advanced stages of cancer. The regression of tumors was also more prevalent in patients below the age of 60 than in those that were older (Brohult *et al.*, 1986). Synthetic analogues of *O*-alkylglycerols have been shown to inhibit the proliferation of cancer cells *in vitro*. They are also cytotoxic against many tumors and leukemias of human origin *in vitro* (Schick *et al.*, 1987). Additionally, synthetic alkylglycerols are active against Lewis lung carcinoma and in grafted tumors in mice when administered *in vivo* (Brachwitz *et al.*, 1987; Pedrono *et al.*, 2004). The role of ether lipids and their analogues in oncological treatment either alone or in combination with other agents continues to be investigated.

Shark Liver Oil and Immunomodulation

As stated earlier, alkylglycerols have been used to enhance immune function following radiation therapy, specifically as a treatment for leukopenia following irradiation (Brohult and Holmberg, 1954). Alkylglycerols may play a role in hematopoiesis and are found in hematopoietic organs such as bone marrow in other organisms (Denaiu *et al.*, 2010). The putative role of alkylglycerols in stimulating the immune response has become the focus of many studies, both *in vitro* and *in vivo*, particularly with regard to their effects on wound healing, immune dysfunction, and inflammation (Pugliese *et al.*, 1998). In particular, the effect of alkylglycerols on the activation of macrophages has been investigated. Both butyl alcohol, a naturally occurring alkylglycerol found in shark liver oil, and dodecylglycerol, a synthetic ether lipid, activate mouse peritoneal macrophages *in vivo*. *In vitro* studies have shown that the activation of macrophages is dependent upon the presence of both B and T lymphocytes, suggesting that there is significant cell-to-cell interaction required for its mechanism of action. Activation of macrophages results in increased phagocytic activity as well as increased extracellular cytolytic activity due to an upregulation in IgG receptor binding of macrophages to target cells (Pugliese *et al.*, 1998).

Subsequent studies have also found that components of shark liver oil can regulate complement levels, natural killer cell activity, and the production of reactive oxygen metabolites by peripheral blood leukocytes in patients suffering from rheumatoid arthritis (Tchorzewski *et al.*, 2002). Peripheral blood mononuclear cells produce a predominantly Th1 type cytokine profile in response to shark liver oil, characterized by upregulation of interferon-gamma (IFN-γ), tumor necrosis factor alpha (TNF-α), and interleukin 2 (IL-2) (Lewkowicz *et al.*, 2005). In a recent study, it had a significant effect on tumor infiltrating lymphocytes and cytokine production in tumor-bearing mice. Shark liver oil has been shown to enhance delayed-type hypersensitivity response against sheep red blood cells as well as increase the number of T lymphocytes infiltrating the tumor. The cytokine production of splenic mononuclear cells shifted to a Th1 type pattern in response to liver oil injection. A decrease in the size of tumors following treatment was noted in this murine study (Hajimoradi *et al.*, 2009).

One of the difficulties in determining the biological activity of various alkylglycerols derived from shark liver is that some of the beneficial effects observed may be due to other components present in the mixtures used in these studies. In many studies carried out over the last 50 years, the purity and chemical composition of the shark liver oil preparations used was highly variable from one study to another. Shark liver oil contains high amounts of squalene and polyunsaturated fatty acids (PUFA) that may contribute to the different effects observed (Denaiu, 2010). Fish oil (from nonshark source) and particularly PUFA were studied extensively and also appear to modulate the immune response and decrease the growth of tumors. In a recent study, shark liver oil, which contains both PUFA and alkylglycerols, was compared to fish oil on its effect on tumor growth and macrophage function in Walker-256 tumor-bearing rats. Tumor growth was inhibited independently by each type of oil and also when both fish oil and shark liver oil were combined in the treatment of rats. The study concluded that the shark liver oil had no additive effect on the action of fish oil and that the mechanism of action was likely unrelated to the regulation of macrophage function. Fish oil enhanced macrophage phagocytic ability, superoxide production, and NO production, whereas shark liver oil only stimulated slight increases in NO production in peritoneal macrophages (Belo *et al.*, 2010).

From a public health point of view, caution should be exercised in drawing conclusions as to the composition of shark liver oil in dietary supplements because such products are not regulated to meet certain acceptable standards with regard to content. It is not known whether commercial preparations of shark liver oil contain comparable levels of bioactive alkylglycerols described in the earlier studies because depending on the manufacturer's extraction process many of the alkylglycerols can be considerably reduced or removed entirely (Pugliese *et al.*, 1998). In many northern European countries, there are commercial preparations of shark liver oil that purport to contain a standardized dose of alkylglycerols for consumption, but these dosages are based on early animal studies investigating the effects of purified alkylglycerols. The recommended dosage of dietary supplements in the United States is not standardized and is relatively high, requiring intake of multiple capsules daily. Furthermore, it has been reported that commercial preparations of shark liver oil contain components that may raise blood cholesterol levels (Hajimoradi *et al.*, 2009). An important alternative to be considered in the continued marketing of shark liver oil as a dietary supplement is its replacement with fish oil shown to be equally biologically active while having a potentially less severe ecological impact. Thus, fish oil represents a viable alternate choice as a source of marine lipid compounds sought for their medicinal value. Furthermore, there is overwhelming evidence in the literature that many of the active components in shark liver oil can be developed and synthetically manufactured and which exhibit similar *in vitro* and *in vivo* effects with respect to cancer and immunomodulation.

Shark Cartilage

Shark cartilage-derived products have been used as a source of food and dietary supplements for many years and more recently as natural remedies in the practice of complementary and alternative medicine. The increasing interest in the application of shark cartilage products in medicine prompted investigators to isolate and characterize components of shark cartilage with an aim to define their bioactive properties. For over two decades, most shark cartilage research has focused largely on its anti-angiogenic properties and its application as an anticancer product in clinical medicine. However, in the last decade, attention has been given to its potential as an immune regulator, which has stemmed from the considerable interest directed to the immunomodulatory properties of natural products of plant and animal origin.

Research on cartilage began in earnest following the studies of Brem and Folkman, who demonstrated that bovine cartilage exhibits anti-angiogenic properties (Brem and Folkman, 1975). The apparent anti-angiogenic properties of cartilage led to speculation as to its potential use in preventing the formation and/or controlling the proliferation of cancerous tumors by limiting the neovascularization of developing tumors (Brem and Folkman, 1975; Prudden and Balassa, 1974). Since only a relatively small yield of cartilage can be recovered from mammalian (nonhuman) tissues per gram of raw material, investigators were prompted to examine shark cartilage for similar properties, because a higher yield of cartilage per unit weight could be obtained given that the entire shark skeleton is made of cartilage. Results from these and subsequent studies have confirmed that certain derivatives of shark cartilage have anti-angiogenic properties against tumors in experimental animal studies (Gonzalez et al., 2001a). Bargahi and Rabbani-Chadegani (2008) have extracted a partially purified protein fraction from shark cartilage consisting two proteins of molecular masses of 14.7 and 16.0 kDa and have shown that both have an inhibiting effect on endothelial cell angiogenesis. Further studies by Rabbani-Chadegani et al. (2008) identified SCP1, a protein of molecular weight 13.7 kDa with potent inhibitory effect on capillary growth.

There have been several studies where the oral administration of powdered, commercial shark cartilage has been tested in an in vivo rabbit cornea assay. The results show that shark cartilage may inhibit neovascularization and fibroblast growth factor-induced angiogenesis (Gonzalez et al., 2001a). In studies where commercially available powdered shark cartilage was administered orally in murine and rat models, oral ingestion inhibited angiogenesis and, although it did not abolish tumor progression, it did delay its development significantly (Barber et al., 2001; Davis et al., 1997). These studies suggest that, under certain experimental conditions, enough of the active anti-angiogenic ingredient in shark cartilage is being absorbed through the gut to have an effect following oral ingestion. Despite these experimental results in some animal models, human trials have not shown any significant improvements in patients' tumor regression or disease status following the consumption of shark cartilage. Several studies have examined the effects of shark cartilage extracts or products on humans either in vitro or in vivo (Bukowski, 2003; Gingras et al., 2000; Hillman et al., 2001; McGuire et al., 1996; Miller et al., 1998). Liquid cartilage extract (LCE) was administered orally along with a placebo to male volunteers, and wound angiogenesis was measured indirectly by endothelial cell density. The results indicated that LCE contains an anti-angiogenic component that is bioavailable to humans following oral administration (Berbari et al., 1999). There have been anecdotal reports of several clinical trials investigating the effect of shark cartilage ingestion on cancer patients' health, some of which elicited wide public attention although their experimental methods did not undergo the scrutiny of peer review. Moreover, the results of these studies have not been corroborated (Ostrander et al., 2004).

Because the marketing of shark cartilage as an anticancer treatment continues in the absence of acceptable scientific evidence as to its effectiveness, several experimentally robust studies have now been initiated to address this controversy. Because Benefin is a well-known shark cartilage product, investigators chose to examine its efficacy in a placebo-controlled clinical trial. Benefin was developed as shark cartilage treatment for cancer following Dr. William Lane's, 1992 publication *Sharks don't get cancer* and remains on the market today despite having had an Food and Drug Administration (FDA) injunction placed against it and its manufacturer, Lane Labs, in 1999 for promoting Benefin as a cancer treatment. In the placebo-controlled trial, no benefit was found with the intake of shark cartilage with respect to cancer progression (Loprinzi *et al.*, 2005). Unfortunately, websites such as the *shark cartilage information exchange* continue to promote it as an alternative anticancer therapy, although statements are cautiously made given the severe skepticism shown by the scientific community (Gingras *et al.*, 2000) to much of claims made by Lane Laboratories. Recently, a study published in cancer research reviewed more than 50 studies and concluded that sharks do get both benign and cancerous tumors and claimed that there is no scientific evidence that supports the ability of cartilage extracts to reach target sites in the body and eradicate cancer cells (Ostrander *et al.*, 2004). Unfortunately, shark cartilage is also marketed as a therapeutic agent for a number of different diseases and is promoted as a potential treatment for other diseases such as arthritis and osteoarthritis and as prophylaxis to prevent disease (Gonzalez *et al.*, 2001).

As shark cartilage has been marketed as a therapeutic agent for a number of other disease states, its use has not been limited to the treatment of cancer (Gonzalez *et al.*, 2001b). However, reliable scientific evidence in support of its therapeutic effectiveness in the treatment of a variety of medical conditions remains lacking. Although several studies have shown that cartilage extracts have analgesic, anti-inflammatory, antioxidant, and anti-angiogenic capabilities under certain experimental conditions, none have clearly established their role in modulating or enhancing basic immune function (Chen *et al.*, 2000; Dupont *et al.*, 1998; Felzenszwalb *et al.*, 1998; Fontenele *et al.*, 1996, 1997; Gingras *et al.*, 2000; Lee and Langer, 1983; McGuire *et al.*, 1996; Miller *et al.*, 1998; Oikawa *et al.*, 1990; Rabbani-Chadegani, 2008; Sheu *et al.*, 1998). Many of these studies focused on the effect of cartilage-derived compounds on inhibition of angiogenesis and tumor progression in cellular and animal models and the compounds were derived from shark cartilage prepared by investigators themselves (i.e., relatively *pure* preparations), not from the commercial shark cartilage products sold as dietary supplements, which in addition to cartilage may contain any number of noncartilage tissue derivatives. Because most investigative studies carried out to define the scope of biological activity of shark cartilage have tested relatively *pure* material prepared for the most part in laboratories and not the commercial preparations of shark cartilage commonly sold over the counter to consumers, and given the increasing number of individuals taking shark cartilage, the lack of studies examining the bioactivity of commercial brands of shark cartilage is significantly relevant when conclusions are drawn as to its activity in bioassays and/or animal studies, which might be misleading based on results obtained using *pure* material.

Clinical Assessment of Shark Cartilage as an Anticancer Agent

A few shark cartilage-derived compounds have entered clinical trials based on their inhibition of angiogenesis demonstrated in various *in vivo* and *in vitro* models. In the last 40 years, over a dozen clinical trials on cartilage-derived anticancer agents have been conducted. However, the results of only seven have appeared as peer-reviewed publications and none have resulted in FDA-approved clinical medicines (Gonzalez *et al.*, 2001a). Agents include commercially available shark cartilage preparations such as Cartilade and Benefin as well as other purified cartilage compounds such as Catrix and Neovastat (Leitner *et al.*, 1998; Puccio *et al.*, 1994; Romano *et al.*, 1985; Rosenbluth *et al.*, 1999). Among them, Neovastat (or AE-941) has undergone several phases of clinical trials and is shown to be safe in humans with various dose effect studies conducted (Batist *et al.*, 2002). A number of trials compared the effectiveness of treatment with oral Neovastat in combination with chemotherapy and radiation therapy with that of treatment with placebo in combination with traditional therapies. In 2002, Neovastat was granted orphan drug status by the FDA for the treatment of renal cell carcinoma. Despite this early promise, the results of phase III clinical trials of this drug were not reported in the scientific literature, and in 2004, the company that manufactures Neovastat withdrew their application for orphan drug status (Falardeau *et al.*, 2001; Latreille *et al.*, 2003). In 2010, another randomized, double-blinded, placebo-controlled phase III trial was conducted for Neovastat in patients with advanced cancer. The trial results indicated no statistically significant difference in survival between groups treated with Neovastat and groups treated with placebo (Lu *et al.*, 2010). Although several shark cartilage-derived compounds remain in the preclinical and clinical pipeline, it remains to be seen if any will prove effective cancer treatments in human clinical trials given the poor record of success observed thus far.

Shark Cartilage and Immunomodulation

Although for many years the main thrust of shark cartilage research was directed toward its anti-angiogenic properties, a few scientists choose to systematically investigate the potential immunomodulatory properties of shark cartilage, an area of research that had largely been ignored. One such study focused on the effect of shark cartilage on the infiltration of lymphocytes in a murine tumor model. Intraperitoneal injection of shark cartilage protein fractions into tumor-bearing mice increased T-cell infiltration into the tumor and significantly increased the CD4/CD8 ratio in tumor-infiltrating lymphocytes. The study also showed that a 15-kDa fraction of shark cartilage enhanced immune response by augmenting delayed hypersensitivity against SRBC in mice (Feyzi *et al.*, 2003). Another study investigated the immunostimulating potential of various preparations of shark cartilage extracts using an *in vitro* murine model. The extracts were potent stimulators of B cells and macrophages isolated from BALB/c mice spleen. The active components were shown to be thermally stable proteoglycans with molecular masses exceeding 100 kDa (Kralovec *et al.*, 2003).

Considering the quantities of shark cartilage sold and consumed as a dietary supplement, it soon became apparent that there was a lack of informative data on its effect on human immune cellular function. As stated earlier, most of the laboratory-based studies were carried out using laboratory-prepared shark cartilage that did not represent or reflect qualitatively the composition of commercial products. The first study to examine the immunomodulatory properties of commercial shark cartilage was undertaken by Simjee (1996), who examined the cytokine (TNF-α) response of cultured human leukocytes to various extracts of several brands of commercial shark cartilage (SCE). The level of detectable TNF-α in culture supernatants of SCE-stimulated leukocytes was significantly higher than that of LPS-stimulated control cultures and peak levels were reached at 24 h of exposure. This study showed that different commercial brands of shark cartilage differed in the amount of extractable protein from the same method of extraction and in the level of biological activity (i.e., cytokine induction) per mg of protein recovered. In addition, when three different types of extracts (a salt-soluble, a phosphate-buffered, and an acid extract) were assayed for activity, the acid extract induced higher levels of cytokine production than the others. Furthermore, cytokine-inducing activity of SCE when compared to that of bovine cartilage extracts, the former induced a significantly higher level of cytokine per mg of extract protein. The study also showed that extracts lost 80% of their biological activity (cytokine induction) following treatment with trypsin and chymotrypsin (i.e., protein-digesting enzymes), indicating that the active component(s) in shark cartilage may be proteinaceous. However, it should be noted that a low level of activity (20%) remained following enzymatic digestion, suggesting that oral ingestion and digestion will not necessarily eliminate all activity and that some level of activity can be expected to be retained, which may be sufficient to stimulate cells locally in the gut or affect immunomodulation systemically upon absorption. In addition, heating SCEs up to 95°C abolished cytokine-inducing activity further suggesting the bioactive component was likely to be protein in nature. A concurrent study showed that SCE-induced TNF-α (detected by ELISA) was a functionally active molecule because it was cytotoxic for WEHI-164 cells. The WEHI-164 is a murine fibrosarcoma cell line widely used in bioassays to demonstrate biological cytotoxic activity, particularly where TNF-α is the effector molecule (Ziegler-Hetibrock, 1986). The assay was used to demonstrate inducible cytotoxic activity in stimulated human leukocytes. A correlation between the level of the secreted cytokine (TNF-α) and corresponding degree of cytotoxicity against target cells was observed. The cytotoxic effect of cartilage-stimulated leukocytes was significantly higher (almost two-fold) than that of LPS-stimulated cultures at concentrations used. Similarly, the level of TNF-α secreted by shark cartilage extract-stimulated human peripheral blood leukocytes (hPBLs) was significantly higher than that induced by LPS stimulation. Supernatant from leukocytes stimulated with culture medium alone was noncytotoxic (Simjee, personal communication) (Figure 16.1).

A follow-up *in vitro* study examining overall cytokine induction in response to stimulation with SCE revealed that cartilage significantly induced the production of Th1-type, inflammatory cytokines, namely, IL-1β, IL-2, TNF-α, and IFN-γ, as well as a potent chemokine, IL-8 (Merly *et al.*, 2007). Many of the diseases (i.e., rheumatoid arthritis) for which shark cartilage products are recommended by commercial manufacturers are caused by and are the result of undesirable inflammatory reactions. The intake of shark cartilage that induces upregulation of inflammatory cytokines is, therefore, counter indicated. However, if active components in shark cartilage that stimulate and/or modulate innate immune response are isolated, they may serve as immunostimulatory agents. A better understanding of the active component(s) of shark cartilage, the scope of their bioactivity, and the mechanisms by which they modulate the immune response of cells can lead to their potential application in clinical medicine. One such application would be topical immunomodulation where an agent that induces monocytes/macrophages to produce Th1 cytokines and promote cell-mediated immunity, especially as a TLR agonist, may be clinically useful in treating viral infections and/or improving wound healing (Hengge and Ruzicka, 2004). Recent studies have identified a bioactive component present in extracts prepared from commercial shark cartilage as collagen type II, alpha 1 protein from the lesser spotted catshark, *Scyliorhinus canicula* (Merly and Smith, 2013). A schematic representation of a putative molecular model for how collagen type II, alpha 1 protein derived from shark cartilage might modulate immune cell activity in peripheral blood leukocytes is presented in Figure 16.2. Collagen type II, alpha 1 protein (purified from fetal bovine articular cartilage) is routinely used to experimentally induce arthritis in a rheumatoid arthritis (RA) model in the mouse. This is referred to as collagen-induced arthritis or CIA. In both CIA and RA models, it is the T-cell recognition of this protein that initiates pathology leading to

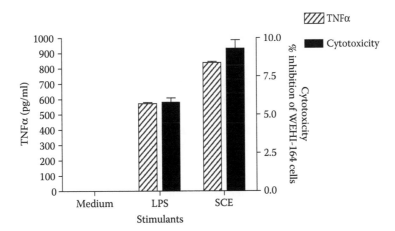

Figure 16.1   TNF-α and cytotoxicity of hPBL culture supernatants. Supernatants from hPBLs stimulated with shark cartilage extract (0.36 mg protein/ml extract), LPS (5 µg/ml) or CM for 24 h were assayed for secreted TNF-α and cytotoxicity against WEHI-164 cells. Cytotoxicity is expressed as percent inhibition of MTT-reducing activity of WEHI-164 cells following treatment with supernatants from stimulated cultures compared to cultures treated with medium alone. TNF-α and cytotoxicity values are presented as the mean ± SEM (n = 6). $P < 0.05$. (Data kindly provided by Dr. S. Simjee, personal communication.)

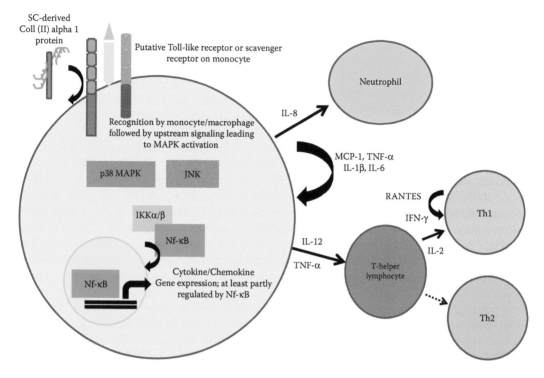

Figure 16.2   A schematic representation of the putative mechanism of the action of immunomodulatory activity of collagen type II, alpha 1 protein derived from shark cartilage in peripheral blood leukocytes *in vitro*. Broken arrow indicates stimulation pathway that does not occur.

arthritic joints. Although it is unknown if these peptides would survive the digestive process *in vivo*, it remains a concern that this protein is at least partially responsible for the inflammatory activity observed *in vitro* for shark cartilage preparations on human leukocytes.

The composition of commercial preparations is not well understood because companies do not readily provide information on source of raw material in terms of shark species, geographic location, age, sex, tissue source, and so on or on how the material is processed into the final commercial product. Because little is known about the composition of these products, the impact that the natural product industry is having on global shark populations has largely been ignored in the literature. The identification of the bioactive component as a collagen protein, a normal constituent of cartilage, suggests that much of the activity observed for these commercial preparations is in fact cartilage derived rather than the result of other tissue contaminants and/or additives. As a result of the earlier study, the lesser spotted catshark was identified as the source of cartilage in the product tested, indicating that this species is being harvested and its tissue in one form or another is making its way into the product marketed as shark cartilage. Interestingly, the label on the commercial product used to prepare the extract for the study claimed the cartilage to be 100% Australian in origin and yet the shark cartilage material was identified as being derived from a shark species not found in Australian waters. This illustrates the discrepancy between the description provided on the product label by the manufacturer and the true source of the commercial preparation which brings into question the reliability of product labeling. Different brands of shark cartilage have been shown to be highly variable in level of bioactivity which is likely due to any number of shark species harvested from different waters. Various shark species are routinely landed and make their way into the global trade in shark products. The lesser spotted catshark is widely distributed from the Northeast Atlantic continental shelf to the Mediterranean where it is abundant and readily available and is not considered of high commercial value. It is not surprising, therefore, that it is being used as a source for these commercial products (Ellis *et al.*, 2009). It brings into question whether this species is specifically being targeted for the production of these supplements.

IMPLICATIONS FOR PUBLIC HEALTH

From a public health point of view, given the complex nature of the gut, one can propose a model by which active component(s) in cartilage that survive the digestive process could potentially induce similar responses in immune cells that line the gastrointestinal tract *in vivo* following oral ingestion of cartilage capsules. Localized inflammatory cytokine production in the gut could lead to gastrointestinal upset as well as exacerbation of symptoms in individuals who suffer from gastrointestinal disease where the underlying pathology involves inflammation (Sanos and Diefenbach, 2013). There is some evidence in the literature of gastrointestinal upset reported as a result of intake of shark cartilage supplementation during clinical trials (Loprinzi *et al.*, 2005). Beyond the gut, one can speculate as to how components of shark cartilage are able to upregulate distinct biochemical pathways in circulating immune cells, so that if active component(s) reached other sites in the body, inflammatory processes could be initiated or worsened, making shark cartilage supplementation a particularly poor choice for people with arthritis and other inflammatory conditions. At least one active component in shark cartilage products is collagen type II, alpha 1 protein which is known to be a major component in the underlying pathology of rheumatoid arthritis. The consumption of shark cartilage is a potentially risky practice in the absence of a comprehensive characterization of bioactive components of cartilage and their immune-stimulating properties and the mechanisms involved. Furthermore, unless components are isolated, purified, and fully characterized with respect to their immunomodulating properties, it is unlikely they can be of practical use in clinical medicine.

ECOLOGICAL CONSEQUENCES

From a conservation and ecological point of view, there is currently very little known of the extent to which the shark cartilage industry is responsible for the global decline of shark numbers and potential loss of species (refer to Cailliet and Ebert chapter 1 in this book for further information of distribution of shark species). This industry may represent a critical cause of shark losses due to direct harvesting by manufacturers. Unfortunately, the sale of shark cartilage provides incentives to commercial fisheries to increase their shark landings and potentially ignore the problem of shark bycatch (known to be a major cause for shark species losses) because there is monetary gain to be had from such activities from manufacturers of shark cartilage.

CONCLUSIONS

Sharks have been used as a marine source of medicinal compounds for centuries. Experimental evidence has identified some compounds that exhibit immunomodulatory activity under certain experimental conditions. Reactivity between shark proteins and human leukocytes has been shown in several studies where shark-derived immune proteins were used successfully to stimulate human cells (Smith *et al.*, 1997). Recent evidence has also shown that shark leukocytes produce substances *in vitro* that exhibit potent cytotoxic activity against human tumor cell lines by inducing apoptosis, suggesting that mediators produced by shark cells are capable of modulating human cellular activity and disease processes (Walsh *et al.*, 2013). Shark liver oil and shark cartilage contain bioactive components both lipid and protein in nature that have immunomodulatory effects. However, the scientific data supporting the use of natural products based on this bioactivity are lacking. With the growth of the natural product industry and Internet marketing, the use of shark-derived compounds with purported health benefits has expanded significantly. With very few exceptions, bioactivity observed under experimental conditions has not been replicated in human trials, nor has the composition of commercial products been standardized in any way. For this reason, caution should be exercised by consumers of these commercial products. Furthermore, no study has been undertaken to assess the extent to which industries that promote these products are contributing to the decline of shark populations. It is likely that their sale plays a substantial role in the increase in global trade of shark products, which presents a survival challenge to shark species and an ecological threat to the oceans which they inhabit.

ACKNOWLEDGMENT

We thank Dr. Shabana Simjee for providing the cytotoxicity data (Figure 16.1) presented in this chapter.

REFERENCES

Baldridge, H.D. Sinking factors and average densities of Florida sharks as functions of liver buoyancy. *Copeia* 1970; 4: 744–754.

Barber, R., Delahunt, B., Grebe, S.K., Davis, P.F., Thornton, A., and Slim, G.C. Oral shark cartilage does not abolish carcinogenesis but delays tumor progression in a murine model. *Anticancer Research* 2001; 21:1065–1069.

Bargahi, A., and Rabbani-Chadegan, A. Angiogeic inhibitor protein fractions derived from shark cartilage. *Bioscience Reports* 2008; 28:15–21.

Batist, G., Champagne, P., and Hariton, C. Dose-survival relationship in a phase II study of Neovastat in refractory renal cell carcinoma patients. (Abstract). *Proceedings of the American Society of Clinical Oncology* 2002; 21:A-1907.

Belo, S.R.B., Iagher, F., Bonatto, S.J., Naliwaiko, K., Calder, P.C., Nunes, E.A., and Fernandes, L.C. Walker 256 tumor growth is inhibited by the independent or associative chronic ingestion of shark liver and fish oil: A response linked by the increment of peritoneal macrophages nitrite production in Wistar rats. *Nutrition Research* 2010; 30:770–776.

Berbari, P., Thibodeau, A., Germain, L., Saint-Cyr, M., Gaudreau, P., Elkhouri, S., Dupont, E., Garrel, D.R., and Elkouri, S. Antiangiogenic effects of the oral administration of liquid cartilage extract in humans. *Journal of Surgical Research* 1999; 87:108–113.

Bhilwade, H.N., Tatewaki, N., and Nishida, H. Squalene as a novel food factor. *Current Pharmaceutical Biotechnology* 2010; 11(8):875–880.

Brachwitz, H., Langen, P., Arndt, D., and Fichtner, I. Cytostatic activity of synthetic *O*-alklyglycerolipids. *Lipids* 1987; 22:897–903.

Brem, H., and Folkman, J. Inhibition of tumor angiogenesis mediated by cartilage. *Journal of Experimental Medicine* 1975; 141:427–439.

Brito, L.A., Chan, M., and Baudner, B. An alternative renewable source of squalene for use in emulsion adjuvants. *Vaccine* 2011; 29:6262–6268.

Brohult, A. Alkoxyglycerols and their use in radiation treatment: An experimental and clinical study. *Acta Radiologica Therapy Physics Biology* 1963; 223:1–99.

Brohult, A., Brohult, J., and Brohult, S. Regression of tumour growth after administration of alkoxyglycerols. *Acta Obstetricia Gynecologica Scandinavica* 1978; 57:79–83.

Brohult, A., Brojult, J., Brohult, S., and Joelsson, I. Reduced mortality in cancer patients after administration of alkoxyglycerols. *Acta Obstetricia Gynecologica Scandinavica* 1986; 65(7):779–785.

Brohult, A., and Holmberg, J. Alkoxyglycerols in the treatment of leukopaenia caused by irradiation. *Nature* 1954; 174(4441):1102–1103.

Bukowski, R.M. AE-941, a multifunctional antiangiogenic compound: Trials in renal cell carcinoma. *Expert Opinion on Investigational Drugs* 2003; 12:1403–1411.

Camin, F., Bontempo, L., Ziller, L., Piangiolino, C., and Morchio, G. Stable isotope ratios of carbon and hydrogen to distinguish olive oil from shark squalene-squalane. *Rapid Communications in Mass Spectrometry* 2010; 24(12):1810–1816.

Chen, J.S., Chang, C.M., Wu, J.C., and Wang, S.M. Shark cartilage extract interferes with cell adhesion and induces reorganization of focal adhesions in cultured endothelial cells. *Journal of Cellular Biochemistry* 2000; 78:417–428.

Clarke, S., Milner-Gulland, E.J., and Cemare, T.B. Social, economic, and regulatory drivers of the shark fin trade. *Marine Resource Economics* 2007; 22: 305–327.

Davis, P.F., He, Y., Furneaux, R.H., Johnston, P.S., Ruger, B.M., and Slim, G.C. Inhibition of angiogenesis by oral ingestion of powdered shark cartilage in a rat model. *Microvascular Research* 1997; 54:178–182.

Denaiu, A., Mosset, P., Pedrono, F., Mitre, R., Le Bot, D., and Legrand, A.B. Multiple beneficial health effects of natural alkylglycerols from shark liver oil. *Marine Drugs* 2010; 8:2175–2184.

Dupont, E., Savard, P.E., Jourdain, C., Juneau, C., Thibodeau, A., Ross, N., Marenus, K., Maes, D.H., Pelletier, G., and Sauder, D.N. Antiangiogenic properties of a novel shark cartilage extract: Potential role in the treatment of psoriasis. *Journal of Cutaneous Medicine and Surgery* 1998; 2:146–152.

Ellis, J., Mancusi, C., Serena, F., Haka, F., Guallart, J., Ungaro, N., Coelho, R., Schembri, T., and MacKenzie, K. *Scyliorhinus canicula*. In: IUCN 2013. IUCN Red List of Threatened Species. Version 2013:1, 2009.

Falardeau, P., Champagne, P., and Poyet, P. Neovastat, a naturally occurring multifunctional antiangiogenic drug in phase III clinical trials. *Seminars in Oncology* 2001; 28(6):620–625.

Felzenszwalb, I., de Mattos, J.C.P., Bernardo-Filho, M., and Caldeira-de-Araujo, A. Shark cartilage-containing preparation: Protection against reactive oxygen species. *Food and Chemical Toxicology* 1998; 36:1079–1084.

Ferretti, F., Worm, B., Britten, G.L., Heithaus, M.R., and Lotze, H.K. Patterns and ecosystem consequences of shark declines in the ocean. *Ecology Letters* 2010; 13:1055–1071.

Feyzi, R., Hassan, Z.M., and Mostafaie, A. Modulation of CD(4)(+) and CD(8)(+) tumor infiltrating lymphocytes by a fraction isolated from shark cartilage: Shark cartilage modulates anti-tumor immunity. *International Immunopharmacology* 2003; 3:921–926.

Fontenele, J.B., Araujo, G.B., de Alencar, J.W., and Viana, G.S. The analgesic and anti-inflammatory effects of shark cartilage are due to a peptide molecule and are nitric oxide (NO) system dependent. *Biological and Pharmaceutical Bulletin* 1997; 20:1151–1154.

Fontenele, J.B., Viana, G.S., Xavier-Filho, J., and de-Alencar, J.W. Anti-inflammatory and analgesic activity of a water-soluble fraction from shark cartilage. *Brazilian Journal of Medical and Biological Research* 1996; 29:643–646.

Gingras, D., Renaud, A., Mousseau, N., and Beliveau, R. Shark cartilage extracts as antiangiogenic agents: Smart drinks or bitter pills? *Cancer Metastasis Reviews* 2000; 19:83–86.

Gonzalez, R.P., Leyva, A., and Moraes, M.O. Shark cartilage as source of antiangiogenic compounds: From basic to clinical research. *Biological and Pharmaceutical Bulletin* 2001a; 24:1097–1101.

Gonzalez, R.P., Soares, F.S., Farias, R.F., Pessoa, C., Leyva, A., de Barros Viana, G.S., and Moraes, M.O. Demonstration of inhibitory effect of oral shark cartilage on basic fibroblast growth factor-induced angiogenesis in the rabbit cornea. *Biological and Pharmaceutical Bulletin* 2001b; 24:151–154.

Hajimoradi, M., Hassan, Z.M., Pourfathollah, A.A., Daneshmandi, S., and Pakravan, N. The effect of shark liver oil on the tumor infiltrating lymphocytes and cytokine pattern in mice. *Journal of Ethnopharmacology* 2009; 126:565–570.

Hallgren, B., and Larsson, S. The glycerol ethers in the liver oils of elasmobranch fish. *Lipid Research* 1962; 3:31–62.

Hengge, U.R., and Ruzicka, T. Topical immunomodulation in dermatology: Potential of Toll-like Receptor agonists. *Dermatologic Surgery* 2004; 30:1101–1112.

Hillman, J.D., Peng, A.T., Gilliam, A.C., and Remick, S.C. Treatment of Kaposi sarcoma with oral administration of shark cartilage in a human herpesvirus 8-seropositive, human immunodeficiency virus-seronegative homosexual man. *Archives of Dermatology* 2001; 137:1149–1152.

Hilvo, M., and Oresic, M. Regulation of lipid metabolism in breast cancer provides diagnostic and therapeutic opportunities. *Clinical Lipidology* 2012; 7(2):177–188.

Kralovec, J.A., Guan, Y., Metera, K., and Carr, R.I. Immunomodulating principles from shark cartilage. Part 1. Isolation and biological assessment *in vitro*. *International Immunopharmacology* 2003; 3:657–669.

Latreille, J., Batist, G., and Laberge, F. Phase I/II trial of the safety and efficacy of AE-941 (Neovastat) in the treatment of non-small-cell lung cancer. *Clinical Lung Cancer* 2003; 4(4):231–236.

Lee, A., and Langer, R. Shark cartilage contains inhibitors of tumor angiogenesis. *Science* 1983; 221:1185–1187.

Leitner, S.P., Rothkopf, M.M., and Haverstick, L. Two phase II studies of oral dry shark cartilage powder (SCP) with either metastatic breast or prostate canceer refractory to standard treatment. (Abstract). *Proceedings of the American Society of Clinical Oncology* 1998; 17:A-240.

Lewkowicz, N., Lewkowicz, P., Kurnatowska, A., and Tchorzewski, H. Biological action and clinical application of shark liver oil. *Polskimerkuriuszlekarski* 2006; 20(119):598–601.

Lewkowicz, P., Banasik, M., Glowacka, E., Lewkowicz, N., and Tchorzewski, H. Effect of high doses of shark liver oil supplementation on T cell polarization and peripheral blood polymorphonuclear cell function. *Polskimerkuriuszlekarski* 2005; 18(108):686–692.

Loprinzi, C.L., Levitt, R., Barton, D.L., Sloan, J.A., Atherton, P.J., Smith, D.J., Dakhil, S.R. *et al*. Evaluation of shark cartilage in patients with advanced cancer: A North Central Cancer Treatment Group trial. *Cancer* 2005; 104:176–182.

Lu, C., Lee, J.J., and Komaki, R. Chemoradiotherapy with or without AE-941 in stage III non-small-cell lung cancer: A randomized phase III trial. *Journal of the National Cancer Institute* 2010; 102(12):859–865.

McGuire, T.R., Kazakoff, P.W., Hoie, E.B., and Fienhold, M.A. Antiproliferative activity of shark cartilage with and without tumor necrosis factor-alpha in human umbilical vein endothelium. *Pharmacotherapy* 1996; 16:237–244.

Merly, L., Simjee, S., and Smith, S.L. Induction of inflammatory cytokines by cartilage extracts. *International Immunopharmacology* 2007; 7:383–391.

Merly, L., and Smith, S.L. Collagen type II, alpha 1 protein: A bioactive component of shark cartilage. *International Immunopharmacology* 2013; 15:309–315.

Miller, D.R., Anderson, G.T., Stark, J.J., Granick, J.L., and Richardson, D. Phase I/II trial of the safety and efficacy of shark cartilage in the treatment of advanced cancer. *Journal of Clinical Oncology* 1998; 16:3649–3655.

Moore, K.S., Wehrli, S., Roder, H., Rogers, M., Forrest, J.M., McCrimmon, D., and Zasloff, M. Squalamine: An aminosterol antibiotic from the shark. *Proceedings of the National Academy of Sciences of the United States of America* 1993; 90:1354–1358.

Oikawa, T., Ashino-Fuse, H., Shimamura, M., Koide, U., and Iwaguchi, T. A novel angiogenic inhibitor derived from Japanese shark cartilage (I). Extraction and estimation of inhibitory activities toward tumor and embryonic angiogenesis. *Cancer Letters* 1990; 51:181–186.

Ostrander, G.K., Cheng, K.C., Wolf, J.C., and Wolfe, M.J. Shark cartilage, cancer and the growing threat of pseudoscience. *Cancer Research* 2004; 64:8485–8491.

Pedrono, F., Martin, B., Leduc, C., Le Lan, J., Saiag, B., and Legrand, P. Natural alkylglycerols restrain growth and metastasis of grafted tumors in mice. *Nutrition and Cancer* 2004; 48:64–69.

Prudden, J.F., and Balassa, L.L. The biological activity of bovine cartilage preparations. Clinical demonstration of their potent anti-inflammatory capacity with supplementary notes on certain relevant fundamental supportive studies. *Seminars in Arthritis and Rheumatism* 1974; 3:287–321.

Puccio, C., Mittelman, A., and Chun, P. Treatment of metastatic renal cell carcinoma with Catrix. (Abstract). *Proceedings of the American Society of Clinical Oncology* 1994; 13:A769.

Pugliese, P., Jordan, K., Cederberg, H., and Brohult, J. Some biological actions of alkylglycerols from shark liver oil. *Journal of Alternative and Complementary Medicine* 1998; 4(1):87–99.

Rabbani-Chadegani, A., Abdossamadi, S., Bargahi, A., and Yousef-Masboogh, M. Identification of low-molecular-weight protein (SCP1) from shark cartilage with anti-angiogenesis activity and sequence similarity to parvalbumin. *Journal of Pharmaceutical and Biomedical Analysis* 2008; 46:563–567.

Romano, C.F., Lipton, A., and Harvey, H.A. A phase II study of Catrix-S in solid tumors. *Journal of Biological Response Modifiers* 1985; 4(6):585–589.

Rosenbluth, R.J., Jennis, A.A., and Cantwell, S. Oral shark cartilage in the treatment of patients with advanced primary brain tumors. (Abstract). *Proceedings of the American Society of Clinical Oncology* 1999; 18:A-554.

Sanos, S.L., and Diefenbach, A. Innate lymphoid cells: From boarder protection to he initiation of inflammatory disease. *Immunology and Cell Biology* 2013; 91:215–224.

Schick, H.D., Berdel, W.E., and Fromm, M. Cytotoxic effects of ether lipids and derivatives in human non-neoplastic bone-marrow cells and leukemic cells *in vitro*. *Lipids* 1987; 22(11):904–910.

Sheu, J.R., Fu, C.C., Tsai, M.L., and Chung, W.J. Effect of U-995, a potent shark cartilage-derived angiogenesis inhibitor, on anti-angiogenesis and anti-tumor activities. *Anticancer Research* 1998; 18:4435–4441.

Simjee, S. Secretion of tumor necrosis factor by human leukocytes stimulated with shark cartilage. Medical Laboratory Sciences, Master of Science. 1996. Florida International University, Miami, FL.

Smith, S.L., Riesgo, M., Obenauf, S.D., and Woody, C.J. Anaphylactic and chemotactic response of mammalian cells to zyomasan-activated shark serum. *Fish Shellfish Immunol* 1997; 7(7):503–514.

Tchorzewski, H., Banasik, M., Glowacka, E., and Lewkowicz, P. Modification of innate immunity in humans by active components of shark liver oil. *Polski Merkuriusz Lekarski* 2002; 13(76):329–332.

Walsh, C.J., Luer, C.A., Yordy, J.E., Cantu, T., Miedema, J., Leggett, S.R., Leigh, B., Adams, P., Ciesla, M., Bennett, C., and Bodine, A.B. Epigonal conditioned media from Bonnethead Shark, *Sphyrna tiburo*, induces apoptosis in a T-cell leukemia cell line, Jurkat E6-1. *Marine Drugs* 2013; 11:3224–3257.

Zasloff, M., Adams, A.P., Beckerman, B., Campbell, A., Han, Z., Luijten, E., Meza, I. *et al*. Squalamine as a broad-spectrum systemic antiviral agent with therapeutic potential. *Proceedings of the National Academy of Sciences of the United States of America* 2011; 108(38):15978–15983.

Zheng, L., Ling, P., Wang, Z., Niu, R., Hu, C., Zhang, T., and Lin, X. A novel polypeptide from shark cartilage with potent anti-angiogenic activity. *Cancer Biology and Therapy* 2007; 6(5):775–780.

Ziegler-Heitbrock, H.W., Moller, A., Linke, R.P., Haas, J.G., Rieber, E.P., and Reithmuller, G. Tumor necrosis factor as effector molecule in monocyte mediated cytotoxicity. *Cancer Research* 1986; 46(11):5947–5952.

Index

Note: Locators followed by "*f*" and "*t*" denote figures and tables in the text

Milton Keynes UK
Ingram Content Group UK Ltd.
UKHW050453071024
449327UK00015B/354

9 780367 378042